CHEMISTRY OF VIRUSES

SPRINGER
STUDY
EDITION

CHEMISTRY OF VIRUSES

Second Edition

C. A. Knight
Department of Molecular Biology and the Virus Laboratory
University of California, Berkeley

Springer-Verlag New York · Wien 1975

This book is affectionately dedicated to Billie, Tom, Susie, and Bob.

Claude A. Knight
Department of Molecular Biology
University of California
Wendell M. Stanley Hall
Berkeley, California 94720

Library of Congress Cataloging in Publication Data

Knight, Claude Arthur, 1914-
chemistry of viruses

1. Virology. 2. Microbiological chemistry.
I. Title. [DNLM: 1. Viruses—Analysis. QW160 K69c]
QR370.K54 1974 576'.64 74-11220

Printed in the United States of America

ISBN 0-387-06772-8 Springer-Verlag New York · Wien
ISBN 3-211-06772-8 Springer-Verlag Wien · New York

Preface

In 1963, the first edition of *Chemistry of Viruses* was published as a contribution to the series on viruses sponsored by Protoplasmatologia. An aim of the first edition was to review some major principles and techniques of chemical virology in a concise manner and to accompany this review with a compilation of pertinent references. It was anticipated that this exercise would be helpful to the author in his teaching and research and, hopefully, would be useful to readers as well.

The literature of virology has grown enormously since then, and it is even more urgent to have a succinct survey. In addition, few authors have attempted to integrate the findings pertaining to the various major classes of viruses (that is, animal, bacterial, and plant viruses) but, rather, have chosen to assemble large monographs dealing in depth with facts and fancies pertaining to specific groups of viruses. Such works are valuable for pursuit of particular topics but fail to yield a brief, integrated view of virology. The present edition of *Chemistry of Viruses* aspires to such a review.

A serious attempt was made to deal concisely with every major topic of chemical virology and to present examples from different classes of viruses. Numerous references are given to original articles and review papers as well as to selected books.

It is hoped that this type of presentation—a compendium of chemical virology with pertinent, selected references—will prove to be a helpful introduction to viruses for neophytes and a convenient reference to veterans.

The author acknowledges with gratitude the contribution of illustrations by several colleagues, who are cited with the illustrations they provided, and the work of E. N. Story in preparing some of the illustrations. He is also indebted to Maureen Rittenberg for her efforts in typing the manuscript.

CONTENTS

I.

Some Events Leading to the Chemical Era of Virology

Near the end of the 19th century, Dutch scientist Martinus W. Beijerinck performed some experiments that were to have far-reaching consequences in science. Working with the sap from leaves of mosaic-diseased tobacco plants, Beijerinck (1898a, 1898b; see also van Iterson et al. 1940) showed that the infectious agent causing mosaic disease was so small that it passed through exceedingly fine bacteria-retaining filters and diffused at a measurable rate through blocks of agar gel. To this unprecedentedly small pathogen, Beijerinck applied the terms "contagium vivum fluidum" (contagious living fluid), or "virus."

As early as 1892, the Russian scientist, Ivanovski, reported filtration experiments with infectious juice from mosaic-diseased tobacco plants, but he was not convinced that his results were valid. In fact, a year after Beijerinck's report, Ivanovski (1899) published a paper on mosaic disease in which he concluded from his experiments that this condition was a bacterial infection. The following excerpt illustrates this point: "Zwar sind die Versuche noch wenig zahlreich und der Prozentsatz der erkrankten Pflanzen gering; doch glaube ich, dass die Bakterielle Natur des Kontagiums kaum zu bezweifeln ist."

In Germany, Loeffler and Frosch reported in 1898 that foot-and-mouth disease could be transmitted to calves by intravenous injection of infective lymph which had been freed of bacteria by passage through a filter candle made of diatomaceous earth (kieselguhr). Experiments involving dilution of the lymph and serial passage virtually eliminated the possibility that the disease could be attributed to a nonreproducing agent such as a toxin. Loeffler and Frosch therefore concluded that the causal agent was able to reproduce in cattle and was so small that it could pass through the pores of a filter that retained the smallest known bacterium. They also suggested that the hitherto elusive agents of such diseases as smallpox, cowpox, rinderpest, and measles might belong to this group of tiny organisms.

During the first 30 years of the 20th century, following the lead given by the work on tobacco mosaic and foot-and-mouth diseases, many infectious agents were tested for their filterability. As a consequence, such diverse diseases as yellow fever, Rous sarcoma of chickens, rabies, infectious lysis of bacteria, cucumber mosaic, potato X disease, and many others were classified in the newly recognized group of ultratiny disease agents, the "filterable viruses." To characterize these newly recognized disease agents better, many studies were made of the effects of various chemical and physical agents on infectivity. The results of these pioneer investigations have been well summarized by Stanley (1938).

While early interpretations of the mechanism of inactivation of viruses by chemical and physical agents were necessarily faulty as judged by more recent knowledge, nevertheless, the results did provide a foundation on which ultimately successful attempts to isolate and purify viruses could be built. For example, it became clear that protein denaturants, oxidizing agents, formaldehyde, strong acids or bases, and high temperatures were inimical to viruses, whereas the milder protein precipitants, low temperatures, and neutral pH could usually be employed without destroying infectivity.

A prelude of what was shortly to come appeared in the experiments of Vinson (1927) and of Vinson and Petre (1929, 1931) on tobacco mosaic virus (TMV). A series of experiments on infectious sap from mosaic-diseased tomato or tobacco plants was summarized by Vinson and Petre (1929) in the following manner:

> We have found that when precipitation of the virus is carried out under favorable conditions, with the proper concentration of safranin, acetone, or ethyl alcohol, the precipitation is almost complete. In each case the precipitate contains practically all of the original activity of the juice, and the virus concentration in the supernatant liquid is no greater than that obtained by diluting a fresh juice sample one thousand-fold. This, together with the fact that the virus is apparently held in an inactive condition in the safranin precipitate and is released when the safranin is removed, makes it probable that the virus which we have investigated reacted as a chemical substance.

In a subsequent publication (Vinson and Petre 1931) the supposed nature of this chemical substance was postulated to be enzymic, largely on the basis of viewing the viral multiplication process as an autocatalytic phenomenon and on experimental hints that the virus might be proteinaceous. The chief clue that the virus might be associated with protein was an observed increase in nitrogen content as the infectious fraction was separated from the bulk of impurities associated with it, although the observations that the infectious principle moved in an electric field and was precipitated by protein precipitants were also consistent with the protein hypothesis.

Interest in TMV increased considerably when Vinson described infectious crystalline preparations of TMV at meetings of the American Association for Advancement of Science in 1928 and 1930, and published the relevant experiments in some detail in 1931 (Vinson and Petre 1931).

These crystalline preparations were obtained by treating infectious tobacco juice with acetone to get a precipitate, which was dissolved in a small amount of water. To this concentrated solution, acetic acid was added to pH 5; then acetone was added slowly with constant stirring until a slight permanent cloudiness appeared. When stored in the icebox, crystalline material often, but not always, separated out. Such crystalline material, when obtained, was described as "moderately active" (infectious), but as a protein preparation it was of dubious purity since about 33 percent was found to be ash (largely calcium oxide). Nevertheless, the finding was acclaimed, somewhat prematurely, in an editorial in the *Journal of the American Medical Association* (1932) in part as follows:

> Possibly the reported successful crystallization of the etiologic factor of mosaic disease of tobacco may be regarded by future medical historians as one of the most important advances in infectious theory since the work of Lister and Pasteur. The announcement of the isolation of a crystallizable pathogenic enzyme necessarily throws doubt on the conception that poliomyelitis, smallpox, and numerous other "ultramicroscopic infections" are of microbic causation. The apparent evidence that a specific protein, which in itself is incapable of self multiplication, may function as a disease germ when placed in "symbiosis" with normal cells seems to furnish experimental confirmation of several highly speculative theories relating to vitamins, hormones, and progressive tissue degenerations.

From the foregoing, it is evident that Vinson and associates contributed substantially to the chemical elucidation of TMV, but fell short of a definitive identification of the infectious agent. Hampered by persistent impurities in the preparations, uncertain biological assays, and variable but great losses of virus, the experiments designed to concentrate, purify, and identify the virus failed to reach fruition.

In 1931 a department of plant pathology was established in the Rockefeller Institute for Medical Research near Princeton, New Jersey. Louis O. Kunkel was brought from the Boyce Thompson Institute for Plant Research at Yonkers, New York, to head the new department. Kunkel felt the time was ripe to add a chemist to the team he was organizing to study plant virus diseases. At this time, Wendell Meredith Stanley (Figure 1), a young organic chemist who had received his doctorate under the tutelage of Roger Adams at the University of Illinois, was working with the noted cell physiologist, W. J. V. Osterhout, at the New York branch of the Rockefeller Institute for Medical Research. Stanley was persuaded to join the Princeton group, and in 1933 began his now-famous studies on TMV.

In preliminary experiments, Stanley worked through previous methods of purification and modified them, especially with respect to the pH used in various steps. Infectivity was closely followed for the first time in the fractionation procedures by use of Holmes' newly developed method of local lesion assay (Holmes 1929). Stanley also took advantage of the pres-

Fig. 1. Wendell Meredith Stanley, 1904–1971.

ence in the Institute of Northrop, Kunitz, Herriott, and Anson, who were engaged in their classic studies on the isolation and properties of crystalline proteolytic enzymes. The proximity of these workers provided, among other things, access to crystalline pepsin, which was used in a crucial experiment of a series on the effect of chemical reagents on viral activity. Stanley (1934b) found that the infectivity of TMV was largely destroyed by pepsin at a pH at which the virus was stable when pepsin was omitted. This result led Stanley (1934) to state, "It seems difficult to avoid the conclusion that tobacco mosaic virus is a protein, or closely associated with a protein, which may be hydrolyzed with pepsin."

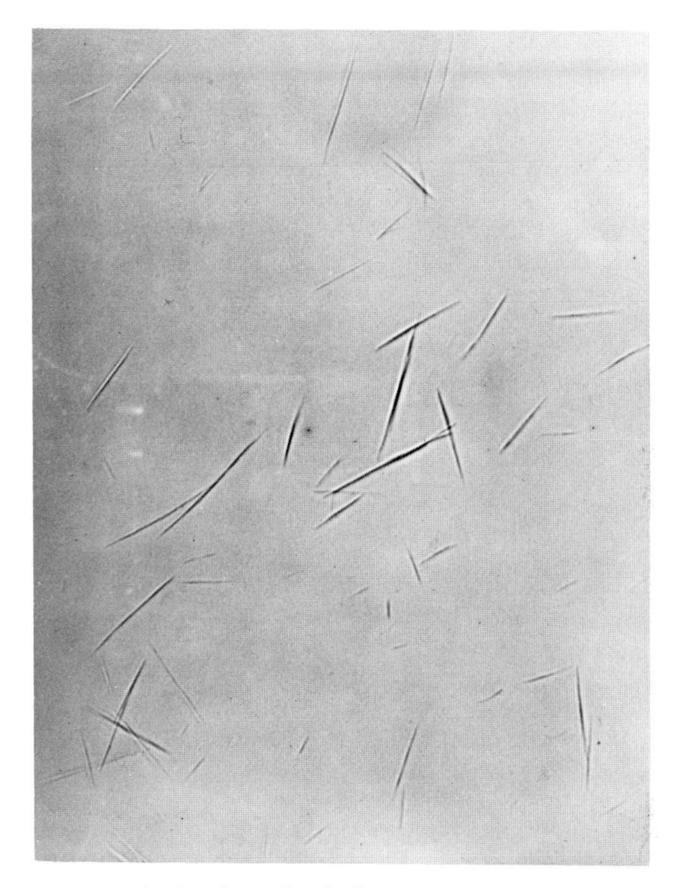

Fig. 2. Crystals of tobacco mosaic virus.

Proceeding, then, with the methods of a protein chemist, Stanley combined repeated precipitation with ammonium sulfate with decolorization by treatment with lead subacetate to obtain high yields of purified virus. Such virus in aqueous solution was crystallized by adding sufficient saturated ammonium sulfate to cause turbidity; then with stirring, adding slowly 0.5 saturated ammonium sulfate in 5 percent acetic acid. Needlelike crystals like those shown in Figure 2 were thus obtained. Such crystals when dissolved were infectious at dilutions as high as 10^9, and the infectivity of the material, in contrast to that of Vinson's preparations, was not lost by as many as ten successive recrystallizations.

From the results of many different tests, the crystalline material appeared to be protein, and preliminary osmotic pressure and diffusion measurements indicated that this protein had an extraordinary molecular weight of the order of several millions. The infectivity of the preparations was shown to depend on the integrity of the protein, and hence infectivity could be considered a property of the protein. Stanley concluded his historic paper published in *Science* (1935) with the statement: "Tobacco-mosaic virus is regarded as an autocatalytic protein, which, for the present, may be assumed to require the presence of living cells for multiplication."

It was inevitable that some details of Stanley's description of the chemical constitution of the virus would need modification. One was the initial report that the virus contained 20 percent nitrogen. Since his own subsequent, more accurate determinations yielded a nitrogen value of about 16.6 percent for the virus, the first reported value has been interpreted to mean that the initial preparations contained about 70 percent ammonium sulfate. However, this possibility was incompatible with other observations, and especially with the simultaneously reported ash content of only 1 percent. Hence, it seems that the earliest nitrogen analyses were faulty, but these were very soon corrected in the detailed paper (Stanley 1936) that followed the announcement in *Science*.

Another discrepancy between earlier and later elementary analyses that persisted for a year or two was the failure to detect any phosphorus in the preparations. However, Bawden and Pirie and associates (1936), who were actively working on plant viruses in England at the same time, reported that three strains of TMV (common, aucuba, and enation mosaic) contained phosphorus and carbohydrate, and that these components were present in ribonucleic acid, which could be released from the virus by heat denaturation. Stanley confirmed this point (1937). Although he at first viewed the nucleic acid as probably not essential for infectivity, he later reversed his judgment, and together with others established that several different plant viruses could be isolated as nucleoproteins.

In this connection, the earlier analyses of a bacterial virus by Max Schlesinger, working at the Institut für Kolloidforschung in Frankfurt, Germany, tend to be overlooked, probably because of the more extensive and definitive studies on TMV. However, Schlesinger (1934) found that a phage preparation that gave strong color reactions for protein and yet gave a negative test for bacterial antigen contained about 3.7 percent phosphorus. This led him to suggest that nucleoprotein might be a major component of bacteriophages, but the proposal lacked the force it would have carried had the presence of purine and pyrimidine bases been demonstrated.

Thus, the chemical era of virology was launched. The impact on research of Stanley's findings was aptly summarized by a pioneer animal virologist, Thomas M. Rivers, when he presented Stanley to receive the gold medal of the American Institute of the City of New York in 1941 (Rivers 1941). His remarks, in part, were as follows:

Stanley's findings, which have been confirmed, are extremely important because they have induced a number of investigators in the field of infectious diseases to forsake old ruts and seek new roads to adventure. As much as many bacteriologists hate to admit it, Stanley's proof that tobacco mosaic virus is a chemical agent instead of a microorganism is certainly very impressive. . . . In fact, the results of Stanley's work had the effect of demolishing bombshells on the fortress which Koch and his followers so carefully built to protect the idea that all infectious maladies are caused by living microorganisms or their toxins. In addition, his findings exasperate biologists who hold that multiplication or reproduction is an attribute only of life. In the midst of the wreckage and confusion, Stanley, as well as others, finds himself unable at the present time to decide whether the crystalline tobacco mosaic virus is composed of inanimate material or living molecules. In fun it has been said that we do not know whether to speak of the unit of this infectious agent as an "organule" or a "molechism."

II.

Purification of Viruses

A. Some General Principles

Each virus poses an individual purification problem that is related to the properties of the virus, the nature of the host, and the culture conditions. Consequently, it is not possible to outline a purification procedure that will work with equal effectiveness for all viruses. Nevertheless, it is possible to describe a few methods and their underlying principles that have led to purified preparations of some viruses, and, hence, that are potentially useful, separately or in combination, for the purification of other viruses. Attention is directed here to comprehensive reviews on the purification of plant and animal viruses (Steere 1959; Sharp 1953; Maramorosch and Koprowski 1967; Habel and Salzman 1969; Kado and Agrawal 1972).

Methods based on centrifugation have come to dominate the techniques of isolating and purifying viruses as well as to characterize viruses, at least in part. When centrifugation is coupled with a variety of other techniques based on different principles, its potential for purification is greatly enhanced. Some of the methods used as adjuncts to centrifugation include precipitation, adsorption, treatment with enzymes, extraction with organic solvents, treatment with antiserum, electrophoresis, and chromatography.

Two basic facts underlie the purification of viruses by whatever method used: (1) all presently known viruses contain substantial quantities of protein and hence are more or less susceptible to protein fractionating techniques; and (2) the sizes and densities of viruses are such that they are not readily sedimented in low gravitational fields, but are generally sedimentable in characteristic ways in high-speed centrifuges at 40,000 *g* or more.

Some general considerations should also be mentioned here. To determine the effectiveness of any purification procedure, it is essential that a suitable quantitative test for virus infectivity be available. For example, if a virus assay is subject to 50 percent variations (which is not uncommon in biological tests), it is difficult to determine in which fraction the virus is contained or the extent to which the purification conditions are destroying virus activity. Thus, an important contributing factor leading to the discovery of the nature of tobacco mosaic virus was the timely development of a local-lesion assay method (Holmes 1929). With this method the infectivities of fractions could be determined with an error of about 10 percent, a value several times as good as that usually achieved by the older dilution-endpoint assay. Later, assays of bacterial and animal viruses were developed that resembled the plant virus assays in the sense that at appropriate concentrations of virus a linear relationship was observed between concentration of virus and numbers of colonies of virus apparent in tests (local lesions on plant leaves in the case of plant viruses and cell plaques for bacterial and animal viruses). Such assays are illustrated in Figure 3.

If a satisfactory measure of virus activity is available, then it is possible to adjust purification conditions to allow for such factors as pH and thermal stabilities of the virus and salt effects. Lacking information on these factors, it is well to begin by working around neutrality and in the cold. Also, the use of 0.01–0.1 M phosphate buffer has proved a good salt medium for several viruses. Salt mixtures such as Ringer's solution are needlessly complex for most viruses; on the other hand, unbuffered "physiological" saline is deleterious to some viruses owing to its tendency to be somewhat acidic in reaction.

Organic buffers have proved superior to inorganic buffers in some biological systems including viral systems. Thus various salts of tris(hydroxymethyl)aminomethane and organic or inorganic acids provide the socalled Tris buffers with a buffering range between pH 7 and 9. Tris buffers, which have been widely used, do not precipitate divalent cations as phosphate buffers may. However, many biological reactions occur optimally between pH 6 and pH 8 and Tris buffers have poor buffering capacity below pH 7.5; moreover, Tris has a reactive primary amine group that can engage in undesirable or even inhibitory reactions. Consequently, considerable use of a series of zwitterionic buffers (Good et al. 1966) has developed. These buffers are mainly amino acid derivatives, many being N-substituted glycines or N-substituted taurines. They were shown to be superior to Tris or phosphate buffers in several important biological reactions (Good et al. 1966). Some commercially available zwitterionic buffers are listed in Table 1.

1. Centrifugation

a. Differential Centrifugation

The sizes of most presently known viruses (10–300 nm in diameter) and their densities are such that the viruses are sedimented from solution in an hour or two in centrifugal fields of 40,000–100,000 × *g*. Such centrifugal fields were achieved in the early years of virus purification with air-driven

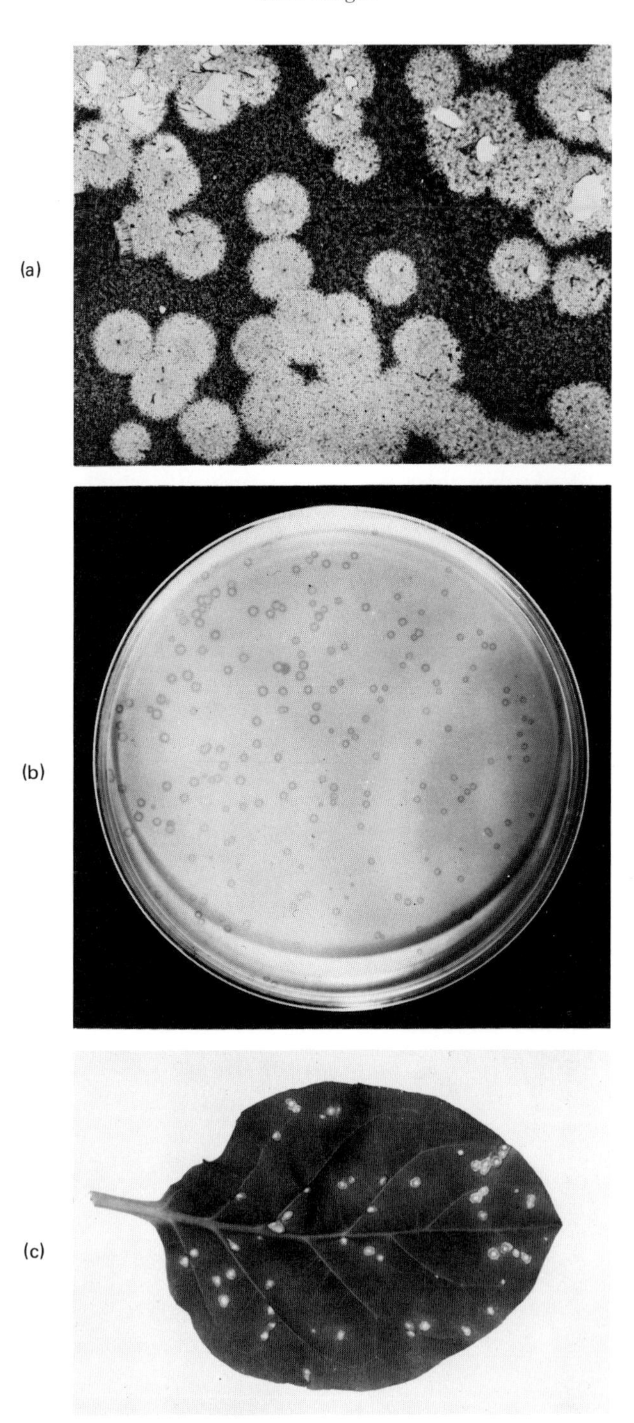
(a)
(b)
(c)

rotors (see Stanley et al. 1959), but these machines have now been largely superseded by electrically driven, commercially available centrifuges.

Some outstanding features of these electrically driven centrifuges are that they are compact, simple in service requirements, and are designed so as to give hours of uninterrupted service with maximum safety to the operator and to the mechanism. The rotors are made of duralumin alloy, a strong substance but relatively light. The same machine accommodates a variety of rotors, most of which hold ten or more plastic tubes in angular holes. The tubes are available in a variety of sizes to match the rotors so that as little as 2 ml or as much as 940 ml can be handled in a run. Both tubes and rotors are sealed, and since the rotor spins in a vacuum, the initial temperatures of rotor and sample change very little during a run. A refrigeration unit around the walls of the vacuum chamber makes it possible to avoid even slight rises in temperature if this is necessary. The force fields obtainable in one commercial model range from 59,000 *g* in the 21,000-rpm rotor, which holds 940 ml, to 198,000 *g* in the 50,000-rpm rotor, which holds 100 ml.

A cardinal feature of the electrically driven centrifuge, aside from its high-speed motor, is the flexible shaft linking rotor with motor. This allows some inexactness in balance of tubes placed opposite one another, for the rotor seeks its own axis of rotation on the flexible shaft.

Some viruses can be obtained in highly purified preparations using only the machine just described. The process of differential centrifugation means simply the application of alternate cycles of low-speed and high-speed centrifugation. This can be done in the same centrifuge, although, commonly, a separate and simpler angle centrifuge is used for the low-speed cycles. The isolation and purification of tobacco mosaic virus by differential centrifugation may be described as follows:

1. Frozen, infected tobacco plants are ground in a meat chopper, 3

Fig. 3. Plaques and local lesions representing infection by three major classes of viruses. Each plaque or spot illustrated represents several hundred cells although each is thought to have originated from the infection of a single cell.

a. Plaques caused by vesicular stomatitis virus in a monolayer of chick fibroblast cells attached to one side of a prescription bottle. After culture under a layer of agar, the overlay is removed and the sheet of cells is fixed and stained with an alcoholic crystal violet solution, rinsed, and air-dried. The background of uninfected cells retains stain, whereas groups of infected cells, which are grossly degraded by the infection, seem clear. (Courtesy F. L. Schaffer.)

b. Plaques caused by phage P22 on a lawn of *Salmonella typhimurium* bacteria. The bacterial lawn is on nutrient agar in a petri dish. The plaques represent groups of lysed cells. (Courtesy J. R. Roth.)

c. Local lesions (necrotic spots) on a leaf of tobacco (*Nicotiana tabacum* L. cv. Xanthi nc) caused by infection with tobacco mosaic virus. Each necrotic spot represents a group of dead cells and the virus is confined to such lesions.

Table 1. Some Zwitterionic Buffers[a]

Name		Useful pH Range[b]	Molarity of Saturated Solution at 0°	Binding of Ca^{2+} and Mg^{2+}	
Common	Chemical			Ca^{2+}	Mg^{2+}
ACES	N-(2-acetamido)-2-aminoethane sulfonic acid	6.0–7.5	0.2	+	+
ADA	N-(2-acetamido)-2-iminodiacetic acid	5.8–7.4	2.5	+	+
BES	N,N-bis(2-hydroxyethyl)-2-aminoethane sulfonic acid	6.4–7.9	3.2	–	–
Bicine	N,N-bis(2-hydroxyethyl) glycine	7.7–9.1	1.1	+	+
Bis-Tris Propane	1,3-bis(tris[hydroxymethyl]-methylamino)-propane	6.0–10.0			
CAPS	Cyclohexylaminopropane sulfonic acid	9.7–11.1			
EPPS	4-(2-hydroxyethyl)-1-piperazine propane sulfonic acid	7.4–8.6			

HEPES	N-2-hydroxyethylpiperazine N′-2-ethanesulfonic acid	6.8–8.2	2.2	–	–
MES	2-(N-morpholino) ethane sulfonic acid	5.5–7.0	0.6	+	+
MOPS	Morpholinopropane sulfonic acid	6.5–7.9			
PIPES	Piperazine-N,N′-bis [2-ethanesulfonic acid]	6.1–7.5		–	–
TAPS	Tris (hydroxymethyl) methylaminopropane sulfonic acid	7.7–9.1			
TES	N-tris (hydroxymethyl) methyl-2-aminoethane sulfonic acid	6.8–8.2	2.6	–	–
Tricine	N-tris (hydroxymethyl)-methyl glycine	7.4–8.8	0.8	+	+

[a]Compiled from Good et al. 1966 and from the March 1974 catalog Sigma Chemical Co., St. Louis, MO 63178, U.S.A.

[b]The pKa values at 20° for these compounds lie approximately midway in the useful range.

percent by weight of dipotassium phosphate is added to the mash, and the mixture is thawed with occasional stirring.

2. The juice is separated from the plant pulp in a basket centrifuge or with a press.

3. The expressed juice is clarified by centrifuging for 10 min in an angle centrifuge at about 8,000 rpm (6,000–8,000 g). The angle pellet is discarded (this contains starch, pigmented material, denatured protein, and so on).

4. The clarified juice is centrifuged at 21,000 rpm (about 59,000 g) for 1 hr. The supernatant fluid is discarded and the virus pellet is covered with 0.1 M phosphate buffer at pH 7 and allowed to soak overnight at 4°. The softened pellets dissolve readily with a little stirring such as that produced by squirting the liquid up and down with a dropping pipet.

5. The virus solution is centrifuged at low speed as in step 3 and the pellet (pigmented material, denatured protein, and so on) discarded.

6. The supernatant fluid is centrifuged at 21,000–40,000 rpm for 1 hr. (The virus becomes increasingly difficult to sediment as it is concentrated, particularly if the salt concentration is lowered.) The supernatant fluid is discarded and the pellets are dissolved again in 0.1 M phosphate buffer at pH 7, and centrifuged at low speed.

7. The alternate low-speed, high-speed runs are continued for four complete cycles. If the virus is to be lyophilized, the last two cycles are made in distilled water, thus removing salt. Although it is probably not necessary with TMV and some of the more stable viruses, refrigeration is employed in the high-speed centrifugation and the material is kept cold throughout the preparative procedure.

A technique for preparative microcentrifugation of viruses and other entities of similar size has been described by Backus and Williams (1953). In this method, pellets obtained with conventional centrifugation equipment are resuspended in 0.01–0.1 ml of diluent and are then transferred and sealed into "field-aligning" glass or quartz capsules. These are suspended in a solvent of suitable density in a standard plastic centrifuge tube and centrifuged in an angle rotor at an appropriate speed. By using supplementary equipment, such as a spectrophotometer, these capsules can also be used for analytical ultracentrifugation of virus preparations (Backus and Williams 1953).

b. Density-Gradient Centrifugation

A powerful adjunct to the conventional differential centrifugation procedure for the isolation and purification of viruses is density-gradient centrifugation (Brakke 1960, 1967; Vinograd and Hearst 1962; Schumaker 1967). Not only can separations be achieved by this method that are impossible in ordinary sedimentation but under appropriate conditions, densities and sedimentation coefficients also can be estimated.

The essence of the density-gradient system is the separation of particles partly or entirely on the basis of their densities in a convection-free medium. There are many modifications of the method, which, however, differ mainly in operational details such as (1) material used to form the gradient, (2) use of preformed gradient or one formed during the sedimentation, (3) gravitational field, or (4) length of time of centrifugation, especially in relation to equilibrium conditions. The results obtained will depend largely upon these factors.

In practice the different modifications of density-gradient centrifugation may be considered to fall into two classes:

1. Rate-Zonal (or Velocity) Density-Gradient Centrifugation [also termed "gradient differential centrifugation" by Anderson (1955)]. In this procedure the virus solution is layered on top of a preformed gradient, such as a sucrose or glycerol density gradient, and centrifuged in a swinging bucket rotor for 0.5–3 hr at about 70,000–170,000 *g*. (The time required for rate-zonal centrifugation is approximately equivalent to that required to sediment the virus completely in ordinary centrifugation in the same gravitational field.) Particles appear in zones according to their sedimentation rates; hence the term "rate zonal."

While density of the sedimenting particles is a primary factor in determining the zones obtained, size and shape of the particles and viscosity of the medium are also involved in these nonequilibrium conditions. Thus in rate-zonal centrifugation, virus particles tend to concentrate in a zone in which, barring interaction, contamination is mainly restricted to particles whose size, shape, and density combine to give them about the same sedimentation velocity as the virus.

An example of the use of the popular rate-zonal method of density-gradient centrifugation is found in the purification of poliovirus by Schwerdt and Schaffer (1956), and may be summarized as follows:

1. Density gradients were set up in 5-ml cellulose acetate tubes by layering 0.7-ml vol of 45, 37, 29, 21, and 11 percent (by weight) sucrose solutions in 0.14 M NaCl. A continuous gradient was established by allowing the tubes to stand at 4° for 12 hr.

2. About 0.7 ml of partially purified poliovirus (butanol extracted, enzyme treated, and two times ultracentrifuged) was layered on top of each density gradient and centrifuged in a swinging bucket rotor at 30,000 rpm (about 70,000 *g*) for 2 hr.

3. A narrow beam of light from a microscope light was shone down through the gradient column; and when viewed at right angles against a dark background, four bands could be distinguished, which were designated, from top to bottom, A, B, C, D, respectively.

4. Each of the four bands was removed in turn, starting with the upper one, A, by puncturing the tube with a hypodermic needle and withdrawing the appropriate volume of liquid into a syringe.

5. Various tests were made on the material of the four bands (Schwerdt and Schaffer 1956; LeBouvier et al. 1957). Virtually all of the infectivity was found in band D, but particles of similar dimensions and serological properties were also found in band C. The particles in band C, however, contained no more than a few percent RNA, whereas those in band D were found to contain 25–30 percent RNA.

A similar separation of two classes of particles that had essentially the same size and shape but differed in nucleic acid content (and hence in density) was made with partially purified Shope papilloma virus, using rate-zonal centrifugation in sucrose or glycerol density gradients (Williams et al. 1960). In conventional differential centrifugation, the two types of particles (nucleic acid containing and nucleic acid free) occur together, but they are nicely separated on the density-gradient column, and it was demonstrated that only the nucleic acid-charged particles are infectious (Figure 4).

Many examples of the applications of rate-zonal centrifugation to plant virus problems are reviewed by Brakke (1960).

2. *Equilibrium (or Isopycnic) Density-Gradient Centrifugation.* If the rate-zonal procedure is continued for a period of hours, most of the particles reach a zone corresponding to their densities (isopycnic position). Thus the zones obtained are essentially the equilibrium ones with respect to densities of particles.

Commonly, however, the concentration gradient is formed either prior to or during centrifugation (Meselson et al. 1957). Inorganic salts—cesium chloride, rubidium chloride, potassium bromide, and so on—have been employed, usually in the concentration range of about 6–9 M, and the establishment of a gradient depends upon the partial sedimentation of these salts in the centrifugal field. The virus solution is introduced either before or after formation of the gradient, and centrifugation is then continued (12–24 hr) until the particles have reached a point in the suspending medium of equal density (isopycnic position).

The original report of Meselson et al. (1957) nicely illustrates the power of the equilibrium density-gradient centrifugation method. They showed that the normal DNA of T2 bacteriophage could be readily distinguished from T2 DNA in which some of the thymine had been substituted by the denser component, 5-bromouracil.

Some other applications of the equilibrium density-gradient centrifugation method include the purification of ØX174 bacteriophage (Sinsheimer 1959a) and the demonstration that its DNA differs significantly in density from the DNA of the host, *Escherichia coli* (Sinsheimer 1959b); the demonstration of differences in density between strains of tobacco mosaic virus (Siegel and Hudson 1959), and between strains of herpes simplex virus (Roizman and Roane 1961); and the purification of potato virus X (Corbett 1961) and of Rous sarcoma virus (Crawford and Crawford 1961). Also,

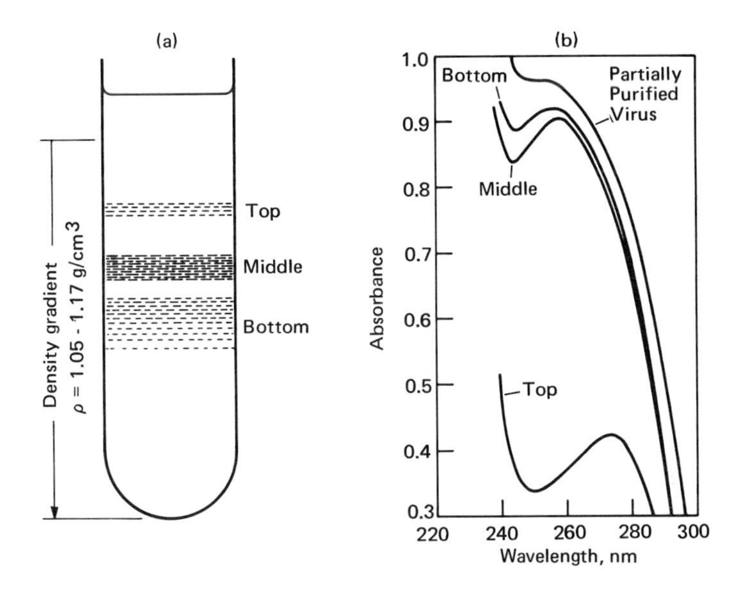

Fig. 4. *a*. Sketch of the bands observed after rate zonal density-gradient centrifugation of partially purified Shope papilloma virus in a glycerol gradient for 2.5 hr at 24,000 rpm in an SW-25 swinging bucket rotor. Material in the top band proved to be very low in DNA, virtually noninfectious, and exhibited particles that appeared hollow by negative staining electron microscopy (see Fig. 33). Material in the middle and bottom bands (bottom material appears to be an aggregate of middle) had much DNA, was highly infectious, and contained particles that appeared filled (dense) in negative staining electron microscopy (see Fig. 33) (Kass and Knight 1965).

b. Ultraviolet absorption curves of partially purified Shope papilloma virus and of fractions obtained from rate zonal centrifugation as sketched in 4*a*. Note that the peak of absorbance of top material is near 280 nm, characteristic of proteins, whereas the peak absorbances of middle and bottom components are near 260 nm, characteristic of nucleic acid and nucleoproteins having more than 5 percent nucleic acid (Williams et al. 1960).

differences in density between standard and "incomplete" influenza viruses have been shown by equilibrium sedimentation in cesium chloride (Barry 1960). Since this technique is almost universally employed in virology, this listing is far from complete.

There has been some concern about the possible deleterious effect of the gradient materials on viruses, especially in the equilibrium method, but thus far most viruses have appeared quite stable in strong cesium chloride and similar salts. While only a milligram or less of material is normally applied per tube for density-gradient sedimentation, it is possible to greatly expand the quantity of material separated by this technique if suitable equipment is employed (see Anderson and Cline 1967).

2. *Enzymatic Treatment*

A property that has aided in the purification of some viruses is the resistance of most of them, unless denatured, to attack by proteolytic enzymes and nucleases. For example, Bawden (1950) and co-workers used snail enzymes to effect a greater release of plant viruses from leaf tissues, or trypsin and chymotrypsin to digest pigment-protein complexes attached to plant viruses. Likewise, Bachrach and Schwerdt (1954) and Schwerdt and Schaffer (1955, 1956) used pepsin and nucleases in purifying poliomyelitis virus. Herriott and Barlow (1952) used treatment with deoxyribonuclease as a step in the purification of T2 bacteriophage in order to destroy host DNA, which might otherwise adsorb to the phage.

3. *Extraction with Organic Solvents*

The purification of poliomyelitis virus is aided by an extraction with *n*-butanol (Bachrach and Schwerdt 1952, 1954). In this procedure, the virus remains in solution in the aqueous layer; lipids, when present, are extracted into the butanol layer, and a significant proportion of the nonviral proteins are denatured and appear mainly in the interphase. A modification of this technique, employing equal parts of *n*-butanol and chloroform, has been used successfully in the purification of some plant viruses (Steere 1956, 1959; Frisch-Niggemeyer and Steere 1961).

Certain fluorocarbons, such as Freon 112 ($FCl_2C\text{-}CCl_2F$) or Genetron 226 ($F_2ClC\text{-}CCl_2F$), alone or mixed with *n*-heptane, have been widely employed in the purification of animal viruses (see Gessler et al. 1956a, 1956b; Philipson 1967b; Brown and Cartwright 1960 for examples and other references). Treatment with the fluorocarbon genesolv-D was an important step in the purification of cytoplasmic polyhedrosis virus of the silkworm (Lewandowski et al. 1969). The fluorocarbon extraction procedure was also used with some success in the isolation and purification of tobacco mosaic and ringspot viruses, but caused loss of infectivity when applied to the isolation of common cucumber mosaic virus (Porter 1956). The success of the fluorocarbon treatment of virus-containing tissue homogenates appears to lie in the efficiency of fluorocarbon emulsions in gathering and holding nonviral proteins and lipids in the organic phase while concentrating the viruses, without significant loss of infectivity, in the aqueous phase.

4. *Precipitation Methods*

An outline of a precipitation method, adapted from Stanley's procedure (1938), used to purify tobacco mosaic virus illustrates some of the details and principles of this method:

1. Infected plants are harvested and frozen. (Many viruses will withstand freezing and thawing in tissues or in crude extracts, whereas normal cellular components are frequently denatured. Hence this step may aid in purification.)

2. The frozen plants are minced in a meat chopper, and 3 percent by weight K_2HPO_4 (in a 50 percent solution) is stirred into the mash, which is then allowed to thaw. The buffer serves a dual purpose: aiding in the extraction of virus and maintaining a suitable pH. (Finer grinding, such as obtained in a roller mill [Bawden 1950] will, in some instances, release considerably more virus than a meat chopper. However, it is often more difficult to purify the virus from such mashes and there is evidence for greater destruction of virus in fine grinding.)

3. The juice is separated from the solids by allowing the mash to drain in the cold through several thicknesses of gauze or cheesecloth followed by further expression of juice from the mash in a canvas bag in a laboratory press. Alternatively, the juice may be separated from the solids by centrifuging the mash in a basket centrifuge.

4. The juice is clarified by passing through about 1 cm of diatomaceous silica (Celite filter-aid) on a Buchner funnel.

5. The virus is precipitated from the clarified juice by addition of 30 percent by weight of ammonium sulfate, and is separated by filtration on a thin layer of Celite.

6. The virus is redissolved by suspending the Celite in 0.1 M phosphate buffer at pH 7 and is separated from the Celite by filtering on a Buchner funnel.

7. The virus is precipitated again by addition of about 11 percent by weight of ammonium sulfate and filtered again on Celite. (Less ammonium sulfate is required to precipitate the virus from purified solutions than from crude juice.)

8. After two or three more precipitations with ammonium sulfate, the last precipitate is dissolved in water and the solution adjusted to pH 4.5, causing precipitation of the virus. (The virus is insoluble at its isoelectric point and for a considerable zone on either side depending on salt concentration.)

9. The virus is filtered on Celite as above and dissolved in water at pH 7. This constitutes the purified preparation of virus. If it is desired to crystallize the virus, saturated ammonium sulfate is added dropwise to the aqueous solution until a cloudiness develops; then 5 percent acetic acid in half-saturated ammonium sulfate is added with vigorous stirring until a lustrous sheen signals the presence of needlelike crystals.

Precipitation with ammonium sulfate has also been used to obtain highly purified crystalline preparations of tomato bushy stunt, tobacco necrosis, southern bean mosaic, turnip yellow mosaic, and squash mosaic viruses (Bawden 1950; Price 1946; Markham et al. 1948; Takahashi 1948).

In some cases magnesium sulfate is substituted for the ammonium salt, and occasionally ethanol is used, either to precipitate the virus or, by selection of proper conditions, to precipitate impurities from the virus solution.

Precipitation by neutralized ammonium sulfate has been helpful in concentrating myxo- and paramyxoviruses. How this step fits into the procedure for purifying simian virus 5, Newcastle disease virus, and Sendai virus is illustrated in the following example (Mountcastle et al. 1971).

1. All three viruses were grown in monolayer cultures of bovine kidney cells on plastic surfaces in reinforced Eagle's medium. The culture medium into which virus was released from the infected cells was harvested 24–42 hr after inoculating the cells.
2. Cell debris was removed by sedimentation at 3,000 *g* for 20 min and virus was precipitated from the supernatant by addition of an equal volume of saturated ammonium sulfate.
3. The precipitate of crude virus was resuspended in Eagle's medium and banded twice in linear 15–40 percent (w/w) potassium tartrate density gradients at 23,000 rpm for 2.5 hr in a Spinco SW 25.1 rotor.
4. The viral band in each case was dialyzed overnight against Eagle's medium or distilled water, and either used immediately or stored at −60°C.

Lowering the pH of virus extracts may precipitate the virus, resulting in concentration and some purification. However, caution is needed in such precipitations because of the sensitivity of some viruses to acidic conditions. Even pH 5 is destructive to many strains of influenza virus. On the other hand, some plant viruses are both stable and soluble at pH 5, whereas host cell constituents are not. Thus, white clover mosaic virus can be purified by adjusting the pH of centrifugally clarified plant sap to 5 with acetic acid and then discarding the resultant precipitate (Miki and Knight 1967). Virus is recovered after this treatment by a few cycles of differential centrifugation.

Precipitation techniques have frequently been used with animal viruses to quickly concentrate virus from dilute solution. For example, the virus in large volumes of allantoic fluid from chick embryos infected with influenza virus was concentrated and partially purified by precipitation in the cold with 25–35 percent methanol (Cox et al. 1947). This procedure has also been applied to the concentration and partial purification of eastern equine encephalomyelitis, MEF poliomyelitis, mumps, and Newcastle disease viruses (Pollard et al. 1949).

Protamine is another precipitating agent used in the concentration and partial purification of animal viruses (Sharp 1953). In some cases this reagent precipitates the virus more or less selectively, whereas in other instances contaminating materials are thrown down and the virus largely remains in the supernatant fluid.

Commercial yeast nucleic acid was found useful in enhancing the acid precipitation of poliovirus (Charney et al. 1961). Previously it had been shown that poliovirus could be precipitated from tissue culture filtrates by acidifying to about pH 2.5 (Mayer et al. 1957; Charney 1957), but acid treatment alone resulted in inexplicably variable degrees of precipitation of the virus. However, essentially complete and reproducible precipitation of the virus is effected by adding 50 μg of commercial sodium ribonucleate per milliliter of tissue culture fluid (containing 0.1–1 μg virus/ml) before lowering the pH to 2.5. After the virus is concentrated in this manner, other methods, such as differential centrifugation, can be used for further purification.

5. *Adsorption Methods*

Adsorption of viruses on, and subsequent selective elution from, various materials has been used with moderate success to purify a few plant and animal viruses, but this principle has seldom been used alone in the concentration and purification of viruses. Historically, however, adsorption techniques were among the first methods employed for concentrating and purifying animal viruses (Stanley 1938), and, with the development of new adsorbents such as ion exchange resins and cellulose derivatives, this approach has new potential.

A common procedure with adsorbing agents is to cause a precipitate to form in the crude virus suspension, upon which the virus is adsorbed and from which it is subsequently eluted, usually by treatment with an appropriate buffer. Included among adsorbing materials of this sort, which have been used especially with animal viruses, are aluminum phosphate, calcium phosphate, calcium sulfate, and ferric and aluminum hydroxides. In addition, kaolin, charcoal, and alumina have been employed as adsorbing agents. Sometimes impurities are adsorbed rather than the virus. Such an example was observed with potato virus X (Corbett 1961) from whose crude preparations pigments and microsomal material were removed preferentially by activated charcoal, permitting subsequent isolation of nonaggregated virus rods of high infectivity by centrifugation on a density gradient.

"Salting out" chromatography on Celite and on calcium phosphate columns has helped in the purification of chicken tumor virus (Riley 1950), tobacco mosaic virus (Tiselius 1954), herpes virus, encephalomyocarditis virus, and others (Philipson 1967a).

Tobacco mosaic virus and related materials, potato virus X, southern bean mosaic virus, rice dwarf virus, and internal cork virus of sweet potatoes have been purified with moderate success on ecteolacellulose, DEAE cellulose, carboxymethyl cellulose, chitin columns, and agar gel columns (von Tavel 1959; Levin 1958; Commoner et al. 1956; Cochran et al. 1957; Townsley 1961; Steere and Ackers 1962; Venekamp 1972). At present, none of these procedures seems likely to replace centrifugation for the routine purification of TMV or other viruses, although they do offer advantages for certain specific objectives.

Similarly, influenza, adenovirus, and T2 phage have been purified on cellulose columns of one sort or another (Laver 1961; Haruna et al. 1961; Taussig and Creaser 1957), and several different animal viruses have been partially purified by adsorption on and elution from ion exchange resins (Lo Grippo 1950; Muller 1950; Muller and Rose 1952; Takemoto 1954; Matheka and Armbruster 1956a, 1956b; Philipson 1967a).

Outstanding success was reported in removing tenacious pigment from southern bean mosaic virus with an ion exchange column (Shainoff and Lauffer 1956, 1957). In one experiment, 490 mg of 520 mg virus applied were recovered when the column was eluted with chloride-phosphate at room temperature at a rate of 100 ml/hr. The pigment could be seen as a brownish discoloration of the resin along two-thirds the length of the column.

Incidentally, the problem of tenacious pigments is rather common to the preparation of plant viruses and the following approaches have been used in attempts to solve it: (1) displacement of the pigment by polyvalent anions (Ginoza et al. 1954); (2) extraction with organic solvents (Steere 1959); (3) treatment with proteolytic enzymes (Bawden 1950; Markham and Smith 1949); (4) column chromatography (Shainoff and Lauffer 1956, 1957); (5) treatment with charcoal (Corbett 1961); and electrophoresis (Goldstein et al. 1967).

The use of magnesium pyrophosphate for the rapid purification of coliphages (Schito 1966) and aluminum phosphate for the purification of influenza viruses (see below) are other successful applications of adsorption techniques. Miller and Schlesinger (1955) used aluminum phosphate-silica gel either in a batch-type process or in a column to adsorb influenza virus directly from infectious allantoic fluid. By adapting this column procedure and by using virus preparations that had already been partially purified by differential centrifugation, Frommhagen and Knight (1959) were able to obtain not only highly purified virus, but also specific host material. This host material is important because it is serologically closely related to one or more components inseparably bound to influenza virus particles (Knight 1944, 1946a).

The following outline describes the use of such a column:

1. A glass column about 5 × 100 cm is charged with aluminum phosphate-silica gel, prepared as described by Miller and Schlesinger (1955).

2. Partially purified influenza virus, obtained from infectious allantoic fluid by one cycle of centrifugation (see section on centrifugation), is diluted with 0.125 M phosphate buffer at pH 6.0 (for an A strain like PR8) or with water (for a B strain like Lee) to a concentration containing 100 chick cell agglutinating units (CCA units) per milliliter. This solution is put

through the column (adsorbing cycle) at a rate of about 4 ml/min. The effluent of the adsorbing cycle emerges with a milky opalescence, possesses almost no hemagglutinating power or infectivity, and can be shown by serological tests to be very similar to normal allantoic material. The effluent can be further purified by repeated passage through the column, and can be concentrated by sedimenting at 44,000 g for 20 min. The virus is eluted from the aluminum phosphate by passing through at about 5 ml/min a volume of 0.25 M phosphate buffer at pH 8.0 equal to one and a half times the volume of solution applied to the column. The eluted virus can be concentrated, if desired, by sedimenting it from the eluate at 44,000 g for 20 min.

The major disadvantages of the adsorption methods mentioned above are that they are usually not sufficiently selective, and the high salt concentrations generally required for elution may be deleterious to the virus, or in any case, often constitute an unwanted component that can complicate subsequent procedures with the virus. However, as illustrated above, adsorption methods, combined with centrifugation or other techniques, may aid substantially in the purification of viruses.

A special application of adsorption is that which involves viruses and red cells. Several animal viruses have been found to adsorb on and cause agglutination of appropriate erythrocytes, and some of these viruses, though not all, can be readily eluted from the red cells (Hirst 1959; Rosen 1964). The specificity of the adsorption in the case of influenza is so great that this phenomenon was an invaluable step in the preparation of highly purified virus from infective chick embryo fluids and from lung tissue homogenates (Knight 1946b), but density-gradient centrifugation has now largely replaced it. In practice, adsorption on and elution from erythrocytes is accompanied by centrifugation procedures, and it is particularly important to subject eluted virus to two cycles of high- and low-speed centrifugation to free it from red cell proteins (mainly hemoglobin) that may enter the preparation during the adsorption-elution process.

6. *Serological Methods*

Occasionally, it is possible to purify partially a virus by serological methods. This technique was successfully applied by Cohen and Arbogast (1950) to some preparations of bacteriophage. These were treated with antiserum to *Escherichia coli*, causing a precipitate of the traces of cellular debris present. This precipitate was removed by centrifugation and discarded, and the virus was freed of serum elements by two cycles of differential centrifugation.

Likewise, antisera to proteins of healthy plants have been employed to help purify certain plant viruses. The problem of adding serum protein contaminants to the virus under purification applies here as it does with phages. However, contamination with serum elements can be minimized by use of partially purified antibodies, that is, the gamma globulin fraction, rather than whole serum. After the complex of plant proteins and their antibodies is precipitated, the virus is separated from excess serum globulin by differential or density-gradient centrifugation. Such a technique was successfully applied by Gold (1961) to the purification of a strain of tobacco necrosis virus and subsequently by Fulton (1967a, 1967b) in the purification of rose mosaic, tulare apple mosaic, and tobacco streak viruses.

7. *Electrophoresis*

Conventional solution electrophoresis has been employed in the purification of southern bean mosaic virus (Lauffer and Price 1947), tobacco ringspot virus (Desjardins et al. 1953), tomato ringspot virus (Senseney et al. 1954; Kahn et al. 1955), influenza virus (Miller et al. 1944), blue-green algal virus LPP-1 (Goldstein et al. 1967), among others. However, this method is limited, largely because of the complexity of the equipment required and the difficulties in sampling. Sampling is facilitated in at least one commercial apparatus with a sampling needle that can be seen and controlled by the operator while looking at the migrating boundaries in the optical system (see van Regenmortel 1972).

Convection and sampling problems are greatly reduced in zone electrophoresis carried out in sucrose density gradients. A procedure of this sort was first used by Brakke (1955) in the purification of potato yellow-dwarf virus, and subsequently was greatly extended by van Regenmortel (1964, 1966, 1972), who used it successfully for separating 20 different plant viruses from contaminating antigenic plant proteins. Several animal viruses have been found to migrate distinctive distances in zone electrophoresis (Polson and Russell 1967), and it is probable that some purification occurs during the process, although this has not been adequately evaluated.

Electrophoresis in solid media such as agar, agarose, or polyacrylamide gels has been effective in the purification of tobacco mosaic virus, potato virus X, and turnip yellow mosaic virus, as well as in the separation of some animal viruses (Townsley 1959; Tiselius et al. 1965; Polson and Russell 1967). However, other purification techniques appear to be more convenient or selective or both.

8. *Partition in Liquid Two-Phase Systems*

Albertsson and co-workers (see Albertsson 1960, 1971) described several two-phase systems of water-soluble polymers, such as dextran-methylcellulose or dextran-polyethylene glycol, in which animal, bacterial, and plant viruses could be selectively concentrated and at least partially purified. Concentration of a virus by this method is based on its low partition coefficient in the systems used, which by proper adjustment of phase volumes permits concentration of most of the virus in a low-volume phase (or at the interface). In practice, concentrations of 10–100 times were obtained in one step or as much as 100–10,000 times in a two-step operation. Purification occurs mainly because impurities distribute differently from the virus. Proteins, for example, distribute rather evenly ($K = 0.5$–1). The purification effect is enhanced by repeated partitioning, as, for example, in a countercurrent apparatus. The usefulness of the method is also increased by combination with other techniques such as fluorocarbon extraction (dextran seems to stabilize sensitive viruses in this treatment) or high-speed centrifugation, with or without a density gradient.

The following one-step example (Albertsson 1960) illustrates the possibilities of this technique:

1. ECHO virus, prototype 7, was grown in monkey kidney tissue culture in Parker 199 medium.
2. To 5,000 ml virus culture was added 64 g of a 20 percent (w/w) sodium dextran sulfate solution and 1,390 g of a 30 percent (w/w) polyethylene glycol solution containing 69.5 g NaCl.
3. The mixture was shaken in a separatory funnel and allowed to stand at 4°C for 24 hr for phase separation.
4. Practically all the virus (as judged by infectivity tests) was found in the clear bottom phase in a volume of 50 ml. Hence the virus had been concentrated 100 times. With respect to purification, the original 5,000 ml of virus culture contained 1090 mg nitrogen and the concentrate was found to have only 39.5 mg nitrogen. Thus more than 96 percent of the nitrogen was removed (much of it in the form of particulate material that collected at the interface) with essentially no loss of infectivity.

The advantages claimed for the liquid two-phase separation are mainly its mildness and simplicity. Its strong point appears to be the ease with which rather great concentrations of virus can be achieved. While significant purification of the virus is also obtained, supplemental methods are usually needed when highly purified virus is the objective. There is also the problem of separating the virus from polymer, but this can be done either by repeated ultracentrifugation, which mainly sediments the virus, or by precipitation of the polymer, which, however, usually adds inorganic salt. The virus can be separated from most of the salt by high-speed centrifugation or the salt can be removed by dialysis or gel filtration.

9. *Criteria of Purity*

Prior to chemical analyses, it is essential to evaluate the purity of a virus preparation. A few definitions are required here. Viruses are infectious agents mainly characterized by their small size (10–300 nm in diameter, or the nonspherical equivalent) and ability to reproduce only in living cells. Traditionally, this is essentially the same as saying that viruses are infectious nucleoproteins or somewhat more complex particles. The traditional view will be followed here despite the suggestion, for which there is considerable justification, that viruses are infectious nucleic acids (Northrop 1961). In these terms, purity means the degree of freedom of viral particles from nonviral components, or, conversely, the extent to which viral particles show gross physicochemical homogeneity. No single test is sufficient to establish this type of purity, but a consistent answer from each of several tests establishes the degree of homogeneity of the preparation in question and hence the reliance to be placed on analytical data and other results obtained with such a preparation.

The degree of homogeneity of a virus preparation with respect to particle size, shape, and density can be evaluated in modern analytical centrifuges (see Stanley et al. 1959; Schramm 1954; Schachman 1959; Markham 1967). Thus a single sedimentation boundary suggests the presence of a single species of particle, two boundaries, two components, and so on.

Furthermore, the nature of the boundary can be significant, for the degree of boundary spreading observed with a homogeneous preparation should be no greater than expected from the diffusion constant, as independently determined. Likewise, the results of the diffusion measurements themselves can provide information regarding the homogeneity of the material. Incidentally, a combination of the results of sedimentation and diffusion measurements permits a calculation of molecular weight, and if supplementary data, such as density or viscosity values, are available, one can estimate the particle radius by application of Stokes' law (Stanley et al. 1959; Schramm 1954; Schachman 1959; Markham 1967).

Another widely applied criterion of purity is electrochemical homogeneity as measured in the electrophoresis apparatus (Stanley et al. 1959; Alberty 1953; Brinton and Lauffer 1959). It can be regarded as good evidence for homogeneity of a virus preparation if the material migrates with a single boundary over the entire pH range within which the virus is stable. This evidence is strengthened if the boundary shows no greater spreading than anticipated from the diffusion constant.

The lower limit of contaminant detectable by either sedimentation analysis or electrophoresis is variable, and is dependent upon the nature of the material and the circumstances of the test. As usually applied in testing virus preparations, these methods cannot be expected to detect less than a few percent of contaminant (Sharp 1953). For many purposes, it is satisfactory to measure purity to this degree, but as the tools for chemical and biological analyses become sharper and sharper, it will be increasingly necessary to remember the limitations of sedimentation and electrophoresis measurements.

The electron microscope can be used to examine directly the physical homogeneity of a virus preparation. Under favorable conditions it is possible to detect an impurity present in a concentration of as little as 1 percent of the virus (Williams 1954). It is obvious, of course, that impurities will escape detection if they have the same size and shape as the virus particles, or if they are below the size resolved by the microscope. Also, particles present in small number but large in mass are easily overlooked, owing to sampling difficulties (Lauffer 1951). Nevertheless, under favorable conditions it appears that the electron microscope is capable of detecting impurities at a level presently unattainable by any other physical method.

Crystallinity, once considered by many as evidence of purity, has fallen into disrepute (Pirie 1940). This is primarily because of the demonstration that crystalline protein preparations may be contaminated by amorphous material, by crystals of other substances, or by reason of containing mixed crystals, that is, solid solutions. Furthermore, not all proteins or viruses will crystallize regardless of their purity. Nevertheless, it is clear that crystallization usually results in purification, and it seldom occurs unless one constituent is predominant and in a native state. Denatured proteins are known to lose crystallizing ability (Putnam 1953). Therefore, the criticism of crystallinity as a criterion of purity is valid, mainly in the sense that this property does not afford a precise means for determining whether or not any contaminant is present, and, if so, how much. If a virus will crystallize, however, it is still a good preliminary indication of purity.

Frequently, immunochemical methods are used to good advantage in testing the purity of virus preparations (Kabat 1943; van Regenmortel 1966; Bercks et al. 1972). With the use of proper antisera one may detect by means of precipitin or complement fixation tests very small amounts of contaminating tissue antigens. Such impurities were demonstrated in early preparations of tobacco mosaic virus, but the most highly purified preparations, such as are now employed in most chemical studies, give no indication by serological means of the presence of normal antigens, even when tested by the extremely sensitive anaphylactic test (Bawden and Pirie 1937).

Use of the immunochemical approach to test for host antigens in preparations of influenza virus led to the discovery of a previously unfamiliar host-virus relationship. Highly purified preparations of the virus, obtained from allantoic fluid of infected chick embryos, and which were homogeneous in the analytical ultracentrifuge under various conditions and in the electrophoresis apparatus over a wide range of pH, reacted strongly in quantitative precipitin tests with antiserum to material isolated from normal allantoic fluid (Knight 1946a). Similarly, highly purified virus isolated from mouse lungs was found to contain an antigenic component characteristic of normal mouse lungs. Since the host antigen material could not be separated from the virus particles by a variety of methods, it was concluded that influenza viruses contain such antigens as integral parts of the virus structure. A similar conclusion was reached by Smith and colleagues as a result of extensive complement fixation tests (1955). Munk and Schäfer found an almost exactly parallel situation for fowl plague and Newcastle disease viruses grown in the chick embryo (1951). Thus, an incorporated host antigen seems to be characteristic of several viruses, especially those whose particles have an envelope structure.

With respect to the quantitative side of immunochemical tests, it should be noted that the quantitative precipitin test will give a figure for the amount of host antigen in the virus preparation (Kabat 1943). When the host antigen is incorporated in or tightly bound to the virus particles, this value may be only approximate, owing to changes in the number and reactivity of antibody-binding sites. Thus, it was calculated that influenza virus particles contain 20–30 percent host antigen (Knight 1946a), and that Newcastle disease virus contains about 42 percent (Munk and Schäfer 1951). If the host antigen and virus particles are not combined, it is possible, of course, to obtain a more precise estimate of the quantity of host antigen present. Furthermore, many antigens are detectable and can be quantitatively measured at 1 percent or less.

The popular gel diffusion serological method has found use in identifying and evaluating the purity of viruses (see, for example, LeBouvier et al. 1957; van Regenmortel 1966).

Finally, the importance of the immunochemical test as a criterion of purity is its independence of the factors basic to other tests, such as size, shape, and electrochemical properties. On the other hand, a limitation of the method that must be remembered is that not all potential impurities are good antigens; furthermore, the antihost constituent serum may not be sufficiently comprehensive to detect all possible contaminants. Nevertheless, the immunochemical method is one of the most powerful and sensitive means for evaluating the purity of a virus preparation.

One of the most exacting tests of homogeneity of proteins is the constant solubility test (Taylor 1953). However, this test is not used widely as a criterion of the purity of viruses because of restrictions imposed by the limited quantity of material available in many cases and technical difficulties in others. From the results of early solubility studies made on purified preparations of tobacco mosaic virus, Loring (1940) concluded that the virus is not a homogeneous chemical substance. Since that time, numerous refined chemical and structural studies made on this virus have demonstrated a surprising uniformity of its properties (Knight 1954; Tsugita et al. 1960), and lead one to conclude that there must have been something anomalous in the solubility studies or the preparations of virus used in these studies. Further work needs to be done along these lines.

In summary, no single criterion of purity is sufficient to establish the homogeneity of a preparation of virus. This must be done by applying critically as many tests as possible (see Knight 1974).

In all of the homogeneity procedures described, it is, of course, essential to try to relate the observed physical particles to the biological activity (infectivity) and to show that the characteristic particles are the biologically active ones. Many such tests of this character were made in establishing the identity of the infectious entity and the characteristic rodlike particles of tobacco mosaic virus (Stanley 1939). Lauffer and colleagues have described rigorous methods for relating biological activity to the physical particles observed in a variety of ways (Lauffer 1952; Epstein and Lauffer 1952; Hartman and Lauffer 1953; Shainoff and Lauffer 1957). Finally, with certain of the T-bacterial viruses it has been possible to make a good correlation between the particles counted with the electron microscope and the number of infectious units found by infectivity measurements (Luria et al. 1951).

III

Composition of Viruses

Several hundred virus diseases are now recognized as such, but the chemical compositions of relatively few viruses have been reported. The main reasons for this situation are that many interesting experiments can be done without precise information about viral composition, and it is not easy to obtain all viruses in a state of purity adequate for analysis. Finally, it is somewhat tedious to perform complete and thorough analyses of viruses even when they can be obtained in adequate amounts and in sufficient purity. Consequently, it is common for investigators to make just those analyses most pertinent to a particular topic under investigation.

Tobacco mosaic virus proved to be a nucleoprotein, that is, a specific combination of nucleic acid and protein. The same characteristic nucleoprotein was obtained from a variety of hosts infected with tobacco mosaic virus (TMV). In each instance the virus nucleoprotein appears to be foreign to its host as judged from its absence in normal plants and its lack of serological relationship to the normal host constituents. This holds true for many other viruses, although among the more complex viruses, those that acquire envelope structures at and bud out through cell membranes typically contain some host cell constituents in their envelopes.

Most viruses give rise during multiplication to occasional mutants, or variants; when isolated, these are found to be of the same general composition as that of the parent virus. It now seems generally true that strains of a given virus have identical proportions of protein and nucleic acid, although exceptions to this rule have been observed in mutants of the more complex viruses.

All types of viruses, including plant, bacterial, higher animal, and insect viruses, contain either ribonucleic acid (RNA) or deoxyribonucleic acid (DNA) (see Table 2). At one time it was thought that some viruses, such as influenza virus, contained both types of nucleic acid, but, as a consequence of more refined methods for purification and analysis of viruses, no virus is presently known that contains both RNA and DNA. This may be contrasted with bacteria including *Rickettsia* and *Bedsonia*, which appear to have both DNA and RNA (see Allison and Burke 1962). Some other comparisons between viruses and bacteria are given in Table 3.

Almost all of the many plant viruses that have been obtained in a highly purified state, as well as some small animal and bacterial viruses, have proved to be simple nucleoproteins. However, some of the larger animal and bacterial viruses also possess lipid, polysaccharide, and other components. It should be noted that a few plant and bacterial viruses have also been found to contain some lipid. The general compositions of some viruses are listed in Table 2.

A. Proteins

The structures of proteins, including those of viruses, are usually complex and hence require a series of studies to characterize them. A division of protein structure into four distinctive types can be made (see Kendrew 1959), each of which can be investigated experimentally:

1. Primary structure: the number of different peptide chains (determined mainly by end group analyses), interchain binding, if any (such as by S-S-bonds), and, most especially, the sequence of amino acid residues. These features of primary structure are illustrated by the formula for bovine insulin as determined by Sanger and associates (Ryle et al. 1955):

S—S

Gly–Ile–Val–Glu–Gln–Cy–Cy–Ala–Ser–Val–Cy–Ser–Leu–Tyr–Gln–Leu–Glu–Asn–Tyr–Cy–Asn

S S

S S

Phe–Val–Asn–Gln–His–Leu–Cy–Gly–Ser–His–Leu–Val–Glu–Ala–Leu–Tyr–Leu–Val—— Cy–*X*

In order to save space, the last portion of the second peptide chain has been represented by *X*, in which *X* stands for -Gly-Glu-Arg-Gly-Phe-Phe-Tyr-Thr-Pro-Lys-Ala.

It can be seen from the formula that insulin has two peptide chains that are joined at two positions by disulfide bonds between cysteine residues, and the sequence of amino acid residues is as indicated. In this formula and in subsequent tables, the abbreviations commonly employed by protein chemists are used. A list of such abbreviations is given in Table 4.

2. Secondary structure: the geometric configuration of the peptide chain (or chains) with special reference to the presence or absence of helical structure. The commonest configuration of peptide chains is probably the so-called alpha helix of Pauling and associates (1951), which is illustrated in Figure 5. It will be noted that a spiral structure is favored by the formation of hydrogen bonds between adjacent -CO and -NH- groups.

3. Tertiary structure: the folding pattern of the peptide chain. Possibilities of folding (often the folding of a helical chain) are illustrated by

Table 2. Approximate Composition of Some Viruses.

Virus	% RNA	% DNA	Protein[a]	% Lipid	% Non-nucleic Acid Carbohydrate	Ref.[b]
Adenovirus		13	87			1
Alfalfa mosaic	19		81			2
Avian myeloblastosis	2		62	35	1[c]	3, 4
Blue-green algal LPP-1		48	52			5
Broad bean mottle	22		78			6
Brome mosaic	21		79			7
Carnation latent	6		94			8
Cauliflower mosaic		16	84			8
Coliphages f1, fd, M13		12	88			9
Coliphages f2, fr, M12, MS2, R17, QB, ZR	30		70			10
Coliphage ØX174, ØR, S13		26	74			9
Coliphages T2, T4, T6		55	40		5[d]	11, 12
Cowpea mosaic	33		67			8
Cucumber 3 (and 4)	5		95			13
Cucumber mosaic	18		82			8
Dasychira pudibunda L.	7		93			14
Encephalomyocarditis	30		70			15
Equine encephalomyelitis	4		42	54		16
Fowl plague	2		68	25	+	17
Herpes simplex		9	67	22	2	18, 18a
Influenza	1		74	19	6	19, 20
Mouse encephalitis (ME)	31		69			21
Pea enation mosaic	29		71			8
Poliomyelitis	26		74			22
Polyoma		16	84			23
Potato spindle tuber	100					24

Potato X	6		94			25
Reovirus	21		79		+	18
Rous sarcoma	2		62	35	1[c]	3, 4
Shope papilloma		18	82			23
Silkworm cytoplasmic polyhedrosis	23		77			26
Silkworm jaundice		8	77[e]			27
Simian virus 5	1		73	20	6	28
Southern bean mosaic	21		79			29
Tipula iridescent		13	82	5		30
Tobacco mosaic	5		95			31
Tobacco necrosis	19		81			32
Tobacco rattle	5		95			33
Tobacco ringspot	40		60			34
Tomato bushy stunt	17		83			35
Tomato spotted wilt	5		71	19	5	36
Turnip yellow mosaic	34		63			37
Vaccinia		5	88	4	3	38
Wild cucumber mosaic	35		65			39
Wound tumor	23		77			26

[a]Rounded figures, often obtained by difference between 100 percent and sum of other components.

[b](1) Green 1969; (2) Frisch-Niggemeyer and Steere 1961; (3) Bonar and Beard 1959; (4) Baluda and Nayak 1969; (5) Brown 1972; (6) Yamazaki et al. 1961; (7) Bockstahler and Kaesberg 1961; (8) Harrison et al. 1971; (9) Ray 1968; (10) Kaesberg 1967; (11) Kozloff 1968; (12) Thomas and MacHattie 1967; (13) Knight and Stanley 1941, Knight and Woody 1958; (14) Krieg 1956; (15) Faulkner et al. 1961; (16) Beard 1948; (17) Schäfer 1959; (18) Joklik and Smith 1972; (18a) Russell et al. 1963; (19) Ada and Perry 1954; (20) Frommhagen et al. 1959; (21) Rueckert and Schäfer 1965; (22) Schaffer and Schwerdt 1959; (23) Kass 1970; (24) Diener 1971; (25) Bawden and Pirie 1938; (26) Kalmakoff et al. 1969; (27) Bergold and Wellington 1954; (28) Klenk and Chopin 1969a; (29) Miller and Price 1946; (30) Thomas 1961; (31) Knight and Woody 1958; (32) Kassanis 1970; (33) Harrison and Nixon 1959a; (34) Stace-Smith 1970; (35) Stanley 1940, DeFremery and Knight 1955; (36) Best 1968; (37) Matthews 1970; (38) Fenner et al 1974; (39) Yamazaki and Kaesberg 1961.

[c]A rough estimate calculated from data of Baluda and Nyak (1969).

[d]Hydroxymethyl cytosine residues of the T-even DNAs are glucosylated but to different extents so that the glucose residues amount to about 4, 5, and 7 percent, respectively, of the DNAs of T2, T4, and T6 (Jesaitis 1956). An average value for DNA and carbohydrate is given here.

[e]About 15 percent of the virus was not accounted for as protein, nucleic acid, or lipid.

Table 3. Some Comparisons of Viruses and Bacteria.[a]

Microorganism	Size, nm (Approx. Diam.)	Chemical Composition	Multiplication	Inhibition by Antibiotics	Staining Characteristics
Bacteria	500–3,000	Complex: numerous proteins (including enzymes), carbohydrates, fats, etc.; DNA and RNA; cell wall contains mucopeptide	In fluids, artificial media, cell surfaces, or intracellularly, by binary fission	Inhibited	Stain with various dyes
Mycoplasmas or PPLOs[b]	150–1,000	Similar to other bacteria but generally possess no cell wall	In media similar to other bacteria but by budding rather than fission	Resistant to penicillins, sulfonamides; sensitive to tetracyclines, kanamycin, etc.	Stain with dyes but poorly
Rickettsia	250–400	Similar to other bacteria	Inside living cells by binary fission; major hosts: arthropods	Inhibited	Stain with various dyes
Chlamydia or *Bedsonia*	250–400	Similar to other bacteria	Inside living cells by binary fission; major hosts: birds and mammals	Inhibited	Stain with various dyes
Viruses	15–250	Mainly nucleic acid (one type) and protein. Some contain lipid and/or carbohydrate in addition	Inside living cells by synthesis from pools of constituent chemicals	Not inhibited	Stain for electron micro scopy with salts of heavy metals

[a]Adapted from Knight 1974.
[b]PPLO is the abbreviation for pleuropneumonia-like organism, the first of this group of wall-less bacteria to be characterized.

Table 4. Names and Abbreviations of Common Amino Acids.

Alanine	Ala	Leucine	Leu
Arginine	Arg	Lysine	Lys
Asparagine	Asn	Methionine	Met
Aspartic acid	Asp	Phenylalanine	Phe
Cysteine	Cys	Proline	Pro
Glutamic acid	Glu	Serine	Ser
Glutamine	Gln	Threonine	Thr
Glycine	Gly	Tryptophan	Trp
Histidine	His	Tyrosine	Tyr
Isoleucine	Ile	Valine	Val

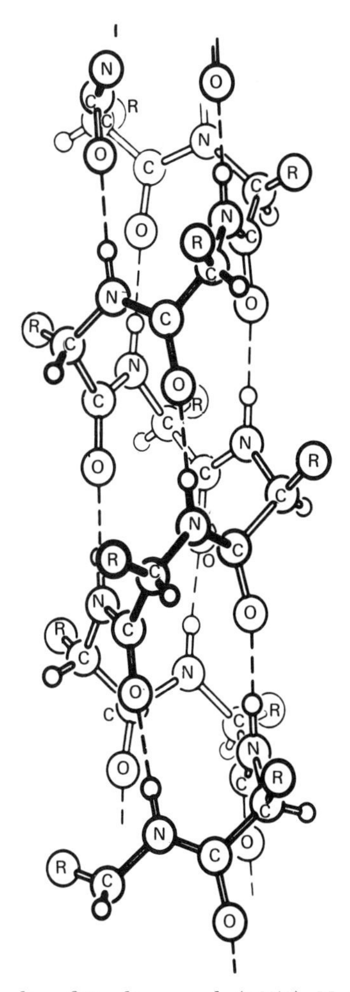

Fig. 5. The α-helix of Pauling et al. (1951). Note the formation of hydrogen bonds, indicated by dashed lines, between -CO and -NH groups. There are 3.7 amino acid residues per complete turn, and a unit residue translation of 1.47 Å giving a pitch of 5.44 Å. (From Kendrew 1959.)

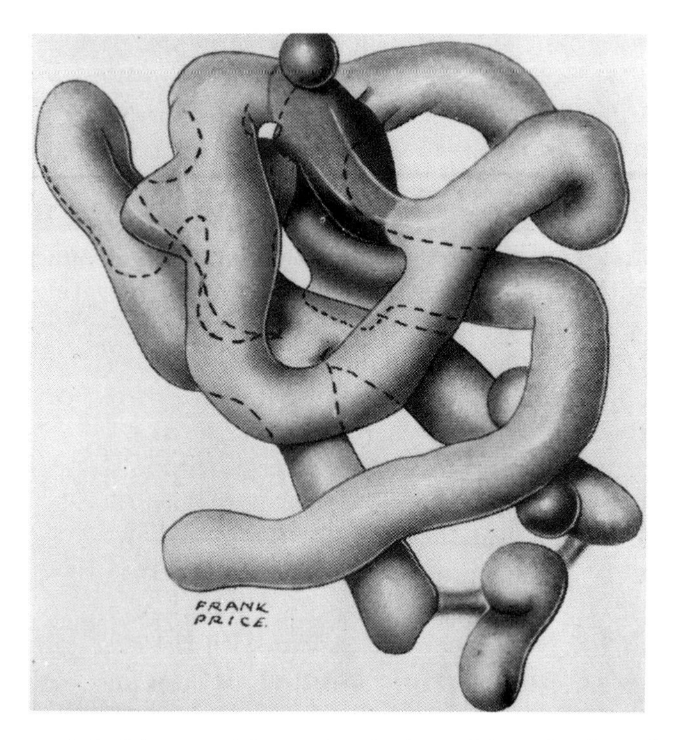

Fig. 6. Model of the myoglobin molecule derived from the three-dimensional electron density map based on x-ray data. The polypeptide chain is represented by solid rods; the side chains have been omitted but if present would fill the regions between the main chains. The small spheres are the heavy atoms used to determine the phases of the x-ray reflections. (From Kendrew 1959.)

the tertiary structure of myoglobin determined from crystallographic studies and represented by the model shown in Figure 6 (Kendrew 1959).

4. *Quaternary structure*: the number and spatial relationship of repeating subunits when these are present. An excellent example of quaternary structure is the architecture of the tobacco mosaic virus particle (see the model in Figure 7).

Primary structure is the stablest of the four types since this kind of structure involves covalent bonds, whereas the others depend on secondary attractions, that is, ionic and hydrophobic interactions and hydrogen bonds. The so-called "denaturation" of proteins results from disruption of secondary bonds and is manifested by the unfolding of the protein and changes in the solubility, charge, hydration, and serological and chemical reactivities. Rupture of secondary bonds affects quaternary as well as tertiary structure, and is the means for disaggregating virus particles, as will be described next.

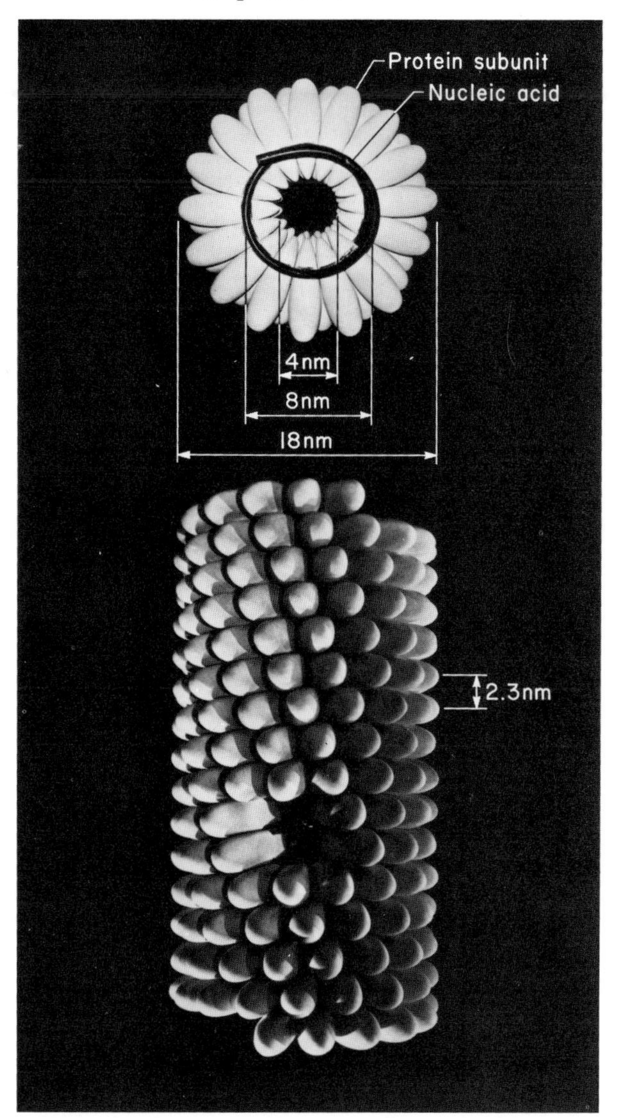

Fig. 7. Model of about one-tenth of the tobacco mosaic virus particle. The protein subunits are schematically illustrated in a helical array about the long axis of the particle. The structure repeats after 6.9 nm in the axial direction, and the repeat contains 49 subunits distributed over three turns of the helix of 2.3 nm pitch. Some of the subunits have been removed in order to show the deeply embedded location of the viral RNA, which, however, is better viewed in cross section as shown at the top of the illustration.

1. *Preparation of Viral Proteins*

A first step in determining primary structure is to obtain an accurate analysis of the amino acid content of the protein. This may be done with hydrolysates of the whole virus, but more reliable results are generally obtainable if the viral protein is first separated from the other viral constituents. This removes nonprotein materials whose acid-degraded products may interfere with the analytical methods, and eliminates spurious glycine that can arise from acid degradation of the nucleic acid (see Smith and Markham 1950). Isolation of proteins before analysis is mandatory for the many viruses that contain more than one species of protein.

Each virus presents a separate problem with respect to appropriate means for disrupting quaternary structure (without breaking covalent bonds) and subsequent isolation of homogeneous protein preparations. Obviously, the most complicated situations arise with such viruses as the myxoviruses (influenza, Newcastle disease, mumps) and bacterial viruses, which possess not one but several discrete protein components, as well as a variety of other chemical constituents. However, it has been found with several viruses that disruption of quaternary structure and resultant release of protein occur upon treatment of virus with acid, alkali, or/and detergent. Also, treatment with salts or with such reagents as urea, phenol, or guanidine hydrochloride have occasionally been used with success. Some of these methods will be summarized.

a. **The Mild Alkali Method (Schramm et al. 1955; Fraenkel-Conrat and Williams 1955)**

Exposure of TMV in the cold to dilute alkali at pH 10–10.5 results in release of protein, which, from its sedimentation and diffusion characteristics, appears to have a molecular weight of 90,000–100,000 (that is, it is either a pentamer or hexamer of the fundamental protein subunit). This material has been called "A protein" (Schramm et al. 1955). At the same time, the RNA of the virus undergoes alkaline hydrolysis to small fragments.

The quality of a good preparation of A protein is apparently so similar to that of the native protein in the virus particle that it can be used, together with viral RNA, to reconstitute (see section on Reconstitution of Viruses) virus rods that are virtually indistinguishable from undegraded virus.

The alkali employed appears not to be critical since sodium hydroxide, borate, carbonate, and glycine buffers have all been used with good results. The amino alcohols, such as ethanolamine, seem to be more effective than alkaline buffers in degrading the virus and hence achieve the desired result in a shorter time (Newmark and Myers 1957).

A convenient alkaline degradation procedure used successfully with TMV and many of its strains is as follows:

An aqueous solution of virus at 10 mg/ml is placed in a cellophane bag and dialyzed at about 4° against 2 liters of 0.1 M carbonate buffer (21.2 g Na_2CO_3 in 2 liters H_2O adjusted to pH 10.5 by addition of solid $NaHCO_3$) for 2–5 days. Alternatively, 0.02 M ethanolamine at the same pH and temperature can be used with a reduction of the dialysis time to 2–4 hr.

Separate undegraded virus by centrifugation of the contents of the dialysis bag at 60,000–100,000 *g* for 1 hr. Discard the pellet and add 1 vol of saturated ammonium sulfate to the supernatant fluid. Sediment the precipitated protein at about 5,000 *g* and redissolve in water. Precipitate twice more with 0.33 saturated ammonium sulfate and dialyze at 4° against several changes of distilled water. Adjust the pH of the dialyzed solution to 7–8 with dilute NaOH and centrifuge at 60,000–100,000 *g* for an hour to remove heavy particles. Store the final solution in the refrigerator, adding a drop or two of chloroform as preservative. The ultraviolet absorption ratio of the maximum to the minimum (280/250) should be about 2 or higher, depending mainly on the tryptophan and tyrosine contents of the protein.

The kinetics of the degradation of the quaternary structure of TMV in mildly alkaline solutions have been studied with the use of the ultracentrifuge and viscometry (Schramm 1947a; Schramm et al. 1955; Harrington and Schachman 1956). Some conclusions of general interest that Harrington and Schachman reached are:

1. The character of the alkaline degradation of TMV changes markedly between 0° and 25°, and different products are obtained at the two temperatures.
2. Contrary to earlier interpretations, some of the intermediates and some of the final products are the result not of degradation of larger components but rather of aggregation of smaller degradation products.
3. TMV seems to possess structural features of such a nature that the protein subunits are rapidly stripped at pH 9.8 from two-thirds of the particle leaving a relatively stable nucleoprotein fragment one-third the initial size of the particle.
4. Some particles of TMV seem to be completely resistant to degradation under conditions that lead to the breakdown of the bulk of the virus.

Conclusion 4 had also been reached earlier by Schramm et al. (1955), and remains a puzzle, along with the seemingly greater stability to alkali of one-third of the particle. Possibly the interaction between protein and nucleic acid is not uniform along the length of the rod, being stronger at one end than at the other.

b. The Cold 67 Percent Acetic Acid Method (Bawden and Pirie 1937; Fraenkel-Conrat 1957)

The proteins of some strains of TMV are too sensitive to alkali to permit their isolation from the virus at high pH values. In such cases the preparation of viral protein may be facilitated by use of the acid degradation procedure developed by Fraenkel-Conrat (1957), and described as follows:

To ice-cold virus solution of 10–30 mg/ml in water is added 2 vol of chilled glacial acetic acid, and the mixture is allowed to stand in an ice bath for 30–60 min, with occasional stirring. The nucleic acid separates out while protein stays in solution. The nucleic acid is removed by centrifuging in the cold and the protein is then dialyzed against several changes of distilled water at 4° for 2–3 days, by which time the protein may have reached its isoelectric range and have come out of solution. If the protein remains soluble, a series of precipitations with ammonium sulfate should be employed for completion of the preparation as described in the mild alkali method above. Otherwise, the isoelectrically precipitated protein is removed from the dialysis bag and pelleted by centrifugation; the pellet is dissolved in distilled water by adjusting to pH 8 with dilute NaOH and the solution is centrifuged at about 100,000 *g* for an hour to remove any undegraded virus or denatured protein. The water-clear supernatant, which contains the protein, may be used as such or dried from the frozen state and stored for use.

The acetic acid method did not work satisfactorily with turnip yellow mosaic virus (Harris and Hindley 1961) because insoluble protein aggregates were obtained. These aggregates were apparently formed by oxidation of SH groups to give S-S linkages between protein subunits. However, by converting the SH groups to carboxymethyl-SH with iodoacetic acid, subsequent aggregation was avoided, and the procedure could be carried out successfully as described above.

A useful modification of the acetic acid method is to substitute formic acid at 37° and to treat for 18 hr (Miki and Knight 1965).

c. The Guanidine Hydrochloride Method (Reichmann 1960; Miki and Knight 1968)

Some viruses are not readily dissociated by treatment with acid or alkali but do respond to protein-denaturing agents such as guanidine hydrochloride and urea. The former has proved more useful with both plant and animal viruses.

Dialyze virus at 10–20 mg/ml against 2 M guanidine HCl at room temperature overnight. During dialysis, the nucleic acid precipitates and the protein remains in solution. The nucleic acid is removed by centrifugation at 5,000 *g* and the supernatant is dialyzed against water for one or two days. The dialyzed material is centrifuged at 100,000 *g* for 2 hr in order to remove incompletely degraded virus. The yield of protein from potato virus X is about 90 percent.

A somewhat modified method is applied to poliovirus (Scharff et al. 1964) consisting of treatment with 6.5 M guanidine HCl at pH 8.3 for 3 hr at

37° followed by separation of protein and nucleic acid by centrifugation on a sucrose density gradient.

d. The Warm Salt Method (Kelley and Kaesberg 1962)

This procedure has been used successfully in preparing protein subunits from alfalfa mosaic virus, a small rodlike plant virus containing about 81 percent protein and 19 percent RNA.

Combine the virus at about 20 mg/ml in 0.01 M phosphate buffer at pH 7 with an equal volume of 2 M NaCl; hold the resultant mixture at 45° for 20 min. (During this time the solution becomes turbid, presumably because the released protein is less soluble than whole virus in M NaCl.) Cool immediately and centrifuge at low speed. The protein is sedimented while the nucleic acid and/or its degradation products remain in the supernatant fluid. Wash the protein pellet with M NaCl and centrifuge again. Dissolve the protein pellet in 0.01 M phosphate at pH 7 containing 0.005 M sodium dodecyl sulfate. Dialyze for 24 hr against distilled water and then precipitate by adding 0.66 vol of saturated ammonium sulfate. Resuspend the precipitate in phosphate buffer containing 0.05 M dodecyl sulfate, and then dialyze for 24 hr against phosphate buffer containing 0.005 M sodium dodecyl sulfate.

The molecular weight of the protein obtained by this method from alfalfa mosaic virus was estimated from sedimentation data to be about 34,000 (Kelley and Kaesberg 1962). The preparation was also found to be serologically active when tested with antiserum to whole alfalfa mosaic virus. This would not have been the case had the tertiary structure been extensively disrupted.

e. The Cold Salt Method (Yamazaki and Kaesberg 1963)

This procedure has been applied successfully to bromegrass mosaic and broad bean mottle viruses with recoveries of 60–80 percent of the viral protein.

The virus, at about 10–15 mg/ml in water, is dialyzed against 1 M $CaCl_2$ at pH 6–7 at 4° for 12 hr. During dialysis a white precipitate of nucleate is formed. The suspension is centrifuged at 5,000 *g* for 20 min. The supernatant is dialyzed against water to remove $CaCl_2$. The dialyzed material contains the soluble protein but practically no nucleic acid.

f. The Phenol Method (Anderer 1959a, 1959b)

Disruption of viruses with phenol is the basis for one of the commonest methods for isolation of viral nucleic acids (see section on Methods for Preparing Viral Nucleic Acids). In the preparation of nucleic acids the protein and other phenol-soluble components are usually discarded in the phenolic layer. However, it has been shown with TMV that protein can be recovered from the phenolic layer, and, by suitable treatment, be restored to a condition which resembles the native state (Anderer 1959b).

To the phenolic layer remaining after separation of the aqueous, RNA-containing layer (see section on The Phenol Method for Preparing Nucleic Acid), add 5–10 vol of methanol and a couple of small crystals of sodium acetate. Remove the precipitate by centrifugation and wash three times with methanol and once with ether. Dry the product in air. To solubilize the air-dried protein, suspend 10 mg in 5 ml of water and heat at 60°–80°, adding enough 0.02 N NaOH to bring the pH to 7.5. The protein should dissolve and remain in solution upon cooling.

The TMV protein is denatured after extraction with phenol, precipitation with methanol, and so on, as described above. However, renaturation is assumed to occur to a large extent when the protein is warmed at about 60° at pH 7–7.5.

g. The Detergent Method and General Conclusions

It was early noted (Sreenivasaya and Pirie 1938) that 1 percent sodium dodecyl sulfate (SDS) disrupts TMV over a wide pH range. Later, when disruption by SDS was coupled with fractional precipitation with ammonium sulfate, Fraenkel-Conrat and Singer (1954) showed that the protein and nucleic acid of TMV could be rather cleanly separated. However, there are at least two disadvantages in the protein prepared by treatment of virus with SDS:

1. The protein strongly binds as much as 15 percent SDS, which seems to introduce only relatively small errors in ultracentrifuge studies (see Hersh and Schachman 1958, for example), but is more serious for other types of investigation. For example, trypsin is inhibited by anionic detergents such as SDS (Viswanatha et al. 1955), and TMV protein prepared by treatment with SDS does not appear to be satisfactory for structural studies dependent on a quantitative cleavage of protein by trypsin (Fraenkel-Conrat and Ramachandran 1959).
2. In addition to effecting the release of protein subunits, SDS tends to degrade the secondary and tertiary structures in a not readily reversed manner. Thus, TMV protein prepared with SDS is insoluble from pH 2 to 10 and does not participate in reconstitution (see section on Reconstitution of Viruses). However, the use of detergents, alone or in combination with other reagents such as phenol, has been invaluable in the disruption of virus particles for the isolation of viral nucleic acids (see Preparation of Nucleic Acids).

In general, all of the methods used to prepare viral proteins probably cause various denaturative changes, some of these reversible and others not. In the case of TMV, Anderer (1959b) has suggested the following criteria for distinguishing between native and denatured protein. Native TMV protein (1) is soluble in neutral aqueous media, (2) aggregates to viruslike rods at pH 5–7, (3) reconstitutes to infectious virus with appropriate viral RNA, and (4) resembles the protein in the virus in amount of TMV antibody it binds. These criteria apply more or less to all viruses.

2. *Analysis of Viral Proteins*

The structure of proteins, as mentioned earlier, can be considered to fall into four main categories, each with its methods of analysis. Main consideration will be given here to the determination of primary structure because this is basic to the other types of structure. Thus the degree of helicity exhibited by the polypeptide chain (secondary structure), the nature of its folding (tertiary structure), and the assembly of protein subunits to form superstructures of characteristic morphology (quaternary structure) are virtually predestined by the sequence of amino acids in the protein chains. Some details of quaternary structure are considered in Sec. IV, Morphology of Viruses.

a. Amino Acid Analyses

In order to determine the composition of a viral protein, it is necessary, as with other proteins, to release the constituent amino acids by hydrolysis. This hydrolysis is usually accomplished by heating under vacuum 2–5 mg of protein in 1 ml of 6 N HCl in a thick-walled, sealed glass tube at 110° for 22–72 hr (see Knight 1964) (some proteins can be satisfactorily hydrolyzed at 120° for 6–24 hr; see Carpenter and Chramback 1962). Tryptophan and cysteine are largely destroyed by these conditions but can be preserved by modifying the hydrolysis medium (see Liu 1972; Liu and Inglis 1972). Alternatively, there are colorimetric procedures for determining tryptophan and cysteine (Anson 1942; Spies and Chambers 1949).

Customarily, two or more different times of hydrolysis are employed for evaluation of the release and recovery of individual amino acids. The highest value observed in a series or by extrapolation of observed values is generally accepted. Thus, as hydrolysis time increases, serine and threonine, and sometimes tyrosine, tend to be proportionately more destroyed; therefore, the contents of these amino acids are usually calculated by extrapolation back to zero time from the values observed at different times of hydrolysis. Conversely, when two or more residues of isoleucine or valine—and, to some extent, leucine—are contiguous, they are less readily released from peptide linkage than other amino acids. Thus even maximum periods of hydrolysis may yield somewhat low values. When the nature of the results suggests this, extrapolation to higher values is generally done by inspection and approximation. The uncertainties connected with such approximations are virtually eliminated if analyses can be made of the tryptic peptides of a protein (see Table 5) because the numbers of residues are

Table 5. Amino Acid Content of Tryptic Peptides of Tobacco Mosaic Virus.

Amino Acid	Tryptic Peptide[a] 1	2	3	4	5	6	7	8	9	10	11	12	Res.[b] Sum	Residue M. W. Sum
Ala	4	0	0	0	0	3	0	3	2	1	0	1	14	995.05
Arg	1	1	1	0	1	1	1	1	2	1	1	0	11	1,718.15
Asn	3	0	0	0	0	1	1	2	0	2	1	0	10	1,141.05
Asp	1	0	0	2	0	2	0	1	2	0	0	0	8	920.68
Cys	1	0	0	0	0	0	0	0	0	0	0	0	1	103.12
Gln	5	1	3	0	0	0	0	1	0	0	0	0	10	1,281.35
Glu	1	0	0	0	0	0	0	3	0	1	0	1	6	774.69
Gly	1	0	0	0	0	1	0	0	0	0	2	2	6	342.33
Ile	3	0	0	0	0	0	0	2	1	3	0	0	9	1,018.49
Leu	4	0	0	0	0	4	0	1	0	2	0	1	12	1,357.98
Lys	0	0	1	1	0	0	0	0	0	0	0	0	2	256.35
Phe	3	0	1	2	0	1	0	0	0	0	1	0	8	1,177.40
Pro	2	0	2	1	0	1	0	1	0	0	0	1	8	776.92
Ser	5	0	2	1	0	0	0	0	0	1	1	6	16	1,391.60
Thr	4	1	1	0	0	2	0	4	1	0	1	2	16	1,617.68
Trp	1	0	1	0	0	0	0	0	0	0	0	1	3	558.62
Tyr	1	0	0	0	1	1	0	0	0	0	1	0	4	652.70
Val	1	2	3	0	1	2	0	1	2	1	0	1	14	1,387.89
Totals													158	17,472.05[c]

[a]The 12 peptides resulting from treatment of tobacco mosaic virus coat protein with trypsin are numbered in order from the N-terminal to the C-terminal of the polypeptide chain.

[b]Res. = Residue. An amino acid residue is the molecular weight of the amino acid less one molecule of water.

[c]The N-terminal amino acid of TMV protein is acetylated. If the molecular weight of the acetyl group and a hydroxyl associated with the C-terminal threonine are added on to the sum of the residues, a value of 17,516.08 is obtained. For most practical purposes, a rounded value of 17,500 can be used for the TMV coat protein.

fewer and systematic errors are less significant. For example, if a peptide has 4 Ala residues, a 3 percent error in its analysis amounts to a negligible ± 0.1 Ala residue, whereas if the whole protein contains 20 Ala residues the error is ± 0.6 residue, which entails an uncertainty of ± 1 residue from the apparent value.

The amino acids resulting from hydrolysis of proteins are generally determined quantitatively in commercial, automatic amino acid analyzers such as that in Figure 8. These machines are an outgrowth of the laboratory models first developed and used effectively in amino acid analyses by S. Moore and W. H. Stein and associates at the Rockefeller University in New York (see Spackman et al. 1958; Spackman 1967).

In operation, something between a few microliters and a milliter or more of hydrolysate is applied at the top of a column of polysulfonic cation exchange resin and then appropriate buffers are pumped through resin under high pressure. In passage through the resin, the various amino acids are repeatedly adsorbed and eluted at rates dependent upon their chemical composition. As the amino acid fractions emerge from the column, they undergo reaction with a ninhydrin solution and the resulting color is measured by passage through a colorimeter whose readings are recorded automatically on a chart such as that shown on the instrument in Figure 8.

The various amino acids are identified by the order in which they emerge from the column as indicated by the successive peaks observed on the recorder chart, and the quantity of each amino acid is obtained by comparison of the areas under the various peaks with the areas obtained with known quantities (for example, 0.1 μM) of standard amino acids. This comparison is made through a series of calculations as specified by the manufacturer of the analyzer, but is facilitated in some instruments by an integrator accessory that automatically integrates the areas under the respective peaks; there is also a system which translates recorder output to a

Fig. 8. Beckman amino acid analyzer. Note the cylindrical glass columns for the ion exchange resin on the left and the chart recorder on the right.

form suitable for subsequent processing in a digital computer (see Hirs 1967). Figure 9 illustrates the peaks obtained in a run made with a standard mixture containing 0.1 μM of each of the commonest amino acids.

The results from analyses made with amino acid analyzers are in terms of micromoles of each amino acid present in the applied sample. For maximum usefulness in structural and genetic analyses, these values are converted to numbers of each amino acid residue present per protein molecule (an amino acid residue is an amino acid minus the elements of water that are lost in the incorporation of the amino acid into a polypeptide chain). This conversion is done as follows.

First, a minimal molecular weight is calculated for the protein. In the absence of any estimate of the molecular size of the protein, this calculation involves a series of tentative assignments whose purpose is to find the lowest level at which the various amino acid residues appear in integral numbers. To start, a value of 1 can be assigned to the amino acid present in lowest amount (or the value of 1 can be assigned to any amino acid characterized by a consistently high recovery from protein hydrolysates) and the number of each of the other amino acid residues can be calculated on the basis of their relative micromolar values. Thus, by trial, a set of residue

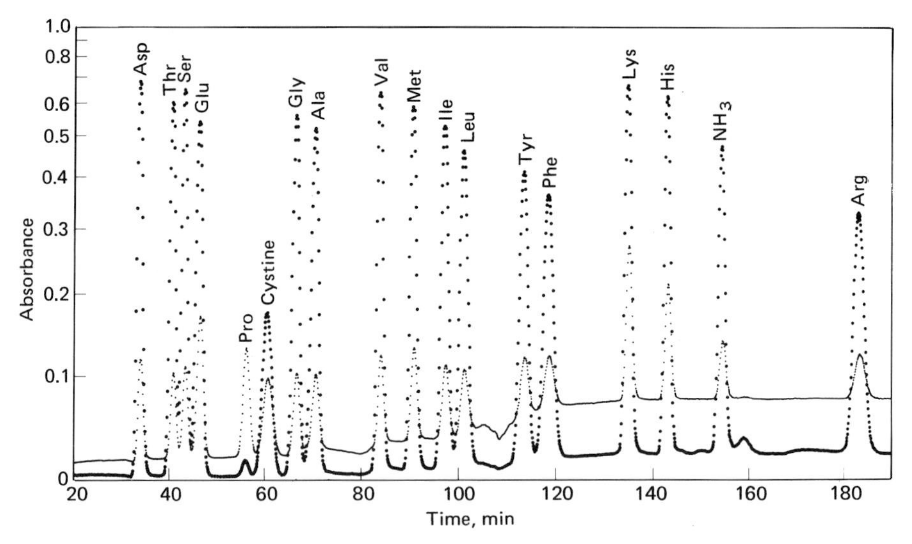

Fig. 9. Tracing from a Beckman amino acid analyzer chart showing the curves obtained with approximately 1 μM of each amino acid. The heavy dots represent the absorbance values at 570 nm and the small dots at 440 nm of the colors obtained in the reaction of each amino acid with ninhydrin. This color is purple for all common amino acids except proline, which yields a yellow product. Note that it took only about 3 hr for a complete analysis of the mixture of amino acids represented here.

numbers can be obtained in which the values are nearly whole numbers for all of the amino acids present. Summation of these residue weights yields a minimal molecular weight for the protein.

At this point, an independent estimate of the molecular size is needed in order to determine the factor by which the minimal numbers of residues must be multiplied to give the actual numbers present per molecule of viral protein. Some procedures to estimate molecular weights of proteins accurately enough for this use include (1) end group analysis (see Sec. IIIA, 2b), acrylamide gel electrophoresis (see Sec IIIA, 2c), agarose gel chromatography (see Sec. IIIA, 2c), and tryptic digestion followed by peptide mapping and counting (see Sec. IIIA, 2c). The molecular weight obtained by one or more of these methods divided by the minimal molecular weight yields a figure whose nearest integer is the factor by which the numbers of residues must be multiplied to give actual numbers per protein molecule and by which the minimal molecular weight must be multiplied to give the actual molecular weight.

In current practice, some estimate of the molecular weight of the protein is usually made early in the process by one of the methods listed above (for example, gel electrophoresis) and this figure is used with results of the amino acid analyses to make the approximations leading to assignment of residue numbers. This type of analysis can be illustrated by data obtained with the protein of tobacco mosaic virus (Table 6).

Similar procedures have been applied to the coat proteins of numerous plant viruses (see Tsugita and Hirashima 1972) and to some bacterial virus coat proteins; analysis of animal virus proteins in terms of amino acid residues has received almost no attention. Some examples of the results obtained with plant and bacterial viruses are given in Tables 7 and 8.

If the data in Tables 7 and 8 are compared with those for nonviral proteins (see, for example, Dayhoff 1972), it appears that viral proteins have ordinary quantities of the common amino acids and that no unusual amino acids have yet been observed in them. Furthermore, the general agreement between results of amino acid analyses made many years ago by microbiological assay (which detects only the L isomers, that is, the form generally present in proteins of all sorts) and more recent results obtained by methods that do not distinguish between optical isomers suggests that the amino acids of viral proteins are primarily if not exclusively the L isomers.

It is also apparent from the values shown in Tables 7 and 8 that viral coat proteins, while containing significant quantities of the basic amino acids arginine and lysine, do not have sufficient quantities of these to place them in the protamine or histone class of proteins. The latter often have been found in sperm and similar nucleoproteins. Actually, the isoelectric points reported for viruses are consistent with the idea that at least some viral proteins are acidic rather than basic. Some examples are given in Table 9.

Table 6. Amino Acid Analysis of Tobacco Mosaic Virus Coat Protein.

Amino Acid	Micromoles Amino Acid Found after Hydrolysis in 6N HC1 at 120°C for				Maximum Micromoles	Number of Residues on Basis of			Best Estimate[c]
	6 hr	24 hr	48 hr	96 hr		Lys=2	Gly=6	Glu=16	
Asp	0.1773	0.1764	0.1755	0.1754	0.1773	18.96	18.21	18.31	18
Thr	0.1470	0.1337	0.1120	0.0953	0.1520[a]	16.26	15.61	15.70	16
Ser	0.1430	0.1119	0.0716	0.0493	0.1540[a]	16.47	15.82	15.91	16
Glu	0.1545	0.1547	0.1547	0.1549	0.1549	16.57	15.91	16.00	16
Pro	0.0799	0.0775	0.0827	0.0799	0.0799[b]	8.55	8.21	8.25	8
Gly	0.0584	0.0584	0.0582	0.0579	0.0584	6.25	6.00	6.03	6
Ala	0.1365	0.1349	0.1360	0.1348	0.1365	14.60	14.02	14.10	14
Val	0.1310	0.1386	0.1386	0.1385	0.1386	14.81	14.23	14.31	14
Ile	0.0744	0.0806	0.0822	0.0812	0.0822	8.79	8.44	8.49	9
Leu	0.1150	0.1185	0.1193	0.1184	0.1193	12.76	12.26	12.32	12
Tyr	0.0378	0.0384	0.0379	0.0354	0.0384	4.11	3.94	3.97	4
Phe	0.0763	0.0783	0.0789	0.0755	0.0789	8.44	8.11	8.15	8
Lys	0.0182	0.0176	0.0187	0.0187	0.0187	2.00	1.92	1.93	2
Arg	0.1045	0.1014	0.1028	0.0981	0.1045	11.18	10.74	10.79	11

[a]Values for Thr and Ser are obtained by extrapolation back to zero time of the plot of micromoles versus time of hydrolysis.

[b]The 48-hr value was discarded as anomalous.

[c]The best estimate is obtained by a series of test calculations. Such calculations were begun here by inspecting the list of maximum micromoles (column 5) of amino acid shown by the analyses and setting the amino acid occurring in smallest amount (Lys) equal to unity. When the residues of the other amino acids were calculated on the basis of Lys as 1, a value of 5.59 Arg was among those obtained. This gives a total of six or seven basic amino residues per protein subunit, and leads to the prediction that about this same number of peptides should result from tryptic digestion of the protein. In fact, nearly twice this number was observed. This suggests setting the Lys at 2, and the results of this assumption are shown in column 6. Inspection of these values shows that there are several that are far from integral and indicates that it might be more accurate to shift to an amino acid which is present in greater numbers of residues than Lys in order to reduce the effect of analytical error on the micromolar unit value. Glycine, which is known to be quantitatively released by and stable to acid hydrolysis conditions, was selected for the next test calculation (set at 6 from the first approximation) and the results yielded are shown in column 7. Values close to integral ones were obtained with almost all amino acids except isoleucine; when a shift was made to Glu as 16, the results were close to those obtained with Gly as 6 (column 8). Hence, the nearest integral values were assigned for each amino acid residue except isoleucine, which was assigned the next highest value on the basis of experience. The results are shown in the last column of the table. [It should be noted that neither histidine nor methionine is present in this protein but tryptophan and cysteine occur and as in proteins in general (see text) must be determined by independent analyses.]

b. Protein End Groups

A protein or peptide chain consists of a linear sequence of amino acid residues with two ends. In the usual manner of writing linear formulas for such structures, the amino acid residue on the extreme left is called the amino terminal, or N terminal, residue, whereas the amino acid residue on the extreme right is called the carboxyl terminal, or C terminal, residue. These terminal groups can be identified by cleaving them chemically or enzymatically from the protein chain followed by application of a method for amino acid analysis.

Knowledge about primary structure of viral proteins accumulated most rapidly with TMV, and the information obtained served as a model for the investigation of other viruses. For some years after Stanley's discovery of the nature of TMV, there was little interest in the primary structure of the virus protein, mainly because most investigators were overwhelmed with the idea of working with a protein whose apparent molecular weight was of the order of 38×10^6. However, the first step in primary structure work, namely, the determination of amino acid content of TMV and related viruses, was taken in Stanley's laboratory by Ross (Ross and Stanley 1939; Ross 1941) and was further developed in the same laboratory by Knight (Knight and Stanley 1941; Knight 1947b).

At the same time, evidence was gradually provided by physicochemical studies on the degradation products obtained from TMV with urea, alkali, and detergents (Stanley and Lauffer 1939; Lauffer and Stanley 1943; Wyckoff 1937; Schramm 1947a; Sreenivasaya and Pirie 1938) that the virus might possess a substructure. This idea was well supported by the classical x-ray diffraction studies that Bernal and Fankuchen (1941) made on TMV and other plant viruses. However, it was not clear from the results of the chemical degradation studies precisely what sorts of bonds were being broken, nor was it possible to define chemically the crystallographic subunits. Therefore the concept of viral protein subunits lay dormant until some years later.

A fresh approach to the question of viral subunits was launched with the attempt to determine the number and nature of the peptide chains in TMV by means of protein end group studies using the enzyme carboxypeptidase A.

Two pancreatic carboxypeptidases, A and B, carboxypeptidase C, found in citrus and a variety of other plants, and carboxypeptidase Y from yeast are known (see Ambler 1972; Hayashi et al. 1973). These differ in the rate at which they release specific terminal amino acids.

Carboxypeptidase A catalyzes the rapid release of terminal Ala, Gln, His, Ile, Leu, Met, Phe, Thr, Trp, Tyr, and Val; the slow release of Asn, Asp, Cys, Glu, Gly, Lys, and Ser; and generally fails to hydrolyze terminal Arg and Pro. In addition to their own refractory response to carboxypep-

Table 7. Amino Acid Residues Per Subunit of Some Plant Virus Coat Proteins.

Amino Acid	Virus									
	Brome Mosaic	Cowpea Chlorotic Mottle	Cucumber Mosaic	Cucumber 3 (Japan)	Cucumber 4	Potato X	Tobacco Mosaic	Tobacco Necrosis[d]	Tobacco Necrosis Satellite[e]	Turnip Yellow Mosaic
Ala	33	27	17	21	18	46	14	13	9	15
Arg	13	8	24	8	9	10	11	14	24	3
Asn							10			3
Asp	10[a]	11[a]	30	20[a]	17[a]	24[a]	8	18[a]	27[a]	7
Asx[b]										1
Cys	1	2	0	0	0	3	1	2	2	4
Gln							9			3
Glu	18[a]	17[a]	20[a]	10[a]	10[a]	19[a]	7	20	18	8
Glx[b]										3
Gly	10	10	16	9	5	13	6	8	8	8
His	4	2	4	1	0	2	0	1	6	3
Ile	8	7	16	7	5	12	9	11	13	15
Leu	15	16	26	18	13	10	12	10	20	17
Lys	13	13	18	4	3	12	2	12	11	7
Met	3	1	8	0	0	8	0	6	4	4

Phe	5	4	7	9	10	12	8	12	11	5
Pro	7	7	18	6	8	18	8	15	4	20
Ser	13	16	32	24	22	17	16	14	12	17
Thr	11	16	17	10	12	29	16	16	25	26
Trp	2	4	1	2	1	6	3	n.d.[c]	n.d.[c]	2
Tyr	5	5	11	4	4	2	4	11	6	3
Val	18	19	22	7	14	14	14	14	13	14
Total	189	185	287	160	151	257	158	197	213	188
Molecular weight	20,300	19,782	32,000	16,940	16,102	26,815	17,493	22,606	24,919	19,979
C-terminal	Arg	n.d.[c]	n.d.[c]	Ala	Ala	Pro	Thr	Ile	Leu	Thr
N-terminal	n.d.[c]	n.d.[c]	n.d.[c]	Acetyl-Ala	Acetyl-Ala	n.d.[c]	Acetyl-Ser	n.d.[c]	n.d.[c]	Acetyl-Met
References[f]	1	2	3	4,5	4	6,7	8-11	12	12	13,14

[a]Expressed as free acid but actually includes free acid and the relevant amide, since asparagine and glutamine were not determined as such.
[b]Uncertain whether residue is amide or free acid.
[c]n.d.: Not determined.
[d]American Type Culture 36 strain (AC 36 TNV).
[e]SV-C, Satellite associated with AC 36 TNV.
[f](1) Stubbs and Kaesberg 1964; (2) Chidlow and Tremaine 1971; (3) van Regenmortel 1967; (4) Tung and Knight 1972a; (5) Funatsu and Funatsu 1968; (6) Tung and Knight 1972b; (7) Miki and Knight 1968; (8) Anderer et al. 1960; (9) Anderer et al. 1965; (10) Tsugita et al. 1960; (11) Funatsu et al. 1964; (12) Uyemoto and Grogan 1969; (13) Harris and Hindley 1965; (14) Peter et al. 1972.

Table 8. Amino Acid Residues Per Subunit of Some Bacterial Virus Coat Proteins.

Amino Acid	Virus			
	Coliphage fd	Coliphage fr	Coliphage f2	Coliphage QB
Ala	9	16	14	15
Arg	0	4	4	7
Asn	0	10	11	8
Asp	3	4	3	7
Asx[a]		1		
Cys	0	2	2	2
Gln	1	5	6	8
Glu	2	6	5	5
Gly	4	9	9	7
His	0	0	0	0
Ile	4	6	8	4
Leu	2	5	8	12
Lys	5	7	6	7
Met	1	2	1	0
Phe	3	5	4	3
Pro	1	5	6	8
Ser	4	11	13	9
Thr	3	9	9	12
Trp	1	2	2	0
Tyr	2	4	4	4
Val	4	16	14	13
Total	49	129	129	131
Molecular weight	5,168	13,736	13,710	14,037
C-terminal	Ser	Tyr	Tyr	Tyr
N-terminal	Ala	Ala	Ala	Ala
References[b]	1	2	3	4

[a]Uncertain whether aspartic acid or asparagine.
[b](1) Asbeck et al. 1969; (2) Wittmann-Liebold and Wittmann 1967; (3) Weber and Konigsberg 1967; (4) Konigsberg et al. 1970.

tidase A when they are carboxylterminal amino acids, Arg, Asp, Cys, Glu, Gly, and Pro when in the penultimate position tend to decrease the rate of cleavage of C terminal amino acids that are normally readily released.

Carboxypeptidase B is effective mainly in the release of basic C terminal amino acids such as Arg and Lys.

Carboxypeptidase C catalyzes the hydrolytic cleavage of almost any C terminal residue, including Pro, although it is inefficient with such combinations as Pro-Pro and Pro-Gly, and also may not function on large polypeptides.

Carboxypeptidase Y catalyzes the release of most amino acids, including proline, from the C terminals of peptides and proteins (Hayashi et al. 1973). Glycine and aspartic acid may be released more slowly than other

Table 9. Electrophoretic Isoelectric Points of Some Viruses.

Virus	Isoelectric Point pH	Reference
Alfalfa mosaic	4.6[a]	Lauffer and Ross 1940
Influenza A (PR8)	5.3	Miller et al. 1944
Shope papilloma	5.0	Beard and Wyckoff 1938 Sharp et al. 1942
Southern bean mosaic	5.9	MacDonald et al. 1949
Tobacco mosaic	3.5	Eriksson-Quensel and Svedberg 1936
Tomato bushy stunt	4.1	MacFarlane and Kekwick 1938
T_2 bacteriophage	4.2	Sharp et al. 1946
Turnip yellow mosaic	3.8	Markham and Smith 1949
Vaccinia	4.5	Beard et al. 1938
Wild cucumber mosaic	6.6	Sinclair et al. 1957

[a]Determined from minimum solubility rather than from electrophoretic measurements.

amino acids. The enzyme has the advantage of retaining activity in 6 M urea, which makes it possible to use it in structural studies on proteins whose C terminals are not exposed unless they are treated with chain-unfolding reagents such as urea.

The reaction of a peptide with carboxypeptidase A is illustrated by the following in which the R groups represent the side chains of common amino acids (such as H, CH_3, benzyl, and so on). The peptide bond at C is split and then, if conditions are favorable, the one at B.

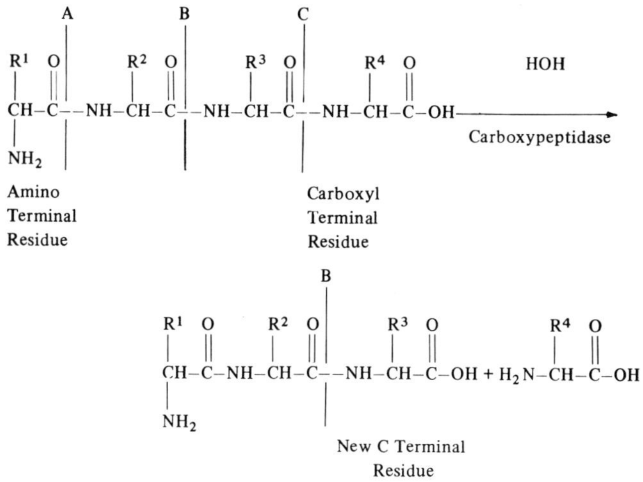

$$\underset{NH_2}{\overset{R^1}{CH}}-\overset{O}{\overset{\|}{C}}\;\overset{A}{|}\;-NH-\overset{R^2}{CH}-\overset{O}{\overset{\|}{C}}\;\overset{B}{|}\;-NH-\overset{R^3}{CH}-\overset{O}{\overset{\|}{C}}\;\overset{C}{|}\;-NH-\overset{R^4}{CH}-\overset{O}{\overset{\|}{C}}-OH \xrightarrow[\text{Carboxypeptidase}]{HOH}$$

Amino Terminal Residue (at A); Carboxyl Terminal Residue (at C)

$$\underset{NH_2}{\overset{R^1}{CH}}-\overset{O}{\overset{\|}{C}}-NH-\overset{R^2}{CH}-\overset{O}{\overset{\|}{C}}\;\overset{B}{|}\;-NH-\overset{R^3}{CH}-\overset{O}{\overset{\|}{C}}-OH + H_2N-\overset{R^4}{CH}-\overset{O}{\overset{\|}{C}}-OH$$

New C Terminal Residue (at B)

Another technique for determining the nature of C terminal amino acid residues is that employing hydrazinolysis (Akabori et al. 1956).

As shown by the following equation, in the hydrazinolysis reaction only the C terminal amino acid comes out as the free amino acid. All others are converted to hydrazides, which have different solubilities than the free amino acids and can be separated by extraction with appropriate organic solvents. The free amino acids can be identified by forming the dinitrophenyl derivatives and subjecting them to two-dimensional chromatography on paper. Elution of the spots and examination in the spectrophotometer provide quantitative values, which, after correction for the significant destruction that occurs during the hydrazinolysis step, are a measure of the C terminal end groups present.

Improved yields by use of a catalyst in the hydrazinolysis step and qualitative and quantitative determination of the free amino acids released by means of column chromatography enhance the usefulness of this technique (see Schroeder 1972).

The hydrazinolysis reaction may be illustrated with a dipeptide as follows:

$$
\begin{array}{l}
\overset{\displaystyle R}{\underset{\displaystyle NH_2}{CH}}-\overset{O}{\overset{\|}{C}}-NH-\overset{R'}{CH}-\overset{O}{\overset{\|}{C}}-OH + NH_2-NH_2 \longrightarrow \overset{\displaystyle R}{\underset{\displaystyle NH_2}{CH}}-\overset{O}{\overset{\|}{C}}-NH-NH_2 + \overset{\displaystyle R'}{\underset{\displaystyle NH_2}{CH}}-\overset{O}{\overset{\|}{C}}-OH
\end{array}
$$

Hydrazide Free Amino Acid

The carboxypeptidase method has been singularly successful with TMV and strains, but has been disappointing with other viruses. For example, when applied to potato virus X, cucumber viruses 3 and 4, southern bean mosaic, tomato bushy stunt, and tobacco ringspot viruses, small, equivocal amounts of several amino acids were released from which no safe conclusions regarding the C terminals or numbers of subunits could be drawn (Knight 1955).

Three possible reasons can be suggested for this result: (1) The C terminal may contain sequences that are incompatible with the specificity of carboxypeptidase A. (2) More vigorous treatment (for example, more enzyme and/or higher temperature) may be needed. (3) The C terminal may be sterically unavailable to the enzyme. The following examples illustrate each of these situations.

C terminal proline is not cleaved by carboxypeptidase A. Potato virus X was found by the hydrazinolysis procedure (Niu et al. 1958; Miki and Knight 1968) to have a proline residue in the C terminal position, thus explaining the negative result with carboxypeptidase A.

Cucumber viruses 3 and 4 when treated with carboxypeptidase A at an enzyme:substrate ratio of about 1:400 at 25° yielded small amounts of several amino acids; and, while alanine seemed to be the major split product, even its concentration was so low as to make the result equivocal (Knight 1955). However, treatment at an enzyme:substrate ratio of 1:25 and at 37° clearly revealed alanine as the C terminal residue and threonine and serine as probable adjacent amino acids (Tung and Knight 1972a). The kinetics of the reaction are shown in Figure 10.

Two cases will serve to illustrate the steric hindrance possibility. No

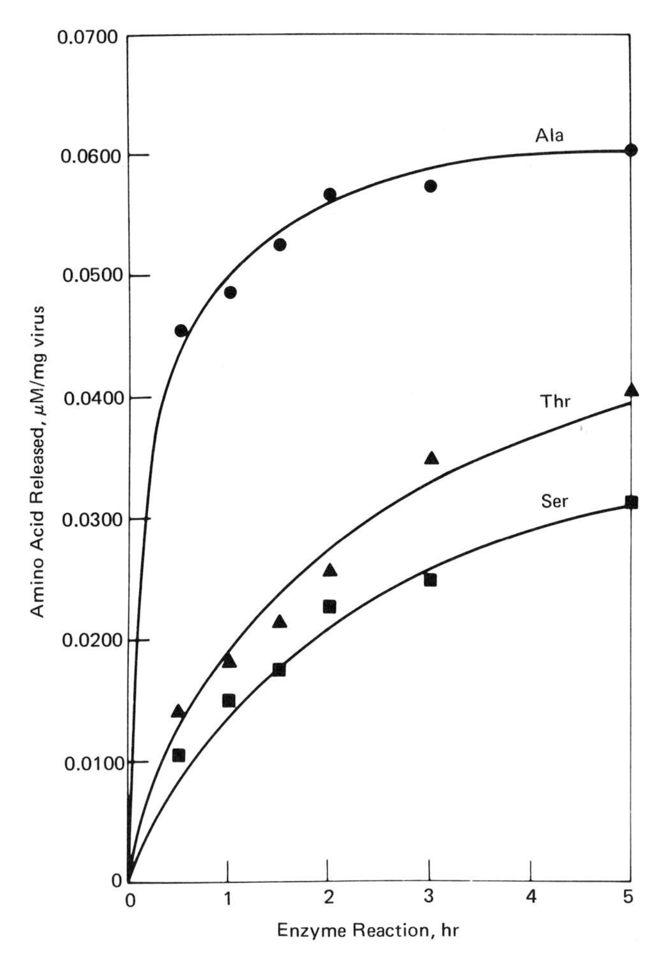

Fig. 10. Release of amino acids from Czech cucumber virus 4 by the action of carboxypeptidase A at an enzyme-substrate ratio of 1:25 at 37°. (From Tung and Knight 1972a.)

free amino acids were produced when intact turnip yellow mosaic virus was treated with carboxypeptidase, but when the isolated protein was employed, four different amino acids were released (Harris and Hindley 1961). From quantitative rate studies of the release, it was found that threonine was the C terminal amino acid, and that after removal of this, carboxypeptidase catalyzed the release from the peptide chain in succession—Ser, Thr, Val, and Asp. Another interesting case is that of the nitrous acid mutant TMV-171 (Tsugita and Fraenkel-Conrat 1960), from which carboxypeptidase caused the release of three amino acids, but from whose isolated protein the enzyme released about 15 residues.

There are several methods for determining the amino terminal (N terminal) residue of proteins and peptides, most of which, however, depend on the presence of an unsubstituted amino group. The two methods that have been used most in determining N terminal amino acid residues of viral proteins are (1) the Sanger fluorodinitrobenzene (FDNB) method (Sanger 1945, 1949; see also Knight 1964) and (2) the Edman phenylisothiocyanate (PTC) method (Edman 1950a, 1950b, 1956; see also Knight 1964).

The principle of the FDNB method is that the free amino group of the N terminal residue in a peptide or protein reacts with FDNB in mildly alkaline solution to give a *N*-dinitrophenyl (DNP) substituted residue. Upon acid hydrolysis, the DNP group remains attached to the N terminal amino acid in most cases, although some hydrolytic cleavage occurs with all DNP amino acids, being most severe with proline, glycine, and cysteine. The formation of the DNP derivative of a dipeptide and its subsequent hydrolysis can be illustrated as follows:

$$O_2N\text{-}C_6H_3(NO_2)\text{-}F + H\text{-}N(H)\text{-}CH(R)\text{-}C(=O)\text{-}NH\text{-}CH(R')\text{-}C(=O)\text{-}OH \xrightarrow{NaHCO_3}$$

1-Fluoro-2,4-dinitrobenzene (FDNB) — Dipeptide

$$O_2N\text{-}C_6H_3(NO_2)\text{-}NH\text{-}CH(R)\text{-}C(=O)\text{-}NH\text{-}CH(R')\text{-}C(=O)\text{-}OH + NaF$$

DNP Peptide

$$\Big\downarrow \text{HOH, HCl}$$

$$O_2N\text{-}C_6H_3(NO_2)\text{-}NH\text{-}CH(R)\text{-}C(=O)\text{-}OH + H_2N\text{-}CH(R')\text{-}C(=O)\text{-}OH$$

DNP Amino Acid — Free Amino Acid

This general formulation becomes specific when R and R′ are replaced with H, CH_3, C_6H_5-CH_2, or other amino acid side chains.

The DNP amino acids are mainly extractable by ether and can thus be readily separated from nonterminal free amino acids, after which they are identified by paper chromatography and comparison with chromatograms of standard DNP amino acids. A standard chromatogram is given in Figure 11. The yellow color of DNP amino acids makes them easily located on chromatograms, and they can be quantitatively extracted and their concentrations determined from their light absorption in a spectrophotometer at 360 nm.

A procedure similar to the DNP method employs 1-dimethylaminonaphthalene-5-sulfonyl chloride ("dansyl" chloride) as the reagent. Dansyl (DNS) derivatives of the amino acids are very resistant to acid hydrolysis and show an intense yellow fluorescence that enables them to be detected at about 1 percent of the concentration needed for DNP amino acids. DNS derivatives can be identified by either electrophoresis or chromatography (see Bailey 1967; Gray 1972). The use of this technique is

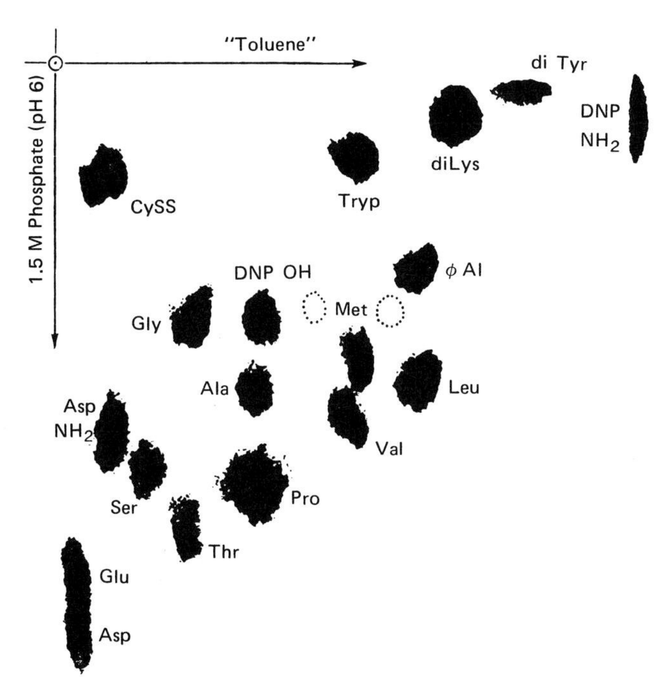

Fig. 11. Two-dimensional chromatogram of a mixture of 16 DNP-amino acids, dinitrophenol (DNPOH) and dinitroaniline (DNP NH_2). Approximately 0.02 μM of each was applied. (Fraenkel-Conrat et al. 1955.)

nicely illustrated by its application to the polypeptides of reovirus (Pett et al. 1973).

The major reactions of the Edman degradation, also called the phenylisothiocyanate (PTC) reaction, resemble the FDNB procedure in that an organic radical is first coupled to the protein at the N terminal in mildly alkaline solution and then the substituted N terminal amino acid is cleaved from the rest of the protein by treatment with acid. A milder acid treatment is used in the PTC method than in the FDNB procedure; hence, in contrast to the latter, the residual protein, shortened by one amino acid residue, is available for further stepwise degradative analysis. As in the case of the FDNB method, the procedure is applicable to peptides as well as to proteins. The chemical steps of the PTC reaction when applied to a peptide can be represented as follows:

$$\underset{\text{PTC}}{C_6H_5{-}N{=}\overset{\overset{S}{\|}}{C}} + \underset{\text{Peptide}}{H_2N{-}\overset{\overset{R}{|}}{CH}{-}\overset{\overset{O}{\|}}{C}{-}NH{-}\overset{\overset{R'}{|}}{CH}{-}\overset{\overset{O}{\|}}{C}{-}NH{-}\overset{\overset{R''}{|}}{CH}{-}COOH} \xrightarrow{\text{pH 8-9}}$$

$$C_6H_5{-}NH{-}\overset{\overset{S}{\|}}{C}{-}NH{-}\overset{\overset{R}{|}}{CH}{-}\overset{\overset{O}{\|}}{C}{-}NH{-}\overset{\overset{R'}{|}}{CH}{-}\overset{\overset{O}{\|}}{C}{-}NH{-}\overset{\overset{R''}{|}}{CH}{-}COOH$$

Phenylthiocarbamyl Peptide (PTC Peptide)

$$\text{PTC Peptide} \xrightarrow{H^+} C_6H_5{-}NH{-}C{=}\,\text{(ring: } HN^+{-}C(R)(H){-}C{=}O{-}S\text{)} + H_2N{-}\overset{\overset{R'}{|}}{CH}{-}\overset{\overset{O}{\|}}{C}{-}NH{-}\overset{\overset{R''}{|}}{CH}{-}COOH$$

Shortened Peptide

2-Anilino-5-thiazolinone

H_2O ↙ ↘ Heat

$$C_6H_5{-}NH{-}\overset{\overset{S}{\|}}{C}{-}NH{-}\overset{\overset{R}{|}}{CH}{-}\overset{\overset{O}{\|}}{C}{-}OH \xrightarrow{H^+} C_6H_5{-}N\,\text{(ring: } N{-}C{=}S{-}NH{-}C(R)(H){-}C{=}O\text{)}$$

3-Phenyl-2-thiohydantoin (PTH)

As shown in the reaction scheme, the N terminal amino acid emerges as a 3-phenyl-2-thiohydantoin (PTH) derivative. The PTH derivative is extracted from the reaction mixture with an organic solvent in which the protein or peptide residue is insoluble. The quantity of PTH derivative can be estimated by reading its absorption in a spectrophotometer at the maximum (usually between 260 and 275 nm) and comparing the result with that of an appropriate standard. Identification of the amino acid present as a PTH derivative can sometimes be made by paper chromatographic comparison with the standard PTH amino acids (Sjöquist 1953), or the PTH derivative can be hydrolyzed in acid to yield free amino acid, which then can be identified by any standard method for amino acid analysis. Various modifications of this procedure are used, including the substitution of potassium cyanate for phenylisothiocyanate (Stark and Smyth 1963; Stark 1972).

The methods just described for the determination of N terminal amino acid depend on the presence of a free amino group on the N terminal residue. However, it appears that many viral proteins have acetylated N terminal amino acids (see Table 10). Two methods have been employed in analyzing such N terminals:

1. The isolated viral protein is digested with pepsin or trypsin and the resulting peptides are fractionated on a strongly acidic cation exchange column. The peptide containing the acetylated amino acid as its N terminal will generally be least basic owing to the neutralization of its N terminal amino group by the acyl substituent. Therefore this peptide elutes from the column in one of the earliest fractions; it can be detected by the Folin colorimetric procedure and analyzed more or less readily by sequencing procedures (see Sec. IIIA, 2d). This approach was used successfully in identifying qualitatively the N terminal amino acids of TMV and cucumber virus 4 coat proteins (Narita 1958, 1959).

2. In the case of viral proteins acetylated at the N terminal, the amount of this terminal amino acid (and hence the molecular weight of the protein) can be estimated from the quantity of acetic acid released upon acid hydrolysis of the protein. The acetic acid can be determined readily by gas chromatography. Thus subunit molecular weights, but not the identity of the acetylated amino acids, were determined for tobacco mosaic and potato X virus proteins (Miki and Knight 1968). Some examples of the terminal amino acid residues found in some viruses are given in Table 9.

c. Protein Subunits

In 1952, Harris and Knight treated TMV with carboxypeptidase and found that more than 2,000 threonine residues were released from each mole of virus. The actual value from repeated determinations (Harris and Knight 1955) was about 2,320. (This figure is based on the now commonly used molecular weight for TMV of 40×10^6. The paper just cited gives a value of 2,900 based on a molecular weight for TMV of 50×10^6.) The key point, however, is that these results indicated that the TMV protein consists of over 2,000 polypeptide chains (protein subunits), and since these peptide chains all terminate in threonine it was correctly assumed, as subsequent data have shown, that they represent identical subunits. The alternative possibilities of a few chains ending in polythreonyl units or a

Table 10. The Protein Subunits of Some Viruses.

Virus	No. of Constituent Proteins	Approximate Molecular Weight of Subunits	N-Terminal	C-Terminal	Ref.[a]
Alfalfa (lucerne) mosaic	1	25,000	Ac-Ser	Arg	1,2
Broad bean mottle	1	20,500		Ala	3
Bromegrass mosaic	1	18,000		Arg	4
Cucumber 4	1	16,000	Ac-Ala	Ala	5,6,6a
Cucumber mosaic	1	24,000			7
Potato X	1	27,000	Ac-X[b]	Pro	8
Shope papilloma	1	40,000		Thr	9
Southern bean mosaic	1	30,000		Ser	10
Sowbane mosaic	1	19,000		Lys	11
Tobacco mosaic	1	17,500	Ac-Ser	Thr	12,13
Tobacco necrosis	1	30,000	Ala	Met	14
Tobacco necrosis satellite	1	20,000	Ala	Ala	15
Tobacco rattle	1	24,000			16
Tomato bushy stunt	1	41,000		Leu	17,18
Turnip yellow mosaic	1	20,000	Ac-Met	Thr	19
White clover mosaic	1	22,500			8

Adeno	9	7,500–120,000		20
Coliphage f2	2	14,000 and 39,000		21
Coliphage ØX174	4	5,000– 48,000		22
Coliphage T4	~28	11,000–120,000		23
Herpes simplex	~24	25,000–275,000		24
Influenza	~7	25,000– 94,000		25–27
Mouse Elberfeld	4	7,300– 33,000		28
Newcastle disease	~6	41,000– 74,000		29
Mouse mammary tumor	5	23,000– 90,000		30
Polio	4	6,000– 35,000		31
Polyoma	6	13,000– 43,000		32
Reovirus	~7	34,000–155,000		33
Rubella	~3	35,000– 62,000		34
Simian 40	6	9,300– 42,000		32
Vaccinia	~30	8,000–200,000		35

[a](1) Hull et al. 1969; (2) Kruseman et al. 1971; (3) Miki and Knight 1965; (4) Stubbs and Kaesberg 1964; (5) Narita 1959; (6) Niu et al. 1958; (6a) Tung and Knight 1972a; (7) Van Regenmortel 1967; (8) Tung and Knight 1972b; (9) Kass 1970; (10) Ghabrial et al. 1967; (11) Kado 1967; (12) Narita 1958; (13) Harris and Knight 1955; (14) Lesnaw and Reichmann 1969; (15) Reichmann 1964; (16) Offord and Harris 1965; (17) Tung and Knight, unpublished (from gel electrophoresis); (18) Niu et al. 1958; (19) Harris and Hindley 1961; (20) Maizel et al. 1968a, 1968b; (21) Hohn and Hohn 1970; (22) Burgess and Denhardt 1969; (23) Laemmli 1970; (24) Spear and Roizman 1972; (25) Compans et al. 1970; (26) Skehel and Schild 1971; (27) Schulze 1972; (28) Rueckert et al. 1969; Stoltzfus and Rueckert 1972; (29) Mountcastle et al. 1971; (30) Nowinski and Sarkar 1972; (31) Jacobson et al. 1970; (32) Hirt and Gesteland 1971; (33) Pett et al. 1973; (34) Vaheri and Hovi 1972; (35) Sarov and Joklik 1972.

[b]X = N-terminal amino acid residue unknown.

single huge polypeptide possessing threonyl side chains on ω carboxyl groups of aspartic or glutamic acid residues were eliminated by appropriate tests (Harris and Knight 1955). Also, application of the newly developed Akabori hydrazinolysis method (1956) confirmed the original conclusion that the virus protein consists of many peptide chains, each terminating in a single threonyl residue (Braunitzer 1954; Niu and Fraenkel-Conrat 1955).

There are several ways of calculating numbers of subunits in a viral protein from terminal amino acid data obtained as described in the preceding section. One of them is illustrated here, with the use of data from TMV analyses.

The molecular weights of TMV and threonine are 40×10^6 and 119, respectively. These molecular weights may be expressed in any units desired for the purposes of calculating relationships between TMV and its C terminal threonine. Thus the same end result will be obtained whether calculations are made using grams, milligrams, or micrograms. If micrograms are used, for example, the weight of one TMV particle (particle or molecular weight) is expressed as 40×10^6 μg, representing 1 μM of TMV. (The absolute weight of one TMV particle is, of course, the particle or molecular weight in grams divided by the number of particles in one mole, that is, by Avogadro's number. Thus the actual weight of one particle of TMV, assuming a molecular weight of 40×10^6, is 6.64×10^{-17} g.)

From 10^4 μg of TMV there were released by carboxypeptidase A 62 μg of threonine. Then

$$\text{Micromoles of threonine released} = \frac{\mu\text{g of threonine found}}{\text{molecular weight of threonine}}$$

$$= \frac{62}{119} = 0.52$$

$$\text{Micromoles of TMV used} = \frac{10^4}{40 \times 10^6} = 2.5 \times 10^{-4}$$

If 2.5×10^{-4} μM of TMV yielded 0.52 μM of threonine, 1 μM of TMV would yield $1 / (2.5 \times 10^{-4}) \times 0.52 = 2{,}080$ μM of threonine. Therefore, there must be 2,080 polypeptide chains (protein subunits) in the TMV particle. Since the virus is 95 percent protein, the molecular weight of each protein subunit is $(0.95 \times 40 \times 10^6) / 2080 =$ about 18,000.

There are many simple viruses like TMV that consist of a strand of nucleic acid ensheathed in a multisubunit protein coat. However, the large, tailed phages and many of the larger animal viruses have more than one species of protein in their particles. In these cases, it is common for each species of protein to be made up of identical subunits. Thus the major head protein of the T-even coliphages consists of about 2,000 identical subunits, each with a molecular weight of 40,000, while another protein, the tail sheath protein, is composed of 144 subunits, each of two species of polypeptide, and so on (Mathews 1971). It is possible in some cases at least to isolate each species of protein from a virus particle and subject it to structural analysis, that is, analyze for amino acid content, do end group analyses, and so on. Often, however, it is useful just to determine the number of different polypeptide species present in the virus and their approximate molecular sizes. This can be done by applying the techniques of electrophoresis in acrylamide gel (Shapiro et al. 1967, 1969; Weber and

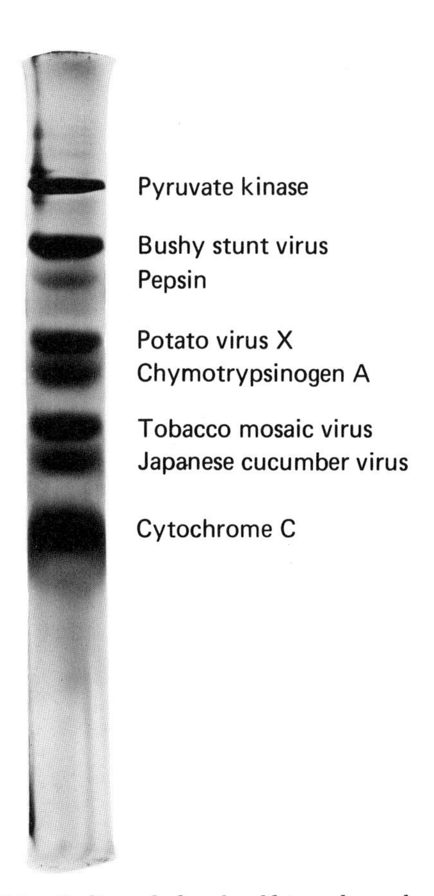

Fig. 12. Sodium dodecyl sulfate-polyacrylamide gel electrophoregram of some proteins ranging in molecular weight from 11,700 (cytochrome c) to 57,000 (pyruvate kinase). Electrophoretic migration, which was in 10 percent gel, is from top toward bottom in the illustration. The bands are visualized by staining in 0.25 percent Coomassie brilliant blue.

Osborn 1969; Dunker and Rueckert 1969) or gel chromatography (also called gel filtration) in agarose (Fish et al. 1969) to dissociated whole virus or to the protein fractions isolated from purified virus by one of the methods described in Sec. IIIA, 1.

In both techniques the molecular weights are determined by comparison of the migration rates of viral polypeptides with those of standard proteins whose molecular weights have been established by various means.

Figure 12 illustrates the migration of several plant virus proteins and standard proteins in SDS—10 percent acrylamide gels. The molecular weights assumed for the standard proteins and those calculated for the viral proteins shown in the gel of Figure 12 are pyruvate kinase, 57,000; tomato bushy stunt virus protein, 41,000; pepsin, 35,000; potato virus X protein, 27,000; chymotrypsinogen A, 25,700; tobacco mosaic virus protein, 17,500; Japanese cucumber virus 3 protein, 16,000; and cytochrome C, 11,700.

All of the plant viruses used in the illustration just given are characterized by a single species of polypeptide comprising the subunits of the viral coat protein. The electrophoresis of the polypeptides of influenza virus, which contains several different species, is illustrated in Figure 13. In this case, the whole virus was dissociated by treatment with SDS and the mixture was applied to the gel for electrophoresis. Duplicate gels are shown in order to illustrate how polypeptides and glycopolypeptides can

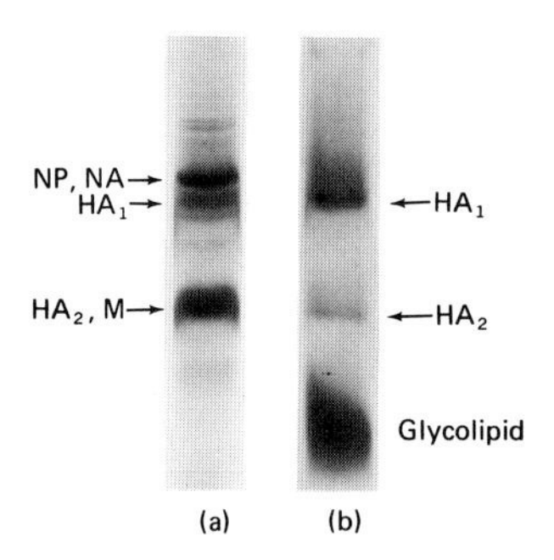

Fig. 13. Polyacrylamide gel electrophoregram of PR8 influenza virus dissociated in 1 percent sodium dodecyl sulfate and run on a 6 percent gel. Gel (*a*) was stained with Coomassie brilliant blue to reveal protein bands, and gel (*b*) was stained with *p*-rosanilin to detect carbohydrate (glycoprotein, two upper bands; glycolipid, lower band).

Abbreviations: NP, nucleoprotein (nucleocapsid subunit); NA, neuraminidase; HA_1, large hemagglutinin component; HA_2, small hemagglutinin component; M, membrane protein of viral envelope.

be distinguished after electrophoresis by application of different stains.

An extraordinarily useful modification of gel electrophoresis is the use of thin gel slabs that permit the side-by-side comparison of many samples (Reid and Bieleski 1968; Studier 1973). Location of protein bands (the procedure is also applicable with modifications to RNAs) is usually accomplished by autoradiography of the dried gels, appropriate isotopes having been previously introduced into the system under investigation. Quantitative distinctions between bands can be enhanced more readily in autoradiography than in the gel staining techniques simply by varying the length of time the recording film is exposed to the gel. Examples of such gels are shown in Figure 13A, which records the proteins extracted from *E. coli* cells infected with various combinations of coliphage P2 and its satellite P4 as well as certain mutants (Barrett and Calendar 1974; Lengyel et al. 1973, 1974). An excellent example of the application of the slab gel technique to the analysis of proteins in an animal virus system is the study by Honess and Roizman (1973) of herpes simplex proteins.

Some caution needs to be exercised with respect to molecular weight values for viral or other proteins obtained by SDS polyacrylamide gel electrophoresis. It has often been assumed that the rate of migration of polypeptides in SDS acrylamide gels depends solely on their molecular size. This assumption appears valid for many proteins. However, a rigorous application of the technique to plant virus proteins (Tung and Knight 1972a, 1972b, 1972c) indicates that the electrophoretic migrations of proteins of similar size in SDS polyacrylamide gels is a closely related function of their molecular weights only when the macromolecules under investigation have the same hydrodynamic shape and charge-to-mass ratio. This situation exists only when standard and test proteins react with SDS in a strictly comparable manner. The data summarized in Table 11 illustrate this point.

The values listed in the first column of the table, which were determined by amino acid analysis and peptide mapping as described earlier, represent the most accurate figures available. They probably deviate from the actual molecular weights by not more than one or two amino acid residues (± 100–200 daltons) and less than that for TMV whose complete amino acid sequence is known. A comparison of the molecular weight determined by other methods with that in the first column shows a good agreement for TMV (± 10 percent accuracy is usually ascribed to the gel electrophoresis and ± 7 percent accuracy is associated with gel chromatography results).

However, focusing on the SDS acrylamide gel values for all of the viruses listed in the table, it is apparent that close agreement with the actual values (column 1) was observed only with TMV and Japan CV3 proteins. The values indicated by gel electrophoresis are spuriously low for Berkeley CV3, Berkeley CV4, and Czech CV4 proteins. It seems likely that

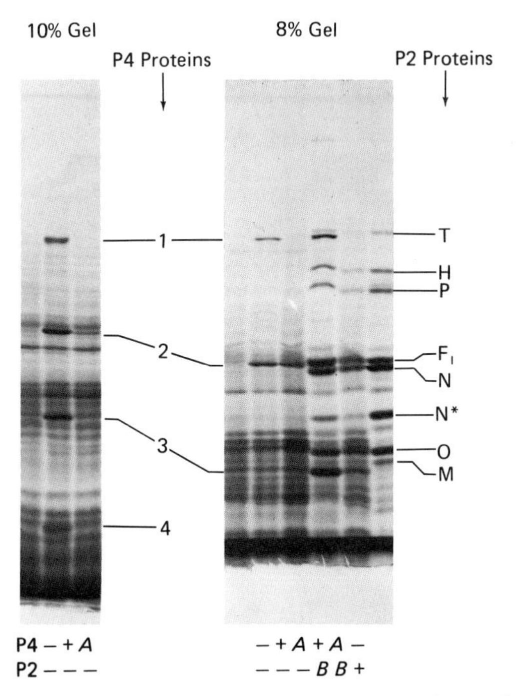

Fig. 13A. Use of slab gel electrophoresis and autoradiography to detect proteins made after infection of *Escherichia coli* bacteria with various combinations of coliphages P2 and P4 and mutants A and B. (Courtesy K. Barrett.)

E. coli C was irradiated with ultraviolet light in order to decrease the synthesis of host proteins. The cells were then infected with P2, P4, or with both together, and labeled with ^{14}C-reconstituted algal protein hydrolysate from 70 to 72 min after infection (*left panel*) or from 40 to 60 min after infection (*right panel*). The incorporation was stopped by adding an excess of cold amino acids and pouring the cells in ice. The cells were collected by centrifugation and lysed by holding in boiling water for 2 min in tris buffer at pH 6.8 containing SDS and mercaptoethanol (Laemmli 1970). The labeled proteins were analyzed in SDS polyacrylamide slab gels. The discontinuous gel system of Laemmli and Maizel as described by Laemmli (1970) was used with a 5 percent stacking gel and an 8 or 10 percent resolving gel. The slab gel apparatus was that of Studier (1973). The gels were dried for autoradiography according to Maizel (1971) and autoradiograms were made with Kodak No-Screen x-ray film.

Symbols: –, no phage; +, wild-type phage as indicated at the left; A or B, phage which contains mutations in genes A or B.

Table 11. Molecular Weights of Coat Protein Subunits of Tobacco Mosaic Virus and of Some Isolates of Cucumber Viruses 3 and 4 Determined by Different Methods.[a]

Virus	Method of Determination			
	Amino Acid Analyses and Peptide Mapping	C-terminal Analysis	SDS-Polyacrylamide Gel Electrophoresis	Agarose Gel Chromatography in Guanidine Hydrochloride
Tobacco mosaic	17,500	18,000	18,000	16,500
Berkeley CV3	17,100	10,100	14,200	16,000
Japan CV3	17,100	25,000	16,000	16,000
Berkeley CV4	16,100	13,300	14,200	16,000
Czech CV4	16,100	16,700	14,200	16,000

[a]Adapted from Tung and Knight 1972a.

these proteins retain enough tertiary structure in the presence of SDS to cause them to migrate at anomalous rates with respect to the standard proteins (and with respect to the proteins of TMV and Japan CV3). Consequently, as judged by SDS gel electrophoresis, the proteins of Berkeley CV3 and Japan CV3 appear to have significantly different molecular weights when, in fact, they are the same, and, conversely, the proteins of Berkeley CV3 and Berkeley CV4 appear to have the same molecular weight when, in fact, they are substantially different.

From these and other data, Tung and Knight have concluded that, as might be expected, the most accurate procedure for determining the molecular weights of viral and other polypeptides is to add up the weights of the constituent amino acid residues. In practice, this means that one determines the minimum number of residues in (and hence a minimum molecular weight of) the polypeptide from careful amino acid analyses and then determines the factor by which the minimum value must be multiplied to give the actual value. This factor is the integral number nearest to the quotient obtained by dividing the molecular weight estimated by gel electrophoresis or gel chromatography by the minimum molecular weight based on amino acid analysis. Alternatively, the factor can be deduced by comparing the number of peptides found on a map of the tryptic digest of the protein with the number expected from the arginine and lysine residues in the minimum molecular weight unit.

Obviously, the less accurate molecular weight values for viral polypeptides (protein subunits) obtainable by gel electrophoresis and gel chromatography are often sufficient and may be much more convenient to obtain since they do not require the chemically homogeneous product required for reliable amino acid analyses. Thus the choice of method for determining molecular weights of viral proteins will doubtless depend on convenience and degree of accuracy sought. In addition, it is clear that information other than molecular weight can be obtained by the various procedures employed, for example, numbers of different polypeptide species present, presence or absence of conjugated carbohydrate moieties, nature of terminal groups of the polypeptide chain, and so on.

Information about the numbers, size, and composition of polypeptides associated with each virus is still fragmentary. However, some data of this sort appear in Table 10.

It can be seen from the data in Table 10 that while there are some large viral polypeptides, most of them fall in the range of 14,000–50,000. As might be expected, the larger, morphologically complex viruses contain several species of polypeptides including some of the larger ones. The rather common occurrence of acylated N terminals is also illustrated.

d. Amino Acid Sequences

Most proteins contain basic amino acids such as arginine and lysine scattered along the length of the peptide chain. This fact, coupled with the marked specificity of the enzyme trypsin for bonds next to basic amino acids, provides a means for cleaving long peptide chains into more readily analyzed fragments.

A convenient survey of the peptides can be made by combining paper electrophoresis and chromatography in a "mapping" procedure. This is illustrated by the diagram of Figure 14, which shows a peptide map obtained after digestion of TMV protein with trypsin. However, the amounts of individual peptides in map spots seldom exceed 200 μg, whereas milligram amounts are usually required for sequential analyses. Therefore, countercurrent extraction or ion exchange chromatography is usually done to separate tryptic peptides for sequential analysis. The separation of the tryptic peptides from TMV protein by ion exchange chromatography is illustrated by the elution diagram shown in Figure 15.

The next step is to determine the amino acid sequences of each peptide. These steps can be illustrated by taking the peptide designated as 11 in Figure 15 and following the sequence determination made by Ramachandran and Gish (1959).

The purity of an aliquot of this fractionated peptide was checked by the mapping procedure. The spot labeled 11 in Figure 14 was the only major spot observed. N terminal analysis by the DNP method indicated that glycine was the N terminal residue. Analysis of the rest of the peptide by acid hydrolysis, formation of DNP derivatives, and so on (instead of the DNP procedure, it is also convenient to subject the hydrolysate to analysis directly in the automatic amino acid analyzer) showed amino acids present in the following molar proportions: Arg 1.00, Asp 1.14, Gly 0.98, Ser 1.00, Thr 1.00, Tyr 0.86. Hence the peptide is a heptapeptide with the formula Gly (Arg, Asp, Gly, Ser, Thr, Tyr). (In the formula the usual convention of

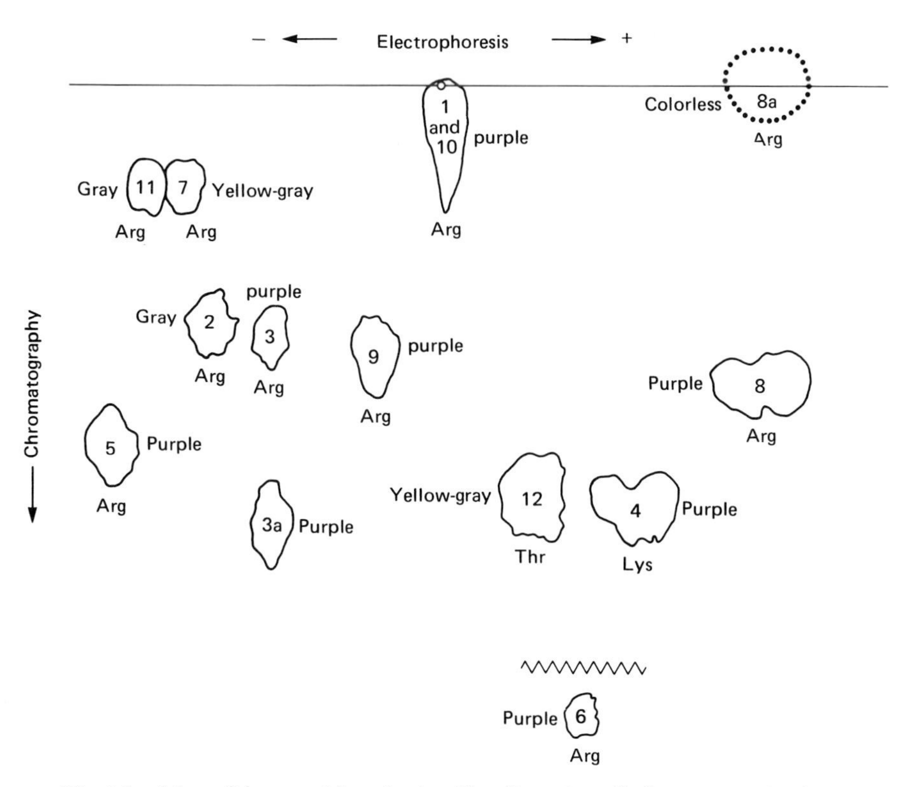

Fig. 14. Map of the peptides obtained by digestion of tobacco mosaic virus coat protein with trypsin. Adapted from Woody and Knight (1959) by deletion of four minor spots and addition of three major ones (tryptic peptides 1, 6, and 10) not shown on the original map. The latter were largely removed by precipitation at pH 4.5 prior to mapping in the experiments of Woody and Knight. Their migration in the mapping procedure was subsequently determined when the individual peptides became available. Peptide 6 travels much farther down the sheet than shown here, and the wavy line above it indicates that it has been brought up to a point more conveniently included in the diagram. Solid lines indicate ninhydrin-reactive spots that gave the colors indicated, and dotted lines signify a ninhydrin-negative but starch-iodine positive spot. The abbreviations for the amino acids—Arg, Lys, and Thr—indicate the nature of the C-terminal amino acid residue in a given peptide. See Table 12 for compositions of the peptides, which are numbered in order of their occurrence from the N-terminal to the C-terminal.

the protein chemists is used in which the amino acids whose sequences are unknown are placed in parentheses.)

Another portion of peptide 11 was treated with the enzyme leucine aminopeptidase, and aliquots were removed at various time intervals and analyzed by the DNP method. (Leucine aminopeptidase catalyzes hydrolytic cleavage of amino acids in a stepwise fashion from the N terminal end

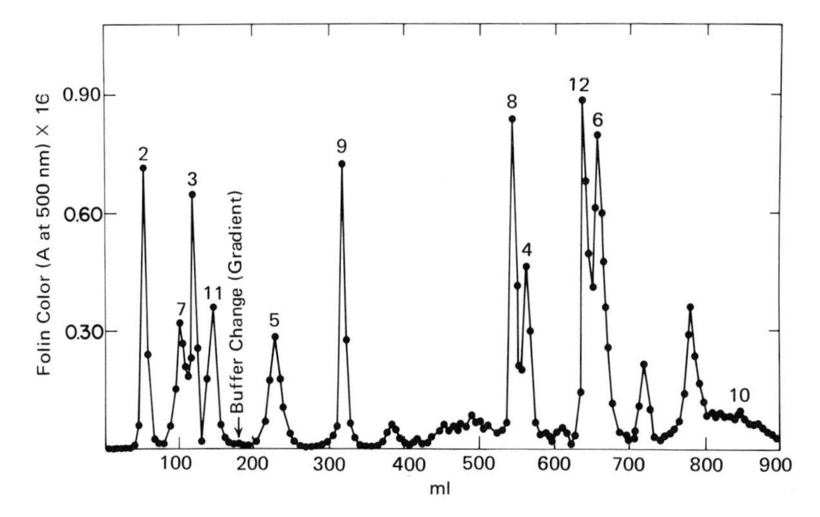

Fig. 15. Separation of tryptic peptides of TMV protein, after removal of material insoluble at pH 4.5 (mostly peptide 1) by passage through a column of Dowex 1-X2. The peptides were detected by their reaction with the Folin reagnet. The numbers are those of the tryptic peptides as listed in Table 12. (From Tsugita and Fraenkel-Conrat 1962.)

of peptide chains.) The results obtained are shown in Figure 16. These results indicate an N terminal order of Gly-Thr-Ser followed by asparagine and tyrosine in unknown order. It will be noted that the enzymatic degradation of the peptide revealed that one of the residues was asparagine rather than aspartic acid. Acid hydrolyses always yield the free acids rather than the amides, so that enzymatic hydrolyses are important in distinguishing between aspartic acid and asparagine and between glutamic acid and glutamine.

Another portion of peptide 11 was hydrolyzed with the enyzme chymotrypsin. Test analyses made by paper chromatography of portions of digest and by use of various indicator sprays indicated that the peptide was split rapidly into two peptides. One of these gave a positive test for arginine and the other a positive color reaction for tyrosine. Amounts of each peptide sufficient for analyses could be obtained by paper chromatography of the chymotryptic digest followed by elution of the separated peptides; a small strip of the chromatogram was reserved for spraying with ninhydrin in order to locate the spots. One of the peptides was found by application of the DNP method to contain equimolar amounts of aspartic acid (asparagine before hydrolysis) and arginine, the aspartic acid being N terminal. Hydrolysis of a portion of this peptide by leucine aminopeptidase yielded asparagine and arginine. Hence the peptide was Asn-Arg. A portion of the second peptide isolated from the chymotryptic digest of peptide 11 was

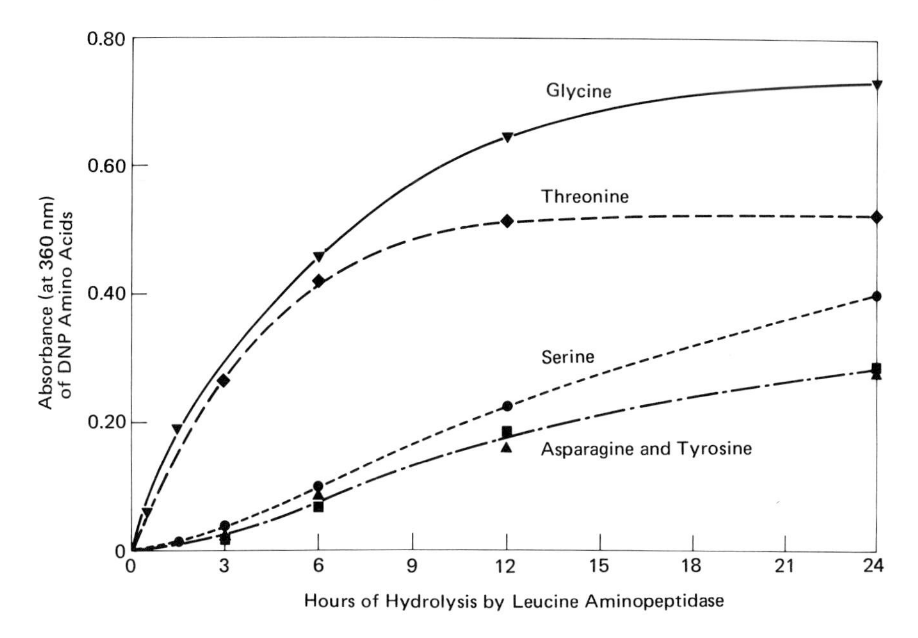

Fig. 16. Release of N-terminal amino acids from TMV tryptic peptide 11 by treatment with leucine aminopeptidase. The amino acids released at various times were identified as their DNP derivatives. (Adapted from Ramachandran and Gish 1959.)

treated with carboxypeptidase. Only tyrosine was released. Analysis of the remainder of the peptide by the DNP method revealed N terminal glycine and about equimolar proportions each of threonine, serine, and glycine. On the basis of these results and those obtained upon treating whole peptide 11 with leucine aminopeptidase, a sequence of Gly-Thr-Ser-Gly-Tyr could be assigned. Upon combination of the analytical results, the complete sequence for peptide 11 was found to be Gly-Thr-Ser-Gly-Tyr-Asn-Arg.

Twelve tryptic peptides were obtained from TMV protein and each was analyzed in a manner just outlined for peptide 11. The results are summarized in Table 12.

The final step in the primary structural analysis of TMV protein was to obtain "bridge peptides" whose sequences overlapped those of the peptides obtained by tryptic digestion. These were obtained by digesting portions of TMV protein with the chymotrypsin, pepsin, and subtilisin, and separating the resulting peptides by the same procedures used for the tryptic peptides. Analyses, or partial analyses, of the amino acid sequences of these new peptides provided information for linking the 12 tryptic peptides in the proper order to give the total sequence for the protein subunit. This procedure may be illustrated as follows.

Table 12. Peptides Obtained from TMV Protein by Tryptic Digestion.

Peptide Sequential Number	Number of Amino Acid Residues and Sequential Locations	Composition and Sequence	Color with Ninhydrin on Paper
1	41 (1-41)	Acetyl-Ser-Tyr-Ser-Ile-Thr-Thr-Pro-Ser-Gln-Phe-Val-Phe-Leu-Ser-Ser-Ala-Trp-Ala-Asp-Pro-Ile-Glu-Leu-Ile-Asn-Leu-Cys-Thr-Asn-Ala-Leu-Gly-Asn-Gln-Phe-Gln-Thr-Gln-Gln-Ala-Arg	Colorless
2	5 (42-46)	Thr-Val-Val-Gln-Arg	Gray
3	15 (47-61)	Gln-Phe-Ser-Gln-Val-Trp-Lys-Pro-Ser-Pro-Gln-Val-Thr-Val-Arg	Purple
3A	Same as 3, except for the preparative artifact of an N-terminal pyroglutamyl residue. See discussion by Gish (1960) of peptide K-0.66-A.		Purple
4	7 (62-68)	Phe-Pro-Asp-Ser-Asp-Phe-Lys	Purple
5	3 (69-71)	Val-Tyr-Arg	Purple

6	19 (72-90)	Tyr-Asn-Ala-Val-Leu-Asp-Pro-Leu-Val-Thr-Ala-Leu-Leu-Gly-Ala-Phe-Asp-Thr-Arg	Purple
7	2 (91-92)	Asn-Arg	Yellow-gray
8	20 (93-112)	Ile-Ile-Glu-Val-Glu-Asn-Gln-Ala-Asn-Pro-Thr-Thr-Ala-Glu-Thr-Leu-Asp-Ala-Thr-Arg	Purple
8A	Same amino acids as in 8. Exact structure unknown. Probably an N-terminal cyclic derivative of 8.		Colorless
9	10 (113-122)	Arg-Val-Asp-Asp-Ala-Thr-Val-Ala-Ile-Arg	Purple
10	12 (123-134)	Ser-Ala-Ile-Asn-Asn-Leu-Ile-Val-Glu-Leu-Ile-Arg	Purple
11	7 (135-141)	Gly-Thr-Gly-Ser-Tyr-Asn-Arg	Gray
12	17 (142-158)	Ser-Ser-Phe-Glu-Ser-Ser-Ser-Gly-Leu-Val-Trp-Thr-Ser-Gly-Pro-Ala-Thr	Yellow-gray

One of the peptides isolated from a chymotryptic digest of TMV protein was identified as Lys-Val-Tyr. There are two Lys residues in TMV protein, and one of them is in the tryptic peptide with a sequence -Lys-Pro-Ser (peptide 3 in Table 12). The other Lys is found in tryptic peptide 4 in a sequence ending in -Asp-Phe-Lys. This latter lysine must, therefore, be the one in the chymotryptic peptide, Lys-Val-Tyr. The only tryptic peptide with a Val-Tyr sequence is Val-Tyr-Arg. Hence the chymotryptic peptide bridges the two tryptic peptides listed in Table 12 as 4 and 5 and establishes their order in the TMV protein subunit.

By similar analyses, all of the tryptic peptides were located to give the total sequence of amino acids for the coat protein of common TMV and subsequently for other mutant strains. The sequences of four such strains are compared in Table 13.

The total number of amino acid residues in the coat proteins of many strains of TMV proved to be the same, 158. However, the coat proteins of some strains appear to reflect both additions and deletions in the viral genome (see Hennig and Wittmann 1972; Tung and Knight 1972a). Two amino acid deletions in the HR protein and their presumed location are indicated in Table 13. Admittedly, the relative evolutionary histories of these two strains are not known, and hence it can be argued that TMV was derived from an HR-like strain rather than the reverse. In that case, TMV protein would be viewed as possessing two additional amino acids rather than HR protein representing two deletions.

A comparison of the amino acid sequences of the fr, f2, and MS 2 bacteriophage coat proteins is shown in Table 14, and the sequence of the smallest known viral coat protein, that of fd phage, is given in Table 15 where it is compared with the protein of the closely related phage ZJ-2.

3. *Function of Viral Proteins*

The mass of most viruses is protein, much of which is located on the exterior of the virus particle. Thus situated, the protein comprises a coat or shell inside of which, or deeply embedded in which, lies the viral genetic material, the nucleic acid. Presumably, in the course of evolution, viruses whose genomes coded for a protein coat possessed survival value superior to those without such a structure. In any case, a noteworthy function of viral protein is the protection of the viral nucleic acid from destruction by nucleases or other degradative agents.

Another important function of viral protein is to mediate the process of infection, often determining the host specificity of a given virus. The basis for this action is that the first step in infection by many animal and bacterial viruses (but apparently not for plant viruses) is the attachment of a virus particle to a receptor site on a cell. A specific viral protein is involved in this attachment. In the case of tailed bacteriophages, the tail fibers, which are protein, serve as specific attachment organs. Only phages whose tail fiber proteins have affinity for receptor sites on the bacterial envelope can attach and initiate infection. Spheroidal phages, such as the RNA-containing R17, f2, and fr phages, appear to possess a specific coat protein essential for initiation of infection even though they have no tails. Likewise, the coat proteins of animal viruses appear to be important in the capacity of these viruses to infect cells.

A striking example of the specificity that animal viral proteins can display is given by poliovirus. This virus when intact has a very restricted host range, namely, primate cells, and this restriction appears to be dependent on the specific affinity between poliovirus coat protein and receptor material on the primate cell surface (Holland 1964). However, when poliovirus RNA is used as the infectious agent, the host range of the virus is vastly expanded (for example, virus production occurs after intracerebral inoculations of mice, rabbits, guinea pigs, chicks, and hamsters) because the RNA gains entrance to cells by an inefficient, nonspecific mechanism. It should be noted that such "unnatural" infections are restricted because whole virus is produced in the cell initially infected by the RNA and whole virus can only attach productively to primate cells.

Viruses are good antigens. That is, when introduced into various animals either by injection or by infection they elicit the production of antibodies. These antibodies can react with viruses in a variety of immunologic and serologic ways (See vol. II, chap. 13 in Fenner 1968; Casals 1967; Matthews 1967). Viral proteins either alone or in some viruses as glycoproteins and lipoproteins are primarily responsible for such properties partly because they comprise a large part of the mass of virus particles and are exteriorly located, and especially because they are better antigens than other constituents of viruses.

Most of the larger and structurally more complex viruses contain enzyme constituents (these should be distinguished from the enzymes coded for by the viral genome but which do not become incorporated into the virus particles). Such enzymes are important protein constituents of viruses that have them. Functionally, they seem to fall mainly into two classes: enzymes that degrade cell envelope or membrane constituents (for example, phage lysozyme and influenzal neuraminidase) and those involved in viral nucleic acid synthesis (for example, the RNA transcriptase of reovirus and the RNA polymerase of Newcastle disease virus and the DNA polymerase called "reverse transcriptase" of Rous sarcoma virus) (Kozloff 1968; Webster 1970; Drzeniek 1972; Shatkin and Sipe 1968; Kingsbury 1972; Baltimore 1970; Temin 1970).

Protein kinases have been detected in the particles of several purified animal viruses, including numerous RNA tumor viruses, some viruses of the influenza and parainfluenza groups, vaccinia virus, and some herpes viruses (see Rosemond and Moss 1973). However, some viruses containing

Table 13. Sequence of Amino Acids in the Coat Proteins of Four Strains of Tobacco Mosaic Virus.[a]

Strain	Sequence
	1 5 10 15
TMV[a]	Acetyl-Ser-Tyr-Ser-Ile-Thr-Thr-Pro-Ser-Gln-Phe-Val-Phe-Leu-Ser-Ser-Ala-Trp-Ala-
D	-Ser- -Val-
U2	Pro- -Thr- -Asn- -Tyr- -Ala-Tyr-
HR	Acetyl-Ser- -Asn- -Thr-Asn-Ser-Asn- -Tyr-Gln- -Phe-Ala-Ala-Val-Tyr-
	20 25 30 35
TMV	Asp-Pro-Ile-Glu-Leu-Ile-Asn-Leu-Cys-Thr-Asn-Ala-Leu-Gly-Asn-Gln-Phe-Gln-Thr-
D	-Leu- -Val- -Ser-Ser-
U2	-Val- -Ile- -Leu- -Asn-Ala-
HR	Glu- -Thr-Pro-Met-Leu- -Gln- —Val-Ser- -Ser-Gln-Ser-Tyr-
	40 45 50 55
TMV	Gln-Gln-Ala-Arg-Thr-Val-Val-Gln-Arg-Gln-Phe-Ser-Gln-Val-Trp-Lys-Pro-Ser-Pro-
D	-Thr- -Gln- -Glu- -Phe-
U2	-Ala-Asp-Ala- -Ser-
HR	-Ala-Gly- -Asp- -Arg- -Asn-Leu-Leu-Ser-Thr-Ile-Val-
	60 65 70 75
TMV	Gln-Val-Thr-Val-Arg-Phe-Pro-Asp-Ser-Asp-Phe-Lys-Val-Tyr-Arg-Tyr-Asn-Ala-Val-
D	-Gly-Asp-Val-Tyr-
U2	Val-Met- -Ala-Ser-Asp-Phe-Tyr- -Ser-Thr-
HR	Ala-Pro-Asp-Gln- -Asp-Thr-Gly- -Arg- -Val-Asn-Ser-Ala-Val-
	80 85 90
TMV	Leu-Asp-Pro-Leu-Val-Thr-Ala-Leu-Leu-Gly-Ala-Phe-Asp-Thr-Arg-Asn-Arg-Ile-Ile-
D	-Ile- -Thr-
U2	-Asn-Ser-
HR	Ile-Lys- -Tyr-Glu- -Met-Lys-

```
            95                      100                     105                     110
TMV   Glu-Val-Glu-Asn-Gln-Ala-Asn-Pro-Thr-Thr-Ala-Glu-Thr-Leu-Asp-Ala-Thr-Arg-Arg-
D                             -Gln-Ser-
U2    Glx-     -Asx-Asx-Glx-Ala-Asx-                -(Asx,Thr,Glx,Glx,Pro,Val,Ile)-
HR    Gln-Thr-Glu-Glu-Gln-Ser-Arg-    -Ser-Ala-Ser-Gln-Val-Ala-Asn-Ala-Thr-Gln-

          115                     120                     125                     130
TMV   Val-Asp-Asp-Ala-Thr-Val-Ala-Ile-Arg-Ser-Ala-Ile-Asn-Asn-Leu-Ile-Val-Glu-Leu-
D                                                             -Val-Asn-
U2                                    -Ala-Ser-               -Ala-
HR                                    -Ser-Gln-   -Gln-Leu-   -Leu-

              135                     140                     145                     150
TMV   Ile-Arg-Gly-Thr-Gly-Ser-Tyr-Asn-Arg-Ser-Ser-Phe-Glu-Ser-Ser-Ser-Gly-Leu-Val-
D     Val-                -Leu-       -Gln-Asn-Thr-           -Met-
U2                        -Met-Phe-       -Ala-Gly-       -Thr-Ala-
HR    Ser-Asx-His-Gly-    -Tyr-Met-   -Arg-   -Glu-       ———Ala-Ile———      -Pro-

                    155         158
TMV   Trp-Thr-Ser-Gly-Pro-Ala-Thr
D               -Ala-         -Ser
U2          -Thr-Thr-         -Thr
HR              -Ala-
```

[a]Adapted from Wittmann-Liebold and Wittmann 1967.

[b]TMV, common (vulgare) strain of tobacco mosaic virus; D, Dahlemense strain; U2, a mild strain; HR, Holmes' ribgrass strain.

Table 14. Sequence of Amino Acids in the Coat Proteins of Three Strains of Bacteriophage.[a]

	1				5					10					15			
fr	Ala-	Ser-	Asn-	Phe-	Glu-	Glu-	Phe-	Val-	Leu-	Val-	Asn-	Asp-	Gly-	Gly-	Thr-	Gly-	Asp-	Val-
f2					-Thr-	Gln-											-Asn-	
MS2					-Thr-	Gln-					-Asp-	Asn-						

		20					25					30					35	
fr	Lys-	Val-	Ala-	Pro-	Ser-	Asn-	Phe-	Ala-	Asn-	Gly-	Val-	Ala-	Glu-	Trp-	Ile-	Ser-	Ser-	Asn-
f2	Thr-																	
MS2	Thr-																	

				40					45					50				
fr	Ser-	Arg-	Ser-	Gln-	Ala-	Tyr-	Lys-	Val-	Thr-	Cys-	Ser-	Val-	Arg-	Gln-	Ser-	Ser-	Ala-	Asn-
f2																		-Gln-
MS2																		-Gln-

	55					60					65					70		
fr	Asn-	Arg-	Lys-	Tyr-	Thr-	Val-	Lys-	Val-	Glu-	Val-	Pro-	Lys-	Val-	Ala-	Thr-	Gln-	Val-	Gln-
f2						-Ile-											-Thr-	Val-
MS2						-Ile-											-Thr-	Val-

			75					80					85					90
fr	Gly-	Gly-	Val-	Glu-	Leu-	Pro-	Val-	Ala-	Ala-	Trp-	Arg-	Ser-	Tyr-	Met-	Asn-	Met-	Glu-	Leu-
f2														-Leu-		-Leu-		
MS2														-Leu-				

					95					100					105			
fr	Thr-	Ile-	Pro-	Val-	Phe-	Ala-	Thr-	Asx-	Asp-	Asp-	Cys-	Ala-	Leu-	Ile-	Val-	Lys-	Ala-	Leu-
f2				-Ile-				-Asn-	Ser-			-Glu-						-Met-
MS2				-Ile-				-Asn-	Ser-			-Glu-						-Met-

		110					115					120					125	
fr	Gln-	Gly-	Thr-	Phe-	Lys-	Thr-	Gly-	Ile-	Ala-	Pro-	Asn-	Thr-	Ala-	Ile-	Ala-	Ala-	Asn-	Ser-
f2			-Leu-	Leu-		-Asp-		-Asn-	Pro-	Ile-	Pro-	Ser-						
MS2			-Leu-	Leu-		-Asp-		-Asn-	Pro-	Ile-	Pro-	Ser-						

			129
fr	Gly-	Ile-	Tyr
f2			
MS2			

[a]Adapted from Wittmann-Liebold and Wittmann 1967; Min Jou et al. 1972.

a protein kinase, such as some of the RNA tumor viruses, do not have phosphate groups in their structural proteins, whereas the structural proteins of some viruses, such as simian virus 40, are all phosphoproteins despite the fact that the virus particles have no kinase (Tan and Sokol 1972). Therefore, the origin and function of these enzymes are unclear, although it has been suggested that production of phosphoproteins may be involved in the regulation of viral transcription.

Table 15. Amino Acid Sequences of The Coat Proteins of Bacteriophages ZJ-2 and fd.[a]

	1				5					10
ZJ-2	Ala	Glu	Gly	Asp	Asp	Pro	Ala	Lys	Ala	Ala
fd										
					15					20
ZJ-2	Phe	Asp	Ser	Leu	Gln	Ala	Ser	Ala	Thr	Glu
fd										
					25					30
ZJ-2	Tyr	Ile	Gly	Tyr	Ala	Trp	Ala	Met	Val	Val
fd										
					35					40
ZJ-2	Val	Ile	Val	Gly	Ala	Ala	Ile	Gly	Ile	Lys
fd						Thr				
					45					50
ZJ-2	Leu	Phe	Lys	Lys	Phe	Thr	Ser	Lys	Ala	Ser
fd										

[a]From Asbeck et al. 1969; Snell and Offord 1972.

B. Nucleic Acids

Nucleic acids are so named because they are acidic substances that were first isolated from the nuclei of cells.[1] It is now known that nucleic acids occur in both the nuclei and cytoplasm of all cells. The two major types of nucleic acid found in nature, ribonucleic acid (RNA) and deoxyribonucleic acid (DNA), both occur in viruses. However, in contrast to bacteria and other organisms, no virus appears to contain both RNA and DNA. The type of nucleic acid present can be determined by qualitative tests for sugar and pyrimidine components since it is only with respect to these constituents that RNA and DNA differ in composition. The detection of deoxyribose and thymine indicate DNA, whereas the presence of ribose and uracil denote RNA (procedures for analysis of these substances are

[1]The properties of nucleic acids in general apply to viral nucleic acids. The interested student may wish to refer to such comprehensive reference works as *Progress in Nucleic Acid Research and Molecular Biology*, J. N. Davidson and W. E. Cohn, editors, New York: Academic Press (published annually since 1963); *Procedures in Nucleic Acid Research,* Vol. 1 and 2, G. L. Cantoni and D. R. Davies, editors, New York: Harper and Row (Vol. 1 in 1966 and Vol. 2 in 1971); *The Chemistry of Nucleosides and Nucleotides,* A. M. Michelson, New York: Academic Press (1963); *Genetic Elements—Properties and Function,* D. Shugar, editor, New York: Academic Press (1967); *Methods in Enzymology—Nucleic Acids*, L. Grossman and K. Moldave, editors, New York: Academic Press (Vol. XII, 1967; Vol. XII, Part B, 1968; Vol. XX, Part C, 1971; Vol. XXI, Part D, 1971; Vol. XXIX, Part E, 1974; Vol. XXX, Part F, 1974.

described by Ashwell 1957; Schneider 1957; Lin and Maes 1967; Burton 1968; Hatcher and Goldstein 1969).

The quantity of nucleic acid, while fairly constant within a given group of viruses, varies considerably among different viruses. The extremes are represented by 0.8 percent RNA in influenza viruses and about 56 percent DNA in coliphage lambda. Viral nucleic acids, like those from other sources, are elongated, threadlike molecules. Some of them are single stranded, some double stranded, and some are cyclic. The amount and type of nucleic acid found in some viruses are given in Table 16.

1. Preparation of Viral Nucleic Acids

Viral nucleic acid is deeply embedded in the protein matrix of the virus particle. Despite this sheltered location, the nucleic acid is accessible to some chemical reagents such as mustards, nitrous acid, formaldehyde, and smaller molecular species in general. Nevertheless, for many experiments, it is desirable to isolate the nucleic acid from the rest of the material. No single procedure has proved universally successful for this purpose. However, reagents noted for an ability to break secondary valence bonds, such as salt linkages and hydrogen and hydrophobic bonds, have been most effective in disaggregating virus particles with release of the nucleic acid.

To best study the properties and function of viral nucleic acids, it has become ever more important to isolate the intact nucleic acid, to the extent that this exists, from virus particles. Three main factors work against this objective: (1) mechanical shearing of the nucleic acid during isolation, (2) chemical degradation at the extreme pH values that favor removal of protein coats, and (3) enzymatic degradation.

Mechanical shearing is a problem primarily with large DNA molecules such as those found in phages. Violent mixing or even forceful pipeting of solutions of phage DNA are sufficient to rupture the molecules (Hershey et al. 1962) and breakage may also be accompanied by denaturation, that is, strand separation, under certain conditions of temperature and salt concentration (Hershey et al. 1963). Therefore, gentle stirring procedures are recommended in the isolation of nucleic acids.

The sensitivities of RNA and DNA to extreme pH values differ somewhat but the structures of both types of nucleic acid may be irreversibly altered at pH values below 3 or above 10. Below pH 3, depurination (cleavage of adenine and guanine) tends to occur with double-stranded molecules. RNA is subject to alkaline hydrolysis above pH 10, and DNA, while resistant to alkaline hydrolysis, may be denatured above pH 12. Consequently, most nucleic acid isolation procedures are performed at intermediate pH values.

Probably the greatest hazard to intact viral nucleic acid is attack by nuclease enzymes, that is, by ribonucleases and deoxyribonucleases.

Traces of nucleases can often be detected even in the most highly purified preparations of viruses; and while several of the methods for isolating nucleic acid include provision for protecting the product from nucleases, none of the procedures is entirely satisfactory in this regard. In principle, use of strong protein denaturants in removing the viral coat protein and releasing nucleic acid will also eliminate accompanying nucleases. In practice, however, unless the removal of denatured protein is complete, traces of nuclease will remain with the nucleic acid and subsequently become renatured and active (Ralph and Berquist 1967).

Since complete removal of denatured protein is difficult to ensure, one needs to combat nuclease activity in preparing viral nucleic acids by starting with the most highly purified virus obtainable, avoiding prolonged procedures, and by adding nuclease inhibitors. Two such inhibitors are the acid clay bentonite (Fraenkel-Conrat, et al. 1961; Singer and Fraenkel-Conrat, 1961) and diethyl pyrocarbonate (Solymosy et al. 1968; Bagi et al. 1970). Some disadvantages of these nuclease inhibitors are that bentonite adsorbs some RNA (Fraenkel-Conrat 1966) and diethyl pyrocarbonate under some conditions seriously inhibits the separation of viral protein and nucleic acid (Bagi et al. 1970).

In addition to the above factors, the salt concentration and nature of cations present can affect the isolation and stability of viral nucleic acids (Ralph and Berquist 1967). For example, at very low ionic strengths (10^{-4} M) strand separations occur in double-stranded nucleic acids and in double-stranded regions of single-stranded nucleic acids (all single-stranded nucleic acids have the tendency to form some double-stranded loops). Such denaturation usually makes nucleic acids more susceptible to degradation by nucleases. Cations such as Mg^{2+} may cause aggregation and loss of nucleic acids, especially of RNA, as do also salt solutions stronger than 1 M.

In summary, the various procedures for preparing intact undenatured viral nucleic acids are generally successful in proportion to their ability to effect a thorough denaturation of viral coat protein and separation of it from nucleic acid under conditions that avoid extremes of mechanical treatment, pH, and salt concentration and that minimize contact with nucleases. Some examples of procedures that apply these principles follow.

a. The Hot Salt Method

This procedure is a modification of the method of Cohen and Stanley (1942) (see Knight 1957; Reddi 1958; Lippincott 1961) that has proved useful in preparing nucleic acid from TMV and strains, although the nucleic acid isolated in this manner is not consistently so infectious as that obtained by detergent treatment or phenol extraction. [Infectivities comparable to the highest obtained by any procedure have been reported by Boedtker (1959) and by Lippincott (1961), using the hot salt method, but

Table 16. Approximate Content and Type of Nucleic Acid in Some Viruses.

Virus	Nucleic Acid per Particle, Daltons	Type of Nucleic Acid[a]	References
Algal N-1	38×10^6	ds-DNA	Adolph and Haselkorn 1971
Adenovirus-2	23×10^6	ds-DNA	Joklik and Smith 1972
Avian myeloblastosis	10×10^6	ss-RNA	Joklik and Smith 1972
Broad bean mottle	1×10^6	ss-RNA	Yamazaki et al. 1961
Brome mosaic	1×10^6	ss-RNA	Bockstahler and Kaesberg 1961
Coliphages f1, fd, M13	1×10^6	ss-c-DNA	Hoffmann-Berling et al. 1966
Coliphages f2, MS2, R17	1×10^6	ss-RNA	Hohn and Hohn 1970
Coliphages ØX174, S13	2×10^6	ss-c-DNA	Thomas and MacHattie 1967
Coliphage lambda	32×10^6	ds-DNA	Thomas and MacHattie 1967
Coliphages T2, T4, T6	130×10^6	ds-DNA	Thomas and MacHattie 1967
Cucumber 3 (and 4)	5×10^6	ss-RNA	Knight and Stanley 1941
Cytoplasmic polyhedrosis	13×10^6	ds-RNA	Kalmakoff et al. 1969
Foot-and-mouth disease	2×10^6	ss-RNA	Rueckert 1971

Fowlpox	200×10^6	ds-DNA	Hyde et al. 1967
Herpes simplex	100×10^6	ds-DNA	Joklik and Smith 1972
Influenza	4×10^6	ss-RNA	Compans and Choppin 1973
Mouse encephalitis	2×10^6	ss-RNA	Rueckert 1971
Newcastle disease	6×10^6	ss-RNA	Blair and Duesberg 1970
Poliomyelitis	2×10^6	ss-RNA	Schaffer and Schwerdt 1959
Polyoma	4×10^6	ds-c-DNA	Kass 1970
Potato X	2×10^6	ss-RNA	Reichmann 1959
Reo, Type 3	15×10^6	ds-RNA	Joklik 1970
Rous sarcoma	10×10^6	ss-RNA	Robinson and Duesberg 1967
Shope papilloma	5×10^6	ds-c-DNA	Kass and Knight 1965
Silkworm jaundice	22×10^6	ds-DNA	Bergold 1953; Bergold and Wellington 1954
Simian 40	4×10^6	ds-c-DNA	Yoshiike 1968
Tobacco mosaic	2×10^6	ss-RNA	Knight and Woody 1958
Tomato bushy stunt	2×10^6	ss-RNA	DeFremery and Knight 1955
Turnip yellow mosaic	2×10^6	ss-RNA	Markham 1959
Wound tumor	16×10^6	ds-RNA	Kalmakoff et al. 1969

[a]ds = double-stranded; ss = single-stranded; c = cyclic.

with temperatures between 90° and 98.5°. However, recoveries of RNA are substantially lower under these conditions and the risk of contamination with undegraded virus, higher.]

1. Hot salt procedure for preparing RNA from TMV and similar viruses. To 0.3 M NaCl in a water bath at 100°C is added enough virus in aqueous solution to give a final concentration of 10–15 mg/ml. After mixing, the mixture is held at 100° for 1 min and then removed to an ice bath. After chilling, the mixture is centrifuged at 5,000–10,000 *g* to remove coagulated protein. The water-clear solution of sodium nucleate can be freed of salt by dialysis against water in the cold or by precipitation two or three times with ice-cold 67 percent alcohol, redissolving the nucleate in water each time. (A final step of high-speed centrifugation may be used, if desired, to pellet traces of insoluble matter.) Yields of about 80 percent are usually obtained.

2. Modified hot salt method for preparation of RNA from tobacco ringspot and turnip yellow mosaic viruses. For success with tobacco ringspot and turnip yellow mosaic viruses, Kaper and Steere (1959a, 1959b) found it necessary to modify the hot salt procedure by reducing the virus concentration and heating time and increasing the salt concentration. Thus, to 2 ml of M NaCl in a boiling water bath is added 1 ml of virus (at 5–10 mg/ml in 0.01 M phosphate buffer at pH 7) and heating is continued for 35 sec with constant mixing. The mixture is cooled immediately in an ice bath and the coagulated protein is removed by centrifugation. The nucleic acid, in the supernatant fluid, can be purified by two precipitations with cold alcohol and high-speed centrifugation as above.

3. Modified hot salt procedure for preparing RNA from influenza and Rous sarcoma viruses. Only about 0.8 percent of influenza virus is RNA and it is difficult to extricate the nucleic acid from the great mass of protein, lipid, and carbohydrate present. A hot salt procedure reported to give very good yields was developed by Ada and Perry (1954). The method has also been used to extract RNA from Rous sarcoma virus (Bather 1957).

The purified, frozen-dried (lyophilized) virus is defatted by extracting twice at room temperature with chloroform-methanol (2:1, v/v) followed by one extraction with *n*-butanol and two washes with ethyl ether. The RNA is obtained by extracting the defatted virus one to three times at 100° with 10 percent (w/v) NaCl, using 20-min extraction periods. The RNA can be freed of salt as above, with alcohol precipitation probably preferable. So far, the nucleic acid thus obtained from influenza virus has proved noninfectious.

A hot salt method has also been used to prepare DNA from phage ØX174 (Guthrie and Sinsheimer 1960; Sekiguchi et al. 1960).

b. Detergent Method

The following procedure, adapted from the method of Fraenkel-Conrat et al. (1957), gives good yields of RNA from TMV and related viruses. The RNA is infectious (Fraenkel-Conrat 1956) and reconstitutes well with protein obtained by acetic acid degradation of the virus (see section on Reconstitution of Viruses).

1. Detergent procedure for isolating RNA from TMV and similar viruses. Virus at 20 mg/ml in water is heated to 55° in a water bath, adjusted to pH 8.8 with dilute NaOH, and mixed with an equal volume of 2 percent sodium dodecyl sulfate (commercial preparations such as Duponol C are satisfactory also) that has also been adjusted to pH 8.8 at 55°. The mixture is allowed to remain in the water bath at 55° for 5 min during which the solution loses its characteristic opalescence, owing to degradation of the virus. After 5 min (greater or less time may be required for different strains of TMV) the mixture is rapidly cooled to room temperature (about 23°C) and 0.5 vol of saturated ammonium sulfate is added. After about 10 min the precipitated protein is removed by centrifugation at 5,000–10,000 g and the clear supernatant fluid is stored at 4° overnight. The RNA precipitates out under these conditions, and the precipitate is packed by centrifugation, redissolved in a small volume of water, and reprecipitated by adding 2 vol of cold alcohol. The alcohol precipitation may be repeated once or twice more and traces of insoluble material may be removed from the final solution of RNA by centrifuging at about 100,000 g for 2 hr with refrigeration. Yields of 60–90 percent are obtained.

2. Modified detergent procedure for isolation of DNA from polyoma virus (Smith el al. 1960). Polyoma virus was isolated from clarified extracts of a mouse embryo tissue culture by differential and density gradient centrifugation. Equal volumes of virus solution and 10 percent sodium dodecyl sulfate at pH 7 are mixed and heated at 65° for 2 hr. After adding enough ammonium acetate to give a final concentration of 0.1 M, the DNA is precipitated by adding 2 vol of ethanol. The precipitate is dissolved in 0.1 M ammonium acetate and reprecipitated by adding alcohol as before. This dissolving and precipitating procedure is repeated once more.

A modified detergent method has also been used in the isolation of DNA from Shope papilloma virus (Watson and Littlefield 1960).

c. Combined Detergent and Hot Salt Method

Some viruses from which low yields of nucleic acid are obtained by either the hot salt or detergent methods alone will give reasonable amounts of nucleic acid by a combined procedure (Dorner and Knight 1953). In this method, 1 vol of 10 percent Duponol C solution (or sodium dodecyl sulfate) is added to 4 vol of aqueous virus at about 10 mg/ml. The mixture is heated in a boiling water bath for 4 min and then cooled in an ice bath. Most of the free detergent is removed by dialysis and then enough 5 N NaCl is added to make the final concentration 1 N with respect to NaCl. This mixture is heated for 3 min at 100°, chilled in an ice bath, and the coagulated protein is removed by centrifugation. Salt is removed by dialysis and the preparation is concentrated by directing a stream of air at the dialysis bag (pervaporation). The concentrate of nucleate can be clarified by centrifugation, or, if desired, the nucleate can be precipitated from the concentrate by addition of 2 vol of cold ethanol.

d. The Phenol Method

The phenol extraction method (Westphal et al. 1952) is perhaps the most generally useful procedure for obtaining nucleic acid from a wide variety of viruses (as well as from tissues). In operation, two layers—a phenolic and an aqueous layer—are obtained and protein is extracted into the phenolic layer (and some in the interface) while nucleic acid (and polysaccharide, if present) goes into the aqueous layer.

Phenol extraction was first used to prepare viral nucleic acid by Schuster et al. (1956), and very soon it was shown (Gierer and Schramm 1956a, 1956b) that RNA thus obtained from TMV is infectious. The initial procedure does not work satisfactorily on all viruses, but in several cases modifications have been developed that have successfully extended the usefulness of the technique. A convenient adaptation of the method for the preparation of RNA from TMV and strains is presented here together with some modifications extending the usefulness of the method.

1. Phenol procedure for preparing RNA from TMV and similar viruses. To the virus solution in 0.02 M phosphate buffer at pH 7 and at a virus concentration of 20–25 mg/ml is added an equal volume of water-saturated phenol. (This is about 80 percent phenol, and it is easily prepared by taking a fresh bottle of commercial reagent grade crystalline phenol and almost filling the bottle with distilled water. The mixture is liquefied by placing in a warm water bath and stirring occasionally. Two layers will be apparent: a large lower layer consisting of the water-saturated phenol and a small upper layer of excess water. If stored at about 4° in the dark glass bottle normally commercially available, the preparation keeps for weeks, and portions of the lower layer are used in the preparation of nucleic acid. Some workers redistill their phenol, add metal-chelating agents such as sodium versenate, and so on, but the author has not found these refinements to be generally necessary.)

The mixture of virus and phenol is stirred on a magnetic stirrer for 10–15 min at room temperature (about 23°C), after which the mixture is separated into two layers by centrifuging at 5,000–10,000 *g* for about 2 min. (The original procedure was carried out at low temperature, and 4° is still used in some cases; in other instances it has been found necessary to use temperatures as high as 50°–60°.) The aqueous (top) layer is drawn off, and about one-tenth its volume of water-saturated phenol is added, and the mixture is stirred again for 3–4 min followed by centrifugation. The aqueous layer is extracted once more with a tenth volume of phenol and then twice with equal volumes of ether (to remove the small amount of phenol which dissolves in the aqueous phase). Residual ether is removed from the aqueous nucleate by two to three precipitations of the RNA with ethanol, which is accomplished by chilling the nucleate solution and adding 2 vol of ice-cold ethanol. If difficulty in precipitating the material is experienced, a drop or two of 3 M sodium acetate at pH 5 can be added. The final precipitate, pelleted by centrifugation, is dissolved in a small volume of distilled water and centrifuged at about 100,000 g for 2 hr. The nucleic acid is not sedimentable under these conditions but a trace of insoluble material is usually removed as a tiny pellet. Yields of the order of 80 percent are commonly obtained.

Conditions similar to those described above (except that usually temperatures around 4° have been employed without evidence that such low temperatures are necessary) have been used successfully to prepare RNA from partially purified poliovirus (Alexander et al. 1958); from potato virus X (Bawden and Kleczkowski 1959); from tobacco rattle virus (Harrison and Nixon 1959b); from cucumber mosaic virus (Schlegel 1960b); from an RNA-containing insect virus, Smithia virus pudibundae (Krieg 1959); and others. The method has also been used to extract DNA from T2, T4, and ØX174 phages (Davison et al. 1961; Rubenstein et al. 1961; Guthrie and Sinsheimer 1960).

e. Phenol-Detergent Method

In some cases it has been found beneficial to make the phenol extraction after the virus structure has been opened up by a detergent such as sodium dodecyl sulfate. Rushizky and Knight (1959) used such a technique to obtain infectious RNA from tomato bushy stunt virus, Huppert and Rebeyrotte (1960) to extract DNA transforming principle from bacteria, Wahl et al. (1960) to prepare ØX174-DNA, and Bachrach (1960) to obtain infectious RNA from foot-and-mouth disease virus. Bachrach's method is given here.

> Virus concentrates are diluted six times in 0.02 M phosphate buffer at pH 7.6 which contains 0.01 percent sodium ethylenediaminetetraacetate (EDTA, or "Versene") and 0.1 percent sodium dodecyl sulfate. The diluted virus is twice extracted at 4° with water-saturated phenol containing 0.01 percent EDTA. Phenol is removed from the final aqueous phase by several extractions with ether. The ether is removed by a stream of nitrogen.

In the standard phenol procedure used for preparation of RNA from TMV and similar viruses, the extraction is now made at room temperature, which is about 20° warmer than used in the original procedure. Even this temperature, however, is not sufficient for some viruses, and it has been necessary to go to about 50°, for example, to extract the RNA from equine encephalomyelitis virus (Wecker 1959).

Another important modification of the phenol method involves the addition of bentonite as an adsorbent for nucleases during the phenol extraction (Fraenkel-Conrat et al. 1961) The infectivity of RNA obtained from TMV by the phenol-bentonite procedure is stabler upon incubation in salt solutions than most preparations made without bentonite (Singer and Fraenkel-Conrat 1961).

f. Guanidine Hydrochloride Method

Bawden and Kleczkowski (1959) reported the preparation of infectious RNA from potato virus X by phenol extraction, but the reproducibility of the results was not good. Therefore Reichmann and Stace-Smith (1959) investigated other procedures and devised a method based on treatment with guanidine that gave consistently 70–90 percent yields of infectious RNA. Their method is as follows:

To virus solution at 5–10 mg/ml is slowly added a sufficient volume of concentrated, recrystallized guanidine hydrochloride and ethylenediaminetetraacetate at pH 8.4 to make a final concentration of 2.5 M in guanidine and 0.005 M in EDTA. After 1 hr the RNA, which is insoluble, and the protein, which is soluble, in this mixture are separated by centrifugation at about 4,500 *g*. The RNA pellet is washed twice with 2.5 M guanidine-EDTA solution and then dissolved in a small volume of water. Further purification of the RNA is accomplished with 2 vol of cold ethanol, a 90-min centrifugation at 75,000 *g*, and dialysis overnight against distilled water.

g. Alkaline Method

As discussed earlier, extremes of pH are generally to be avoided in the preparation of nucleic acids. However, DNA is fairly resistant to alkaline media up to pH 12, and this fact was taken advantage of in isolating the nucleic acid of a nuclear polyhedrosis virus of the silkworm (*Bombyx mori*) (Onodera et al. 1965).

Freshly prepared virus particles are suspended in a small amount of 0.1 M Na_2CO_3–$NaHCO_3$ buffer (pH 10) 0.1 M in sodium citrate. This is dialyzed against a large volume of the same buffer-citrate mixture at 4° for two days. The dialyzed material is centrifuged at 40,000 *g* for 30 min to remove insoluble material. Solid ammonium sulfate is added to the supernatant fluid to a concentration of about 25 percent (wt/vol). The resulting precipitate of protein is removed by centrifugation and the supernatant containing the nucleic acid is dialyzed at 4° against 0.15 M NaCl–0.15 M sodium citrate. The DNA can be further purified by precipitation with 2–3 vol of cold ethanol or by passage through a column of methylated albumin.

In general, it may be stated that isolated viral nucleic acids are stabler than at first supposed. The primary cause of the instability is apparently traces of nucleases, and if these are absent, the nucleic acids maintain their integrity on storage and withstand temperatures considerably above room temperature. In short, nucleic acids are not intrinsically chemically unstable, nor are they particularly thermolabile.

2. *Analysis of Viral Nucleic Acids*

Nucleic acids may be considered to be polynucleotides, that is, chain-like molecules in which the links are nucleotides. This is illustrated in Figure 17, which depicts the essential features while showing only a very small segment of nucleic acid.

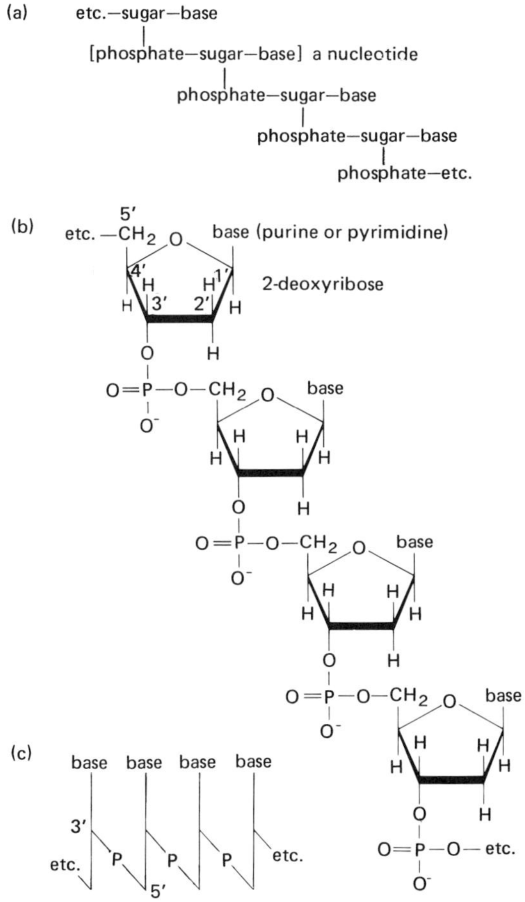

Fig. 17. Composition of a small segment of nucleic acid indicating the order of arrangement of the three major components (sugar, base, and phosphate). *a*. General scheme of sugar-phosphate backbone structure. *b*. Chemical structure of sugar-phosphate backbone of DNA showing numbering of atoms in the sugar (deoxyribose) and phosphodiester linkages between nucleotides. *c*. An abbreviated way to indicate oligonucleotides. (Adapted from Knight 1974.)

Nucelotides are named according to the purine or pyrimidine base they contain. In the case of DNA, which contains deoxyribose rather than ribose as in RNA, this is indicated in the naming of the nucleotides by appending the prefix deoxy. Thus, for RNA, the nucleotides are adenylic acid, guanylic acid, cytidylic acid, and uridylic acid; for DNA they are deoxyadenylic acid, deoxyguanylic acid, deoxycytidylic acid, and thymidylic acid (the deoxy prefix is not necessary for the thymine-containing nucleotide since there is no natural counterpart in the ribonucleic acid series).

As indicated in Figure 17, all nucleotides are built up from three simpler components: phosphate, sugar, and a purine or pyrimidine base. Nucleosides are made from sugar and a purine or pyrimidine base. Nucleosides are thus chemically closely related to nucleotides, and removal of phosphoric acid (by hydrolysis) from a nucleotide yields a nucleoside. Conversely, nucleotides can be viewed as nucleoside phosphates. The common ribonucleosides are adenosine, guanosine, cytidine, and uridine. The comparable deoxyribonucleosides are deoxyadenosine and so on, except that deoxy is commonly omitted from the name of the nucleoside consisting of deoxyribose and thymine (thymidine) since thymidine is characteristic of DNA only.

The purine bases commonly found in nucleic acids are adenine and guanine and the pyrimidines are cytosine, uracil, and thymine, the latter occurring only in DNA. In the DNA of certain phages, 5-hydroxymethylcytosine or 5-hydroxymethyluracil is found in place of cytosine. Formulas for some of these bases are as follows:

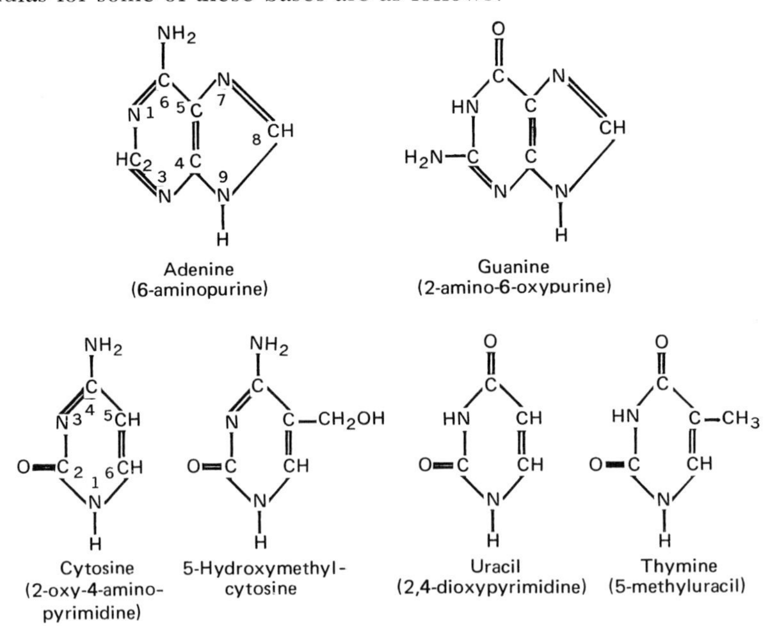

The sugar components of RNA and DNA are D-ribose and 2-deoxy-D-ribose, respectively. These sugars in the β configuration (see the following formula) are attached to purines and pyrimidines in nucleic acids (and also in nucleotides and nucleosides). Both sugars are found in nucleic acids (and in nucleotides) in the furanose ring form (oxygen ring between carbons 1′ and 4′), whereas the free sugars occur mainly in the pyranose form (oxygen ring between carbons 1′ and 5′). Structural and abbreviated formulas may be written for these sugars as follows:

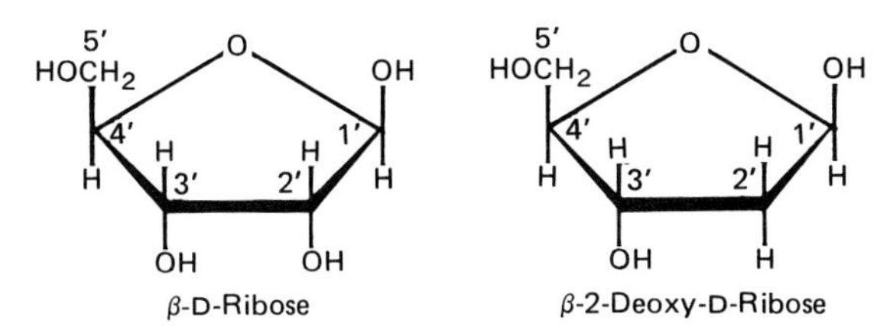

It will be noted in the formulas for nucleic acid components that regular numbers were used to denote positions in the purine and pyrimidine rings, whereas prime numbers were used to indicate positions in the sugars. This convention applies to nucleosides, nucleotides, and nucleic acids in order to make the distinction between derivatives involving the sugar and those affecting the bases.

In nucleic acids, including those of viruses, the nucleotides are uniformly linked through $3' \rightarrow 5'$ phosphate diester bonds between the sugar moieties as illustrated in the bit of DNA shown in Figure 17. This means that all DNA molecules share a common deoxyribose-phosphate structure; similarly all species of RNA have a common ribose-phosphate backbone. Consequently, aside from differences in molecular size, the biological specificity of nucleic acids depends entirely on the purine and pyrimidine bases, or, more precisely, on the sequence in which these bases occur along the sugar-phosphate chain.

An important step in characterizing nucleic acids is to determine the proportions of purines and pyrimidines present. This is commonly done by hydrolyzing the nucleic acid and separating the resultant components of the hydrolysate by either of three methods: paper chromatography or thin layer chromatography, paper electrophoresis, or column chromatography. Concentrations of discrete fractions separated by one of these methods are usually determined by ultraviolet spectrophotometry. Examples of some of the procedures are given below.

a. Determination of the Base Ratios in RNA by Acid Hydrolysis and Paper Chromatography

Numerous studies have been made on the hydrolysis of nucleic acids and the separation of nucleic acid constituents (see, for example, Fink and

Adams 1966; Lin and Maes 1967). An early and still useful procedure for determining base ratios in RNA is that of Smith and Markham (1950). Mild acid hydrolysis is used, which releases the purines in the form of free bases and the pyrimidines in the form of their nucleotides.

Viral RNA, isolated by one of the methods described in the previous section, is hydrolyzed at a concentration of 10 mg RNA/ml in 1 N HCl for 1 hr at 100° (boiling water bath). Twenty μl (equivalent to 200 μg of RNA) of hydrolysate is applied to Whatman No. 1 filter paper with a micropipette in such a way as to give a rectangular spot about 0.5 × 5 cm. Separation of the components is effected by either ascending or descending chromatography in 70 percent tert. butanol-water, 0.8 N with respect to hydrochloric acid (70 ml tert. butanol, 13 ml constant boiling HCl, and 17 ml H_2O). After chromatography, the papers are air-dried at room temperature, and the spots are located by examination with an ultraviolet light and marked with a pencil. Starting at the point of application and going in the direction of the solvent movement, the spots will be found in the following order: guanine, adenine, cytidylic acid, and uridylic acid. The sample spots are cut out, as well as paper blanks of approximately the same size next to the sample spots. Each paper cutout is placed in a test tube together with 5 ml of 0.1 N HCl and eluted by standing at room temperature overnight or by shaking for 2 hr. The absorption of each solution is read in a spectrophotometer at the wavelength of maximum (or near maximum) absorption for the compound in question. The amount of each component can be calculated by use of the proper extinction coefficients such as those given by Sober (1970). Some data of this sort are given in Table 17 for the compounds obtained from RNA as above, and for the compounds obtained by alkaline hydrolysis of RNA or acid hydrolysis of DNA.

During hydrolysis in 1 N HCl, about 5 percent of the pyrimidine

Table 17. Ultraviolet Absorption Data for Some Nucleic Acid Constituents Near Wavelengths of Maximum Absorption.[a]

Substance	pH	Wavelength, nm	Molar Extinction Coefficient, $\times 10^{-3}$
Adenine	1	262.5	13.2
Guanine	1	248	11.4
Cytosine	1	276	10.0
5-Hydroxymethylcytosine	1	279	9.7
Thymine	4	265	7.9
Uracil	0	260	7.8
Adenylic acid	1	257	15.0
Guanylic acid	1	257	12.2
Cytidylic acid	2	279	13.0
Uridylic acid	1	262	10.0

[a]Compiled from Sober 1970.

nucleotides are hydrolyzed to nucleosides. Hence for more accurate values using this method, the cytidylic and uridylic acid figures can be corrected upward by 5 percent (Markham and Smith 1951) and the adenine value downward by the same amount (cytidine arising from partial hydrolysis of cytidylic acid migrates to the same area as adenine in tert. butanol-HCl. Uridine occupies an area between cytidylic and uridylic acids and hence does not affect values for other components). A more precise correction can be made by actual determination of the amount of cytidine in the adenine spot. This is done by using observed absorption values at two wavelengths, standard absorption values, and applying simultaneous equations (Loring and Ploeser 1949). Assuming a similar conversion of uridylic acid to uridine permits complete correction and accounts very well for the nucleic acid components in terms of phosphorus recovery (deFremery and Knight 1955).

The hydrolysis in 1 N HCl is capable of releasing the nucleic acid components from whole virus as well as from isolated nucleic acid (Dorner and Knight 1953). Hence it is not necessary to isolate nucleic acid in order to determine the proportions of purines and pyrimidines present. However, this analysis is affected somewhat by the relative proportions of protein and nucleic acid, and gives most accurate results on the viruses containing 10 percent or more RNA.

b. Determination of the Base Ratios in RNA by Alkaline Hydrolysis and Paper Electrophoresis

Another convenient method for determining the base composition of RNA is by hydrolizing the nucleic acid in dilute alkali at low temperature, separating the resulting four nucleotides by paper electrophoresis, and determining the quantity of each nucleotide by spectrophotometry on the material eluted from the paper (see Smith 1955; Crestfield and Allen 1955a, 1955b). A useful procedure is as follows.

Two mg of RNA is hydrolyzed in 0.1 ml of 0.4 N NaOH at 37° for 24 hr. Ten μl aliquots of this hydrolyzate is applied to buffer-moistened Whatman 3 MM paper in an approximately 2-cm streak. Any of a number of types of electrophoresis apparatus may be used (See Smith 1955; Crestfield and Allen 1955a; Rushizky and Knight 1960b). The buffer used is 0.05 M formate at pH 3.5 (prepared by adding 6.4 g ammonium formate and 10.3 g formic acid, 88–90 percent, to 6 liters of water). Electrophoresis is performed at a voltage gradient of 6 v/cm for about 15 hr. (With proper cooling, a higher voltage can be used and the separation accelerated.) After drying the paper in air, the nucleotides can be located by examination with an ultraviolet light, marked, cut out (with appropriate blanks), eluted in 5-ml portions of 0.01 N HCl, and measured in a spectrophotometer. At pH 3.5 the nucleotides are found in the following order, starting from the cathode side of the paper: cytidylic acid, adenylic acid, guanylic acid, and uridylic acid.

c. Determination of the Base Ratios in DNA by Acid Hydrolysis and Paper Chromatography

Analyses of DNA are based on methods of acid hydrolysis that release the purines and pyrimidines as free bases. Either 70 percent perchloric acid or 88 percent (or 98 percent) formic acid are usually employed. Wyatt (1955) suggests the use of formic acid for best recoveries of the various bases (including the somewhat labile 5-hydroxymethylcytosine), but hydrolyses must be made in sealed bomb (thick-walled) tubes and the pressure from decomposition of formic acid is conveniently released, after hydrolysis and cooling, by heating a small area at the top of the tube until a little hole blows open. The tube may then be safely and fully opened. If it is desired to get base analyses on whole virus without isolation of the nucleic acid, then 70 percent perchloric acid is recommended. The following is a possible procedure based on these observations.

DNA is placed in a pyrex glass bomb tube and enough 88 percent formic acid is added to give a concentration of 2 mg DNA/ml formic acid. The tube is sealed and heated at 175° for 30 min. After cooling, pressure is released as described above; the tube is opened and the hydrolysate is evaporated to dryness in vacuo. The residue is taken up in a small volume of N HCl to give a concentration equivalent to 10–20 mg/ml of the original DNA. Twenty μl of hydrolysate is placed on Whatman No. 1 paper and chromatographed in isopropanol-HCl-water (170 ml isopropanol, 41 ml concentrated HCl, 39 ml H_2O). The migration of the bases in increasing distance from the origin is in the order guanine, adenine, cytosine, and thymine. In cases where 5-hydroxymethylcytosine is present instead of cytosine, it will be found in the cytosine position. Location of the spots and elution and spectrophotometry are carried out as above.

d. Determination of the Nucleotide Ratios in ^{32}P-Labeled RNA by Alkaline Hydrolysis and Column Chromatography

An example of analysis of a phage RNA using column chromatography can be drawn from the studies of coliphage β (Nonoyama and Ikeda 1964). Coliphage β was grown in *E. coli* K12 bacteria in the presence of ^{32}P so that this was incorporated in the phage RNA. Radioactive phage RNA extracted from the virus by the phenol procedure was mixed with carrier yeast RNA (use of carrier enables analysis of minute amounts of viral nucleic acid) and the mixture was hydrolyzed to nucleotides by treatment with 0.3 N NaOH for 18 hr at 37°. The hydrolysate was neutralized with 0.3 N HCl and loaded on a Dowex column (formate type, 1 $\times$ 2). The nucleotides were separately eluted with a gradient of formic acid (0–4 N) and the radioactivities of the issuing fractions were measured in an automatic gas-flow counter. From these data and the assumption of equivalent labeling of the different nucleotides, the composition of the RNA could be calculated.

e. Determination of Base Ratios in DNA from Buoyant Density and Thermal Denaturation Values

By examination of many different samples it has been determined that the buoyant density of DNA in cesium chloride is directly proportional to its guanine plus cytosine content (the buoyant density of a substance is equivalent to the density of solution at the equilibrium position to which the substance sediments in a density gradient). Deviations from this linear relationship occur only if the purine or pyrimidine bases are substituted, that is, if the DNA contains bases other than adenine, thymine, guanine, or cytosine. Such cases are rare. Hence, by density-gradient centrifugation of viral DNA, together with a marker DNA of known density, data are obtained enabling the calculation of the viral DNA. These data are obtained by use of a centrifuge equipped to record the positions of the sedimenting species from their ultraviolet absorbancies. A detailed description of the technique and an illustrative calculation of density are given by Mandel et al. (1968). From a curve representing the best fit of measurements made on 51 DNA samples, Schildkraut et al. (1962) developed the relation

$$(GC) = \frac{\rho - 1.660 \text{ g/ml}}{0.098}$$

where (GC) is the mole fraction of guanine plus cytosine and ρ is the buoyant density of the DNA in CsCl.

Similarly, a linear relationship exists between the molar percentage of guanine plus cytosine in DNA and the denaturation or "melting" temperature (T_m) of the nucleic acid. From observations on 41 samples of DNA, Marmur and Doty (1962) developed the relation: (GC) = (T_m-69.3) 2.44, where (GC) is the mole percentage (note that mole percentage = mole fraction × 100) and T_m is in degrees centigrade in a solvent containing 0.2M Na^+. The absorbance of the DNA solution at 260 nm is measured as a function of temperature, and T_m is taken at the midpoint of the increase in absorbance (hyperchromic rise). Details for the performance of such measurements are given by Mandel and Marmur (1968).

The errors associated with both the density and thermal denaturation procedures appear to be small and both procedures can be performed with microgram amounts of DNA. A comparison of some results obtained by these procedures and those obtained by chemical analysis are illustrated in Table 18.

An example of the use of such data as those in Table 18 is as follows. Using either the buoyant density or thermal denaturation temperature value for guanine plus cytosine listed for herpesvirus DNA in Table 18, and applying the molar equivalence rule that applies to the bases of double-stranded DNA (see Sec. f), one can readily calculate that the molar percentages of bases in this DNA are 16 percent adenine, 16 percent thymine, 34 percent guanine, and 34 percent cytosine.

Table 18. Guanine Plus Cytosine Content of Some Viral DNAs.

Virus	Molar Percentages of Guanine plus Cytosine: From Chemical Analysis	From Buoyant Density	From Thermal Denaturation Temp.	References[a]
Coliphage T3	50	53	49	1,2
Coliphage T7	48	51	48	1,2
Coliphage lambda	49	51	47	1,2
Adenovirus-2	58	57	57	3
Herpes	74	68	68	4,5
Shope papilloma	48	50	49	6

[a](1) Schildkraut et al. 1962. (2) Marmur and Doty 1962; (3) Piña and Green 1965; (4) Ben-Porat and Kaplan 1962; (5) Russell and Crawford 1963; (6) Watson and Littlefield 1960.

f. Proportions of Nucleotides in Some Viral Nucleic Acids

There are different ways of expressing the results of the base analyses made on nucleic acids. The commonest are (1) an arbitrary basis such as (a) moles of base per total of 4 moles, or (b) any one of the bases is set equal to 1 (or 10) and the values of the other bases are calculated accordingly; (2) moles percent, that is, moles base per 100 moles total bases; (3) moles base per mole phosphorus. Method 3 is probably to be preferred since it permits a ready evaluation of the recovery of the bases (there should be a total of 1 mole of bases per mole of phosphorus). However, this requires phosphorus analyses to be made, and sometimes there is not enough sample to make the desired replicate base determinations and phosphorus analyses too. Finally, since there is one base per nucleotide, results may obviously be expressed interchangeably in terms of moles base or moles of nucleotide. The compositions of some viral nucleic acids are summarized in Table 19 in terms of mole percent of nucleotides.

Several points about the compositions of viral nucleic acids as listed in Table 19 may be noted in passing. Nucleic acids of plant and mammalian tissues often contain small amounts of 5-methyldeoxycytidine (Hall 1971). However, such methylation rarely appears among viral nucleic acids. Two exceptions are noted in Table 19: 5-hydroxymethylcytosine found in the T-even coliphages and 5-hydroxymethyluracil observed in the DNA of *B. subtilis* phage SP8. In fact, on the basis of current information unusual purines or pyrimidines are quite uncommon in viral nucleic acids. One interesting variation is the occurrence of deoxyuridylic acid rather than thymidylic acid in the DNA of *B. subtilis* phage PBS2.

Inspection of the molar proportions of nucleotides for different viruses (Table 19) indicates considerable variation among viruses in this regard. Two examples, the DNAs of coliphage T3 and of *Salmonella* phage P22, appear to have equimolar proportions of all four constituent nucleotides. This is fortuitous and does not mean that the four nucleotides occur repetitively in tandem (old tetranucleotide hypothesis). In contrast, the nucleic acids of potato virus X and of white clover mosaic virus have unusually large proportions of adenine; those of herpes simplex and pseudorabies, similarly big proportions of guanine; those of turnip yellow mosaic and wild cucumber mosaic viruses, extraordinary amounts of cytosine; and coliphages of the f1 group and influenza virus have lopsided proportions of thymine and uracil, respectively. The nucleic acids of most of the other viruses listed in Table 19 have undistinctive compositions.

While there are distinctive differences in the compositions of most of the nucleic acids of different viruses, such distinctions are seldom demonstrable with strains of the same virus. For example, the composition given for TMV suffices for its various strains, and one composition can be given for five strains of influenza A virus, one for the T-even coliphages, one for coliphages of the f1 series, one for three types of poliovirus, and so on. It should be remembered, however, that the present methods of analysis have an accuracy of about ± 3 percent, which, for example, is equivalent to about ± 50 nucleotides for any of the four nucleotides of TMV-RNA and of course proportionately larger for the bigger nucleic acids. This analytical situation should be viewed in the context that a difference in one nucleotide may be biologically significant.

The nucleic acids of viruses containing either double-stranded DNA or double-stranded RNA exhibit a molar equivalence of bases first noted in some DNAs by Chargaff and associates (1955):

$$\frac{\text{Adenine}}{\substack{\text{Thymine}\\ \text{(or uracil)}}} = \frac{\text{Guanine}}{\substack{\text{Cytosine}\\ \text{(or 5-HMC)}}} = \frac{\text{Purines}}{\text{Pyrimidines}} = 1$$

These regularities have definite implications concerning the structure of DNA (and of RNA) and were instrumental in the development of the Watson-Crick (1953a) double helix model.

g. Polynucleotide End Groups and Other Structural Features

In the case of proteins, primary structure analysis involves determination of numbers of chains in the protein molecule and the sequence of amino acid residues in the chain or chains. This concept, somewhat modified, can be carried over into nucleic acid structure, and, in the case of viruses, involves determination of the number of polynucleotide chains per virus particle and the sequence of nucleotides in the chain or chains.

With respect to these analyses, two points can be stated at the outset: (1) It is very common for viruses to have a single molecule or chain of nucleic acid per virus particle, but there are several instances, particularly in the case of viruses with double-stranded RNA in which the nucleic acid is segmented, that is, occurs in 10 to 15 discrete pieces per particle. (2) Sequencing of nucleic acids at present is more difficult than sequencing

Table 19. Amount of Nucleic Acid and Nucleotide Ratios of Some Viral Nucleic Acids.[a]

Virus	Type NA[b]	Daltons NA per particle, × 10^{-6}	Approximate Moles Nucleotide per 100 moles					
			Ap[c] or dAp	Gp or dGp	Cp or dCp	Up or dUp[d]	Tp	5 HMdCp[e] or 5 HMdUp
Adenovirus 2	DNA	23	21	29	29		21	
Avian myeloblastosis	RNA	10	25	29	23	23		
B. subtilis PBS2	DNA	190	36	14	14	36[d]		
B. subtilis SP8	DNA	120	28	22	22			28[e]
Broadbean mottle	RNA	1	27	25	19	29		
Brome mosaic	RNA	1	27	28	21	24		
Coliphages f1, fd, M13	DNA	1	24	20	21			35
Coliphages f2, fr, M12, MS2, R17, β	RNA	1	23	26	26	25		
Coliphages ØX174, ØR, S13	DNA	2	24	25	19		32	
Coliphage Qβ	RNA	1	22	24	25	29		
Coliphages T2, T4, T6	DNA	120	33	17			33	17[e]
Coliphage T3	DNA	—[f]	25	25	25		25	
Coliphage lambda	DNA	30	26	24	24		26	
Coliphage T5	DNA	77	30	20	20		30	
Coliphage T7	DNA	24	26	24	24		26	
Cucumber 4	RNA	2	26	26	19	29		
Cytoplasmic polyhedrosis	RNA	13	29	21	21	29		
Foot-and-mouth disease	RNA	2	26	24	28	22		
Fowlpox	DNA	200	32	18	18		32	
Herpes simplex	DNA	81	16	34	34		16	
Influenza	RNA	4	23	20	24	33		
Mouse encephalitis	RNA	2	25	24	24	27		

Newcastle disease	RNA	6	24	24	23	29	
Poliomyelitis	RNA	2	29	24	22	25	
Polyoma	DNA	4	26	24	24		26
Potato X	RNA	2	32	22	24	22	
Pseudorabies	DNA	55	14	36	36		14
Reo Type 3	RNA	15	28	22	22	28	
Rous sarcoma	RNA	10	25	28	24	23	
Salmonella P22	DNA	28	25	25	25		25
Shope papilloma	DNA	5	26	24	24		26
Silkworm jaundice	DNA	22	30	20	20		30
Simian 40 (SV40)	DNA	5	26	24	24		26
Sindbis	RNA	—	29	26	25	20	
Tipula iridescent	DNA	156	34	16	16		34
Tobacco mosaic	RNA	2	28	24	22	28	
Tobacco necrosis	RNA	2	28	26	22	26	
Tobacco necrosis satellite	RNA	0.4	28	25	22	25	
Tobacco ringspot	RNA	2	24	25	23	28	
Tomato bushy stunt	RNA	2	26	28	21	26	
Turnip yellow mosaic	RNA	2	23	17	38	22	
Vaccinia	DNA	160	30	20	20		30
White clover mosaic	RNA	2	33	16	23	28	
Wild cucumber mosaic	RNA	3	18	16	40	26	
Wound tumor	RNA	16	31	19	19	31	

[a]Adapted from Knight 1974.
[b]NA is nucleic acid.
[c]Ap is adenylic acid, dAP is deoxyadenylic acid, and so on (see text for naming of nucleotides).
[d]*B. subtilis* phage PBS 2 is unusual in that it has deoxyuridylic acid in place of thymidylic acid, which is characteristic of DNA.
[e]5HMdCp is 5-hydroxymethyldeoxyytidylic acid which is found in coliphages T2, T4, and T6; and 5HMdUp is 5-hydroxymethyldeoxyuridylic acid found in *B. subtilis* phage SP8.
[f]–is unreported.

Fig. 18. Drawing of a segment of the tobacco mosaic virus particle with the protein subunits removed from the top two turns of the protein helix but maintaining the configuration of the RNA strand as it would be if the protein were there. The fit of the RNA in a helical groove of the protein subunits is indicated and individual nucleotides are denoted on the RNA strand as bead-like objects. It will be noted that there are about three nucleotides per protein subunit. (From Klug and Caspar 1960.)

proteins and, while extensive progress has been made, such sequencing is far behind that of proteins.

Progress in the chemical characterization of viral nucleic acids has been most pronounced with certain plant and phage nucleic acids. Therefore, these nucleic acids will be mainly used to illustrate some techniques and principles applicable to the determination of structure.

1. Number of nucleic acid molecules per virus particle. From the molecular weight of tobacco mosaic virus (about 40×10^6) and an RNA content of about 5 percent (Knight and Woody 1958), it can be calculated that each TMV particle contains 2×10^6 daltons of RNA. Likewise, it can be calculated that a polynucleotide chain of about this molecular weight would just occupy the length of a 300-nm rod if it were located at the radius shown in Figure 18 and followed the helical pitch of the protein subunits, as it appears to do (Franklin et al. 1959; Hart 1958; Schuster 1960a). The total length of such a fiber would be 3,300 nm.

The critical question, of course, is whether there is a single fiber of RNA or several molecules, perhaps subunits, regularly spaced along the length of the TMV particle. Studies on hot salt preparations of TMV-RNA, using light-scattering measurements, yielded a molecular weight of 1.7×10^6 daltons for the isolated nucleic acid (Hopkins and Sinsheimer 1955). Similarly, Boedtker (1959), starting from highly monodisperse preparations of TMV, obtained fairly homogeneous preparations of RNA by a modified hot salt method (dilute virus, 90° heating for 1–3 min), and the molecular weight reported for this RNA, as determined from both light-scattering and sedimentation-viscosity measurements, was $1.94 \pm 0.16 \times 10^6$. Other light-scattering investigations were made by Friesen and Sinsheimer (1959) on TMV-RNA prepared by either the detergent or phenol procedures. A weight average molecular weight of 2×10^6 was found for both types of preparation, and infectivity was associated with this material and not with the smaller components that appeared upon storage of the RNA.

The physical properties of the RNA obtained by the phenol extraction procedure were also investigated extensively by Gierer (1957, 1958a, 1958b, 1958c). In sedimentation studies a well-defined, high molecular weight component was observed that accounted for the bulk of the RNA, the rest appearing on the sedimentation diagrams as smaller, polydisperse products. The observed sedimentation coefficient was about 31 Svedberg units and, after applying a viscosity correction, a molecular weight for the RNA of 2.1×10^6 was calculated. When such nucleic acid was treated with ribonuclease, the kinetics of degradation was found to be as expected for the random splitting of a single-stranded structure (Gierer 1957).

Thus, such data indicate that the RNA isolated from TMV has a molecular weight equivalent to the entire RNA content of a virus particle, and that it occurs in the form of a single strand. A question unanswered at this stage was: Is the RNA a uniform polynucleotide chain in which all the 6,400 nucleotides are linked by covalent bonds, or might there be polynucleotide subunits, joined perhaps by hydrogen bonds, to form a single strand? Gierer found (1959, 1960) that TMV-RNA could be heated at 70° for 10 min or at 40° in 36 percent urea for 30 min without the RNA strands breaking down. Since both of these treatments are known to be disruptive to hydrogen bonds, it may be concluded that these are not linking polynucleotide subunits together in TMV-RNA but TMV-RNA is rather a single, large, polynucleotide strand. More recently, studies of the electrophoretic migration of TMV-RNA in polyacrylamide gels yielded results consistent with a viral RNA strand of about 2×10^6 daltons (Bishop et al. 1967).

The RNAs of several phages as well as of animal viruses of the polio and mouse encephalitis types appear to occur in their respective virus particles in one piece, like TMV. The same holds true for DNA in a wide variety of viruses. However, there are some exceptions to the unitary genome structure. For example, the double-stranded RNA of reovirus occurs in 10 segments (Shatkin et al. 1968; Millward and Graham 1970). A similar situation is found with the double-stranded RNAs of cytoplasmic polyhedrosis and wound tumor viruses (Kalmakoff et al. 1969). In addition, the single-stranded RNA of influenza virus occurs in segments (Barry et al. 1970) and there may be other examples. Two consequences of a segmented genome are that it provides a basis for unusually high recombination in mixed infection (genotypic mixing) and it provides individual gene segments of nucleic acid that can probably be sequenced and the data subsequently related to specific gene products.

2. *End group determinations.* Names are given to the two ends of polynucleotide chains of nucleic acids just as they are for ends of the polypeptide chains of proteins. Thus, in formulas depicting nucleic acid structure, the left side of the linear array of nucleotides is customarily referred to as the 5′-end of the structure, while the right side is called the 3′-end. This nomenclature is based on the occurrence of a free hydroxyl on the 5′-carbon and 3′-carbon atoms, respectively, of the terminal nucleotides. This can be illustrated in the following abbreviated structure for RNA:

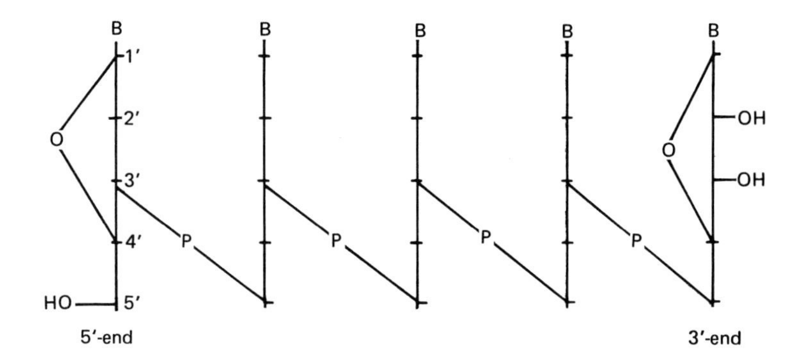

Inspection of this structure shows that the ribose residues (indicated by vertical lines topped with a purine or pyrimidine base, B) are joined by 3′-5′ linkage through phosphate (P). The 5′-OH is involved in this linkage in every case except for the terminal residue on the left. This terminal is therefore recognizable on the basis of its free 5′-OH as the 5′-end. Similarly, the 3′-OH is involved in the formation of phosphodiester linkages everywhere except in the ribose residue on the extreme right. The free 3′-OH there marks this residue as the 3′-end.

Two important features of the ends of nucleic acid chains are often investigated: the presence or absence of terminal phosphate groups and the nature of the purine or pyrimidine base on the terminal nucleotides.

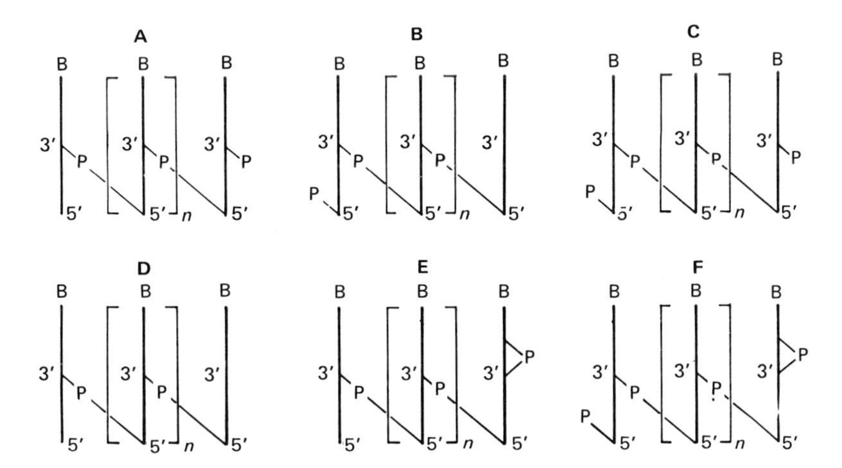

Fig. 19. Schematic representation of six chain end situations for RNA. Using the shorthand notation of Fig. 17*c* for the polynucleotide chain, B stands for purine or pyrimidine base, P for phosphate group, vertical lines for ribose residues, and *n* for number of nucleotides. (From Gordon et al. 1960.)

Some possibilities with respect to terminal phosphate groups, for example in TMV-RNA, are illustrated in Figure 19. It can be seen that treatment of RNA having a structure like that represented by A, B, or C, with an appropriate phosphatase enzyme (see Fraenkel-Conrat and Singer 1962), should release either one or two moles of inorganic phosphate per mole of RNA. By quantitative analysis of the released phosphate, which on a small scale is best done with ^{32}P-tagged RNA, the mean size of the RNA chains can be calculated. Subsequent mild alkaline hydrolysis, which breaks internucleotide links to the right of the phosphate groups as they are depicted in Figure 19, would yield one nucleoside per mole of RNA, and this can be separated from the nucleotides by paper electrophoresis and identified by comparison with standard compounds. Similarly, if the RNA should have one of the structures depicted by B and D, an estimate of the mean chain size and nature of the terminal residues can be obtained by determining the quantity and nature of nucleoside (and nucleoside diphosphate in the case of structure B) produced when the RNA is degraded with dilute alkali.

Other combinations of enzymatic and chemical treatments can provide information about the other end of the RNA chain, and such procedures can also be used to determine whether such structures as E and F are present. A structure similar to B except that it has a 5′-triphosphate group is common in certain phage RNAs. Alkaline hydrolysis of this structure would yield a nucleoside tetraphosphate that would have properties quite distinct from those of all the other digestion products and could be readily separated and identified.

The points just made regarding ends of RNA also apply in principle to DNA except that DNA is not subject to alkaline degradation and some of the enzymes active on DNA are different from those active on RNA. Also there is the problem of two strands in double-stranded DNA and these will need to be separated for some analyses. Finally, there is one structure not shown in Figure 19 that does occur with DNA, that is, a cyclic structure in which the two ends are joined. Essentially, in such a case, no split products are obtained when the nucleic acid is treated with phosphatases and exonucleases (exonucleases are phosphodiesterases that attack polynucleotide chains only or preferentially at ends and liberate mononucleotides therefrom. In contrast, endonuclease enzymes catalyze cleavage of mononucleotides from various locations in the middle of the chain). The cyclic single-stranded DNA of coliphage ØX174 constitutes such a structure (Fiers and Sinsheimer 1962a, 1962b) and cyclic double-stranded DNA is common in DNA-containing tumor viruses (Crawford 1968).

In the initial studies on TMV-RNA using ^{32}P-labeled RNA and treatment with prostatic phosphomonoesterase, Gordon et al. (1960) reported the release of one inorganic phosphate per 3,000–5,000 nucleotides. This was interpreted to mean that TMV-RNA had at the most one monoesterified end. Later results obtained with an *E. coli* phosphomonoesterase and specially purified RNA indicated considerably less than 1 mole of phosphorus per mole of RNA (Fraenkel-Conrat and Singer 1962; Gordon and Huff 1962). This combined with the evidence by Fraenkel-Conrat and Singer that snake venom diesterase (which catalyzes cleavage to the left of each internucleotide phosphate group in the formulations of Fig. 19) caused the release of considerably less than one mole of nucleotide diphosphate from TMV-RNA, leads to the conclusion that there is no monoesterified phosphate in TMV-RNA.

Assuming on the basis of the above that formula D of Fig. 19 represents the correct structure for TMV-RNA, it can be seen that mild alkaline hydrolysis should yield one nucleoside per mole of RNA from the 3′-end (right side of Fig. 19D), the rest of the RNA being converted to nucleoside 2′- or 3′-phosphates. Conversely, degradation of the RNA with venom diesterase should yield one nucleoside per mole of RNA from the 5′-end of the molecule, the rest of the RNA being converted to nucleoside 5′-phosphates.

However, the task of separating one nucleoside from approximately 6,400 nucleotides and identifying it is a formidable one. The key to the problem was the use of TMV-RNA highly labeled with ^{14}C (this is done by growing virus-infected plants in a chamber containing $^{14}CO_2$). When the degradative methods just outlined were applied to TMV-RNA, and the products were separated by paper electrophoresis and paper chromotography, it was found that in each case about 1 mole of adenosine was released per 2×10^6 daltons of RNA (Sugiyama and Fraenkel-Conrat 1961a, *pppA + HO–[NA: B B B, p p] —Kinase→ ppA + *p–[B B B, p p]

ATP NA ADP Phosphorylated NA

1961b; Sugiyama 1962). Thus it appears that TMV-RNA has the structure represented by formula D of Figure 19 and that adenine is the base at both the 5′- and 3′-ends of the molecule.

However, it appears that the 5′-ends of TMV-RNA (and of brome mosaic virus RNA as well) may not be as uniform as the 3′-ends (Fraenkel-Conrat and Fowlks 1972). This was discovered by application of a neat labeling technique, that is, treatment of the nucleic acid with a polynucleotide phosphokinase (Richardson 1965), which transfers phosphorus from labeled adenosine triphosphate to the 5′-end of the nucleic acid:

Up to 1 mole of phosphate per mole of RNA was transferred in this manner to TMV-RNA, confirming the earlier conclusion that the 5′-end is unphosphorylated. The product was then degraded to nucleotides by alkali and the digest was analyzed for content of radioactive nucleoside 3′, 5′-diphosphates after separation of these by paper electrophoresis (or were analyzed for radioactive nucleoside 5′-phosphates separated from snake venom digests of the labeled RNA). Both methods indicated terminal heterogeneity with molar proportions of A:U:G:C = 54:17:18:11. As yet there is no evidence that this variability of the 5′-terminal of TMV-RNA has a serious effect on the biological activity, whereas oxidation of ribose in the 3′-terminal destroys most of the infectivity of the RNA (Steinschneider and Fraenkel-Conrat 1966).

The RNA of turnip yellow mosaic virus is another example of a viral nucleic acid with an unphosphorylated 5′-end (Suzuki and Haselkorn 1968). However, the 5′-ends of several phage RNAs have been found to terminate in triphosphate (Glitz 1968; de Wachter and Fiers 1969; Young and Fraenkel-Conrat 1970). Since nucleic acids appear to be synthesized from the 5′-end toward the 3′-end and nucleotide triphosphates are used in the syntheses, the presence of 5′-triphosphate ends as in the phage RNAs is expected. The absence of phosphate at the terminals of some of the plant virus nucleic acids may indicate that certain plant cells have more phosphatase activity than some other types of cells.

In any case, it appears that some viral nucleic acids as isolated from mature virus particles have phosphorylated ends and others do not. A special case of those that do not are the nucleic acids with cyclic structures.

As just indicated, failure to detect release of significant amounts of phosphate when RNA is treated with *E. coli* phosphatase suggests that the terminals are not phosphorylated. This point can be checked with respect to the 5′-terminal, and at the same time the nature of the terminal base can be established by phosphorylation of the 5′-OH of the terminal nucleoside using the Richardson kinase (Richardson 1965). Phosphorylation places a radioactive label on the terminal and subsequent alkaline hydrolysis yields one radioactive nucleoside diphosphate per RNA molecule and all the rest of the RNA as nucleoside monophosphates. The nucleoside diphosphate can be separated from the nucleotides and identified by electrophoresis (also called ionophoresis) in two dimensions (for details of methodology, see Dahlberg 1968; Brownlee 1972). Likewise, if the 5′-terminal is naturally phosphorylated, as those of several phage RNAs are, alkaline hydrolysis will yield a distinctive product that can be identified in the same manner. Uniformly ^{32}P-labeled RNA is highly desirable for this analysis because of the small amount of terminal compound compared with the bulk of the RNA. (Such compounds can be detected by ultraviolet spectrophotometric methods, but it should be noted that the sensitivity of the isotope technique employing ^{32}P is almost 1,000-fold greater than that with spectrophotometry.) Label can be introduced by growing virus in a medium containing ^{32}P-phosphate, or, if a specific replicase is available as it is for some phage RNAs, it is possible to synthesize a radioactive complementary copy by using unlabeled RNA as template in the presence of α-^{32}P-phosphate-labeled nucleotide triphosphates as substrates for the replicase (the three phosphate groups of nucleotide triphosphates are designated α, β, and γ, starting with α as the first one attached to the 5′-carbon).

Terminal 3′-groups can also be identified by special labeling procedures. An example of such a procedure for analyzing the 3′-end of a polynucleotide chain is the periodate oxidation-borohydride reduction method (Glitz et al. 1968; Leppla et al. 1968). This procedure requires an unphosphorylated terminal; if the end nucleoside is phosphorylated, the terminal phosphate is removed in a preliminary treatment with phosphatase. The adjacent 2′, 3′-hydroxyl groups on the ribose of the terminal nucleoside (these hydroxyls constitute the only such readily oxidized pair in the molecule) are oxidized to aldehyde groups by treatment with periodate. The aldehyde groups are next reduced and simultaneously tagged by treatment with tritiated borohydride. Finally, the terminal nucleoside derivative, conveniently referred to as a nucleoside trialcohol, is released by alkaline hydrolysis and separated from the nucleotides derived from the rest of the RNA molecule by paper or column chromatography procedures that permit its identification.

The main steps of the procedure can be illustrated as follows:

ROCH$_2$ O A
OH OH
$\xrightarrow{IO_4^-}$
ROCH$_2$ O A
HC=O HC=O
$\xrightarrow{NaBH_4^*}$

ROCH$_2$ O A — (NaOH →) — HOCH$_2$ O A + Nucleotides

CH_2^* CH_2^* CH_2^* CH_2^*

OH OH   OH OH

Adenosine
Trialcohol

R stands for all of the RNA molecule except for the 3′-terminus, and A is adenine. The asterisk indicates location of the radioactive label (tritium). One atom of tritium is transferred from the borohydride to each alcohol group formed by reduction of the dialdehyde. A modification of this procedure that yields a somewhat less stable radioactive product is to substitute ^{14}C semicarbazide ($NH_2\text{-}NH\text{-}\overset{\overset{O}{\|}}{C}\text{-}NH_2$) for the tritiated borohydride (Steinschneider and Fraenkel-Conrat 1966), which then yields a radioactive semicarbazone.

Application of the method described above to TMV-RNA and to the RNAs of coliphages f2 and MS2 indicated that the 3′-terminal in all cases is adenosine (Glitz et al. 1968). The same result was obtained with the RNA of phage R17. But, as shown in Table 20, which illustrates the kind of data obtained, some distinctive situations were found with the double-stranded RNAs of three viruses with segmented genomes. Although analyses were made on individual segments of the genomes, the results were essentially the same within ± 2 percent; hence the table shows the results as though each virus contained only a single molecule of nucleic acid per particle as the R17-RNA does. A striking feature of the results shown in Table 20 is that the RNAs of cytoplasmic polyhedrosis and wound tumor viruses appear to have two different 3′-terminal nucleosides, uridine and cytidine. Since these appear in approximately equivalent amounts, it has been concluded that one strand of each RNA segment terminates in uridine while the complementary strand terminates in cytidine (Lewandowski and Leppla 1972).

Dahlberg (1968) has developed a procedure for analyzing the 3′-end of RNA that is uniformly labeled with ^{32}P. This procedure depends on the fact that in a complete T_1 ribonuclease (T_1 RNase) digest of RNA, the only product not susceptible to attack by alkaline phosphatase is the oligonucleotide derived from the 3′-terminal end of the RNA since this is unphosphorylated in most RNAs. (T_1 RNase catalyzes hydrolysis next to guanylic acid residues only. Oligonucleotide means a small polynucleotide segment.) Thus, a T_1 RNase digest of RNA is electrophoresed on DEAE paper providing a spread of oligonucleotide spots that are treated in situ with alkaline phosphatase. Charges will generally be different on the phos-

Table 20. The 3′-Terminals of some Viral RNAs as Determined from Incorporation of Label in Terminal Trialcohols Formed in the Periodate Oxidation-Borohydride Reduction Procedure.[a]

RNA from	Percent of Total Tritium Label in Nucleoside Trialcohols			
	U-triAlc[b]	G-triAlc	A-triAlc	C-triAlc
Reovirus	3		1	96
Cytoplasmic polyhedrosis virus	50		2	48
Wound tumor virus	52	1.5	6	40
R17 phage	3	1.5	94	1.5
Tobacco mosaic virus	1		99	

[a]Adapted from Lewandowski and Leppla 1972; Glitz et al. 1968.
[b]U-triAlc is uracil trialcohol, G-triAlc is guanine trialcohol, and so on.

phatase-treated oligonucleotides except for the terminal one. This will be the only one without any guanine, which migrates in the same way when subjected to electrophoresis at right angles to the first run but in the same solvent. The terminal oligonucleotide usually appears by itself on a diagonal drawn across the paper from the origin and can be eluted and digested separately with alkali, pancreatic RNase and venom nuclease. Electrophoresis in two dimensions of the digestion products provides data from which the nucleotide content and sequence can be deduced, including the 3′-terminal residue. Some data on terminal residues of a few viral RNAs are given in Table 21.

Table 21. Terminal Groups of Some Viral RNAs.

Virus	5′-End	3′-End	Reference[a]
f2, MS2, R17 coliphages	pppG. . .	. . .A_{OH}	1–3
Qβ coliphage	pppG. . .	. . .A_{OH}	1
Satellite necrosis	ppA. . .	. . .C_{OH}	4,5
Tobacco mosaic	A. . .	. . .A_{OH}	6,7
Turnip yellow mosaic	A. . .		8

[a](1) Dahlberg 1968; (2) Wachter and Fiers 1969; (3) Glitz 1968; (4) Wimmer et al. 1968; (5) Wimmer and Reichmann 1969; (6) Sugiyama and Fraenkel-Conrat 1961a; (7) Sugiyama and Fraenkel-Conrat 1962; (8) Suzuki and Haselkorn 1968.

h. Nucleotide Sequences

Viral nucleic acids, like viral proteins, are too large to analyze from end to end by stepwise degradation procedures. Methods of detecting structural units (amino acids in the case of proteins and nucleotides in the case of nucleic acids) as they are split from the macromolecule are not sensitive enough to permit analysis of single molecules and hence thousands must be used. This requires substantial purity of the starting material and synchrony in the cleavage. Such synchrony can be achieved for a number of residues, but then begins to fail and is often accompanied by internal splits of the polynucleotide chain that yield spurious ends.

Consequently, the basic approach is to partially but specifically degrade the molecules into smaller fragments of various sizes. After purification, the sequences of these smaller fragments are established by further degradative procedures, often enzymatic, and identification of the products. By using enzymes of different specificities, and sometimes by partial rather than complete digestion, polynucleotide segments that overlap can be obtained. With these, progressively more of the oligonucleotides can be arranged in their correct order in larger segments of nucleic acid until the total sequence is deduced. Analyses of oligonucleotides for their nucleotide composition are made by alkaline and enzymatic digestion procedures.[2]

Some years ago, a two-dimensional procedure for separating small segments of RNA (one to ten nucleotides long) was developed (Rushizky and Knight 1960a, 1960b, 1960c; Rushizky et al. 1961; Rushizky 1967). The procedure was called two dimensional because it involves paper electrophoresis in one dimension on a large sheet of filter paper followed by chromatography in the second dimension on the same paper. The resulting reproducible spread of nucleotides and oligonucleotides on the paper was termed a "map" (some investigators subsequently called it a "fingerprint"). Such a map is shown in Figure 20. In the mapping procedure pancreatic ribonuclease (ribonuclease A) was employed at first to split the RNA into nucleotides and oligonucleotides. In subsequent experiments, a micrococcal nuclease (Reddi 1959; Rushizky et al. 1962a) and a fungal ribonuclease, ribonuclease T_1 (Miura and Egami 1960; Reddi 1960; Rushizky et al. 1962b), were used. An outline of the mapping procedure as it was applied to TMV-RNA is as follows.

To 1 ml of aqueous solution of TMV-RNA at about 8 mg/ml is added 0.3 ml of pancreatic ribonuclease (RNase) solution at 1 mg/ml and 0.02 ml of M sodium phosphate at pH 7.1. The mixture is allowed to stand at about 23° (room temperature) for 6–8 hr. To determine the precise amount of RNA being analyzed, two 25-μl aliquots are removed from the RNase digest, diluted with N KOH to 10 ml, and, after standing for 24 hr, read in the spectrophotometer at 260 nm. The initial concentration of RNA is calcu-

[2]Many procedures pertinent to analysis of oligonucleotides are described in detail in *Methods of Enzymology*, Vol. 12, Part A, Section II, L. Grossman and K. Moldave, editors, New York: Academic Press (1967).

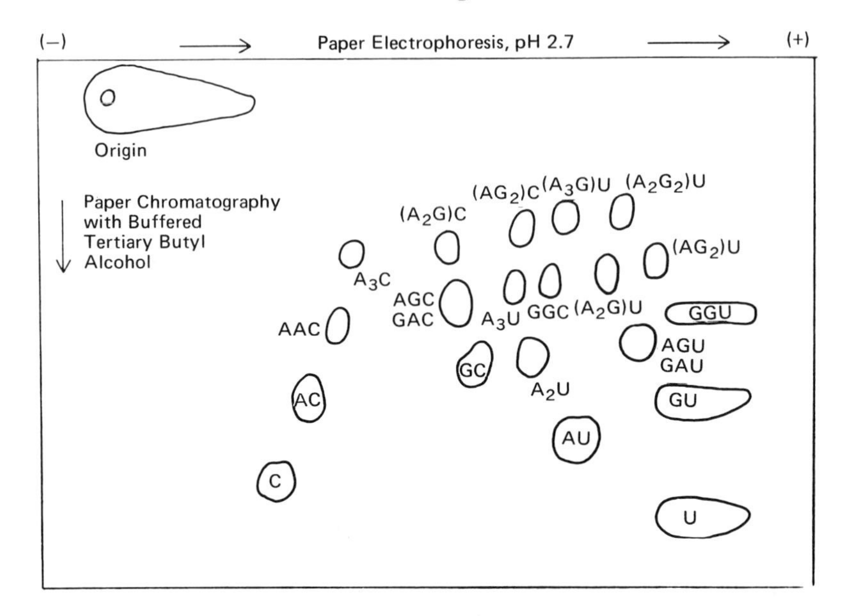

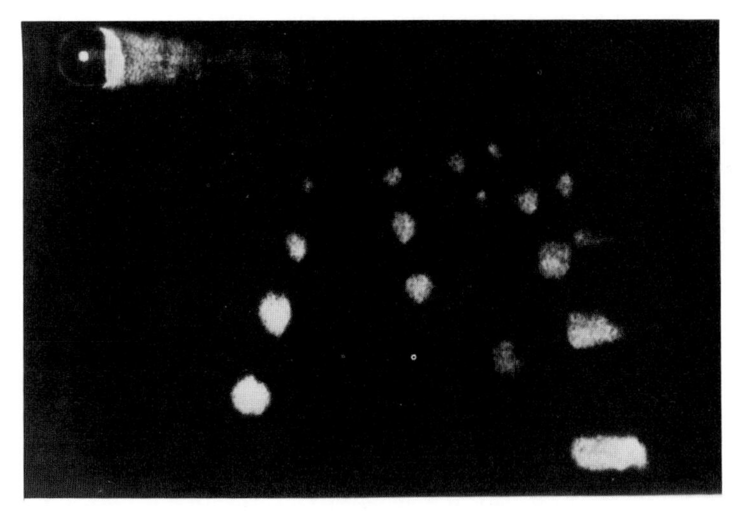

Fig. 20. Contact print map (*bottom*) and key to map (*top*) of pancreatic ribonuclease digestion products obtained from the RNA of the M strain of tobacco mosaic virus. (Maps for the RNAs of other strains and for many RNAs, including yeast RNA, show the same spots; however, they may differ in intensity, reflecting quantitative differences in runs of oligonucleotides). For convenience in labeling, the abbreviations used in the key omit the p normally used to designate phosphate groups. Thus Cp (cytidylic acid) becomes C, ApCp becomes AC, and so on. (From Rushizky and Knight 1960c.)

lated from the relationship that 1 mg of TMV-RNA/ml in N KOH has absorbance at 260 mn of 32.53 (Rushizky and Knight 1960c).

To fractionate the oligonucleotides in the RNase digest, 0.5-ml aliquots (equivalent to about 3 mg of RNA) are applied to buffer-moistened Whatman 3MM paper (46 × 57 cm), and a small application of picric acid (a visible electrophoresis marker that moves slightly faster than any component in the RNase digest) is also applied to the same edge but at the opposite corner of the paper. The buffer used is ammonium formate prepared by adding 7.0 ml of 98 percent formic acid to 2.5 liters of water and adjusting the pH to 2.7 with concentrated ammonium hydroxide. The paper electrophoresis is performed at 350 v (6 v/cm) for 17–20 hr with the point of application of sample near the cathode chamber. The run is finished when the picrate marker reaches the level of the buffer in the anode chamber. A Durrum-type electrophoresis apparatus was used by Rushizky and Knight (1960b). The paper is dried in a current of air at room temperature turned 90° from the direction used in electrophoresis, and serrated at the edge opposite the band of material in order to permit descending chromatography with runoff.

Chromatography is performed with a solvent consisting of equal parts of the electrophoresis buffer, adjusted to pH 3.8 with concentrated ammonium hydroxide, and tertiary butanol (the pH of the mixture at the glass electrode, without solvent correction, is about 4.8). For best results, chromatography should be carried out in a tank thoroughly saturated with solvent vapor for about 36 hr at approximately 25°.

After chromatography, the paper is dried in a current of air at room temperature, and the spots are located with an ultraviolet lamp. A record can also be secured by the contact printing method of Smith and Allen (1953), thus providing a map similar to that in Figure 20. Identification of the spots is by comparison with the positions of spots on a standard map on which the compounds had been identified by elution and analysis using enzymatic and chemical degradations (Rushizky and Knight 1960a).

For quantitative analysis, the spots, located under ultraviolet light, marked with a pencil, and cut out, are eluted in 5–10 ml of 0.01 N HCl at room temperature overnight. An appropriate paper blank is cut from each level of spots and treated in the same manner. The sample eluates are then read against an eluate of the proper paper blank in a spectrophotometer at 260 nm. The spectrophotometer readings are converted into quantities of compound by use of published extinction values for mononucleotides, or, in the case of the oligonucleotides, by use of extinction values calculated from composition using the extinctions of the component nucleotides. In such calculations, allowance is usually not made for the hypochromic state of the oligonucleotides since the error thus introduced is in most cases not great.

Some sequences found in TMV-RNA are summarized in Table 22. The

Table 22. Some Oligonucleotide Sequences Found in Tobacco Mosaic Virus Ribonucleic Acid.[a,b]

Dinucleotides	Trinucloetides	Tetranucleotides	Pentanucleotides
ApAp	ApApAp[c]	ApApApCp	(ApApApGp)Up
ApCp	ApApCp	ApApApGp	(ApApGpGp)Up
ApGp	ApApGp	ApApApUp	
ApUp	ApApUp	CpCpCpGp	
CpAp[c]	ApCpCp	UpUpUpGp	
CpCp	ApCpGp	(ApApCp)Gp	
CpGp	ApGpCp	(ApApGp)Cp	
CpUp	ApGpGp	(ApApGp)Up	
GpAp[c]	ApGpUp	(ApApUp)Gp	
GpCp	ApUpGp	(ApGpGp)Cp	
GpGp[c]	ApUpUp	(ApGpGp)Up	
GpUp	CpApGp	(ApCpCp)Gp	
UpAp	CpCpCp[c]	(ApCpUp)Gp	
UpCp	CpCpGp	(ApUpUp)Gp	
UpGp	CpUpGp	(CpCpUp)Gp	
UpUp	CpUpUp	(CpUpUp)Gp	
	GpApCp		
	GpApUp		
	GpGpCp		
	GpGpUp		
	GpUpUp		
	UpApGp		
	UpCpCp		
	UpCpGp		
	UpGpGp		
	UpUpGp		
	UpUpUp[c]		

[a]Data taken from Rushizky and Knight 1960c; Rushizky et al. 1961, 1962a, 1962b.
[b]The abbreviations are as indicated in the general section on nucleic acids. Where the composition is known but the sequence is not, parentheses are used.
[c]These sequences were deduced from higher oligonucleotides, whereas most of the compounds listed were actually isolated and identified as such after enzymatic digestion of TMV-RNA.

theoretical permutations (P) of the four common RNA nucleotides in which any nucleotide can occupy any position is given by $P = 4^n$ in which n is the number of nucleotide units in the oligonucleotide. Thus 4^2 dinucleotides are possible, 4^3 trinucleotides, and so on. As shown in Table 22, all 16 possible dinucleotide sequences have been found in TMV-RNA, as well as 27 of the possible 64 trinucleotide sequences. The compositions of relatively few tetranucleotide or higher fragments have been determined, although even now it is clear that a great variety occurs.

One of the findings arising from this early analysis of viral RNA sequences is that all RNAs yield the same pattern of oligonucleotides after complete digestion with pancreatic ribonuclease. However, definite quantitative differences are readily demonstrated, as will be described in the section on chemical differences between strains of a virus.

The determination of nucleotide sequences in a segment of nucleic acid larger than those just described for TMV can be illustrated by an example drawn from a study of the RNA of bacteriophage R17 (Jeppesen 1971). The nucleic acid of phage R17 was labeled by growing the bacteria infected with this virus in a radioactive medium, that is, a medium to which ^{32}P phosphate was supplied. The phage was isolated and purified by a combination of precipitation with ammonium sulfate and differential centrifugation. The RNA was isolated from the purified phage by extraction with sodium dodecyl sulfate and phenol, and analyzed by the following procedure.

About 20 μg of ^{32}P-labeled R17 RNA (approximately 20 μCi in radioactivity) is digested with 1 μg of ribonuclease T_1 for 30 min at 37° in 3 μl of 0.01M tris-HCl buffer, pH 7.4 containing 1 mM EDTA. The digestion mixture is then separated into oligonucleotides of various sizes and compositions by a two-dimensional procedure of electrophoresis and chromatography such as developed by Brownlee and Sanger and associates (see Brownlee 1972 for extensive details) and applied as follows.

The digest is applied to a 3 cm × 55 cm strip of cellulose acetate and subjected to electrophoresis in 7 M urea buffered with 5 percent (v/v) acetic acid and pyridine at pH 3.5 until the blue and pink marker dyes (xylene cyanol and acid fuchsin, respectively) separate by approximately 15 cm. The oligonucleotides, which can be detected by a portable Geiger counter, extend from the pink spot to about 3 cm behind the blue spot. The oligonucleotides are transferred from the cellulose acetate to a DEAE cellulose thin layer chromatography plate by placing the former on top of the latter and then a pad of Whatman 3 MM paper wet with water is placed on the cellulose acetate. The strips are pressed evenly together by placing a glass plate on top. Water from the paper pad passes through the cellulose acetate carrying the nucleotides with it into the DEAE cellulose where they are held by ion exchange.

Chromatography is performed on the thin layer plate by developing with a 3 percent mixture of partially hydrolyzed (10 min in 0.2 M NaOH at 37°) RNA dissolved in 7 M urea. (The use of carrier oligonucleotides in the chromatography of other oligonucleotides is called homochromatography.) The oligonucleotides from the partial digest of carrier RNA saturate the DEAE groups and displace the radioactive phage oligonucleotides. The latter then travel along in series of fronts with the nonradioactive oligonucleotides in accordance with size, which governs affinity for the DEAE groups. The smaller oligonucleotides are displaced by the larger ones and thus move more rapidly on the thin layer.

After the chromatography is completed, the thin layer is dried and the spots are located by autoradiography. The sort of separation achieved in such a two-dimensional procedure is illustrated diagrammatically in Figure 21.

Oligonucleotides located by autoradiography can be individually removed from the thin layer and eluted from the DEAE cellulose with 30 percent (v/v) triethylamine carbonate at pH 10. The process of sequencing then involves treating aliquots of the isolated oligonucleotides with different enzymes, analyzing the resultant products, and deducing the sequence from the data obtained. The process of stepwise deduction of sequence can be illustrated for the oligonucleotide (a) (Figure 21) obtained from a ribonuclease T_1 digest of ^{32}P-labeled phage R17 RNA. Enzymes used to digest oligonucleotide (a) and the products found in each digest are summarized in Table 23.

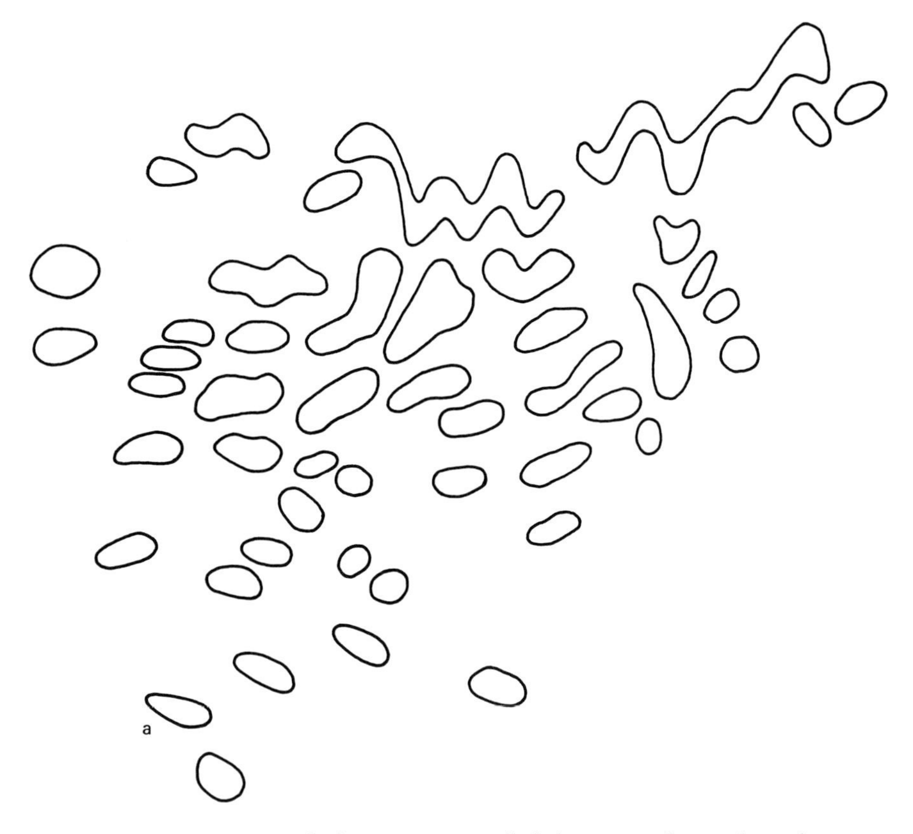

Fig. 21. Diagram of the spots revealed by autoradiography of a two-dimensional thin layer chromatogram. The material fractionated was a ribonuclease T_1 digest of ^{32}P-labeled phage R17 RNA. The oliognucleotide from the spot labeled (*a*) on the diagram is the one whose analysis is described in the text. (Adapted from Jeppesen 1971.)

Owing to the known specificity of ribonuclease T_1 (and confirmed by analysis) the 3′-terminal nucleotide of oligonucleotide (a) is Gp, which is the only guanylic acid residue in the oligonucleotide. Since $\dot{G}$p (see Table 23 for explanation of $\dot{G}$p) was found in the ribonuclease A digest of the CD oligonucleotide in which only bonds next to cytidylic acid are split, the 3′-terminal sequence is -C-Gp. A product of ribonuclease U_2 action is (U_4,C)Gp, which enables extension of the deduced 3′-terminal sequence to Pu-U-U-U-U-C-Gp, where Pu stands for purine. Since there is only one G in the oligonucleotide, all of the remaining purines must be A, and therefore the sequence is -A-U-U-U-U-C-Gp.

Among the ribonuclease A products of the CD oligonucleotide (Table 23, column 2) there is (A-A-U,U_3)Cp, which, because it was derived by ribonuclease action limited to C residues, must be preceded by a C to give (-C-A-A-U,U_3)Cp. This sequence appears to overlap the previously de-

Table 23. Products Obtained by Enzymatic Digestion of Oligonucleotide (a) from Phage R17 RNA.[a]

Ribonuclease A Products[b] (Molar Proportions)	Ribonuclease A Products from CD Oligonucleotide (Molar Proportions)	Ribonuclease U_2 Products (Molar Proportions)
2 A-A-Up	1 (A-A-$\dot{U}$,$\dot{U}$)A-A-Cp	1(U_4,C)Gp
1 A-A-Cp	1(A-$\dot{U}$,$\dot{U}_2$)Cp	<1(U_2,C_2)A-Ap
1 A-Up	1 (A-A-$\dot{U}$,$\dot{U}_3$)Cp	<1(U_2,C_2)Ap
1 Gp	1 $\dot{G}$p	<1 U-U-A-Ap
3 Cp	1 Cp	<1 U-U-Ap
6 Up		1(C,U)Ap
		1 A-Ap
		1 Ap

[a]Adapted from Jeppesen 1971.

[b]The respective enzymatic specificities yielding from oligonucleotide (a) the products listed in the three columns of the table may be summarized as follows. Ribonuclease A: The oligonucleotide is cleaved to the right of each *cytidylic acid* and *uridylic acid* residue. Ribonuclease A on CD oligonucleotide: A particular carbodiimide reacts with uridylic and guanylic acid residues to form carbodiimide derivatives (CD products) indicated by $\dot{U}$ and $\dot{G}$ in the Table. $\dot{U}$ residues are resistant to digestion by ribonuclease A. Therefore, cleavages of CD oligonucleotide (a) occur only to the right of *cytidylic acid* residues. Ribonuclease U_2: This enzyme cleaves to the right of purine (adenylic and guanylic acid) residues, but purine-pyrimidine sequences are split more readily than purine-purine sequences. Partial splitting of purine-purine bonds accounts for the less than molar yields of four compounds in column 3 of the table. In the formulas of the table, the nucleotides are indicated by the first letter of their names and terminal phosphoric acid residues by p. The linkage of nucleotides by the conventional 3′-5′ phosphodiester bond is represented by a hyphen when the sequence is known and a comma when the sequence is unknown. Unknown sequences adjacent to known sequences are placed in parentheses. (These arrangements are in accordance with recommendations of the IUPAC-IUB Commission on Biochemical Nomenclature as summarized in *Journal of Molecular Biology*, **55**, 299–310, 1971.)

duced one and combining them yields -C-A-A-U-U-U-U-C-Gp. Looking for oligonucleotides that might overlap this sequence, it is observed that there are three ribonuclease U_2 products terminating in -A-Ap. Since none of the ribonuclease A products contain A-A-A, it appears that the A-Ap observed must come from the 5′-terminal of oligonucleotide (a). Of the remaining two ribonuclease U_2 products with terminal -A-Ap, only (U_2,C_2)A-Ap fits compositional requirements for overlapping the nine-nucleotide segment above.

The U_2C_2 can be arranged in six different ways (U-U-C-C, C-U-U-C, U-C-U-C, C-C-U-U, C-U-C-U, U-C-C-U) but only the three with C on the 3′-end would fit requirements for the overlap. Two of these possibilities are eliminated because they would have produced U-Cp or U-U-Cp upon degradation of CD oligonucleotide (a) by ribonuclease A and these compounds are not among the products. This leaves U-U-C-C as the proper sequence for U_2,C_2 and since the oligonucleotide containing this was obtained by action of ribonuclease U_2, it must be preceded by an A. This brings the known sequence at the 3′-end to -A-U-U-C-C-A-A-U-U-U-U-C-Gp.

The only remaining CD product which can overlap with this is (A-U,U_2)Cp whose sequence must therefore be U-A-U-U-Cp, which adds a U to the 3′-sequence just given. The CD oligonucleotide not yet appearing in the sequence is (A-A-$\dot{U}$,$\dot{U}$)A-A-Cp, which, in order to account for the three remaining ribonuclease U_2 oligonucleotides—A-Ap, U-U-A-Ap, and (C,U)Ap (the latter is involved in an overlap)—must have the sequence A-A-U-U-A-A-Cp. Therefore, the total sequence of the 21 nucleotides of oligonucleotide (a) from the above is A-A-U-U-A-A-C-U-A-U-U-C-C-A-A-U-U-U-U-C-Gp.

From this example it should be apparent that sequencing viral RNAs ranging from 3,200 to over 20,000 nucleotides is a formidable task and not likely to be undertaken even for the smallest viral RNAs unless the information to be gained is highly important. However, sequencing of small segments of RNA in order to elucidate initiation and termination signals for translation or to characterize enzyme attachment sites are projects that may warrant the effort required.

Analyses similar to those used for RNA have been employed to determine the sequence of the single-stranded ends (cohesive ends) of the mainly double-stranded DNA of lambdoid phages (phages similar to coliphage lambda) such as Ø80 (Bambara et al. 1973). It appears that the cohesive ends, comprising 12 nucleotides, of the DNAs of phages lambda and Ø80 are identical. This structural feature permits through base pairing the formation of interesting mixed dimers between the two phage nucleic acids. The sequences of the complementary strands are

(5′-3′): -A-G-G-T-C-G-C-C-G-C-C-C-OH
(3′-5′): HO-T-C-C-A-G-C-G-G-C-G-G-G-

A fragment of bacteriophage ØX174 DNA(a single-stranded DNA), 48 nucleotides long, has also been sequenced by using enzymatic and electrophoretic techniques similar to those used on RNA (Ziff et al. 1973). In this work good use was made of the T4-induced enzyme, endonuclease IV, which cleaves DNA to yield cytidine 5′ phosphate terminals.

i. Two Ways to Compare Nucleotide Sequences Without Sequencing: Nearest Neighbor Analysis and Hybridization

If there is an *in vitro* method available for the synthesis of nucleic acids from radioactive substrates (specifically, nucleoside triphosphates containing $\alpha^{32}P$), it is possible to compare nucleic acids in terms of the frequencies with which various nucleotide pairs (doublets) occur (since most nucleic acids contain four different nucleotides, the total number of different nucleotide pairs is $4^2 = 16$). This procedure, which has been termed "nearest neighbor analysis," was initially proposed by Josse et al. (1961).

The basic plan of the procedure is to use the nucleic acid whose analysis is desired as a template ("primer") for synthesis by a polymerase enzyme of complementary strands containing radioactive phosphorus at specific points in accordance with which radioactive triphosphate was used in the substrate mixture. Thus four successive syntheses are performed in which all four of the usual nucleotides are present, but in each case a different deoxyribonucleoside triphosphate contains the ^{32}P marker:

Reaction 1: ppp*A, pppG, pppC, pppT + template + polymerase
Reaction 2: pppA, ppp*G, pppC, pppT + template + polymerase
Reaction 3: pppA, pppG, ppp*C, pppT + template + polymerase
Reaction 4: pppA, pppG, pppC, ppp*T + template + polymerase

The mechanism of synthesis with DNA polymerase is that the substrates are 5′-deoxyribonucleoside triphosphates and these are linked into polynucleotide chains by esterification with the 3′-OH of adjoining nucleotides (see Figure 22). After synthesis is complete, the product is digested enzymatically to yield 3′-nucleoside monophosphates. In effect, the ^{32}P goes into the product in one nucleotide and comes out in digested product attached to the neighboring nucleotide (see Figure 22).

The nucleotides liberated by enzymatic digestion are separated by paper electrophoresis and the ^{32}P content of each is estimated in an appropriate counter. From the data the frequency with which any nucleotide occurs next to any other can be calculated. This then provides a measure of the frequency with which each of the 16 possible dinucleotides occurs in the nucleic acid in question.

This type of analysis appears to give patterns of dinucleotide frequency that are reproducible and characteristic for different nucleic acids and hence are useful in comparing nucleic acids. The examples given in Table 24 are drawn from the more extensive compilations of Subak-Sharpe et al.

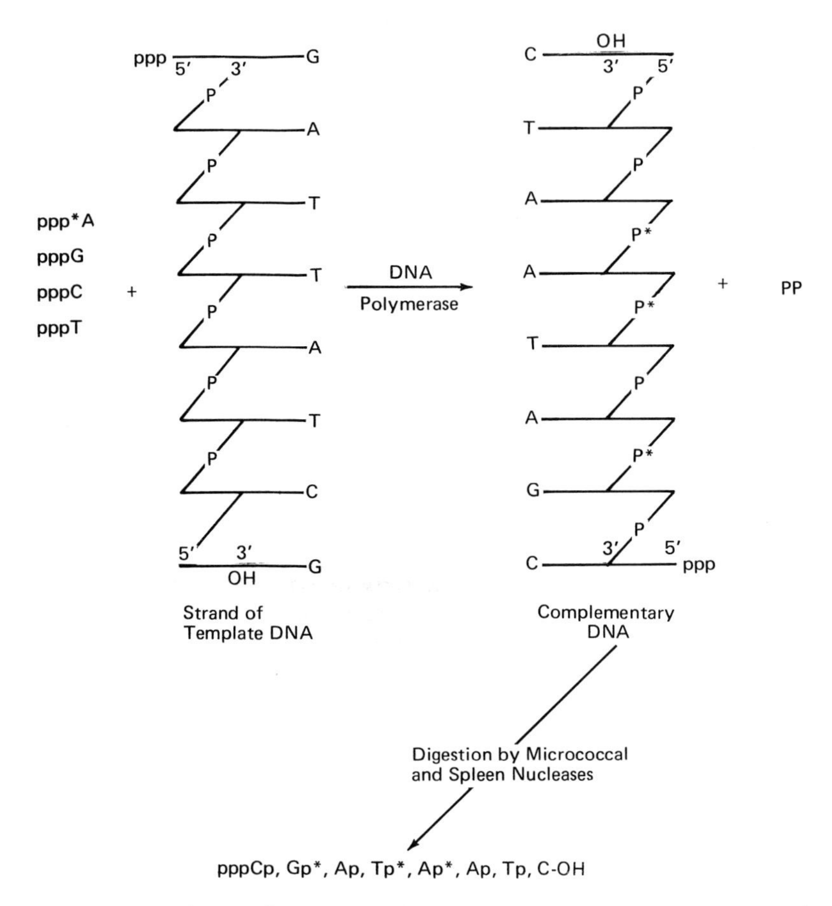

Fig. 22. Synthesis of new DNA from deoxyribonucleoside 5′ triphosphates. Radioactive phosphorus is indicated by p*. The new strand has a composition complementary to that of the template (when the template is double-stranded DNA, both strands serve as template for the polymerase). From the knowledge that each phosphate built into the new DNA is attached to the 5′ position of the entering nucleotide but leaves upon enzymatic digestion attached to the 3′ position of the neighboring nucleotide (the other partner of the phosphodiester linkage), the following dinucleotides can be deduced from the nature of the radioactive nucleotides in the enzymatic digest of the polymerase product shown above: GpA, TpA, and ApA.

Table 24. Some Doublet Frequencies[a] for DNA from Viruses and Other Sources.[b]

Doublet	Polyoma	Shope pap.	Vaccinia	Herpes	Phage λ	T1	T4	T5	ØX174	*E. coli* 2	Hamster BHK 21
ApT	56	57	69	68	70	64	64	65	59	72	53
TpA	50	45	58	68	47	48	55	62	46	47	50
GpC	64	70	54	65	72	76	70	78	80	72	75
CpG	22	30	72	65	64	70	53	58	59	64	13

[a]In parts per thousand normalized to correspond with DNA containing 50 percent (G + C).
[b]Adapted from Subak-Sharpe et al. 1966.

(1966). Even with only four of the 16 doublets represented in Table 23, it can be seen that there are substantial similarities between some of the T phages [Josse et al. (1961) found even closer correspondence, as would be expected on chemical, morphologic, and serologic grounds between the doublet frequencies for T2, T4, and T6], but also distinct differences among the several groups of viruses represented in the table. Comparison of the doublet frequencies of viral and host DNAs [see Table 24 and the more detailed results of Subak-Sharpe et al. (1966); and Bellett (1967)] leads to the conclusion that the smaller DNA-containing viruses exhibit doublet frequencies resembling those of mammalian animal cells (compare the values in Table 24 for the DNAs of polyoma and Shope papilloma viruses with those listed for hamster BHK 21 cells), whereas the DNAs of such large viruses as herpes and vaccinia, like those of phages, are more similar to bacterial DNA, such as that of *E. coli.* While the doublet frequencies do not provide adequate evidence to establish phylogenetic relationships, it has been suggested that the observed data are consistent with the idea that some bacterial and animal viruses evolved from their hosts, whereas others, such as herpes and vaccinia viruses may have had an external origin, perhaps from bacteria.

Limited nearest neighbor analyses have also been made on viral RNAs, including those of tobacco mosaic and turnip yellow mosaic viruses and of phage MS2 (Fox et al. 1964). Such analyses have also been especially useful in sequencing the RNA of some small phages (Bishop et al. 1968; Billeter et al. 1969).

In summary, comparisons of nucleotide doublets by the nearest neighbor technique can be very useful. However, such comparisons are in no way a substitute for the more arduous linear sequencing method outlined earlier. For example, as Josse et al. (1961) indicated, analytical errors as little as 1 percent, which are not unlikely, mean that even the moderate-size phage lambda DNA (molecular weight 20×10^6) could differ from a related DNA in 1,000 nucleotide sequences without any differences being detected by the nearest neighbor analyses.

The molecular hybridization technique is a method for obtaining information about nucleic acid sequences that are very much longer than the doublets of the nearest neighbor method. The basis for this procedure is as follows.

Hydrogen bonds between complementary bases (A:T or U, and G:C) of the two strands of double-stranded nucleic acid (Figure 24) are broken by heating (and also by certain chemicals such as alkali or dimethyl sulfoxide), but the bonds are reformed by slow cooling ("annealing" or "renaturation") or removal of added chemical in the case of chemically induced destruction of hydrogen bonds. The separation of nucleic acid strands by heat is called molecular melting or denaturation, the latter being the more general term for strand separation by whatever cause. If strands of a second species of nucleic acid are added to those of the originally denatured one during the annealing process, strands of the second nucleic acid may compete with the original ones in the reforming of double stranded structures. This occurs only if substantial nucleotide sequences are the same or very similar in the two species of nucleic acid. If such homology exists, "hybrids" may be formed between complementary strands of DNA or between complementary strands of DNA and RNA; thus, the term hybridization is applied to the process.

Experimentally, either agar gels, or more commonly now, nitrocellulose filters, are used to immobilize denatured strands of nucleic acid. Hybrids can be formed if complementary structures are brought in contact with such fixed, denatured nucleic acid. (Appropriate nitrocellulose membranes bind denatured DNA, DNA-DNA, and DNA-RNA hybrids, but not free RNA or undenatured DNA.) In addition, there are procedures in which columns containing hydroxyapatite or other substances are employed. In almost all cases, quantitation is achieved by having one of the reacting species radioactively tagged. Details of various methods are given by Raskas and Green (1971) and Bøvre et al. (1971). An evaluation of the specificity of hybridization reactions is given by McCarthy and Church (1970).

Some examples of situations in which homologous sequences of nucleic acid or lack of them can be demonstrated by molecular hybridization include an evaluation of the degree of similarity of the nucleic acids of viruses that are thought to be similar, for example, comparison of the DNAs of the many types of adenovirus; investigation of the amount of replicating viral nucleic acid present at different times after infection; estimation of the number of viral genomes incorporated into host nucleic acid; determination of the presence or absence of virus-specific messenger RNA (mRNA); and discrimination between types of mRNA present at various times after infection.

An example involving simian virus 40 (SV40) can be cited here of the use of molecular hybridization to demonstrate the presence and number of copies of SV40 DNA integrated into the DNA of mouse cells (Westphal and Dulbecco 1968). Such integration of viral genome is thought to be associated with the transformation of normal cells to tumorous cells. In order to achieve greater sensitivity in the hybridization test, instead of seeking direct hybridization between viral DNA and mouse cell DNA, RNA complementary to SV40 DNA (cRNA) was employed. Two advantages of using cRNA rather than the direct approach with viral DNA are that higher specific radioactivities can be readily obtained in the cRNA and the use of cRNA eliminates the possibility of self-annealing of denatured viral DNA. Reconstruction experiments indicate that as few as three or four viral DNA molecules per cell can be detected by hybridization with cRNA (this is equivalent to less than two parts of viral DNA in a million parts of cellular DNA).

cRNA is synthesized *in vitro* from highly radioactive nucleoside triphosphates using SV40 DNA as primer and a DNA-dependent RNA polymerase. Hybridization is carried out essentially by the method of Gillespie and Spiegelman (1965). The circular, supercoiled SV40 DNA molecules are treated with deoxyribonuclease for 60 min at 30° to convert them into linear, circular strands with some breaks in one or another of the strands. This DNA is poured into 2 vol of boiling water and boiled for 15 min, thus effecting strand separation. The denatured DNA is chilled in ice and adjusted to contain 0.9 M NaCl–0.09 M Na citrate (called 6× SSC, 1 SSC being 0.15 M NaCl–0.015 M Na citrate). The DNA is then slowly passed through a Millipore membrane filter to which much of the denatured DNA attaches. In order to check the variability of results, replicate filters are employed with comparable aliquots of DNA.

The filters are next incubated at 66° for 22 hr in vials containing 1 ml 6X SSC, tritiated cRNA obtained as described above, 1 mg yeast RNA carrier, and sodium dodecyl sulfate at a concentration of 0.1 percent (the latter appears to reduce background counts by reducing nonspecific attachments of the radioactive cRNA). The filter-containing vials are gently shaken in a water bath during incubation. After incubation, during which hybridization is expected to occur when possible, the filters are removed from the vials and washed with 50 ml of 2X SSC using suction filtration. After this the filters are treated with 20 μg/ml of RNase A and 10 units/ml RNase T_1, for 60 min at 37°. Following this treatment, which is designed to remove RNA complexed to the DNA over short regions and hence of uncertain specificity, the filters are washed again, dried, and counted in a scintillation counter. The amount of DNA on each filter to which the counts need to be related is determined after counting by the colorimetric diphenylamine reaction.

In the investigation by Westphal and Dulbecco (1968) it was found that the DNAs of two different lines of mouse cells that had been transformed by SV40 (SV3T3-47 and SV3T3-56) fixed by hybridization different amounts of the SV40 specific cRNA. The counts per minute per 100 μg DNA above the backgrounds observed with the DNAs of untransformed cells were, respectively, 800 and 264 (these are the means from counts of 10 to 12 filters each). From calibration tests in which known numbers of SV40 DNA molecules had been added to cell DNAs, it was determined that 40 cpm from attached cRNA was equivalent to one SV40 DNA; therefore, the counts per minute noted above represent about 20 and 7 SV40 DNA molecules, respectively, present in the DNAs of the two mouse cell lines that had been transformed by the virus. In these same experiments, it was demonstrated by hybridization that the SV40 DNA was in the nucleus rather than in the cytoplasm of transformed cells.

Another example may be cited, without giving the experimental details, to illustrate the many useful applications of molecular hybridization. Lacy and Green (1967) investigated the hybridization reactions between the DNAs of six members of the weakly oncogenic (oncogenic means tumor inducing) adenovirus group consisting of serological types 3, 7, 11, 14, 16, and 21. It was found that these viral DNAs are closely related, apparently sharing 70–100 percent of their nucleotide sequences. However, the DNAs of the weakly oncogenic adenoviruses apparently differ substantially from those of the strongly oncogenic types 12 and 18, for the results of the hybridization tests indicated that the two groups showed only 11–22 percent homology.

j. Secondary and Higher Structure of Nucleic Acids

The term "secondary structure" will be used here as it was in connection with proteins to mean geometric configuration, with special reference to the presence or absence of helical, hydrogen-bonded structures; similarly, folding of the polynucleotide chain can be considered "tertiary structure."

A secondary structure for DNA was proposed by Watson and Crick (1953a, 1953b) that seemed at once compatible with data on the composition of DNA, general chemical features of DNA, and x-ray diffraction data (Wilkins et al. 1953). The validity of this structure has been confirmed by many experiments over the succeeding years, and it is now widely accepted for DNA from many sources.

This DNA structure is briefly described as a dyad, or duplex, of right-handed helical chains each coiled around the same axis but with antiparallel nucleotide sequences (sequences running in opposite directions). Such an arrangement is shown diagrammatically in Figure 23. The two chains are held together by hydrogen bonding between complementary pairs of bases, one base of each pair being a purine and the other a pyrimidine. Thus, as shown in Figure 24, adenine pairs with thymine and guanine with cytosine. In terms of DNA composition, this should be reflected in A/T and G/C ratios of unity. It will be further noted (Figure 24) that two hydrogen bonds can readily form between adenine and thymine on adjacent strands, but three can form between guanine and cytosine.

One of the first examples of DNA to be shown to give an x-ray diagram consistent with the Watson-Crick structure was that of T2 coliphage (Wilkins et al. 1953). At first, the results of chemical analyses seemed not to support this conclusion. The double helix structure requires equimolar amounts of adenine and thymine and of guanine and cytosine for proper base pairing. Early analyses (Wyatt and Cohen 1952) showed T2 DNA to contain 33.2 moles adenine, 35.2 moles thymine, 17.9 moles guanine, and 13.6 moles 5-hydroxymethylcytosine per 100 moles of bases. The agreement between adenine and thymine was fair, but the guanine/5-HMC ratio was seriously off. With the development of the double helix theory, interest in analyses grew, which led to refinements in procedure, and in this case especially, a recognition of the lability of 5-HMC under common hydrolytic conditions. When these factors were adjusted for, both the A/T and G/5-HMC ratios were found to be close to unity (Wyatt and Cohen 1953). The molar equivalence of purines to pyrimidines in double-stranded viral DNAs is illustrated by many examples in Table 19.

The double helical secondary structure of DNA imparts noteworthy properties to the particle that distinguish it from a single-stranded structure. Primarily, the double-stranded structure has greater rigidity and order than the single-stranded one, and this is reflected in hydrodynamic behavior, optical properties, and chemical reactivity. Thus the transition from helical structure to the less ordered random coil structure assumed by separated strands (or vice versa, since the process is more or less reversible) can be followed by:

1. Sedimentation behavior (helical form sediments slower, that is, has lower sedimentation coefficient).
2. Viscosity (helical form is more viscous than random coil). The properties described in (1) and (2) apply to helical and denatured DNAs in dilute salt in the middle pH range; under these conditions, the separated strands of denatured DNA collapse to a globular form that sediments faster and has a lower viscosity than undenatured DNA. If conditions are employed to keep the separated strands of denatured DNA extended, they may sediment slower and have a higher viscosity than the helical structure (see Studier 1965).
3. Optical rotation (helical form has "handedness" so that it acquires optical activity above that inherent in its components, such as the sugar. Thus optical activity is proportional to helicity).
4. Ultraviolet absorption (helical form absorbs less than random coil at 260 nm because of stacking of bases. The state of reduced ultraviolet absorption characteristic of an hydrogen-bonded, ordered structure is called "hypochromicity." An increase in absorption is then called hyperchromy and a decrease, hypochromy).
5. X-ray diffraction (helical forms have distinctive x-ray diffraction characteristics, of which the absence of meridional reflections is outstanding).

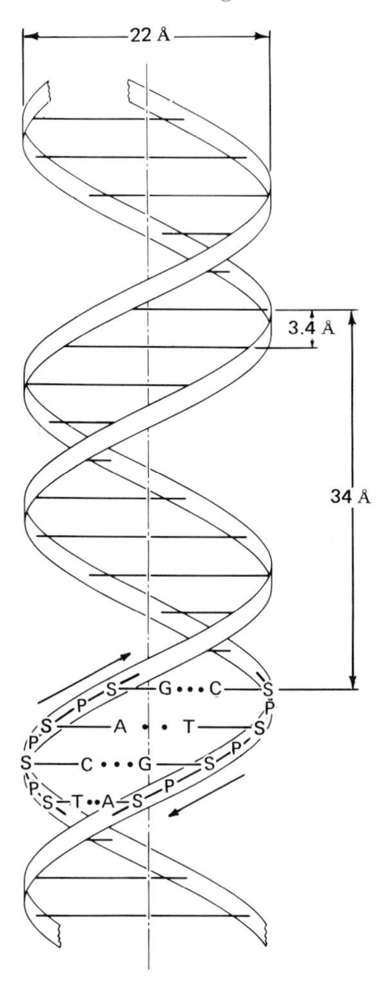

Fig. 23. Diagrammatic sketch of the structure of DNA (modified from Watson and Crick 1953b) by indication of the components in a segment: P, phosphate; S, sugar (2-deoxyribose); G, guanine; C, cytosine; A, adenine; T, thymine. The two ribbons represent the sugar-phosphate backbones of the two helical strands of DNA, which, as the arrows indicate, run in opposite directions, each strand making a complete turn every 34 Å. The horizontal rods symbolize the paired purine and pyrimidine bases. There are ten bases (and hence ten nucleotides) on each strand per turn of the helix. The nature of the base pairs and the number of hydrogen bonds between them are shown in the detailed central segment. The vertical line marks the fiber axis. (From Knight 1974.)

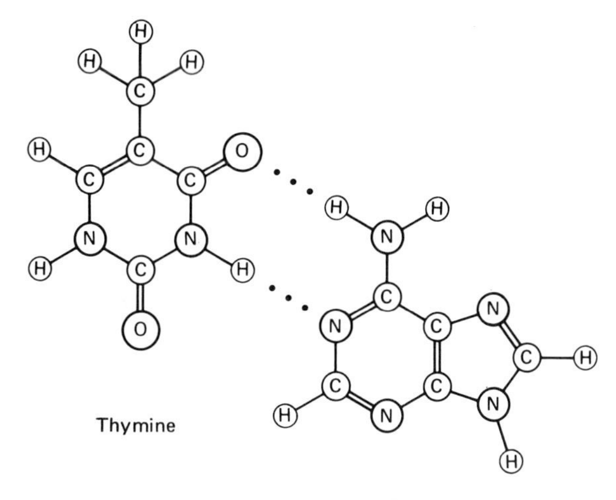

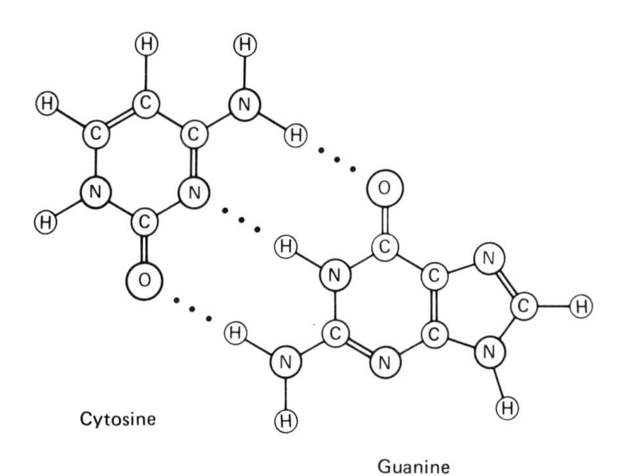

Fig. 24. Illustration of common hydrogen bonding that occurs between specific purine and pyrimidine bases in nucleic acids.

6. Chemical reactivity (keto and amino groups involved in hydrogen bonding do not titrate like free groups, and the reaction of amino groups with reagents like formaldehyde or nitrous acid is inhibited by hydrogen bonding).

7. Chromatography (double-stranded nucleic acid adheres to hydroxyapatite at salt concentrations and temperatures at which single-stranded molecules readily elute).

8. Electron microscopy (under appropriate conditions the thicker strands of double-stranded nucleic acid can be distinguished from those of single-stranded nucleic acid).

The nucleic acids of most viruses are linear structures but some, such as those of DNA tumor viruses and of some phages, are circular (see Table 16). The circular DNAs of animal tumor viruses tend to occur not only in circles but also in supercoiled forms or twisted circles. The significance of these unusual structures is not clear.

Turning now to RNA, it has been noted that the RNA of TMV is held in a helical configuration in the intact virus, and follows the pitch of the protein subunits at a radius of 40 Å from the long axis of the particle. In the TMV particle, the RNA therefore may be represented as a helix with a diameter of 80 Å and a pitch of 23 Å. Since this pitch is too large to permit hydrogen bonding between bases on successive turns of the helix, the helical structure must be stabilized simply by its position in the helical groove formed by the protein subunits (Klug and Caspar 1960). This situation is illustrated in Figure 18, a model drawing based on the results of the x-ray studies described earlier showing a segment of the TMV particle from which the protein subunits of the last two turns are removed. The nucleotides are indicated by the little discs in the RNA chain.

In contrast to the regular helical form assumed by TMV-RNA in the virus particle, the high molecular weight RNA isolated from the virus by the methods described earlier behaves in solution at low temperature and low salt concentration as though it were a flexible but tight random coil (Haschemeyer et al. 1959; Boedtker 1959; Gierer 1960) arising from a single polynucleotide strand. However, the strictly random coil concept is almost surely too simple to account for the observed physical properties of such isolated RNA in solution. The marked changes of properties of the RNA in different ionic media and at different temperatures strongly suggest the ready formation of secondary valence bonds under one set of conditions and the rupture of these bonds under another set (Boedtker 1959; Haschemeyer et al. 1959). Since there is no evidence for combination of separate RNA strands, it must be assumed that it is possible to form intramolecular bonds. To explore this possibility, Doty et al. (1959) applied to TMV-RNA some of the tests for helix-coil transition listed above.

In taking TMV-RNA in 0.1 M phosphate at pH 7 from 10° to 70°, Doty et al. (1959) found a 32 percent increase in absorption in the ultraviolet at 260 nm. This change was at least 95 percent reversible upon cooling. At room temperature, treatment of the RNA with 6 M urea caused half of the increase in absorbance observed in the thermal experiment.

In another test, the reaction of the RNA with formaldehyde at different temperatures was followed spectrophotometrically. This test was based on the observation by Fraenkel-Conrat (1954) that treatment of TMV-RNA with 1–2 percent formaldehyde at pH 6.8 caused a gradual increase in the ultraviolet absorbance at the maximum as well as a shift of 3–5 nm toward higher wavelengths. This effect seems to depend upon the presence of free amino groups in adenine, guanine, and cytosine, and is illustrated by the results shown in Table 25. In the experiments of Doty et al., an increase in reactivity of TMV-RNA with formaldehyde at 45° as compared with 25°

Table 25. Effect of Formaldehyde on the Ultraviolet Absorption of Viruses, Proteins, Nucleic Acids, and Some Nucleic Acid Constituents.[a]

Material	Approximate Increase of Maximum Absorption,[b] %	Approximate Shift in Wave Length of Maximum, nm
Nucleic acids and constituents:		
TMV-RNA (prepared by hot salt method)	+29	+3
TMV-RNA (prepared by detergent method)	+28	+3
Liver-RNA	+19	+4
Yeast-RNA (commercial)	+24	+3
Thymus-DNA	None	None
Adenine	+23	+5
Adenosine	+19	+5
Adenylic acid	+22	+5
Guanylic acid	+5	+5
Cytidylic acid	+16	+3
Thymine	None	+3
Uracil	None	+1
Uridine	None	None
Uridylic acid	None	None
RNA-Containing Viruses:		
TMV	+3	+3
TMV (after 24 hr in 1% sodium dodecyl sulfate)	+18	+3
Tomato bushy stunt virus	+15	+4
Turnip yellow mosaic virus	+15	+3
Tobacco ringspot virus	+23	+3
DNA-Containing Viruses and Proteins:		
Shope papilloma virus	None	None
T2-coliphage	−3	None
T2-coliphage (after 24 hr in sodium dodecyl sulfate)	None	None
TMV protein	None	None
Bovine serum albumin	None	None
Ovomucoid	None	None

[a]Adapted from Fraenkel-Conrat 1954.

[b]Solutions containing the equivalent of about 0.025–0.05 mg of nucleic acid per milliliter in 0.1 M phosphate buffer at pH 6–8 were treated with 1–2 percent formaldehyde for 12 hr at 40° or 48 hr at 23°. The same maximum was reached at the two time and temperature levels.

(after 50 min) was about 19-fold, whereas a control mixture of the appropriate nucleotides showed only a sixfold increase.

The results cited so far definitely support the concept that in dilute, neutral salt at low temperature, portions of the TMV-RNA chain are bound to other portions of the same molecule by hydrogen bonds, presumably of the base-pairing sort, but these results do not answer the question of whether the hydrogen bonding is random or occurs in definite regions in such a manner as to provide helical segments. Evidence for helicity was obtained by Doty et al. (1959) by studying the effect of temperature on the optical rotation of TMV-RNA. If the RNA were devoid of any regular secondary structure, that is, if it were a random coil, its only optical activity would be that of the ribose, which has asymmetric carbon atoms. However, if a significant portion of the RNA had helical structure, the contribution to optical rotation, as judged from known helical structures, might be substantial. Furthermore, such optical rotation should be largely abolished by treatments that convert helical structures to random coils. It was found that the specific rotation of TMV-RNA decreased about 160° in going from a temperature of 8° to about 75°. Most significantly, the optical rotation-temperature and optical density-temperature profiles can be shown to coincide by adjusting the ordinate scales as shown in Figure 25. Thus the decrease in optical rotation of TMV-RNA with rise in temperature is approximately congruent with the increase in optical density observed.

From these results, Doty et al. conclude that the hydrogen bonding in TMV-RNA occurs in definite areas and results in helical segments. Some support for such helical structure is provided by the x-ray pattern obtained by Rich and Watson (1954), which shows some of the characteristics of the patterns obtained with helical DNA. An estimate of the extent of these helical regions was made by Doty et al. by comparing the maximum variation with temperature of the specific rotation (or optical density) of TMV-RNA with the maximum variation observed with the completely helical model, polyadenylic acid-polyuridylic acid. The resulting conclusion is that about 50–60 percent of the nucleotides in TMV-RNA are involved in helical regions. It is further postulated, from experiments with polyribonucleotides and by analogy with DNA, that the predominant base-pairing is probably between adenine and uracil and between guanine and cytosine. Unmatched bases are also predicted and the resulting structure, illustrated in part in Figure 26, consists of a number of imperfect helical loops with randomly coiled regions interspersed. Such a structure would be compatible with the observed hydrodynamic properties of TMV-RNA and would especially account for variations in physical properties with changes in environment. A somewhat similar model has been deduced from the results of x-ray studies made on RNA from ascites tumor cells, *E. coli*, and yeast by Timasheff et al. (1961). They concluded that their RNA was represented by short, rigid, double helical rods about 50–150 Å long joined by small flexible single-stranded regions.

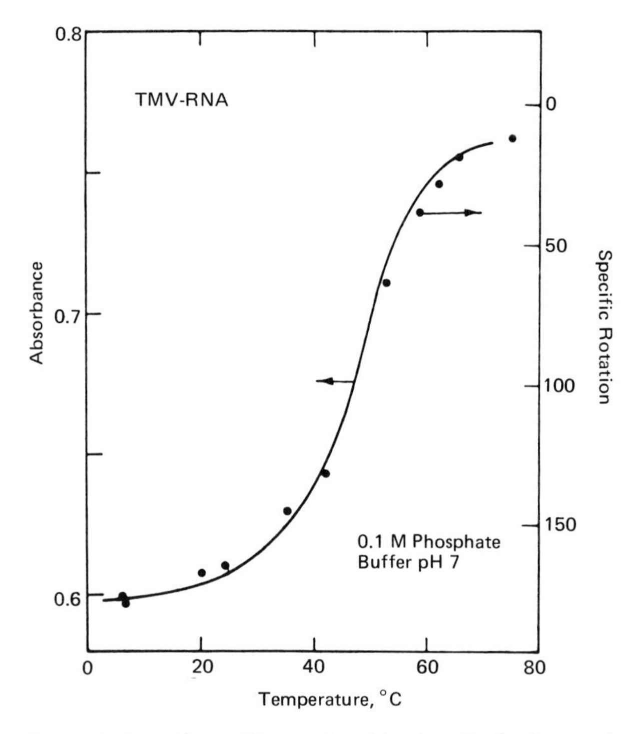

Fig. 25. The variation of specific rotation (*dots*) and adsorbance (*solid line*) of TMV-RNA with temperature. (From Doty et al. 1959.)

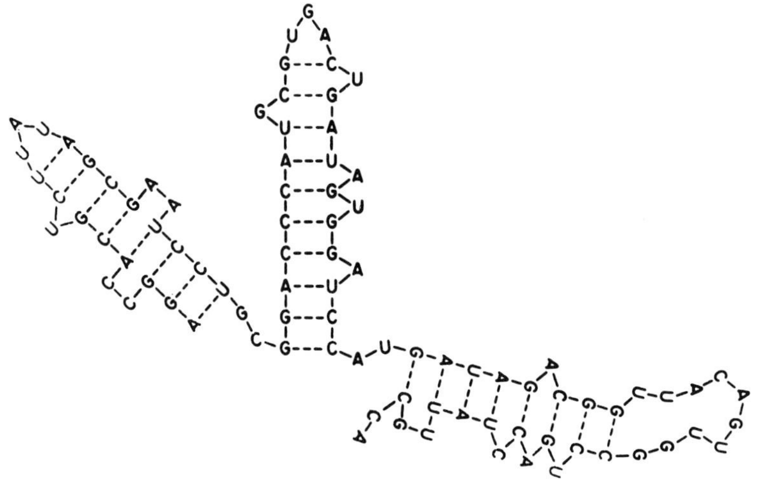

Fig. 26. A possible model for a segment of TMV-RNA in dilute neutral salt solution at low temperature. A, C, G, and U are abbreviations for the nucleotides: adenylic, cytidylic, guanylic, and uridylic acids, respectively. (From Doty 1961.)

Viral nucleic acids, whether single stranded, double stranded, or cyclic, generally appear to occur in unbroken strands. In contrast to this generality there are some RNA viruses whose nucleic acid appears to occur in segments. Thus the double-stranded RNA genomes of reovirus and of cytoplasmic polyhedrosis virus of the silkworm occur in ten segments (Shatkin et al. 1968; Millward and Graham 1970; Lewandowski and Millward 1971) and that of the wound tumor virus of clover in 12 segments (Kalmakoff et al. 1969; Reddy and Black 1973). (A viral genome may be defined as the total ensemble of genes associated with a virus.) Separation of the RNA segments of the cytoplasmic polyhedrosis virus by electrophoresis in acrylamide gel is illustrated in Figure 27. Similar segments but consisting of single-stranded RNA seem to characterize influenza viruses (Duesberg 1968; Pons and Hirst 1968), and possibly RNA tumor viruses (Vogt 1973).

It should be emphasized that, in general, the physical properties of nucleic acid within a virus particle may or may not be the same as those of the extracted nucleic acid treated as a hydrodynamic entity. This was demonstrated by Bonhoeffer and Schachman (1960) using four viruses, two of which contained DNA and two RNA. A comparison of the ultraviolet absorption spectra before and after degradation with sodium dodecyl sulfate with the spectra obtained by heating the degradation mixtures was used as a measure of the degree of hypochromicity (and hence of hydrogen-bonded structure) of the nucleic acid within the virus particles and upon release from them. In the case of the DNA viruses, Shope papilloma virus and T6 coliphage, no change in secondary structure upon release of the nucleic acid could be detected. Upon heating the degraded viruses, however, the ultraviolet absorptions increased 30–35 percent, from which it can be assumed that the DNA is present, in each of the cases, within the virus as well as upon release, in the form of the classical double helix.

On the other hand, a definite decrease in ultraviolet absorption occurred upon release of RNA from TMV, as shown in Figure 28. Heating the degradation mixture restored the absorption to the level of the undegraded virus. This result confirms the deduction made from the x-ray and other data (see above) that the spacing of RNA in the TMV particle precludes base-base interaction, which, however, occurs intramolecularly upon release of the nucleic acid. Thus the TMV-RNA appears to go from the protein-imposed helical configuration of the intact virus illustrated in Figure 18 to an intramolecularly, partially hydrogen-bonded structure such as shown in Figure 26.

Upon degradation of the bushy stunt virus, a slight decrease in absorbance was noted. Heating the degraded virus caused a 23 percent increase. From these facts it was concluded that the RNA within bushy stunt virus has some secondary structure and that more is acquired upon release of the RNA from the particle.

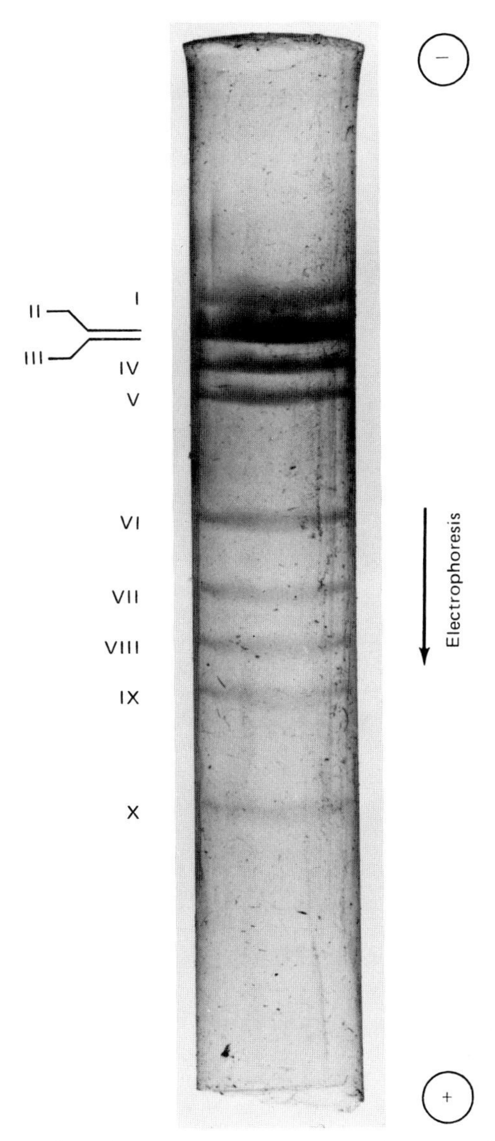

Fig. 27. Polyacrylamide gel electrophoresis of the double-stranded RNA genome of cytoplasmic polyhedrosis virus of the silkworm. There are ten segments of RNA, and all except segments II and III separated under the conditions of electrophoresis used for this experiment (3 percent polyacrylamide gel, pH 7.5, stained with methylene blue). (Courtesy B. L. Traynor.)

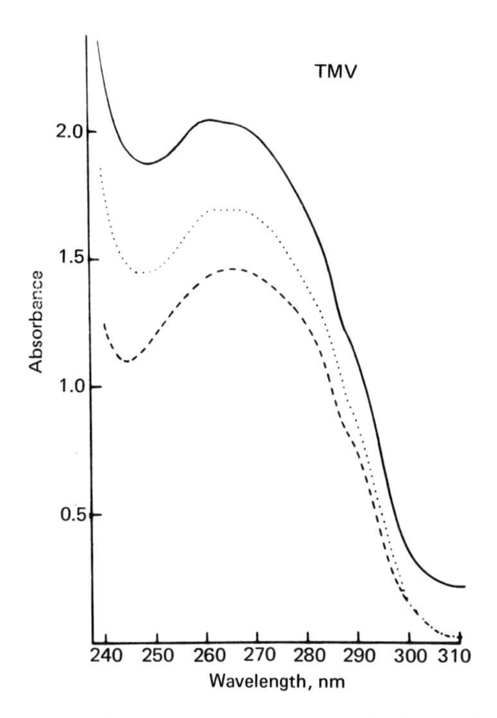

Fig. 28. Ultraviolet absorption spectra of TMV before and after degradation by sodium dodecyl sulfate into RNA and protein. Solid and dotted curves are for the intact nucleoprotein, the former giving the observed optical density values and the latter the values after application of light-scattering corrections. The dashed curve was obtained after degradation of the virus into RNA and protein. (From Bonhoeffer and Schachman 1960.)

Therefore, at present it appears that the nucleic acid within viruses can have (1) no independent secondary structure (TMV); (2) extensive secondary structure (Shope papilloma virus and T6); (3) partial secondary structure (bushy stunt virus).

Finally, it has been suggested by Fresco et al. (1960) that the unpaired bases in such semihelical RNA structures as illustrated in Figure 26 constitute a tertiary structure since they are held in loops or folds in fixed relation to the rest of the structure. It is further suggested that their conformation might provide new possibilities in coding and information transfer that are not inherent in a structureless single strand or in a perfect base-paired helix. These suggestions remain open for further investigation.

Investigations have also been made to determine whether the segmented state of certain RNA genomes mentioned above might be an artifact of preparation or represent the state of the nucleic acid in the viral particle as well. Application of the periodate oxidation-borohydride reduction procedure described in Sec. 2, g2 showed the same number of 3′ terminal groups for the RNA within the virus particles as in the isolated nucleic acid. Hence it appears that the nucleic acid in reovirus, cytoplasmic polyhedrosis virus, and influenza virus particles is segmented to the same extent as the nucleic acid isolated from such particles (Millward and Graham 1970; Lewandowski and Millward 1971; Lewandowski et al. 1971).

3. *Function of Viral Nucleic Acids*

That nucleic acids are the genetic material of viruses is now widely accepted. This view did not appear suddenly but evolved through a series of observations made over a period of years. The following are noteworthy examples.

a. A Suggestive Idea from Bacterial Transformation

In 1944 Avery et al. reported experiments in which it was shown that minute amounts of DNA extracted from Type III pneumococci were able, under appropriate cultural conditions, to induce the transformation of unencapsulated R variants (characterized by rough colonies) of pneumococcus Type II into fully encapsulated S cells (characterized by smooth colonies) of pneumococcus Type III. Evidence that the active transforming agent was actually DNA, unaided by protein, polysaccharide, or any other substance, mounted over the years, together with reports of transforming principles in other bacterial systems (see reviews by Zamenhof 1957; Hotchkiss 1957; Hotchkiss and Gabor 1970). Since the presence or absence of capsules was known to be genetially determined, the potential significance of DNA in bacterial genetics was made apparent by transformation phenomena. Moreover, other hereditary characteristics are now known to be transferred in this manner.

b. A Hint from the Chemical Analysis of Spontaneous Mutants of Tobacco Mosaic Virus

At a time when proteins were considered largely responsible for the biological properties of viruses, enzymes, and certain hormones, some mildly disturbing results were obtained upon analysis of the protein coats of spontaneous mutants of tobacco mosaic virus (Knight 1947a). At least one strain was found whose protein appeared to have the same composition as that of common TMV, although this strain caused markedly different symptoms in infected Turkish tobacco. Therefore, it was suggested that the primary change responsible for mutation might be in the nucleic acid of the virus. However, it was not certain that the amino acid analyses were accurate enough to have detected a small but significant difference between mutants (although this was later shown to be true), and there was the possibility that the proteins of the two mutants were identical in composition but differed in sequence of amino acids. Consequently, these findings had less effect than might have been expected on the direction of thought about the role of viral nucleic acids.

c. RNA Shown Essential for Plant Virus Duplication

A few years after the work of Avery et al. (1944) on the pneumococcal transforming principle, Markham and Smith (1949) isolated and crystallized a new plant virus from turnip, which they called turnip yellow mosaic virus. This virus proved to be homogeneous in the Tiselius electrophoresis apparatus, but had two distinct components as judged from sedimentation studies. The major component, comprising 70–80 percent of the material by weight, and the minor component, comprising the balance of the material, had sedimentation coefficients of 106 S and 49 S, respectively. In other physicochemical properties, the two components were virtually identical save that the major component (called "bottom component" from sedimentation behavior) contained about 37 percent RNA, whereas the minor component ("top component") had essentially none. Significantly, the particles containing RNA were found to be highly infectious while those lacking RNA were noninfectious.

These findings could be interpreted to mean that only the combination of protein and nucleic acid is infectious, or that nucleic acid alone is essential for infectivity. Markham (1953) took the latter view in a paper presented at Oxford in April 1952, in which he said, "The role of the protein constituent of plant viruses is undoubtedly very important, but there is some evidence that the nucleic acid is in fact the substance directly controlling virus multiplication."

d. Role of DNA in Infection by T Phages

The idea of the hereditary primacy of viral nucleic acid received a great stimulus from Hershey and Chase's study (1952) of the process of infection by *E. coli* by coliphage T2. Using phage whose protein was labeled with ^{35}S and whose DNA contained ^{32}P, they showed that at least 80 percent of the phage sulfur (and hence most of the protein) remained on the outside of infected cells, whereas only 21–35 percent of the phosphorus (representing DNA) remained outside. The bulk of the protein was mechanically removed at this stage, and yet the cells went ahead and produced T2 phage. Furthermore, 30 percent or more of the parental phosphorus was found in the progeny phage in contrast to less than 1 percent of the sulfur. From these and other facts, it was proposed that the DNA probably exercises the genetic function of the phage, and the protein of a mature phage particle acts as a protective coat for the DNA and is responsible for the adsorption of the phage to the bacterium and the injection of the DNA into the cell.

e. Infectious Nucleic Acid from Tobacco Mosaic Virus

The crowning evidence that nucleic acid is the prime germinal substance of viruses was obtained when it was shown that TMV-RNA is infectious. This was demonstrated after the Hershey-Chase experiment, but a few years prior to the report that bacterial protoplasts (bacteria whose cell walls had been enzymically removed), in contrast to whole bacteria, could be directly infected with phage nucleic acid.

Fraenkel-Conrat (1956) reported that RNA preparations obtained from TMV by treatment of the virus with sodium dodecyl sulfate (see Detergent Procedure in section on Methods for Preparing Viral Nucleic Acids) was infectious though apparently devoid of characteristic virus particles, and that this infectivity was abolished by treatment of the preparation with ribonuclease. At about the same time, Gierer and Schramm (1956a, 1956b) described similarly infectious TMV-RNA preparations they had obtained by extracting the virus with phenol, according to the method of Schuster et al. (1956) (see The Phenol Method in the preparations section). The infectivity of the Gierer-Schramm preparations was also sensitive to ribonuclease; in addition it was shown to sediment much more slowly than virus, to be relatively unaffected by anti-TMV serum, and to be considerably more sensitive to elevated temperature than intact virus. Also, the level of protein in the infectious RNA preparations was found to be very low. These points were confirmed by Fraenkel-Conrat et al. (1957).

The indication from these pioneer experiments that viral nucleic acid is the genetic material of viruses has been repeatedly verified in many ways and is now taken for granted. Thus, in viruses, as well as in higher organisms such as protists, plants, and animals, a major function of nucleic acid is as a repository of genetic information. In addition, and in contrast with higher organisms, viral nucleic acid, when it is RNA, often acts as its own messenger RNA (mRNA). When viral nucleic acid is DNA, it cannot perform this function directly but serves as a template from which mRNA is transcribed. The nucleic acids found in mature virus particles, whether RNA or DNA, serve as templates for their own replication.

C. Lipids

Viruses containing lipid include representatives of all major types of viruses (see Table 26), although lipid components are much commoner among animal viruses than they are with bacterial or plant viruses.

Lipid-containing viruses share three common properties:

1. *Particle morphology.* The virus particles usually exhibit a nucleoprotein core surrounded by a membranous envelope composed of lipid, protein, and sometimes carbohydrate through which glycoprotein structures called spikes project (see Figure 31).

Table 26. Some Lipid-Containing Viruses.

Virus	Percent Lipid	Lipid Constituents	Reference[a]
Avian myeloblastosis	35	Partly phospholipid	1, 2
Equine encephalomyelitis	54	Phospholipids, cholesterol, triglycerides	3
Fowl plague	25	Phospholipid, cholesterol	4, 5
Fowlpox	27	Phospholipid, cholesterol, triglycerides, fatty acids	6
Herpes simplex	22	Phospholipid	7
Influenza	18	Phospholipids, tryglyceride, and cholesterol	8
Potato yellow dwarf	20	Phospholipids, sterol and possibly other lipids	9
Pseudomonas phage Ø6	25	Phospholipid	10
Pseudomonas phage PM2	10	Phospholipid	11
Rous sarcoma	35	Partly phospholipid	12
Simian virus 5 (SV5)	20	Phospholipids, cholesterol, triglyceride	13
Sindbis	29	Phospholipids, cholesterol	14
Tipula iridescent	5	Phospholipid	15
Tomato spotted wilt	19	Not yet known	16
Vaccinia	5	Phospholipid, cholesterol, triglycerides	17

[a] (1) Bonar and Beard 1959; (2) Allison and Burke 1962; (3) Beard 1948; (4) Schäfer 1959; (5) Gierer 1957; (6) Randall et al. 1964; (7) Russell et al. 1963; (8) Frommhagen et al. 1959; (9) Ahmed et al. 1964; (10) Vidaver et al. 1973; (11) Espejo and Canelo 1968; (12) Vogt 1965; (13) Klenk and Choppin 1970; (14) Pfefferkorn and Hunter 1963; (15) Thomas 1961; (16) Best 1968; (17) Joklik 1966.

2. *Mechanism of maturation and release of virus particles.* Nascent virus particles mature at plasma, vesicular, or nuclear membranes through which they are then released by an extrusion or budding process.

3. *Sensitivity to lipid-degrading agents.* Most lipid-containing viruses disintegrate and lose infectivity upon treatment with organic solvents (for example, ether or methanol-chloroform), certain detergents (for example, deoxycholate), or lipolytic enzymes (for example, phospholipase A). Vaccinia virus and certain iridescent insect viruses, which contain small amounts of lipid, are exceptions to this rule, but logical exceptions because their lipids serve little or no structural function, nor do they play a vital role in the infectious process.

Several different kinds of lipids have been identified among the fatty substances extracted from viruses and include cholesterol, triglycerides ("neutral fat"), and such phospholipids as sphingomyelin, phosphatidylcholine, phosphatidylserine, phosphatidylethanolamine, and phosphatidylinositol. The structures of some of these compounds are illustrated in Figure 29.

As noted above, most lipid-containing viruses have envelope structures that are acquired upon budding from a membrane. Abundant analytical evidence supports the assumption that much if not all of the viral lipid is obtained directly from the cell membrane in the budding process (Wecker 1957; Frommhagen et al. 1959; Kates et al. 1961; Franklin 1962; Klenk and Choppin 1969b, 1970). The chemical relationship between viral envelope lipids and cell membrane lipids is especially well illustrated by the studies of Klenk and Choppin (1969, 1970) with the paramyxovirus, simian virus 5 (SV5), and the membranes of different cells in which this virus was cultured. The data reproduced in Table 27 show how the quantities of different types of SV5 lipids parallel those of the membranes of two types of kidney cells in which the virus was grown.

Phospholipids predominate among the lipids found in cell membranes and this is reflected in the composition of viral lipids, including those of SV5 (Table 27). The comparison between cell membrane and viral lipids is sharpened by comparing the contents of individual phospholipids as is done for SV5 in Table 28. As shown in the table, there are marked differences in content of individual phospholipids between membranes from mouse kidney and hamster kidney cells. These differences are reflected in the compositions of SV5 lipids from virus grown in the two types of cells. Data of this sort mean that the same virus grown in different cell strains can have lipid of diverse compositions.

How are diverse lipids fabricated into viral envelopes? This is not yet clear. Cholesterol may be dissolved in the other fatty substances, but the various phospholipids appear to be coupled to protein and polysaccharide to form specific lipoprotein and glycolipid complexes. The precise nature of the linkages involved in these complexes and in their fabrication into an

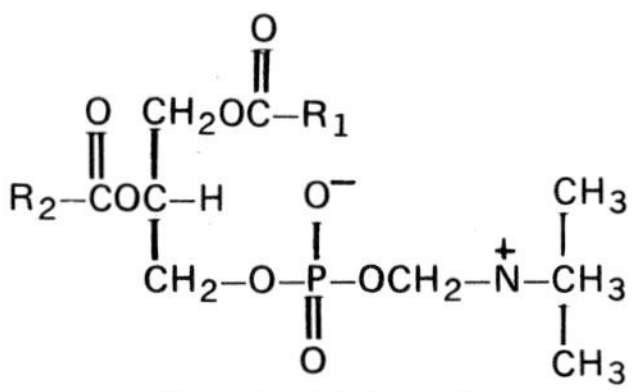

Phosphatidyl choline

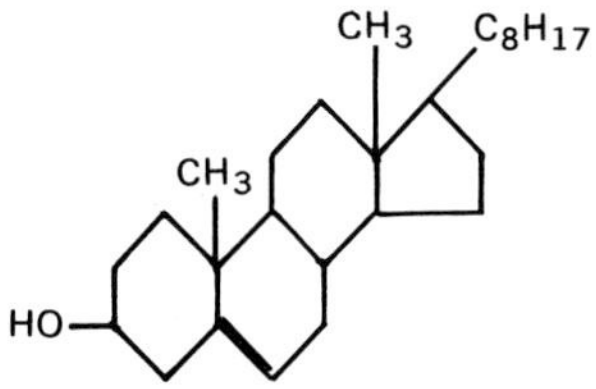

Cholesterol

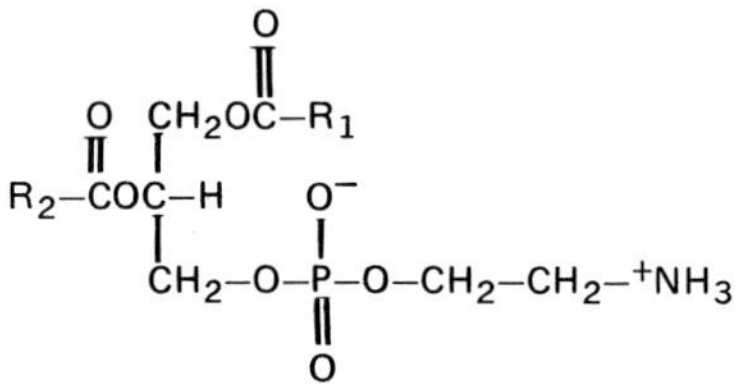

Phosphatidyl ethanolamine

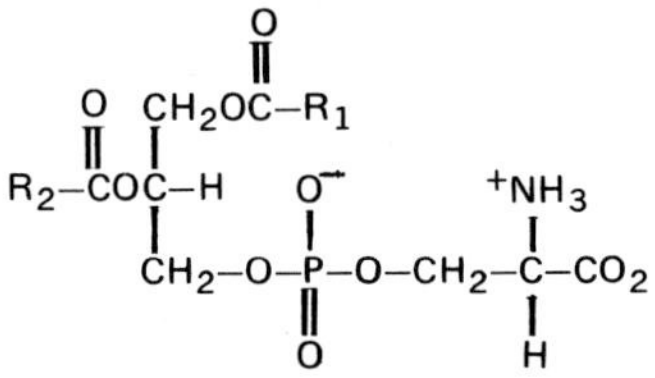

Phosphatidyl serine

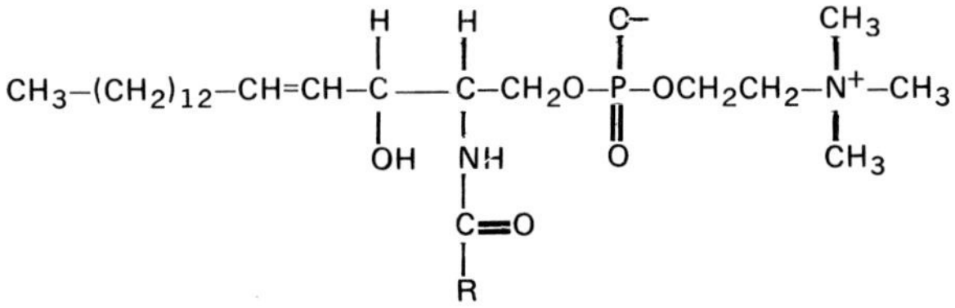

Sphingomyelin

Fig. 29. Structural formulas for some phospholipids (phosphatides) and cholesterol. In these formulas R_1 is typically the hydrocarbon chain of a saturated fatty acid, while R_2 is a similar chain for an unsaturated fatty acid. Usually these fatty acids contain 16 or 18 carbon atoms.

Table 27. Lipid Content of Simian Virus 5 (SV5) and of Plasma Membranes from Monkey Kidney (MK) and Baby Hamster Kidney (BHK21-F) Cells in Which the Virus Was Grown[a]

Source	Total Lipid Percent of Dry Weight	Percent of Total Lipid			
		Phospholipid	Triglycerides	Cholesterol	Cholesterol Esters
MK membranes	28.5	55.0	5.1	23.0	0.3
SV5 from MK cells	20.0	50.9	5.0	29.0	0.9
BHK21-F membranes	30.7	60.0	2.8	20.5	1.3
SV5 from BHK21-F cells	21.0	57.0	3.0	18.6	2.7

[a]From Klenk and Choppin 1969b.

Table 28. A Comparison of the Phospholipid Contents of Simian Virus 5 (SV5) with Those of Plasma Membranes of Monkey Kidney (MK) and Baby Hamster Kidney (BHK21-F) Cells.[a]

Source	Percent of Total Phospholipid				
	Sphingomyelin	Phosphatidyl Choline	Phosphatidyl Inositol	Phosphatidyl Serine	Phosphatidyl Ethanolamine
MK membranes	11.8	32.1	—[b]	17.2	38.8
SV5 from MK cells	12.2	25.2	2.9	17.9	40.3
BHK21-F membranes	24.2	49.5	10.0	5.1	11.2
SV5 from BHK21-F cells	30.0	38.5	10.5	5.2	15.6

[a]From Klenk and Chopin 1969b.
[b]None detected.

envelope remains to be elucidated, but probably involves some of the same sorts of interactions that characterize enzyme-substrate complexes. In short, there is little evidence for primary covalent linkages between protein and lipid moieties in lipoproteins and the combination seems to depend on steric fit and upon interactions between nonpolar hydrophobic residues and between polar or charged groups. In addition there is probably significant hydrogen bonding in which water molecules have a bridging function.

1. Preparation of Viral Lipids

Lipids tend to be less soluble in aqueous media than the other constituents of viruses and more soluble in organic solvents. Generally, they are also more susceptible to air oxidation and to temperature effects. Consequently, lipids are extracted from frozen-dried (lyophilized) virus samples with organic solvents at moderate temperatures and often in an atmosphere of nitrogen. It should be noted that some lipolytic enzymes are solvent activated, an effect that increases with temperature. This suggests that extraction at room temperature is generally desirable.

Since in many cases lipids appear to occur in lipoprotein complexes and water plays some part in this union, it appears that dehydrating organic solvent should help to rupture the lipid-protein linkage. Hence such polar solvents as methanol and ethanol are usually included in the initial solvent of a several step procedure. However, since many lipids are not very soluble in such solvents, a more nonpolar solvent such as chloroform or diethyl ether is often included. A commonly employed solvent system of this sort is chloroform-methanol, approximately 2:1 (v/v). An example using such a mixture in the extraction of lipid from SV5 is as follows (Klenk and Choppin 1969a).

In order to extract total lipid, lyophilized virus is extracted with chloroform-methanol-water (65:25:5) (10 ml/50 mg dry weight) twice for 20 min at room temperature and once for 20 min under nitrogen using boiling solvent. To the combined extracts is added 1/6 vol of water, and the mixture is separated into aqueous and organic phases. If gangliosides (complex lipids composed of sphingosine, fatty acid, one or more sugars, and neuraminic acid) are present, they go into the aqueous phase while all other lipids remain in the organic phase. The latter includes virtually all of the SV5 lipid and the solvent in this fraction is removed under nitrogen in a rotary evaporator to give total lipids.

2. Analysis of Viral Lipids

The analysis of lipids obtained in the manner described in the previous section is fairly complex. The techniques employed include column chromatography, thin layer chromatography, gas-liquid chromatography, phosphorus analysis, and sometimes infrared spectroscopy. A detailed description of such methodology is given by Kritchevsky and Shapiro (1967); the details of the work on SV5 are described by Klenk and Choppin (1969a). A summary sketch of the SV5 analysis follows.

A sample of the total lipid fraction, extracted from SV5 as described in the previous section, was applied to a silicic acid column. Neutral lipids (a term applied to cholesterol and its esters, free fatty acids, and triglycerides) were separated from phospholipids by elution first with chloroform, which yielded neutral lipids, and then with methanol, which eluted phospholipids. The chloroform eluate was evaporated and the residue dissolved in hexane, which then was applied to a Florisil column (Florisil is a synthetic magnesium silicate). Chromatography on the Florisil column separated components of the neutral lipid, mainly cholesterol and its esters and triglycerides. The triglycerides were identified and quantitated as hydroxamic acids while cholesterol and its esters were also determined colorimetrically by another procedure. The methanol eluate from the silicic acid column was dried and dissolved in 2:1 chloroform-methanol. Aliquots of this solution were used for phosphorus determinations, and the different phospholipids were identified and quantitated by a combination of gas chromatography and quantitative two-dimensional thin layer chromatography. The results of such analyses are summarized in Tables 27 and 28.

3. *Function of Viral Lipids*

The lipids occurring in viral envelopes have been termed "peripheral structural lipids" (Franklin 1962). Extraction of these lipids with organic solvents or detergents or digestion of them with lipases results in considerable degradation of the viral particles. Clearly, such lipid components are essential for maintaining the structure of virus envelopes. The reason for loss of infectivity when viral lipids are removed is doubtless associated with the inability of disrupted virus to attach and penetrate because these steps in infection depend largely on viral surface structures. Specific attachment is especially important in infections by animal and bacterial viruses. How removal of lipid alters the infectivity of plant viruses is not yet apparent.

D. Carbohydrates

Carbohydrates are found in all viruses since all viruses have a nucleic acid component. Nucleic acids, as indicated in a previous section, contain one of two carbohydrates, ribose or deoxyribose. However, some viruses also contain nonnucleic acid carbohydrate. This has been observed in two general situations: (1) glucose residues attached to pyrimidine in the DNA of certain bacterial viruses, and (2) polysaccharide coupled with protein (that is, glycoprotein) and lipid (glycolipid) in the envelope structures possessed by some animal and plant viruses. Quite a few enveloped viruses exhibit surface projections called spikes, and these are often glycoprotein in composition, especially in the case of viruses showing the capacity to agglutinate red cells.

Carbohydrate appears in the T-even (T2, T4, T6) bacterial viruses in the form of glucose or gentiobiose in O-glycosidic linkage with the 5-hydroxymethylcytosine (5-HMC) of the viral DNA (Sinsheimer 1960; Lehman and Pratt 1960; Kuno and Lehman 1962). The T-even phages contain essentially the same quantity of 5-HMC in their DNA components; yet the amount of glucoside is distinct for each. Glucose occurs in the proportions of about 0.8, 1.0, and 1.6 moles per mole of HMC for T2, T4 and T6, respectively (Jesaitis 1956; Lichtenstein and Cohen 1960). The glucose is uniformly distributed in the case of T4 where each HMC is glucosylated, but in T2 there is some unsubstituted HMC, some monoglucosylated, and a small amount of diglycosylated-HMC, the latter represented by the glucose disaccharide gentiobiose. In T6, an unsymmetrical distribution of glycosyl units is also found, but here about two-thirds of the glucose is present as gentiobiose.

There is evidence that the degree of glucosylation of the T phage DNA is an inherited trait, although in crosses the trait "glucose content" does not segregate symmetrically as a simple Mendelian character (Sinsheimer 1960). For example, the progeny of a T2 and T4 cross were all found to have the T4 glucose content, whereas the recombinants of a T2 × T6 cross were found to have the glucose content of either T2 or T6, although many more were found to have the glucose content of T2 and the host range of T6 than vice versa.

As indicated above, it has been found that viruses with an envelope structure often contain some of their proteins in the form of glycoproteins. Also, some of the lipid of viral envelopes may be present as glycolipid. The widespread occurrence of such nonnucleic acid carbohydrate constituents among viruses is illustrated by examples given in Table 29. In the table specific viruses are listed but in each case the group of viruses to which the example belongs is also indicated because the characteristics of the example are likely to apply throughout the group, and some of the groups are very large. The carbohydrate components of viral glycoproteins and glycolipids are complex polysaccharides usually fabricated from fucose, galactose, glucosamine, and mannose. Some viral glycoproteins, such as those from Sindbis and vesicular stomatitis viruses, contain sialic acid (Burge and Huang 1970). Sialic acid is the group name for a series of acylated derivatives of neuraminic acid:

$$\begin{array}{ccccccccccccccccc}
 & & \text{O} & & & & \text{OH} & & \text{H} & & \text{H} & & \text{OH} & & \text{OH} & & \\
 & & \| & & & & | & & | & & | & & | & & | & & \\
\text{HOOC} & — & \text{C} & — & \text{CH}_2 & — & \text{C} & — & \text{C} & — & \text{C} & — & \text{C} & — & \text{C} & — & \text{CH}_2\text{OH} \\
 & & & & & & | & & | & & | & & | & & | & & \\
 & & & & & & \text{H} & & \text{NH}_2 & & \text{OH} & & \text{H} & & \text{H} & &
\end{array}$$

Neuraminic Acid

(5-amino-3,5-dideoxy-D-glycero-D-galactononulesonic acid)

The simplest sialic acid is *N*-acetylneuraminic acid.

Although detailed structural analyses are yet to be made of viral glycoproteins, it seems likely, by analogy with other better studied glycoproteins, that the mode of linkage between protein and carbohydrate is by glycosidic bonds between carbohydrate chains and asparagine, serine, and threonine residues of the protein (Neuberger et al. 1972).

1. *Preparation of Viral Carbohydrates*

General methods have not been developed for the isolation of viral carbohydrates as they have for viral nucleic acids and proteins. Instead, it

Table 29. Some Viruses Containing Nonnucleic Acid Carbohydrate.

	Carbohydrate-containing Constituent		
Virus	Glycolipid	Glycoprotein	Reference[a]
Herpes simplex virus (a herpesvirus)	?	+	1
Influenza virus (an orthomyxovirus)	+	+	2
Murine leukemia virus (an oncornavirus)	?	+	3
OC 43 (a human coronavirus)	?	+	4
Potato yellow dwarf virus (a plant rhabdovirus)	?	+	5
Simian virus 5 (SV5) (a paramyxovirus)	+	+	2
Sindbis virus (a togavirus)	−	+	6
Vesicular stomatitis virus (an animal rhabdovirus)	+	+	7

[a](1) Roizman and Spear 1971; (2) Compans and Choppin 1971; (3) Nowinski et al. 1972; (4) Hierholzer et al. 1972; (5) Knudson and MacLeod 1972; (6) Schlesinger and Schlesinger 1972; (7) Knudson 1973.

has usually seemed sufficient to obtain qualitative and quantitative values for these constituents, and even such analyses have often been neglected. However, a growing realization of the importance of glycoproteins in many animal viruses may lead before long to the development of general procedures for isolating viral carbohydrates in undegraded forms.

The investigation of avian tumor virus glycopeptides represents a step in this direction. For example, Lai and Duesberg (1972) used the following technique to isolate avian tumor virus glycopeptides which were estimated to contain less than 10 percent protein: Purified tumor virus was disrupted by treating it at 37° for 30 min with 1 percent sodium dodecyl sulfate in the presence of 0.05 M mercaptoethanol. A precipitate of proteins and glycoproteins from the disaggregated virus was obtained by addition of 5 vol of ethanol; this precipitate was subsequently dissolved in a solution of 0.1 percent SDS–0.1 M tris buffer at pH 8. Alternatively, the viral proteins and glycoproteins were extracted from the virus by treatment with water-saturated phenol; they were recovered from the phenol phase by precipitation with 5 vol of ethanol in the presence of 2 M ammonium acetate. After washing twice with 75 percent ethanol, the precipitate was dissolved in the tris buffer–0.1 percent SDS noted above. By digesting the proteins in this mixture with pronase (a 48-hr treatment with pronase at 1 mg/ml followed by a second 48-hr treatment with pronase at 0.5 mg/ml) the proteins, including those of the glycoproteins, were largely degraded. The glycopeptides could be separated from this mixture by gel filtration chromatography on Sephadex G-50. When the viral protein had been labeled with ^{3}H amino acids before applying the procedures outlined above, it was found that over 90 percent of the label eluted from the Sephadex as amino acids or small peptides rather than in the glycopeptide fraction. This suggests that the procedure just described could be employed to prepare viral polysaccharides containing only a small amount of protein still attached.

2. *Analysis of Viral Carbohydrates*

The results of chemical and spectrophotometric analyses made by Taylor (1944) on influenza viruses provided the first indication that any highly purified virus contained nonnucleic acid carbohydrate. These early assays also indicated that the viral carbohydrate might contain galactose, mannose, and glucose. Later, more extensive studies based on colorimetric, chromatographic, and spectrophotometric analyses indicated that galactose, mannose, glucosamine, and fucose are constituents of the influenzal carbohydrates (Knight 1947b; Ada and Gottschalk 1956; Frommhagen et al. 1959). Since then, sialic acid and galactosamine have been added to the list of viral polysaccharide constituents (Strauss et al. 1970; McSharry and Wagner 1971).

The total carbohydrate of a virus can be estimated colorimetrically by the orcinol reaction as follows [based on Marshall and Neuberger (1972)]:

1. Dissolve 3–4 mg of dry virus in 1 ml of 0.1 N NaOH in a 10-ml glass-stoppered bylinder or test tube.
2. Add 8.5 ml of orcinol-H_2SO_4 reagent (a fresh mixture of 7.5 vol of 60 percent H_2SO_4 and 1 vol of 1.6 percent orcinol in H_2O) and mix well.
3. Place the loosely stoppered cylinder together with cylinders containing reagent and 1 ml of 0.1 N NaOH and other cylinders containing various total amounts (from 50 to 200 μg) of carbohydrates (standard solution containing equal amounts of fucose, galactose, and mannose) in a water bath at 80°C.
4. After 15 min, cool the tubes in tap water and take readings in a spectrophotometer at 505 nm.

This test will give only a crude approximation owing to small but variable contributions to the color by other constituents of the virus and by a failure to get an appropriate color yield from amino sugars. The best estimate of total nonnucleic acid carbohydrate is obtained by the summation of analyses for the individual carbohydrates.

The analysis of individual carbohydrates in viral glycoproteins and glycolipids involves a variety of methods ranging from colorimetric analyses through chromatography (paper, thin layer, column, and gas-liquid) (see Marshall and Neuberger 1972; Clamp et al. 1972). In any case, the analyses must be preceded by or include in them hydrolysis of the polysaccharides. Hydrolysis of carbohydrates has many of the features and cautions of protein hydrolysis, and these must be taken into consideration if incomplete hydrolysis is to be avoided, on one hand, and destructive hydrolysis is to be minimized, on the other. Such problems are thoroughly discussed by Marshall and Neuberger (1972).

Interference in the carbohydrate analysis can be reduced if glycopeptides are first isolated from the virus by gel electrophoresis or chromatography. Seven different carbohydrates (fucose, mannose, galactose, glucose, galactosamine, glucosamine, and neuraminic acid) were identified by gas-liquid chromatography as constituents of the nonnucleic acid carbohydrate of vesicular stomatitis virus (McSharry and Wagner 1971). Gas-liquid chromatography has attracted considerable interest for analyzing viral carbohydrates but there are technical difficulties in the method. One of them is in getting quantitative derivitization of the individual carbohydrates and another is in eliminating or reducing spurious peaks (background). Balanced against these problems are the exquisite sensitivity of the procedure (in the nanomole region) and the added specificity that can be achieved if the chromatography is coupled to a mass spectrometer or to a counter (gas-liquid radiochromatography).

3. *Function of Viral Carbohydrates*

Glucosylation of T-even phage DNAs appears to be essential for phage survival in certain bacterial strains where the glucosyl residues appear to confer resistance to degradation by nucleases. This resistance mechanism seems to be peculiar to the T-even phages and certain *E. coli* cells since other nonglucosylated phages multiply and produce infectious progeny in these bacteria.

The specific function of carbohydrates in the glycoproteins and glycolipids of enveloped viruses is not known. However, the spikes of ortho- and paramyxoviruses are glycoproteins and these are of two types, one of which constitutes the hemagglutinin of these viruses and the other a neuraminidase enzyme (see Compans and Choppin 1971). The potential importance of the carbohydrate moiety in hemagglutination is indicated by the loss of hemagglutination capacity concomitant with the cleavage of reducing sugar by a specific glycosidase enzyme (Bikel and Knight 1972). It can also be surmised because of their location in the surface of virus particles that the carbohydrates of viral envelopes play a role in the attachment and penetration of these viruses in the course of infecting cells and probably also in their exit from cells. It is not yet clear with regard to release of enveloped viruses from infected cells to what extent the viral envelope is determined by host genome and viral genome, respectively. An interesting aspect of this question is the observation that with avian tumor viruses, the glycopeptides of all viruses released from transformed cells are larger than those of viruses released from normal cells (Lai and Duesberg 1972).

Glycoproteins also are heavily involved in the immunological reactions of enveloped animal viruses. For example, antiserum to influenza hemagglutinin has potent virus-neutralizing capacity (Schild 1970), and the glycoprotein of vesicular stomatitis virus appears to be the specific antigen that induces the synthesis of and reacts with viral neutralizing antibody (Kelley et al. 1972).

E. Polyamines and Metals

In addition to protein, nucleic acid, lipid, and carbohydrate, some other substances are found in small amounts in highly purified preparations of certain viruses. Most of these minor components are probably adventitious elements. For example, many cells contain significant amounts of polyamines (Tabor et al. 1961; Cohen 1971), and these cations are strongly attracted to the phosphoryl anions of viral nucleic acids, where they remain to become a part of the mature virus particle in those cases in which low particle permeability and other relationships are favorable. Thus, putrescine, $H_2N\text{-}CH_2\text{-}CH_2\text{-}CH_2\text{-}CH_2\text{-}NH_2$, and spermidine, $H_2N(CH_2)_4NH$-

$(CH_2)_3NH_2$, were found by Ames and associates (1958, 1960) in T2 and T4 phages in amounts sufficient to neutralize about half of the DNA charge. A similar situation was reported by Kay (1959) for bacteriophage 3 of *E. coli* 518.

In the case of the T2 and T4 phages, which are not normally very permeable to cations, it was shown that the putrescine is associated with the DNA inside the phage head and that this internal putrescine could not be displaced with ^{14}C-containing putrescine on the outside nor by Mg^{++} and Ca^{++}. These closely adhering polyamines are thought to be the same as Hershey's A substances (1957), which are injected along with DNA in the course of infection by T2 and T4. The lack of specificity of putrescine and spermidine was shown by growing the host cells in a medium rich in spermine, $H_2N(CH_2)_3NH(CH_2)_4NH(CH_2)_3NH_2$, a polyamine not normally present in these bacteria. The mature phages isolated from these cells were found to contain spermine rather than putrescine or spermidine (Ames and Dubin 1960).

Hence the polyamines appear to play no specific role in phages, a view which is supported by the absence of polyamines in T3, T5, and P22 phages whose permeabilities to cations presumably allow displacement of polyamines by metallic cations during isolation and purification of the phages. Furthermore, it was shown that a permeable mutant of T4 could be isolated with or without spermidine with no change in biological properties. Nevertheless, it is possible that polyamines or metallic cations may assist in the folding of DNA in the process of phage assembly. Moreover, these cations may be essential to some other stage or stages of phage biosynthesis although the mechanism of such effects is presently obscure (Cohen and Dion 1971).

Polyamines occur in the virions of herpesvirus and of influenza and Newcastle disease viruses (Gibson and Roizman 1971; Bachrach et al. 1974). Traces have also been reported in several plant viruses while amounts sufficient to neutralize about a fifth of the charges of the viral RNA have been found in turnip yellow mosaic, turnip crinkle, and broad bean mottle viruses (Ames and Dubin 1960; Johnson and Markham 1962; Beer and Kosuge 1970).

As many as 14 metallic cations have been found in plant virus preparations, some of them loosely bound to protein and others more tightly bound to RNA (Pirie 1945; Loring et al. 1958; Wacker et al. 1963; Johnson 1964). These cations can be largely removed by treatment with a chelating agent without significantly reducing infectivity. It is doubtful if the remaining few atoms of tightly bound metal are of crucial importance.

The general conclusion about both organic and inorganic cations is that they bind randomly to protein and nucleic acid in amounts dependent on the environment and relative affinities of the ions involved. Such binding, especially to the nucleic acid, may well affect the conformation and function, but specific effects have yet to be elucidated.

F. Summary: Composition of Viruses

There are many viruses in nature whose mature particles consist solely of nucleic acid and protein. There are numerous other more complex viruses that contain, in addition to nucleic acid and protein, lipid, nonnucleic acid carbohydrate, and a variety of other minor constituents. Nucleic acid and protein are properly emphasized because these constituents play a predominant role in the structure and function of viruses, although some of the minor constituents may in specific cases, as indicated in the preceding sections, be very important. Finally, nucleic acid is recognized as the one indispensable constituent of all viruses (some viruses may consist of nucleic acid alone) because it is the genetic material and is capable of inducing infection by itself.

IV.

Morphology of Viruses

The chemical constituents described in the previous chapter are found in particles of diverse size and shape in the various viruses isolable from animals, bacteria, plants, and fungi. Despite the diversity of size and shape of different viruses, the size and shape of any one virus tend to be much more uniform than do the cells of a bacterium. This uniformity is reflected in the fact that many viruses can be crystallized whereas bacteria cannot. Some examples of virus crystals are shown in Figure 30. Note that a single virus crystal contains millions of virus particles as is nicely illustrated in the electron micrograph obtained by Steere and Williams (1953) of a partially dissolved crystal of tobacco mosaic virus. Thus, although a simple virus particle may consist of hundreds of molecules of protein and one or more molecules of nucleic acid, large populations of these particles often behave as though they were just molecules, crystallization of particles being one manifestation of this characteristic (behavior of virus particles in hydrodynamic tests such as electrophoresis or sedimentation is also molecular in character).

Each virus has a characteristic size and shape. The range in size for viruses as a group is from about 20 nm in diameter for minute virus of mice to about 300 nm for a poxvirus (some elongated plant and bacterial viruses exceed this upper limit in one dimension; for example, beet yellows virus is about 10 × 1,250 nm).

Few distinctive shapes have been observed among viruses, and most viruses fall into one or another of three general groups characterized by (1) spheroidal particles (also called spherical or isometric particles); (2) elongated particles; and (3) combination particles, such as a tailed bacteriophage that may have a spheroidal head and an elongated tail. The sizes and shapes of viral particles in some distinctive groups of algal, animal, bacterial, insect, and plant viruses are given in Tables 30 to 34.

Although some attempt was made in these tables to group viruses according to recommendations of international committees concerned with virus classification and nomenclature, the purpose of the tables is not to deal with virus classification, but rather to illustrate the distribution of sizes and shapes among distinguishable classes of viruses. In order to treat viruses in groups, the dimensions assigned must necessarily encompass the

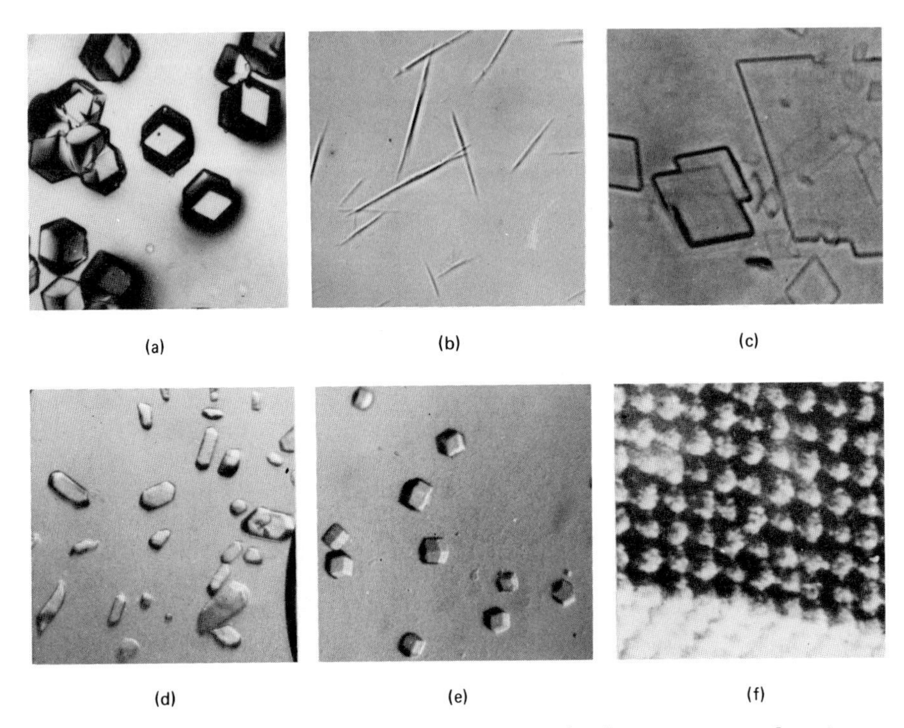

Fig. 30. Crystals of some viruses. *a*. Tomato bushy stunt virus; *b*. tobacco mosaic virus; *c*. Southern bean mosaic virus; *d*. poliovirus; *e*. polyoma virus; *f*. electron micrograph of a portion of a crystal of tobacco necrosis virus showing orderly array of virus particles. (*a* and *b*, courtesy W. M. Stanley; *d*, courtesy F. L. Schaffer; *e*, courtesy W. T. Murakami; and *f*, courtesy R. W. G. Wyckoff.)

range found in the group and thus suffer in precision for individual viruses. However, when precision in dimensions is required, it can be obtained from the references accompanying each table. Another caveat about virus dimensions is that some viruses are more plastic than others; enveloped viruses are most apt to be pleomorphic and to exhibit a range of sizes and shapes. Some of these points will be illustrated in electron micrographs of different viruses where it will also be evident that viruses occur in many sizes but in relatively few shapes.

Evidence concerning the size and shape of virus particles was obtained by indirect methods for some years prior to the common availability of the electron microscope. Some of these methods are still very useful. Thus, estimates of size can be obtained by ultrafiltration, diffusion measurements, gel chromatography, and light scattering, while indications of shape are readily obtained by flow birefringence or viscosity measurements; density

Table 30. Sizes and Shapes of Some Blue-Green Algal Viruses.[a]

Virus	Diameter or Dimensions (nm)	Shape
Anacystis, Synechococcus[b] (AS) AS-1	Head 90 Tail 23 × 244	Spheroidal head and elongated tail
Lyngbya, Plectonema, Phormidium (LPP) LPP-1, LPP-2	Head 59 Tail 15 × 20	Spheroidal head and short tail
Nostoc (N) N-1	Head 55 Tail 16 × 110	Spheroidal head and elongated tail
Synechococcus, Microcystis (SM) SM-1	88	Spheroidal with collar and possibly a very short tail

[a]Compiled from Brown 1972.
[b]These viruses are named according to the algal genera the viruses infect; hence, the names in the table are generic names of some susceptible blue-green algae. The algal viruses contain linear, double-stranded DNA. (See also Padan and Shilo 1973.)

alone, or a composite indication of size, shape , and density can be determined by various centrifugation techniques. Applications of some of these methods were described in the section on Purification of Viruses.[1]

The most versatile and direct method for determining the size and structure of virus particles is by electron microscopy. Many techniques are available that enhance the usefulness of the electron microscope beyond its ability to resolve objects down to about the 1 nm level in contrast to the approximately 200 nm resolving power of the light microscope. Some of these techniques increase contrast between virus particles and the plastic film of the microscope mount, some minimize the tendency of particles to collapse when exposed to osmotic and surface tension forces, and others limit the destructive effects of beams of electrons used to illuminate the field under examination. For descriptions of these methods and their applications, see Kay (1961), Huxley and Klug (1971), Dalton and Haguenau (1973), Williams and Fisher (1974); for reviews, see Horne (1967) and Milne (1972).

The development of electron microscopy, coupled with chemical and physical analyses, revealed various features of virus particles that might be called ultrastructural details. Many such structural components have been given names (Caspar et al. 1962; Lwoff and Tournier 1966); the commoner terms and their synonyms will be briefly presented here.

[1]Detailed descriptions can be obtained in such works as *Methods in Virology*, Vol. 2, K. Maramorosch and H. Koprowski, editors, New York: Academic Press (1967).

Table 31. Sizes and Shapes of Some Diverse Groups of Animal Viruses.[a]

Virus	Diameter or Dimensions (nm)	Shape
A. DNA-Containing Vertebrate Viruses		
Adenoviruses:	70–90	Spheroidal with projecting fibers
Avian adenoviruses		
Gallus-adeno-like (GAL)		
Chicken-embryo-lethal-orphan (CELO)		
Bovine adenoviruses		
Canine adenoviruses		
Infectious canine hepatitis virus (ICH)		
Human adenoviruses		
31 serological types		
Murine adenoviruses		
Ovine adenoviruses (may be same as bovine strains)		
Porcine adenoviruses		
Simian adenoviruses		
Herpesviruses:	100–150	Spheroidal with envelope
Group A		
B virus of monkeys, equine abortion, equine respiratory disease, feline rhinotracheitis, herpes simplex (types 1 and 2), infectious bovine rhinotracheitis, infectious laryngotracheitis, owl monkey herpes, marmoset herpes, squirrel monkey herpes		
Group B		
Cytomegalovirus, varicella-zoster		
Group C		
Burkitt lymphoma, herpesvirus ateles, herpesvirus saimiri,		

herpesvirus sylvilagus, Lucké frog tumor, Marek's disease of chickens		
Papovaviruses	43–53	Spheroidal
K virus of mice		
Papilloma viruses		
Bovine, canine, rabbit, human (wart)		
Polyoma of mice		
Vaccuolating viruses: rabbit, simian (SV40)		
Parvoviruses (picodnaviruses)	18–22	Spheroidal
Adeno-associated viruses (AAV)		
Hamster osteolytic viruses		
Latent rat viruses (Kilham rat virus (RV), X14, H-1, H-3)		
Minute virus of mice (MVM)		
Poxviruses		
True poxviruses	230 × 300	Brick shaped with core, lateral bodies, outer membrane with whorled surface filaments
Vaccinia-variola group		
Alastrim, cowpox, ectromelia, monkeypox, rabbitpox, vaccinia, variola (smallpox)		
Fibroma-myxoma group		
Hare fibroma, rabbit fibroma, rabbit myxoma, squirrel fibroma		
Birdpox group		
Canarypox, fowlpox, pigeonpox, turkeypox		
Sheeppox group		
Goatpox, lumpyskin disease, sheeppox		
Orf group		
Orf, bovine papular stomatitis, pseudocowpox		
Ungrouped poxviruses	150 × 200	Ovoid, with surface filaments
Molluscum contagiosum, swinepox, Yaba monkey tumor		

Table 31. Sizes and Shapes of Some Diverse Groups of Animal Viruses.[a] (cont.)

Virus	Diameter or Dimensions (nm)	Shape
B. RNA-Containing Vertebrate Viruses		
Arenaviruses Lassa, Lymphocytic choriomeningitis (LCM) Tacaribe hemorrhagic fever (several viruses)	50–150	Spheroidal with envelope
Coronaviruses Avian infectious bronchitis (IBV), several human respiratory viruses, mouse hepatitis, and rat pneumonotropic	70–120	Spheroidal with envelope
Diplornaviruses Orbiviruses (bluetongue group) African horse sickness, bluetongue, Changuinola, Chenuda, Colorado tick fever, epizootic hemorrhagic disease of deer, Eubenagee, Irituia, Palyam, simian virus SA-11, Tribec, Wad Medani	60–80	Spheroidal with inner and outer capsids
Reoviruses Avian reoviruses (5 serological types) Mammalian reoviruses (3 serological types)		

Myxoviruses	100–350	Spheroidal with envelope and projecting spikes; filamentous forms are also common
Metamyxoviruses		
pneumonia virus of mice		
respiratory syncytial (RS)		
Orthomyxoviruses	90–120	
Equine influenza, fowl plague, human influenza types A, B, and C, swine influenza		
Paramyxoviruses	120–450	
Mumps, Newcastle disease of chickens, parainfluenza (cattle, man, mice) types 1-4, simian virus 5 (SV5)		
Pseudomyxoviruses	120–300	
Canine distemper, measles, rinderpest		
Oncornaviruses	About 100	Spheroidal with envelope
Leukosis (leukemia) viruses		
Avian leukosis viruses (ALV)		
Lymphomatosis (RPL-12 and other strains), erythroblastosis (AEV), MC29, myeloblastosis (AMV), Rous-associated viruses (RAV-1, RAV-2, etc.)		
Murine leukemia viruses (MLV)		
Friend, Graffi, Gross, Kirsten, Moloney, Rauscher, etc.		
Mammary tumor viruses (MTV)		
Bittner, human (?)		
Miscellaneous oncornaviruses		
Feline, hamster, human (?), monkey, rat, reptile, etc.		
Sarcoma viruses		
Avian sarcoma viruses		
Fujinama sarcoma,		
Rous sarcoma (RSV) (several strains)		
Murine sarcoma viruses		
Harvey, Kirsten, Maloney		

Table 31. Sizes and Shapes of Some Diverse Groups of Animal Viruses.[a] (cont.)

Virus	Diameter or Dimensions (nm)	Shape
Picornaviruses	20–30	Spheroidal
Enteroviruses		
Encephalomyocarditis		
Columbia SK, encephalomyocarditis (EMC), mengo, mouse Elberfeld (ME)		
Human		
Coxsackie A, Coxsackie B, ECHO, polio		
Mouse encephalomyelitis		
Theiler's virus		
Simian enteroviruses (multiple serotypes)		
Rhinoviruses		
Human, other animals		
Unclassified picornaviruses		
Foot-and-mouth disease (FMDV), vesicular exanthema of swine (VE)		
Rhabdoviruses	75 × 130–230	Bullet-shaped or bacilliform with envelope
Bovine ephemeral fever, Chandipura,		
Flanders-Hart Park, Kern Canyon,		
Lagos, Marbur, Oregon sockeye disease, rabies, vesicular stomatitis, and others		
Togaviruses	40–60 for alpha- and flavoviruses and 100 for Bunyamuera supergroup	Spheroidal with envelope
Alphaviruses (Arbovirus Group A)		
Bebaru, Chikungunga,		
eastern equine encephalitis (EEE), Mayaro, Mucambo,		
O'nyong nyong, Pixuna, Ross River, Semiliki forest,		
Sindbis, Venezuelan equine encephalitis (VEE), Western equine encephalitis (WEE)		

Flavoviruses (Arbovirus Group b)
Dengue, diphasic meningoencephalitis, Japanese encephalitis, louping ill, Powassan, St. Louis encephalitis, West Nile, yellow fever
Bunyaviruses (Bunyamwera Supergroup)
Bunyamwera, California encephalitis, Inkoo
Miscellaneous togaviruses
Lactic dehydrogenase (LDH) of mice, phlebotomous fever, rubella

[a]Adapted from Melnick 1971, 1972. See also Joklik and Smith 1972, pp. 747–754.

Table 32. Sizes and Shapes of Some Bacterial Viruses.[a]

Virus	Diameter or Dimensions, nm		Shape
	Head	Tail	
A. Phages with contractile tails:			
Alcaligenes faecalis A6	90	16 × 110	Spheroidal head and rodlike tails ("spheroidal" includes oblong, octahedral, icosahedral, etc.)
Bacillus subtilis SPO1	90	30 × 210	
Escherichia coli E1	75	17 × 210	
Escherichia coli T2, T4, T6	65 × 95	25 × 110	
Lactobacillus 206	72	16 × 138	
Myxococcus xanthus MX1	75	25 × 100	
Proteus hauseri 78	61	16 × 89	
Pseudomonas aeruginosa PB1	75	20 × 140	
B. Phages with noncontractile tails:			
Escherichia coli lambda	54	10 × 150	Spheroidal head and rod-like or flexuous tail
Escherichia coli T1	50	10 × 150	
Escherichia coli T3, T7	60	10 × 15	
Escherichia coli T5	65	10 × 170	

Table 32. Sizes and Shapes of Some Bacterial Viruses.[a](cont.)

Virus	Diameter or Dimensions, nm		Shape
Pseudomonas Pc	65	10 × 160	
Staphylococcus 6	40 × 92	10 × 300	
Streptococcus 3ML	40 × 55	9 × 100	
Typhoid 1	75	9 × 180	
Typhoid S1 BL	50	10 × 130	
C. Tailless phages:			
1. With apical structures	27		Spheroidal
Escherichia coli α3, ØX174, ØR, S13			
2. With apical structures but enveloped	60		Spheroidal with envelope
Pseudomonas PM2			
3. Without apical structures[b]			Spheroidal
Caulobacter crescentus Cb23r	22		
Escherichia coli f2, fr, MS2, Qβ, R17	24		
Pseudomonas aeruginosa 7s	25		
D. Filamentous phages:			
Escherichia coli f1, fd, M13	6 × 800		Filamentous
Pseudomonas aeruginosa Pf	6 × 1300		
Salmonella typhimurium If1, If2	6 × 1300		

[a]Compiled from Bradley and Kay 1960; Bradley 1971; Joklik and Smith 1972, p. 829. The phages are listed by name of host bacterium followed by designation of phage.

[b]The phages of group C3 contain RNA; all of the others are DNA phages.

Table 33. The Sizes and Shapes of Some Insect Viruses.[a]

Virus	Diameter or Dimensions, nm	Shape
A. Occluded viruses (occur in inclusion bodies):		
1. Granulosis viruses		Rodlike (occur in inclusions called granules or capsules)
Armyworm	62 × 412	
Codling moth	51 × 314	
Spruce budworm	40 × 270	
2. Polyhedrosis viruses		
Nuclear		Rodlike (occur in inclusions called polyhedral bodies)
Gypsy moth	18 × 280	
Silkworm	40 × 280	
Western oak looper	62 × 332	
Cytoplasmic		Spheroidal with surface projections (occur in inclusions called polyhedral bodies)
Monarch butterfly	67	
Silkworm (CPV)	69	
Spruce budworm	70	
3. Insect poxviruses		Brick shaped with core and lateral body (occur in inclusions called spherules)
Amsacta pox	250 × 350	
Melolontha pox	250 × 400	
4. Beetle viruses		
Melolontha (cockchaffer) spindle disease	250 × 370	Ovoid (occur in spindle-shaped or ovoid inclusions)
B. Nonoccluded viruses:		
1. Iridescent viruses	150	Spheroidal
Chilo, Sericesthis, and *Tipula iridescent* viruses (CIV, SIV, TIV)		

Table 33. The Sizes and Shapes of Some Insect Viruses.[a](cont.)

Virus	Diameter or Dimensions, nm	Shape
2. Miscellaneous viruses		
Acute bee paralysis	28	Spheroidal
Antherea	50	Spheroidal
Densonucleosis virus of Galleria (DNV)	20	Spheroidal
Sac brood of bees	28	Spheroidal
Sigma virus of *Drosophila*	70 × 200	Bullet-shaped with envelope

[a]Compiled from Smith 1967; Smith 1971; Bellet 1968; Bergoin and Dales 1971; Vago and Bergoin 1968; Kurstak 1972. It should be noted that there are numerous plant viruses, especially of the rhabdovirus type, that multiply in both plants and insects and are not listed here (see Knudson 1973).
The cytoplasmic polyhedrosis viruses contain double-stranded RNA and some of the miscellaneous nonoccluded viruses such as that of sac brood of bees contain RNA; all of the rest appear to contain DNA.

Table 34. Sizes and Shapes of Some Diverse Groups of Plant Viruses.[a]

Virus	Diameter or Dimensions, nm	Shape
Alfalfa (lucerne) mosaic	18 × 18; 18 × 36; 18 × 48; 18 × 58	Pleomorphic: 3 bacilliform particles and 1 spheroidal particle
Beet yellows	10 × 1250	Flexuous rods
Brome mosaic	25	Spheroidal
Also: broad bean mottle virus, cowpea chlorotic mottle		

Carnation latent Also: Cactus 2, chrysanthemum B, pea streak, potato virus M, potato virus S, red clover vein mosaic	15 × 620–700	Bent rods
Cauliflower mosaic Also: Carnation etched ring, dahlia mosaic	50	Spheroidal
Clover wound tumor Also: Maize rough dwarf, rice dwarf	70	Spheroidal
Cowpea mosaic Also: Bean pod mottle, broad bean stain, radish mosaic, red clover mosaic, true broad bean mosaic	30	Spheroidal
Cucumber mosaic Also: Cucumber yellow mosaic, tomato aspermy	30	Spheroidal
Pea enation mosaic	28	Spheroidal
Potato virus X Also: Cactus X, clover yellow mosaic, hydrangea ringspot, white clover mosaic	13 × 480–540	Flexuous rods
Potato virus Y Also: Bean common mosaic, bean yellow mosaic, beet mosaic, clover yellow vein, Columbian datura, cowpea aphid-borne mosaic, henbane mosaic, pea mosaic, potato virus A, soybean mosaic, tobacco etch, watermelon mosaic (S. Africa)	15 × 730–790	Flexuous rods

Table 34. Sizes and Shapes of Some Diverse Groups of Plant Viruses.[a] (cont.)

Virus	Diameter or Dimensions, nm	Shape
Potato yellow dwarf	50–100 × 200–300	Bacilliform with lipid containing envelope
Also: Lettuce necrotic yellows, eggplant mottled dwarf, maize mosaic, Russian winter wheat mosaic, sowthistle yellow vein		
Prunus necrotic ringspot	25	Spheroidal
Also: Apple mosaic, rose mosaic		
Southern bean mosaic	30	Spheroidal
Tobacco mosaic	18 × 300	Tubular rods
Also: Cucumber green mottle mosaic, cucumber yellow mottle mosaic, odontoglossum ringspot, ribgrass mosaic, Sammons opuntia, sunn hemp mosaic, tomato mosaic		
Tobacco necrosis	28	Spheroidal
Tobacco rattle	22 × 50–102 and 22 × 170–210	Tubular rods of two characteristic lengths
Also: Pea early browning		

Tobacco ringspot	30	Spheroidal
Also: Arabis mosaic, grapevine fanleaf, raspberry ringspot, strawberry latent ringspot, tomato black ring, tomato ringspot		
Tomato bushy stunt	30	Spheroidal
Also: Artichoke mottle crinkle, carnation Italian ringspot, pelargonium leaf curl, petunia asteroid mosaic		
Tomato spotted wilt	70–80	Spheroidal
Turnip yellow mosaic	30	Spheroidal
Also: Andean potato latent, belladonna mottle, cacao yellow mosaic, dulcamara mottle, eggplant mosaic, ononis yellow mosaic, wild cucumber mosaic		

[a]Examples were taken mainly from Harrison et al. 1971 in which a group is defined as "a collection of viruses and/or virus strains, each of which shares with the type member all or nearly all the main characteristics of the group." All of the viruses listed contain single-stranded RNA except the clover wound tumor group, which has double-stranded RNA, and the cauliflower mosaic group has double-stranded DNA instead of RNA. The above list is an alphabetic arrangement according to type member of a group. For additional details about listed viruses see the semiannual compilations, *Descriptions of Plant Viruses*, from 1970 on, Gibbs et al. eds. These compilations are issued jointly by the Commonwealth Mycological Institute, Ferry Lane, Kew, Surrey, England, and the Association of Applied Biologists. Orders should be addressed to Central Sales Branch, Commonwealth Agricultural Bureaux, Farnham Royal, Slough SL 2 3BN, England.

The mature (structurally complete), potentially infectious virus particle is called a *virion*. Virus, or virus particle, are synonyms, although in one usage the term "virus" embraces all phases of the viral life cycle rather than just the mature virus particle. *Capsid* is a term given to the protein built around and closely associated with the viral nucleic acid, the combination of the two being called *nucleocapsid, nucleoprotein* (*NP*), or *core*. Synonyms for capsid are *protein coat* and *protein shell*. *Structure units* are the identical protein molecules that make up the capsid; they are also known as *protein subunits*. *Capsomers* are the capsid substructures distinguishable in the electron microscope. They may be individual protein subunits or more often represent small clusters (for example, two, five, or six) of subunits; capsomers are also called *morphologic units*. Viruses that mature at cell membranes may acquire a structure consisting of lipid, protein, and carbohydrate that surrounds and encloses the nucleocapsid, and hence is called *envelope* (*peplos* has also been suggested for this structure but has not been widely adopted). Projections from the surface of a virus particle, especially from the surface of enveloped viruses, are called *spikes* and occasionally *peplomers*. A schematic diagram of three types of virus particles showing some of these structural features is given in Figure 31.

A basic feature of virus morphology is that a virus particle is in many instances composed of numerous identical protein subunits and one or a few molecules of nucleic acid. Also, the shape of a virus particle is usually determined by the virus protein since this comprises most of the mass of the particle, and the configuration and interactions of protein subunits are essentially fixed by the amino acid sequences they possess. A combination of the data issuing from chemical, x-ray, and electron microscopic analyses with principles of symmetry from solid geometry and model building has led to the conclusion that there are two basic designs generally used in nature in the fabrication of virus particles from protein subunits: helical tubes and icosahedral shells (see Horne and Wildy 1961; Caspar and Klug 1962).

In a particle showing helical symmetry, the protein subunits are arranged in a regular helical array perpendicular to the long axis of a particle. This arrangement may result in a tubular structure such as in the tobacco mosaic virus particle (see model, Figure 18 and 7) or a flexuous strand as in the shell of elongated plant and bacterial viruses (for example, potato virus X and coliphage fd, Figure 35) or in the elongated but folded nucleoprotein components of animal viruses such as influenza, vesicular stomatitis, and Sendai viruses (Figure 34A). (Note that the helical nucleoprotein of influenza virus is enclosed in a spheroidal envelope which, though made of repeating units, cannot be readily classified in terms of symmetry.)

Icosahedral symmetry (a form of cubic symmetry) exhibited by many spheroidal virus particles requires that there be specific axes of symmetry (five-, three-, and twofold) about which the particles can be rotated to give a series of identical appearances.

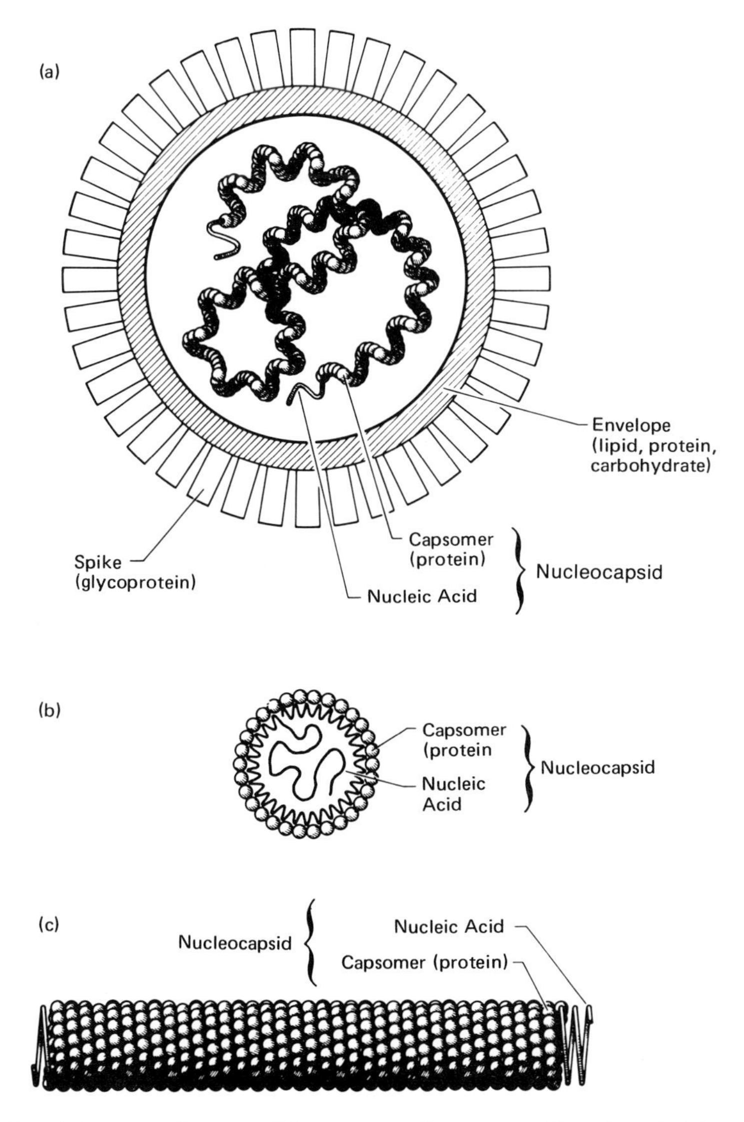

Fig. 31. Schematic diagram of three types of virions. *a*. Enveloped virion with helical nucleocapsid. *b*. Spheroidal virion. *c*. Elongated virion.

It would be an oversimplification to state that the structures of virus particles exhibit either helical symmetry or icosahedral symmetry, for some virus particles have very complex structures. Nevertheless, it is remarkable that the particle structures of many viruses can be interpreted in terms of helical and icosahedral symmetries. Some viruses exhibit both, as for example, a tailed bacteriophage whose head may show icosahedral symmetry and the tail, helical symmetry. In terms of icosahedral symmetry it can be predicted that spheroidal viruses will have specific numbers of morphologic units. Some examples of the classes according to number of morphologic units, and some viruses possibly illustrating the classes, are given in Table 35. The numbers of protein subunits are also given in the table as a reminder that the units visualized in the electron microscope (morphologic units) usually consist of more than one protein subunit. In the Caspar and Klug concept of icosahedral viruses, the protein subunits may be thought to occur in groups of five (pentamers) and six (hexamers), as the

Table 35. Possible Numbers of Morphologic Units and Subunits in Virus Particles Having Icosahedral Symmetry.[a,b]

No. of Morphologic Units	No. of Subunits	Grouping of Subunits in Forming Morphologic Units[c]	Virus Example
12	60	12 pentamers	Coliphage ØX174
32	180	12 pentamers 20 hexamers	Broad bean mottle, cowpea chlorotic mottle, cucumber mosaic, turnip yellow mosaic
42	240	12 pentamers 30 hexamers	Arabis mosaic, tobacco ringspot
72	420	12 pentamers 60 hexamers	Human wart, polyoma, simian virus 40, Shope papilloma
90	180	90 dimers	Tomato bushy stunt, turnip crinkle
92	540	12 pentamers 80 hexamers	Reovirus, wound tumor
162	960	12 pentamers 150 hexamers	Herpes simplex, varicella
252	1,500	12 pentamers 240 hexamers	Adenovirus, infectious canine hepatitis

[a]From Knight 1974.

[b]There are classes of icosahedral particles other than those listed here, but they were omitted for lack of virus examples to illustrate them. See Caspar and Klug (1972) for a detailed discussion.

[c]These groupings of subunits are conceptual and may or may not coincide with the actual situation. For example, coliphage ØX174 seems to have four different protein components rather than 60 copies of one, and the precise numbers and morphologic arrangement of the four proteins remain to be worked out. Similarly, adenovirus has several different protein components, of which the major coat constituent, the hexon, probably consists of three polypeptides, which, moreover, are not identical.

examples in Table 35 indicate. The number of subunits per particle is 60 or some multiple of 60.

Finally, it should be noted that while the concepts of symmetry can be very important in studies of virus fine structure, molecular structures, as Caspar and Klug indicated, are not built to conform to exact mathematical concepts, but rather to satisfy the condition that the system be in a minimum energy configuration. Moreover, with modern techniques of electron microscopy, one can obtain considerable information about virus structures without any knowledge of symmetry in the mathematical sense.

In electron microscopy of viruses, contrast between particle and mount was greatly enhanced by introduction of a shadowing technique (Williams and Wyckoff 1945) in which the particles are coated obliquely with metal vapors in vacuo. This technique is tremendously useful in enhancing the contrast between virus particles and the medium on which the particles are supported, but the metal coating often obscures surface details. An exception is the Shope papilloma virus, shadowed particles of which were observed to show regular arrays of knobs (Figure 32) (Williams 1953b). This appears to represent the first direct observation of morphologic units, each of which is now thought to be composed of five or six protein subunits (see Table 35).

Fig. 32. Micrograph of a cluster of air-dried, uranium-shadowed particles of Shope papilloma virus showing regularly arranged surface knobs. (From Williams 1953b.)

A major advance in visualizing morphologic units as well as other structural features of virus particles occurred with Huxley's (1957) demonstration of the central hole in the TMV particle with a "negative staining" technique. This method, elaborated by Brenner and Horne (1959), was subsequently used extensively by Horne and associates (see Horne 1962) and is now universally employed. It may be briefly described as follows:

A 2 percent solution of phosphotungstic acid (PTA) is brought to neutrality or slightly above by the addition of N KOH. Equal volumes of virus (usually about 10–100 μg/ml in water or ammonium acetate) and PTA are mixed and transferred to a carbonized electron microscope grid from which much of the applied drop is removed with a small strip of filter paper. The grid is allowed to dry and then is examined in the electron microscope. Another method for applying the virus-phosphotungstate mixture is by spraying from an atomizer, giving a very fine mist. The advantage of this technique is that one can get isolated fields (spray droplets), the particles of which are more or less representative of the whole population and are contained within a single field. There are other variations of the technique, including washing of the mounts after application of the virus or virus-phosphotungstate mixture in order to remove excessive salts or small molecules, the virus generally adhering more firmly to the mount than the smaller molecules. Also, uranyl acetate or uranyl formate is sometimes substituted for phosphotungstate, especially if there is any evidence that the virus is unstable in phosphotungstate as is, for example, alfalfa mosaic virus (Gibbs et al. 1963).

With negative staining, the PTA, under the usual conditions, does not adhere specifically to the virus particles as it would in positive staining (which can be done under appropriate conditions). Rather, as the mount dries, the PTA drains down the virus particles and deposits on the particles and on the supporting mount in such a way as to reflect the topography and internal hollow regions of the particles. The micrographs presented here to illustrate the structure of different viruses were made by the negative staining technique.

A. Nonenveloped Spheroidal Viruses

Some spheroidal plant, bacterial, and animal viruses of various sizes are illustrated in Figure 33. Morphologic units are discernible in most of the particles, in some more clearly than in others. The supposed numbers of such units are indicated in Table 35. The viruses illustrated in Figure 33, as is also the case with those shown in Figures 34 to 37, are representative of dozens of other viruses (see Tables 30 to 34 for a partial listing).

Comparison of the particles shown for coliphages ØX174 and Qβ illustrates an interesting difference between these viruses. Both have spheroi-

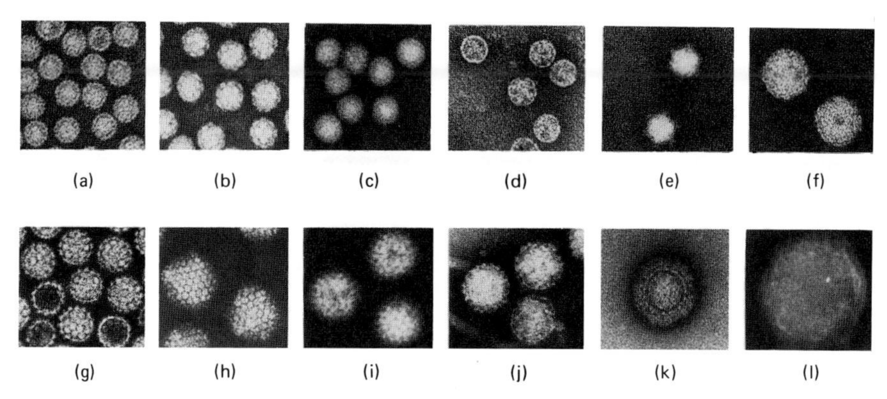

Fig. 33. Some nonenveloped spheroidal viruses. *a*. brome mosaic virus; *b*. turnip yellow mosaic virus; *c*. poliovirus; *d*. Qβ coliphage; *e*. ØX174 coliphage; *f*. cauliflower mosaic virus; *g*. Shope rabbit papilloma virus; *h*. adenovirus-5; *i*. wound tumor virus of sweet clover; *j*. cytoplasmic polyhedrosis virus of the silkworm, *k*. reovirus; *l*. *Tipula iridescent* virus. The virions in the top row are about 30 nm in diameter except for that of cauliflower mosaic, which is about 50 nm; the virions shown on the bottom row range about 50–130 nm in diameter. All mounts were prepared by the negative staining technique (see text). Note the morphologic units exhibited by some virions and especially the apical knobs on the ØX174 particles. (Courtesy R. C. Williams and H. W. Fisher.)

dal particles, but apical knobs are discernible on the ØX174 particles and not in those of Qβ. One or more of the knobs on the ØX174 particles may serve in the specific attachment of this and similar viruses to bacterial cells susceptible to infection by these phages.

Adenoviruses are among the larger viruses (about 80 nm in diameter) and consequently the faces of its icosahedral particles are more clearly evident than in smaller viruses of this shape. The particles have been studied extensively, and it is known that there are 252 morphologic units in the coat protein; these fall into two structural groups termed hexons and pentons. There are 240 hexons (each hexon consists of six protein subunits in a regular cluster) comprising most of the protein coat (capsid) of the virion. The hexons are polygonal discs about 7–8 nm in diameter with a central hole about 2.5 nm across (See Figure 33A) and each hexon is bounded by six other morphologic units. The 12 pentons are situated at the 12 vertices of the icosahedron and each is bounded by five morphologic units. The pentons serve as base structures to which fibers, called penton fibers, are attached. Each penton fiber is about 2×20 nm and terminates in a spherical knob about 4 nm in diameter. These structures are often invisible in the electron dense PTA medium employed in negative staining, but in areas where the PTA matrix is less dense they can be discerned as shown with one particle in Figure 33A. The penton fibers are important in the serologic and hemagglutinating activities of adenoviruses and may also

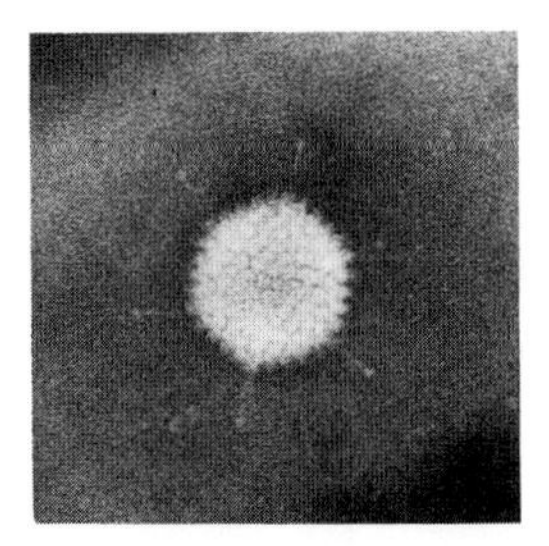

Fig. 33A. An adenovirus virion showing some hexon and penton units (Courtesy R. C. Williams and H. W. Fisher.)

serve as attachment organs in initiating the process of infection (Horne 1973).

It is not uncommon to find particles in preparations of various viruses that, although otherwise closely resembling intact particles in size and shape, are lacking in nucleic acid. Such empty particles exhibit dark centers on electron micrographs, presumably reflecting the ability of PTA to flow readily through empty viral shells and puddle beneath the particles in larger amounts than under complete particles. This is illustrated in the micrograph of Shope papilloma virus in Figure 33. It should be noted in this connection that phosphotungstate may cause a proportion of initially full particles to leak out their nucleic acid (Milne 1972), since a much smaller percentage of empty particles is observed with sensitive viruses when uranyl acetate is employed as the negative stain.

The particles of diplorna viruses such as reovirus are distinctive in containing segmented, double-stranded RNA, as well as for having two protein shells, an outer and an inner one. The morphologic units of both shells are arranged according to icosahedral symmetry. The outer shell appears as a ring in the micrograph of reovirus shown in Figure 33. The outer shell can be digested away with chymotrypsin to leave the inner nucleocapsid, or "core." However, it is not yet clear whether other diplornaviruses have double capsids. The particles of wound tumor virus of sweet clover (WTV) (Figure 33) and of silkworm cytoplasmic polyhedrosis virus (CPV) approximate the size of reovirus cores rather than whole particles and are also similar to cores in possessing RNA transcriptase activity (Lewandowski and Traynor 1972). Of course it is possible that the outer capsids of WTV and of CPV are more readily lost in isolating these viruses than is that of reovirus. This is especially a possibility with CPV, which is usually extracted from polyhedral bodies at rather high pH.

The particles of the iridescent insect viruses (the term "iridescent" comes from the fact that diseased tissues as well as gelatinous pellets of purified virus obtained by centrifugation are iridescent when examined by reflected white light) are the largest presently known nonenveloped viruses; they have a diameter of about 150 nm and are clearly icosahedrons (Williams and Smith 1958). Negatively stained particles, such as those of *Tipula iridescent* virus shown in Figure 33, exhibit a hexagonal outline on micrographs and the protein coat appears as a membranous one- or two-layer structure.

A general point can be made here concerning the relationship between protein and nucleic acid in viruses. There is no evidence for covalent linkage of these substances in any type of virus; nevertheless, the secondary attractions between protein and nucleic acid tend to result in specific configurations. With respect to spheroidal viruses, the nucleic acid is not just randomly packed in a protein shell. For example, x-ray analyses made on turnip yellow mosaic virus (Klug et al. 1966) indicate a regular interlacing of the nucleic acid with the protein subunits far enough below the surface of the particle to protect the nucleic acid from outside degradative agents. The crude sketch of Figure 31 is intended to suggest this relationship as opposed to the simple bag-of-nucleic-acid concept. A similar structure has been deduced for broad bean mottle virus (Finch and Klug 1967). However, specific details of protein-nucleic acid association for spheroidal viruses, as well as for most viruses, are yet to be elucidated.

B. Large, Enveloped Spheroidal and Elongated Viruses

Numerous, large (70 nm in diameter or greater) animal and plant viruses share the feature of maturing at cell membranes (nuclear, vesicular, cytoplasmic) through which they bud, acquiring an envelope structure in the process. The envelope is composed of both host and viral components, the protein tending to be virus specific, while the lipid and perhaps carbohydrate may be characteristic of the host membrane. Quite often, discernible protuberances called spikes (see Figure 31) are apparent in negatively stained preparations of virions. In cases where they have been most thoroughly studied (ortho- and paramyxoviruses), the spikes are rod-shaped structures about 4–5 $\times$ 8–14 nm and appear to be glycoproteins (Compans and Choppin 1973).

Two viruses whose morphologies exemplify numerous spheroidal enveloped viruses are illustrated in Figure 34. They are influenza (a myxovirus) and Rous sarcoma virus (an oncornavirus); some other viruses belonging in these groups are listed in Table 31. Such enveloped viruses tend to be plastic and thus exhibit pleomorphism, which is illustrated in Figure 34 with the influenza virions shown there. While a myxovirus tends to be spheroidal in shape, its nucleoprotein constituent is usually an elongated

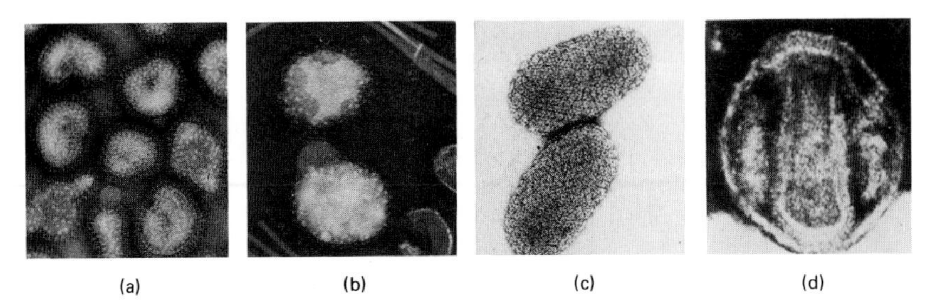

Fig. 34. Three large enveloped viruses and a poxvirus. *a*. influenza virus; *b*. Rous sarcoma virus. *c*. sowthistle yellow vein virus; *d*. vaccinia virus (a poxvirus). Note the pleomorphism and peripheral spikes of the influenza virions. The vaccinia virus particle has been partly stripped with detergent in order to reveal core and lateral bodies. (*a* and *b*, courtesy R. C. Williams and H. W. Fisher; *c*, courtesy D. Peters and *d*, courtesy K. B. Easterbrook.)

structure with helical symmetry that exists in a folded or coiled state within the envelope. A segment of nucleoprotein (nucleocapsid) released from the paramyxovirus, Sendai virus, is illustrated in Figure 34A. This same sort of structure has also been associated with the nucleocapsids of rhabdoviruses.

In both ortho- and paramyxoviruses the nucleocapsid is composed of a single polypeptide species associated with single-stranded RNA in an elongated helical structure similar to that for Sendai nucleocapsid in Figure 34A. However, a basic difference is that the nucleocapsid of paramyxoviruses appears to exist in a single helical structure, whereas that of the orthomyxoviruses may occur in or readily dissociate into several segments (Compans and Choppin 1973). There are hints that such a segmented nucleocapsid may also occur in oncornaviruses (Tooze 1973).

Among the enveloped viruses, herpesviruses are unique in having an icosahedral rather than a helical nucleocapsid. This group of large spheroidal, enveloped viruses has many members, including some with oncogenic properties (see Table 31) (Roizman and Spear 1971).

Another type of large enveloped virus is represented by the group called rhabdoviruses, which has representatives among both animal and plant viruses. Vesicular stomatitis virus, which has bullet-shaped particles (rod with one end rounded and the other planar), is representative of numerous other animal viruses (Table 31) (Hummeler 1971). Such bullet-shaped particles have been observed also with plant viruses, but a bacilliform shape (rod with both ends rounded) seems more characteristic of undegraded particles of these viruses as illustrated by the virion of sowthistle yellow vein virus shown in Figure 34 (see also Knudson 1973). Some other bacilliform viruses in the plant series are listed in Table 34 under potato yellow dwarf, the morphological prototype of this group.

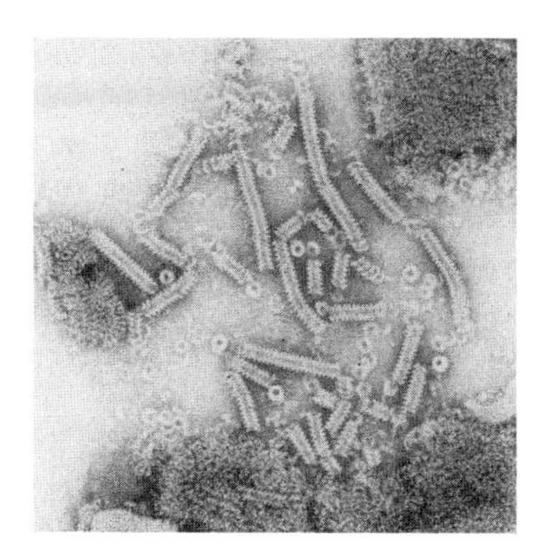

Fig. 34A. A segment of the helical nucleocapsid of Sendai virus. Compare with Fig. 35*c* and 35*d*. (Courtesy R. C. Williams and H. W. Fisher.)

C. Brick-Shaped Viruses

Poxviruses are the largest and most complex of the animal viruses; their virions are usually described as brick- or loaf-shaped. Whether isolated from insects, birds, or mammals (see Tables 31 and 33) (Dales 1973), a basic structural pattern is observed in the virions; they have a highly convoluted, tubular, lipoprotein outer membrane, an internal protein-nucleoprotein core (sometimes called nucleoid), and proteinaceous lateral bodies. These latter features are illustrated by the micrograph of vaccinia virus in Figure 34. In addition to the double-stranded DNA and associated protein, the cores of vaccinia virus enclose four enzymes: a RNA polymerase (transcriptase), a nucleotide phosphohydrolase, and two deoxyribonucleases (DNases)—one an exonuclease and the other an endonuclease. While the function of the lateral bodies is not definitely known, they may serve as inhibitors of the viral DNases since both DNases show elevated activities if the lateral bodies are removed from cores by treatment with a proteolytic enzyme (Dales 1973). Thus the lateral bodies could restrain the activity of the DNases in the vaccinia virions but upon removal during the course of infection might release them to attack host cell DNA.

D. Elongated Viruses

The two basic elongated structures of virions observed thus far are tubular and filamentous particles. They have been noted for several bacterial and plant viruses (Table 32 and 34). Two examples of each type of structure are shown in Figure 35.

Tobacco mosaic virus is the best known and most thoroughly studied rod-shaped virus. The structure of TMV virions was rather well understood by the time negative staining was developed, so this technique only served to confirm the morphology already established by chemical and x-ray studies. Since the approach used for TMV is a classical one for deducing structure of rodlike particles, it will be briefly sketched here.

There was evidence that TMV protein was a single species that occurred in about 2,000 identical subunits (molecular weight about 18,000) per virion of 40×10^6 daltons (Harris and Knight 1955). It was further known that the RNA of TMV was a single-stranded molecule with a molecular weight of about 2×10^6 and about 3,300 nm long, which ran the length of the TMV rod (Hart 1958; Gierer 1957). Important information missing at this time were the arrangement of the protein subunits and the spatial relationship that protein and nucleic acid took with respect to one another. This was supplied by study of the low-angle x-ray scattering patterns yielded by concentrated gels of purified TMV (Watson 1954; Franklin et al. 1957, 1959; Caspar 1956). The x-ray data indicated that the protein subunits of the virus are arranged in a helical array about the long axis of the virus rod; that there is a central hole about 4 nm in diameter so that the rod is actually a tube; that there are regions of high and low density in the particle at specific radii; and, by comparison of radial density distributions of complete and nucleic acid-free particles, that the nucleic acid is not in the center of the tube but is intermeshed with the protein subunits at a radius of about 4 nm. Some of these points are evident from the radial density distribution diagrams shown in Figure 36. As indicated in the figure, density distribution curves similar to that of TMV were also obtained with three strains of this virus and for cucumber virus 4; it will be noted that the curves all show maxima at the same radii and differ mainly in

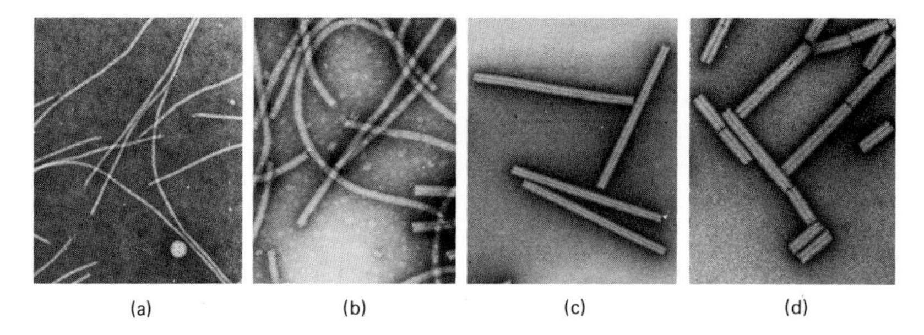

Fig. 35. Some elongated viruses. *a*. coliphage fd; *b*. potato virus X; *c*. tobacco mosaic virus; *d*. tobacco rattle virus. The particles of phage fd and of potato virus X are too long to be shown in their entirety at the magnification used here. (Courtesy R. C. Williams and H. W. Fisher.)

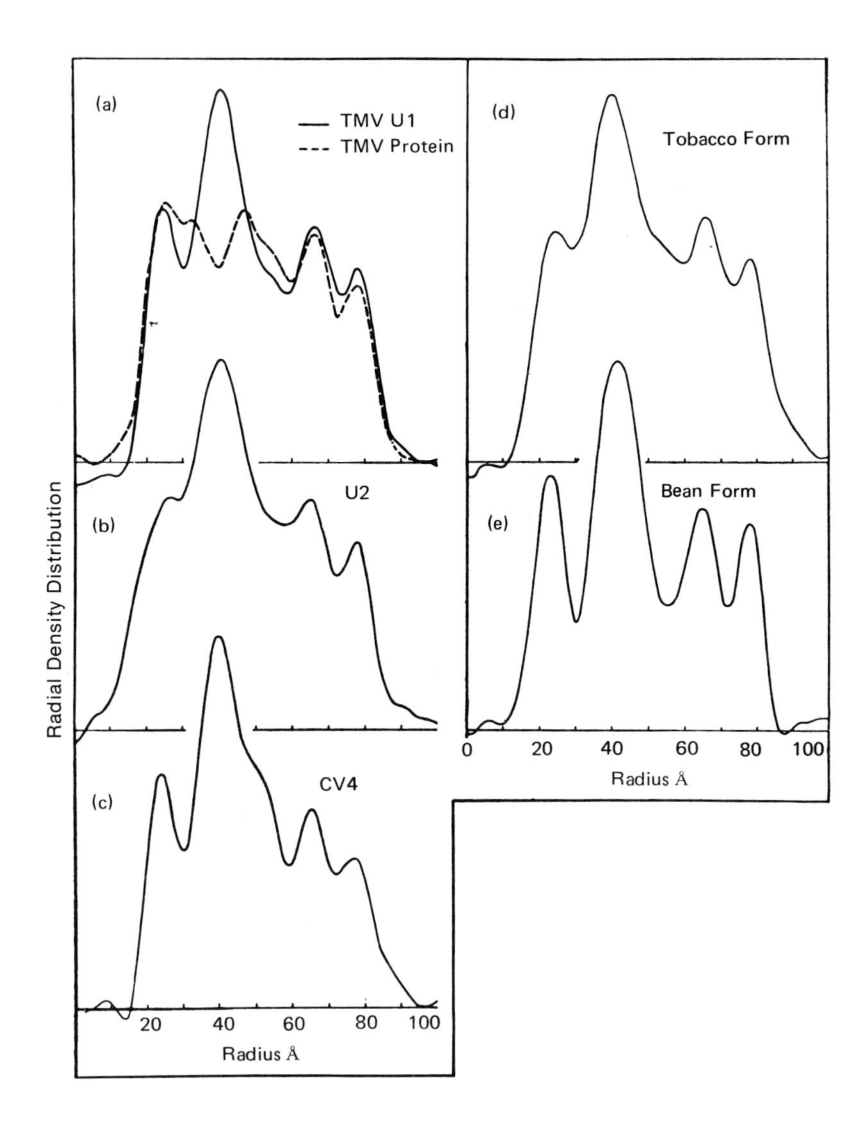

Fig. 36. The cylindrically averaged, radial, electron density distributions of tobacco mosaic virus, some of its strains, cucumber virus 4 (CV4) and TMV protein. The curves show the difference between the electron density of the particles and that of water plotted as a function of radial distance from the particle axis. TMV U1 is common TMV (also called vulgare and wild type) and U2 is a mild strain of TMV (Siegel and Wildman 1954). The strains represented in (*d*) and (*e*) originated in Nigerian cowpea (Bawden 1958). (From Klug and Caspar 1960; see also Caspar 1956; Franklin et al. 1957.)

quantitative respects, which probably represent slight differences in packing of material.

Putting all the evidence together, a model of the TMV particle can be constructed illustrating the helical arrangement of protein subunits in the TMV shell and the manner in which the RNA strand intermeshes with the protein subunits and assumes the helical configuration of the subunits (Figures 7 and 18).

In Figure 35 the central hole is evident in the virions of TMV and of tobacco rattle virus. Cross striations also delineate the helical array of subunits in all of the elongated virions. It will be noted that the filamentous viruses exhibit flexuous shapes rather than the straight form shown by the elongated viruses with greater cross-sectional diameters.

As with other viruses, including the isometric ones, the protein subunits of elongated viruses are associated with the nucleic acid by noncovalent bonds. However, in some cases the stability of this structure is very great; for example TMV has been reported to retain infectivity in extracts at room temperature for 50 years (Silber and Burk 1965).

E. Tailed Viruses

Some bacterial viruses are characterized by spheroidal particles, some by filamentous particles, and many are combinations in which head and tail structures are evident. In the latter case, head capsid may exhibit icosahedral symmetry and the tail helical symmetry. Among the tailless phages there is at least one known, *Pseudomonas* PM2, which has a lipoprotein envelope (Espejo and Canelo 1968; Silbert et al. 1969) that appears to fit closely around an icosahedral capsid. This phage is unusual also in being the only tailless phage possessing double-stranded DNA (which happens to be circular).

The head sizes of different tailed phages vary considerably and the shape ranges from almost spherical to oblong. The head houses the nucleic acid (apparently always double-stranded DNA), while the tail serves as an attachment organ in the initial step of infection and a tube through which the DNA travels in a subsequent step (penetration). Some tails are short, some long, some straight, and some curved; they vary tremendously in complexity, especially with regard to possession or not of accessory structures such as collars, base plates, spikes, tail fibers, and so on. Many of these features are illustrated in Figure 37 and characterize numerous phages, some of which are listed in Table 32.

Tailed viruses have also been observed as infectious agents of blue-green algae (Table 30); two of these are illustrated in Figure 37. The N-1 algal virus (Adolph and Haselkorn 1971), as can be seen in the figure, resembles long-tailed bacteriophages, especially those with contractile sheathed tails. The SM-1 algal virus (MacKenzie and Haselkorn 1972)

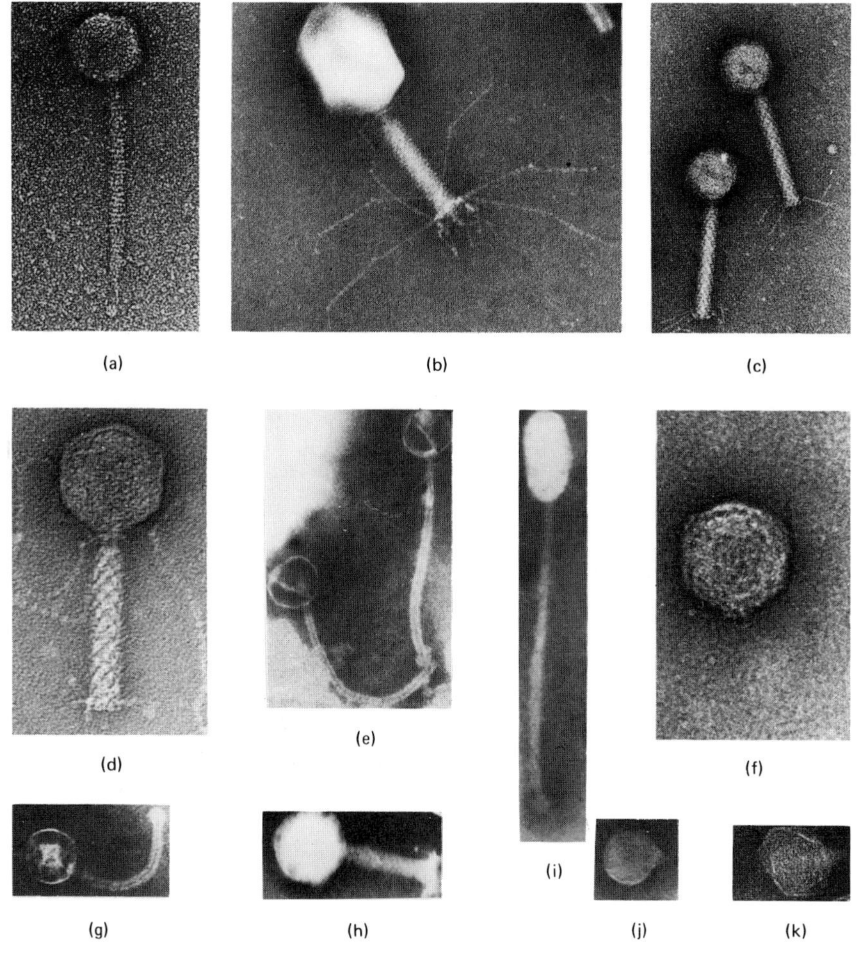

Fig. 37. Some tailed bacteriophages and algal phages. *a*. coliphage lambda; *b*. coliphage T4; *c*. coliphage P2; *d*. N-1 algal virus (from *Nostoc muscorum*); *e*. *staphylococcus* phage 77; *f*. SM-1 algal virus; *g*. *pseudomonas* phage Pc; *h*. typhoid phage Vi 1; *i*. *staphylococcus* phage 6; *j*. a *brucella* phage; *k*. coliphage T7. Mounts were all prepared for electron microscopy by the negative staining technique. (*a, b, c,* and *k*, courtesy R. C. Williams and H. W. Fisher; *d* and *f*, courtesy R. Haselkorn; and the rest, courtesy D. E. Bradley and D. Kay.)

resembles the short-tailed phages; it has an icosahedral head capsid from which there protrudes a collar and a short appendage that could be a tail.

F. Encapsulated Viruses

There are two morphologically different classes of insect viruses that may be called occluded or nonoccluded, depending on whether they typically appear in their mature form in special inclusion bodies or not (Table 33). The nonoccluded virus of *Tipula paludosa* (the crane fly or daddy longlegs), called *Tipula iridescent* virus, is illustrated in a micrograph in Figure 33. However, most insect viruses appear to occur in their mature form in characteristic inclusion bodies. These inclusion bodies are generally crystalline protein packages that contain one or more virus particles. Some of these packages are called polyhedral bodies, and are found characteristically in either nuclei or cytoplasm of infected cells; the diseases associated with them are correspondingly termed nuclear polyhedroses and cytoplasmic polyhedroses. The occluded virions of nuclear polyhedroses are generally rod-shaped, while those of the cytoplasmic polyhedroses are spheroidal and have icosahedral capsids. Hundreds of virus particles are occluded in the crystalline protein matrix of each polyhedral body whether nuclear or cytoplasmic; they can be released by treatment with dilute alkali. For example, the cytoplasmic polyhedrosis virus of the silkworm is released from polyhedra by holding the polyhedral bodies at 25° in 0.1 M NaCl and 0.05 M Na_2CO_3 at pH 10.6 for 1 hr (Lewandowski et al. 1969).

In the insect diseases called granuloses, the inclusion bodies are called granules or, more frequently, capsules. Some distinctions between capsules and polyhedral bodies are

1. Shape of the inclusion bodies: polyhedral bodies occur in a variety of shapes depending on the polyhedrosis involved and have been described as dodecahedral, tetrahedral, rectangular, hexagonal, and crescent-shaped; capsules are usually described as ovoid or egg-shaped in outline although some cubic capsules have been reported.
2. Size: the polyhedral bodies vary in size both in the same and in different polyhedroses but in general they are much larger than capsules and range from 500 to 15,000 nm in diameter, whereas the range of sizes of capsules is more of the order 119 to 350 nm wide and 300–511 nm long.
3. Number of virus particles occluded: hundreds or thousands of virions may be found in polyhedral bodies but on the average only one virion occurs in a capsule.

Thin sections can be made of polyhedral viruses and capsules which upon electron microscopy reveal the dispersion of virus particles (Figure 38). Two concentric membranes can be observed surrounding each virus particle in nuclear polyhedral bodies and capsules but not in cytoplasmic polyhedral bodies. The membranes, when present, are termed inner or intimate membrane (next to the virion) and outer membrane. Their precise functional relationship to the virions is not yet clear. A cross section of a nuclear polyhedral body showing the occluded cabbage looper virus particles and a similar section of a capsule showing a meal moth virus are shown in Figure 38.

As indicated in Table 33 there are at least two other types of occluded insect viruses. One of them occurs in inclusion bodies called spherules and the occluded virus appears to be a poxvirus (Bergoin and Dales 1971); the other is a beetle virus found in peculiar spindle-shaped or ovoid inclusions (Vago and Bergoin 1968).

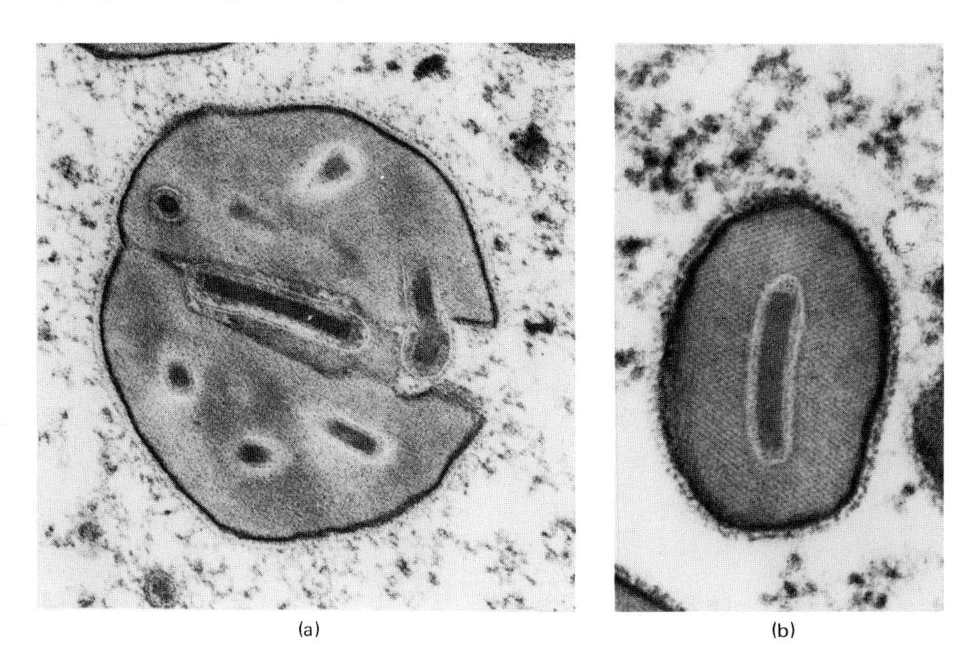

Fig. 38. Two types of occluded insect viruses. *a*. Thin section of a polyhedral body from the nuclear polyhedrosis of the cabbage looper (*Trichoplusia ni*). Bits of the randomly oriented viral rods are apparent in the section with a complete rod discernible in the center of the section. *b*. Thin section from a capsule of the granulosis of the meal moth (*Plodia interpunctella*) showing the single virus particle embedded in a crystalline capsule. (*a*, courtesy M. D. Summers; *b*, courtesy H. J. Arnott and K. M. Smith.)

V.

Action of Chemical and Physical Agents on Viruses

The reaction of viruses with chemical and physical agents seemed rather complex and mysterious some years ago when the structure of viruses and the basic features of the process of infection were poorly understood. Now, the details may still be intricate and incompletely defined, but the main facts relating chemical and physical treatments of viruses to the biological activity of viruses are simple and clear:

In order for a given infectious virus particle to remain fully infectious, the chemical structure of its nucleic acid must not be irreversibly harmed and the nucleic acid must be capable of release from the virion in a form that can react normally with transcription-translation apparatus (enzymes, attachment factors, and so on). A prediction of this formulation is that it should be possible to inactivate viruses in two general ways: (1) by changes in the nucleic acid that render it partly or wholly nonfunctional in the central dogma scheme (self-replication, transcription, translation) or (2) by alteration of the protein coat or other structures of the virion (for example, tail fibers of a phage, RNA polymerase of a poxvirus) in such a way as to prevent delivery of the viral nucleic acid into a functional area of the cell. Both types of inactivation have been detected and both can be caused by heat, radiation, and chemicals.

In addition to the two general types of inactivation that chemical and physical treatment of viruses can produce, they can also produce noninactivating, heritable changes in the nucleic acid. This is called mutation, which will be considered below. Actually, moderate treatments with some agents cause mutation while harsher treatments cause inactivation; in practice, a combination of the two effects is often observed.

A. Inactivation of Viruses

The primary characteristic that makes a virus a virus is its infectivity, that is, its ability to cause the production of progeny like itself. Therefore, the term "inactivation of viruses" is used here to mean the abolishment of infectivity even though it is possible for chemical and physical treatments to affect other characteristics of viruses instead of, or as well as, infectivity. Conversely, it is often possible to eliminate infectivity without destruction of antigenic or serologic reactivity (a fact used in the production of one type of vaccine), and virus particles that have lost the capacity to reproduce fully sometimes still can induce virus inhibitory substances (interferons), cell fusion, enzyme production, and oncogenic transformation of cells (Potash 1968; Kleinschmidt 1972; Watkins 1971a, 1971b; Mathews 1971; Kajioka et al. 1964; Rubin 1965). These examples are explicable on the basis that many of the properties of viruses are expressed by their protein coats and/or envelopes, and that partial transcription (or translation) of viral genomes is known to occur. Thus the nucleic acid of a virion can be altered to the point that production of complete virus is blocked but several viral functions can still be performed.

When inactivation of viruses is reduced to the two simple terms outlined above—alteration of the exterior of the virion or of the nucleic acid—much of the earlier mystery of inactivation is dispelled, and specific explanations for the effect of heat, radiations, chemicals, and so on can be sought in a logical, systematic way. There are practical consequences of this view too. For example, if one is interested in producing a vaccine constituted from an inactivated virulent strain of virus, it is clear that selection of an agent that will attack mainly the nucleic acid is better than an agent that inactivates by altering the viral attachment sites. First, the type of agent that mainly attacks nucleic acid preserves more of the antigenic structure of the exterior of the virus and hence should elicit a better immune response. But, equally important, the nucleic acid is inactivated in this case, whereas when inactivation ensues from altering virus surface structure, the nucleic acid may be intact and could by some fortuitous circumstances (for example, local changes in pH or salt concentration, presence of a latent helper virus) be released from its shell to initiate a devastating infection.

To carry the example further, even use of an agent such as formaldehyde, which can get through the virion superstructure and react efficiently with the nucleic acid, does not guarantee inactivation of the virus to a safe level. Formaldehyde reacts with proteins as well as with nucleic acids and prolonged reaction causes cross-linking between various groups of the protein (so-called tanning or membrane forming effect) (Fraenkel-Conrat and Olcott 1948). Consequently, it can become more difficult for formaldehyde to penetrate the viral shell, and Gard (1957) demonstrated with poliovirus that as inactivation with formaldehyde proceeds, the residual infectivity of the virus becomes progressively more resistant to inactivation. Another point in respect to the inactivation of viruses is that there are some ways of inactivating viruses that are reversible in the cell, such as inactivation with moderate doses of ultraviolet light (see section on Inactivation of Viruses by Radiations).

A uniformity in the kinetics of inactivation of all types of viruses was recognized some time before it was clear that there were two general modes of inactivation of viruses (Luria and Darnell 1967, Chap. 7). Thus, regardless of whether the inactivating agent is chemical or physical, loss of infectivity often follows an exponential decay law based on the simple first-order reaction formula

$$V/V_0 = e^{-kt}$$

where V is infectivity at time t, V_0 is initial infectivity, and k is the rate constant. If the logarithm of infectious titer is plotted against time, a straight line is observed (Figure 39). This means that a constant fraction of the virions undergoes a change, causing loss of infectivity in each unit of time, and that one such independent change is sufficient to inactivate a virion (one-hit theory). In viral structural terms, infectivity can be lost, for example, by a single change in a viral protein involved in the initiation (for example, attachment) step of infection, by a single change in the RNA

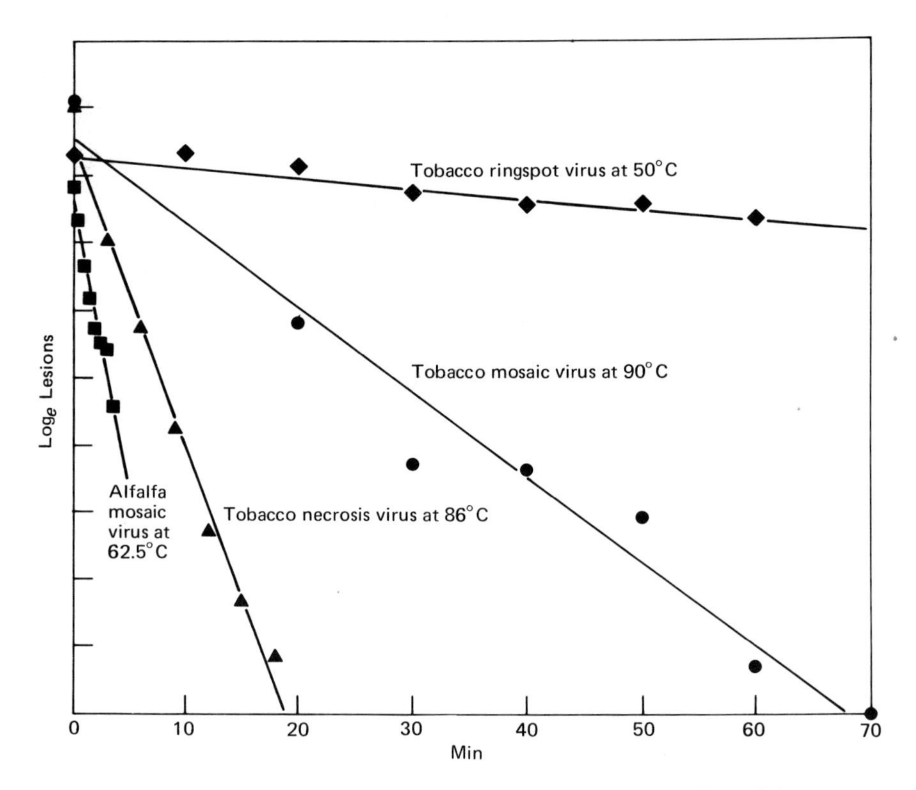

Fig. 39. Thermal inactivation rate of some plant viruses. (Adapted from Price 1940.)

polymerase of a virion containing such an enzyme, or a single change in the viral nucleic acid. The same type of kinetics applies to the production of mutants if the rule is restricted to events affecting the viral nucleic acid.

1. *Inactivation of Viruses by Heat*

The infectious quality of most viruses persists quite well in the cold and especially well below freezing such as at the temperature of dry ice (−70°C) or of liquid nitrogen (−196°C). However, there is a great variation in the lability of different viruses at elevated temperatures. At one extreme the infectious half-life of enveloped viruses such as the myxoviruses and RNA tumor viruses may be only an hour at 37°, while at the other extreme some plant viruses such as tobacco mosaic virus are known to have maintained some infectivity stored in plant sap at about 20°–25°C for 50 years and can stand 80°–90° for 10 min with only a moderate loss of infectivity (Silber and Burk 1965; Price 1940).

Much of the loss of infectivity of viruses at temperatures between 25° and 70°C can doubtless be attributed to changes in protein components of the viruses since proteins in general are more readily denatured by heat than are nucleic acids. Furthermore, nucleic acids, located as they always are in the interior of virions, receive substantial protection from thermal and other degradative assaults. Above 70°C, strand separation occurs with double-stranded nucleic acid and finally random breaks occur in the sugar-phosphate backbone of all types of nucleic acids. Thus, at high temperatures both protein and nucleic acid can be seriously and often irreversibly damaged. In addition, the lipid-containing envelopes of some viruses are doubtless degraded at moderate to high temperatures.

In the context of the process of infection, heat-induced changes in configuration of viral surface protein (including glycoprotein and lipoprotein) could be expected to result in one or more of the following consequences: (1) prevent the specific attachment of virions to cell receptor sites (and thus block the initial step of infection); (2) inactivate virion-associated enzymes needed for virus replication; (3) hinder removal of coat protein and release of viral nucleic acid.

It should be noted that heat inactivation is modulated by other environmental conditions such as the presence of extraneous proteins and divalent cations such as Mg^{++} and Ca^{++}. Conversely, the presence of protease and nuclease enzymes can be expected to increase inactivation observed with heat, especially at moderate temperatures.

2. *Inactivation of Viruses by Radiations*

In their effect on matter, electromagnetic radiations can be considered to fall into two classes: ionizing and nonionizing. Gamma rays and x-rays are common ionizing radiations used in virology, while ultraviolet light is the predominant nonionizing radiation employed. Ionizing radiations provide much higher energy than nonionizing radiations; for example, 0.1 Å x-rays have about 25,000 times the quantum energy of ultraviolet light at 2,537 Å. However, ionizing radiations show little selectivity, their effect on a virus being governed by little more than the atomic density encountered in various portions of the virion. In contrast, the purine and pyrimidine rings of nucleic acids strongly absorb nonionizing ultraviolet light (proteins absorb this radiation too, but much more weakly and mainly in proportion to their aromatic amino acid content). The main inactivating effect of either type of radiation is assumed to be in the viral nucleic acid. Reviews of techniques for using ionizing and nonionizing radiations in virology have been made by Ginoza (1968) and Kleczkowski (1968).

Both types of radiations inactivate viruses by direct effects, and ionizing radiations may also inactivate by indirect effects stemming from the production in aqueous media of hydrogen and hydroxyl-free radicals and of peroxides (Luria and Exner 1941; Watson 1950). An example of the indirect effect is that phages can be inactivated by free radicals, the mechanism appearing to be a radical attack on the phage tail causing a premature release of DNA from the phage head (Dewey 1972). The direct inactivating action of an ionizing radiation such as x-rays involves rupture of covalent bonds in protein or, more likely, with moderate doses, in nucleic acid (Freifelder 1965, 1966; Summers and Szybalski 1967; Lauffer et al. 1956). In the latter instance, strand scission is more serious with single-stranded nucleic acid than with double-stranded nucleic acid since rupture of complementary strands in the same vicinity is required in order to disrupt double helical structures, whereas a break anywhere severs single-stranded nucleic acid. This is reflected in the approximately tenfold greater efficiency of x-rays in inactivating viruses containing single-stranded nucleic acid as opposed to those containing double-stranded nucleic acid.

In contrast to ionizing radiations, ultraviolet light can inactivate viruses without breaking polynucleotide chains (chain breaks, which also are inactivating, can occur at doses of 10^4 ergs/mm^2 or higher). Inactivation by moderate doses of ultraviolet light appears to be attributable to one or more of three observed effects of ultraviolet light on pyrimidines in polynucleotide chains (McLaren and Shugar 1964): (1) covalent bonds may be formed between adjacent thymine residues in DNA or between uracil residues in RNA to form so-called thymine dimers or uracil dimers, respectively; (2) hydration may occur at the C5-C6 double bond of pyrimidines to form 5-hydro-6-hydroxy derivatives (these are thought to be especially important in the inactivation of RNA-containing viruses); and (3) some cross-linking may occur, probably involving pyrimidines, between complementary chains of double-stranded nucleic acids.

Thymine dimer formation may be shown as follows, the horizontal straight line representing the sugar-phosphate backbone of a strand of nucleic acid containing adjacent thymine residues:

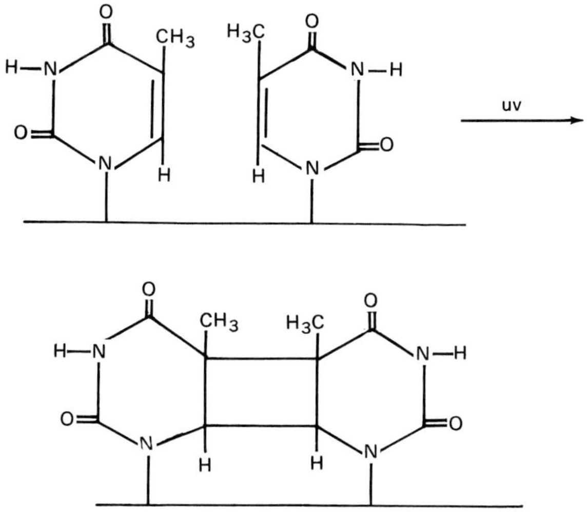

The formation of a uracil hydrate may be represented as follows:

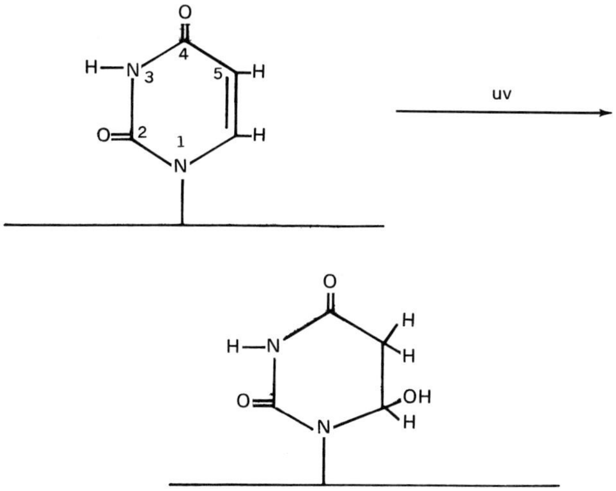

It has been shown that dimers can be removed from bacterial DNA by two types of enzymatic mechanisms, and it is supposed that these may also function in plant and animal cells since reversal of ultraviolet radiation damage has been observed under conditions analogous to those used to accomplish reversal in bacteria. One reactivation system is photoreactivation (Kelner 1949; Dulbecco 1950; Setlow 1968) in which light at 3,500–4,500 Å in concert with an enzyme is able to split thymine dimers (Setlow 1968; Howard-Flanders 1973). Photoreactivation by such a light-activated enzyme system may not occur with some RNA-containing viruses such as tobacco mosaic virus, apparently because of interactions between protein and nucleic acid; however, infectious RNA from TMV can be photoreactivated following treatment with ultraviolet light (Bawden and Kleczkowski 1959; Rushizky et al. 1960).

The second type of enzymatic reversal of ultraviolet damage is called dark repair (Setlow 1968; Howard-Flanders 1973). The enzyme system involved in this is not light activated and operates by excision of the thymine dimer and a few adjacent nucleotides. Other enzymes (DNA polymerase and polynucleotide ligase) replace the excised nucleotides restoring the strand to its state before the ultraviolet treatment.

Another type of radiation inactivation is the photodynamic action that accompanies exposure of viruses to certain dyes in the presence of visible light. Toluidine blue, methylene blue, flavins, and acridines are examples of dyes that have been employed. Since oxygen is required for, and reducing agents protect against, the effect, it appears that photodynamic inactivation is dye-mediated photooxidation (Oster and McLaren 1950; Appleyard 1967; Orlob 1967; Schaffer and Hackett 1963; McLaren and Schugar 1964). In some cases, it appears that an important effect of photodynamic action is modification of guanine residues of nucleic acids (Singer and Fraenkel-Conrat 1966).

3. *Inactivation of Viruses by Chemicals*

The main chemical agents that inactivate viruses were known many years before the detailed structures of virions and the mechanisms of virus infections were known (see review by Stanley 1938). They include enzymes, protein denaturants, oxidizing agents, acids and bases, and agents affecting primary amino groups such as formaldehyde and nitrous acid. The mechanism of inactivation by each of these agents can be readily predicted on the basis of the principle stated earlier: that infectivity may be lost by damage to the exterior of the virion of a sort that interferes with the early stages of infection (attachment, penetration, and so on) or by injury to the viral nucleic acid that prevents its complete functioning. Some treatments can cause both types of damage. Knowledge of the vulnerability of the major constituents of virions to various chemical agents is also important in predicting which mechanism is probably involved in inactivation of a given virus. Some pertinent observations of this sort follow.

a. Enzymes

Virions of different types vary in their resistance to inactivation by enzymes depending on individual structural characteristics.

Phospholipases often inactivate enveloped viruses (Franklin 1962; Simpson and Hauser 1966; Cartwright et al. 1970) by attacking the phospholipid components of the envelope and presumably disorganizing the structure required for attachment and penetration. Likwise, treatment of enveloped viruses with proteases removes glycoprotein spikes with concomitant loss of infectivity, which is probably also associated with inability of such altered viruses to attach to cells (Cartwright et al. 1970; Chen et al. 1971).

Nonenveloped viruses tend to be quite resistant to inactivation by proteases (trypsin, chymotrypsin, pepsin, papain, bromelain, pronase, and so on) unless the coat protein is denatured by some means. Poliovirus illustrates such resistance to enzymes (and to a range of pH values) by passing unscathed through the human digestive tract to establish infection in intestinal cells. However, another picornavirus similar to poliovirus, coxsackievirus type A9, does appear susceptible to partial degradation by the enzyme pronase (Herrmann and Cliver 1973). The coat proteins of some other simple viruses can be somewhat degraded by proteases and yet retain infectivity. For instance, the C-terminal threonine residues can be removed by carboxypeptidase from all of the coat protein subunits of tobacco mosaic virus without reducing infectivity of the virus (Harris and Knight 1955); and a peptide of over 2000 daltons can be digested off the coat protein subunits of potato virus X by trypsin without loss of infectivity (Tung and Knight 1972b). However, some nonenveloped viruses, such as poxviruses and reoviruses, are inactivated by proteolytic digestion. These viruses possess outer shells that are susceptible to degradation by proteases but their cores are resistant (Joklik 1966; Dales and Gomatos 1965). Even though cores (which contain the viral nucleic acid) may penetrate cells, they appear unable to carry out virus replication.

It may be noted that a kind of pseudo-inactivation can occur in which infectivity of a virus is reduced simply by formation of a complex between virus and enzyme (Stanley 1934a; Loring 1942; Arimura 1973). Upon dissociation of such complexes, infectivity is recovered. Inactivation can also be a secondary effect of proteolytic attack on the exterior of virions. For example, viruses in which removal of some coat protein exposes the viral nucleic acid to nucleases will be inactivated even when coat protein is not involved in initiating infection (for example, plant viruses).

The nucleic acids of most viruses are fairly well shielded from degradation by nucleases by coat proteins, envelopes, or both. There appear to be a few exceptions in which the coat protein subunits have a loose structure with spaces between them large enough for enzyme to penetrate. A strain of the small spheroidal cucumber mosaic virus is an example of this kind; it loses infectivity in low concentrations of ribonuclease at pH 7.2 (Francki 1968). Obviously, if the protein coat of any virus is breached by some physical or chemical agent, the nucleic acid is liable to attack by nucleases.

b. Protein Denaturants

Some chemical reagents are deliberately used to strip the protein coat from viruses in order to isolate the nucleic acid or to determine the number and nature of polypeptides in a virion. When the purpose is to isolate the nucleic acid for further studies, precautions must be taken to avoid degradation by nucleases (deoxyribonucleases and ribonucleases for DNA and RNA, respectively). Phenol is the most commonly used reagent to dissociate virions for the isolation of nucleic acid, although the anionic detergent, sodium dodecyl sulfate (SDS), is also employed, or the two together. This has been discussed in the section on preparation of nucleic acids.

SDS is also used to dissociate the proteins of virions into their constituent polypeptide chains which then can be separated as SDS-protein complexes on acrylamide gels as described in an earlier section.

SDS is effective in solubilizing viral envelopes and hence can be used to disaggregate the more complex virions as well as the simple ones. Nonionic detergents (for example, Nonidet-P40, Tween 20, Tween 80, Sterox SL) and sodium deoxycholate, another anionic detergent, are also used to disrupt enveloped viruses (Appleyard et al. 1970; Nermut 1972; Webster and Darlington 1969; Stromberg 1972). They preserve the structure of protein constituents better than ether, chloroform, and some other organic solvents. A list of useful detergents in virology is given in Table 36.

Guanidine and urea are good reagents for disrupting hydrogen bonds, which are abundant in virus particles and important in maintaining their structure. Practically all proteins are denatured by these reagents, often irreversibly.

c. Nitrous Acid

Nitrous acid can react with the primary amino groups of either protein or nucleic acid. The primary amino groups in proteins other than those at the N-terminal of the polypeptide chain, are the epsilon amino groups of lysine residues. Consequently the degree of reactivity of viral proteins with nitrous acid and the occurrence of one type of inactivation will depend considerably on the lysine content of the viral coat protein(s). In contrast, there are thousands of amino groups per molecule of any viral nucleic acid since three out of four of the constituent purines and pyrimidines have one. In the case of isolated viral nucleic acid, the accessibility of these amino groups to nitrous acid will be governed by the degree of hydrogen bonding present between complementary bases. This factor also applies to the reaction between nitrous acid and nucleic acid located inside virions to which must be added the possible complications of protein-nucleic acid interactions and the ability of nitrous acid to penetrate virions. In any case, many of the amino groups of isolated nucleic acids as well as those of nucleic acids in virions undergo reactions with nitrous acid that cause inactivation or mutation.

One of the the first reactions of infectious RNA to be studied was the effect of nitrous acid on infectivity (Schuster and Schramm 1958). The chemical reaction expected is the classical van Slyke (1912) oxidative deamination of primary amines: $R\text{-}NH_2 + HNO_2 \rightarrow R\text{-}OH + N_2 + H_2O$. The constituents of RNA containing primary amino groups are adenine, guanine, and cytosine (the same bases are involved in DNA with the exception that 5-hydroxymethylcytosine is substituted for cytosine in the T-even phages), and these deaminate as shown below to give hypoxanthine, xanthine, and uracil, respectively:

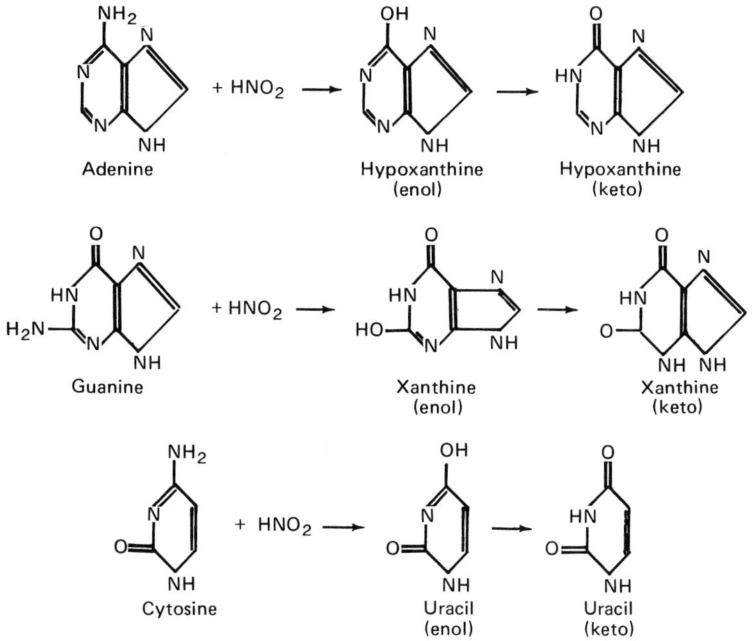

Adenine + HNO_2 → Hypoxanthine (enol) → Hypoxanthine (keto)

Guanine + HNO_2 → Xanthine (enol) → Xanthine (keto)

Cytosine + HNO_2 → Uracil (enol) → Uracil (keto)

Schuster and Schramm (1958) showed that the deamination products indicated above were produced in TMV-RNA by treatment with nitrous acid between pH 4.1 and 4.3. These products were identified by removing samples of RNA at various times during the nitrous acid treatment, hydrolyzing, separating the products by paper chromatography, and determining the quantity of each product on the chromatograms by elution and

Table 36. Some Surfactants (i.e., Surface-active Agents or Detergents) Representative of Groups Used in Disassembling Virions.[a]

Category	Name	Chemical Formula
Anionic	1. Sarkosyl NL-97 (An N-acyl sarcosinate)	$CH_3(CH_2)_{10}C(=O)-N(CH_3)-CH_2-COO^- \; Na^+$
	2. Sodium deoxycholate (A steroid)	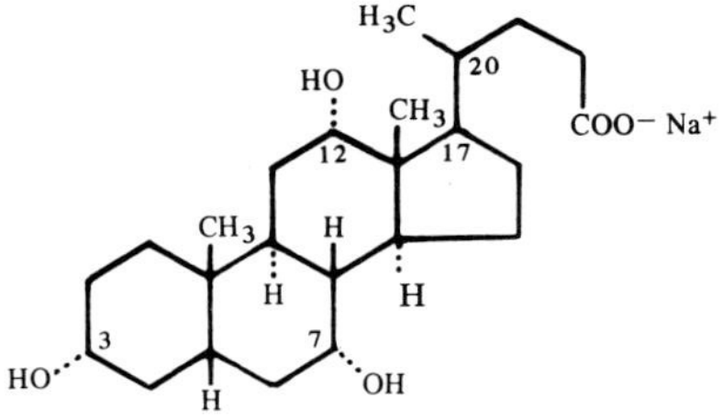
	3. Sodium dodecyl (lauryl) sulfate (An alkyl hydrogen sulfate)	$CH_3(CH_2)_{10}CH_2OSO_3^- \; Na^+$
Cationic	1. Cetab (An alkyl quaternary ammonium salt)	$(CH_3(CH_2)_{15}N(CH_3)_3)^+ \; Br^-$
	2. Hyamine 1622 (An alkyl aryl quaternary ammonium salt)	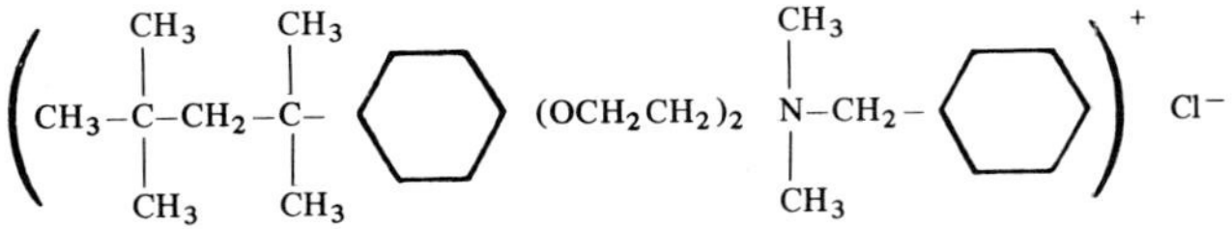

Nonionic	1. Alfol 16-18 (An alkylpolyethoxy alcohol)	$CH_3(CH_2)_{15\text{-}17}(OCH_2CH_2)_{12}OH$
	2. Neodol 25-12 (An alkylpolyethoxy alcohol)	$CH_3(CH_2)_{11\text{-}14}(OCH_2CH_2)_{12}OH$
	3. Sterox 67-K (An alkylpolyethoxy alcohol)	$CH_3(CH_2)_{13\text{-}14}(OCH_2CH_2)_{11}OH$
	4. Trycol TDA-12 (An alkylpolyethoxy alcohol)	$CH_3(CH_2)_{12}(OCH_2CH_2)_{12}OH$
	5. Nonidet P-40 (An alkylarylpolyethoxy alcohol)	Same as Triton X-100
	6. Triton X-100 (An alkylarylpolyethoxy alcohol)	$CH_3(CH_2)_7C_6H_4(OCH_2CH_2)_{9\text{-}10}OH$
	7. Tween 80 (Polyethoxy sorbitan monooleate)	$CH_2-OOC(CH_2)_7CH{=}CH(CH_2)_7CH_3$; CH; $CH(OCH_2CH_2)_x-OH$; $CH(OCH_2CH_2)_y-OH$; CH; $CH(OCH_2CH_2)_z-OH$ (ring closed through O)

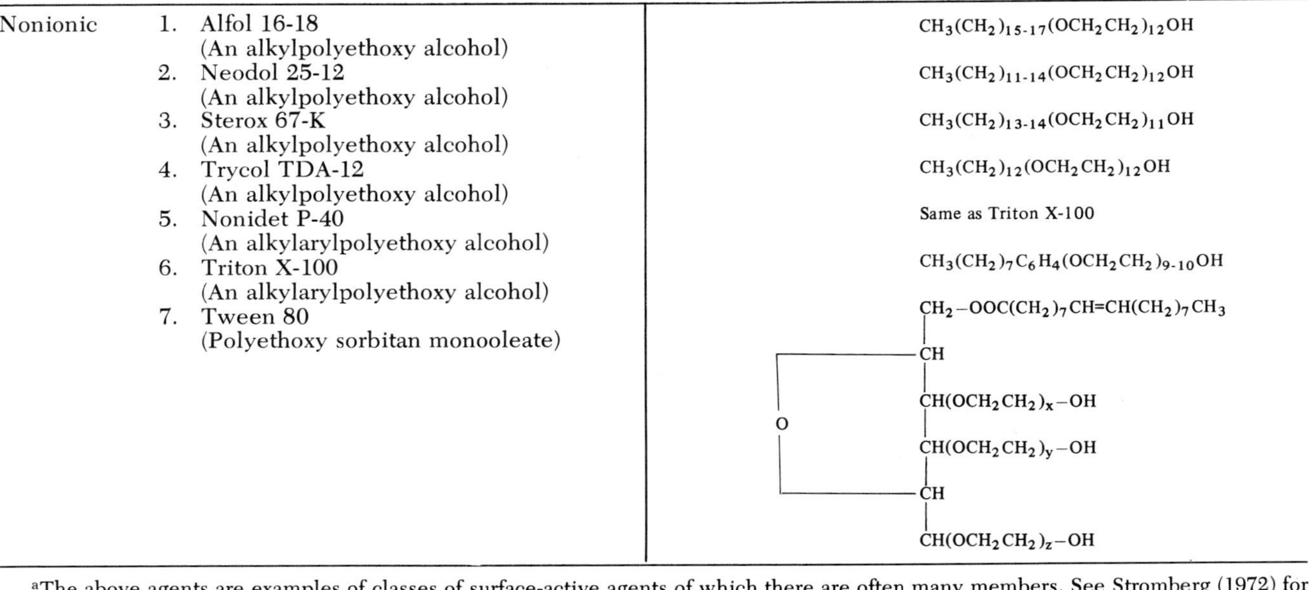

[a]The above agents are examples of classes of surface-active agents of which there are often many members. See Stromberg (1972) for other examples. Additional data about major classes of surface-active agents can be obtained from the manufacturers: Hyamines and Tritons, Rohm and Haas Co., Philadelphia 19105; Sarkosyls, Ciba-Geigy Corp., Ardsley, N.Y. 10502; Tweens, Atlas Chemicals Division of ICI America, Inc., Wilmington, Del. 19899; Alfols and Steroxes, Monsanto Co., St. Louis 63166. Most surfactants are not single chemical compounds but mixtures. Thus, as indicated in the table, the number of ethoxy groups or the length of a carbon chain may vary a little. The manufacturers strive for uniformity in their products so that the mixture usually contains a predominant amount of one compound and small amounts of other members of the homologous series. In the case of surfactants with fatty acid residues such as the Tweens, formulations with different properties are provided by changing the fatty acid (e.g. the fatty acid residue of Tween 20 is lauric acid, that of Tween 40 is palmitic acid, and that of Tween 80 is oleic acid). In addition, Tweens may vary in numbers of ethoxy groups incorporated and the numbers of ethoxy groups attached to different carbons in any one Tween formulation are not constant. Hence, in the formula given for Tween 80, the variable numbers of ethoxy groups are indicated by the subscripts x, y, and z. In summary, no one structural formula represents the product; instead Tween 80 is best described as a mixture of oleate partial esters of sorbitol and sorbitol anhydrides condensed with approximately 20 moles of ethylene oxide for each mole of sorbitol and its anhydrides.

spectrophotometry. In this manner it was shown that as adenine, guanine, and cytosine decreased, hypoxanthine, xanthine, and uracil increased proportionately. Significant deamination could be demonstrated after 2–4 hr at pH 4.1 and after 10–20 hours at pH 4.3 (Figure 40).

The following evidence led Schuster and Schramm (1958) to conclude that inactivation of TMV-RNA actually depends on the deamination of the purine bases and cytosine. In control experiments run at the same pH, temperature, and salt concentration but without addition of nitrous acid, the infectivity of the RNA was not lost. Hence loss of infectivity is associated with the nitrous acid treatment. Next, the possibility that nitrous acid may have caused splits in the RNA chain was investigated. No change in sedimentation coefficient could be detected after inactivation of the RNA with nitrous acid, nor could dialyzable split products be found. Hence, the most reasonable assumption is that the observed loss of infectivity is due to deamination of the bases in the RNA.

As shown in Figure 40, the inactivation of TMV-RNA with time is much more rapid at pH 4.1 than at 4.3. In both cases, the curves are of the first-order, one-hit type, indicating that deamination of a single nucleotide base is sufficient to inactivate (the aberration of the pH 4.3 curve at the beginning is due to the fact that the first two or three samples fell outside the range in which lesion numbers are proportional to RNA concentration).

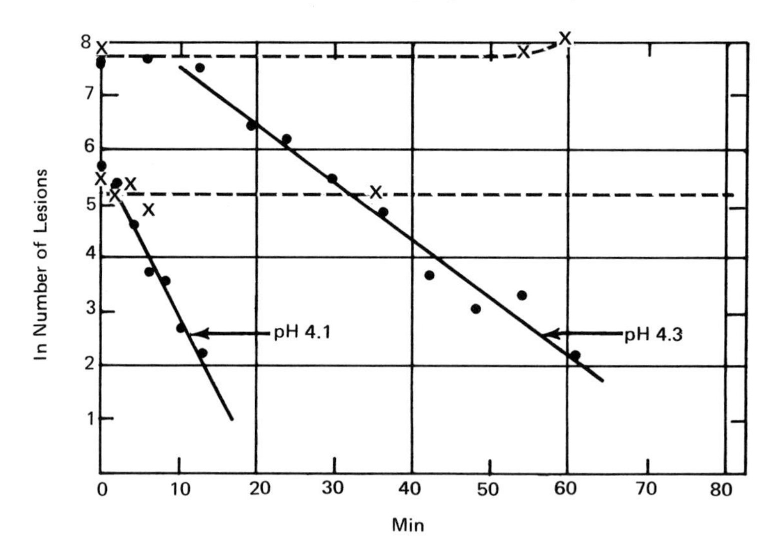

Fig. 40. Loss of infectivity of TMV-RNA with time upon treatment with nitrous acid at pH 4.1 and 4.3, respectively. Solid lines represent the nitrous acid treated samples, and dotted lines the controls held at the same pH levels and salt concentrations but not treated with nitrous acid. (Adapted from Schuster and Schramm 1958.)

According to first-order kinetics, if an average of one nucleotide base out of N nucleotides must be deaminated to produce inactivation of the RNA, then the loss of infectivity with time is given by

$$I_t = I_0 \, e^{-N\alpha t}$$

where I_t = infectivity at time t, I_0 = infectivity at time zero, and α = moles of nucleotide deaminated per minute. By changing from the exponential to the logarithmic expression and rearranging, the relation becomes

$$\ln I_0 - \ln I_t = N\alpha t$$

Then, by selecting the time when infectivity has been reduced to 37 percent (1/e), $\ln I_0 - \ln I_t = 1$ and the relation $N = (1/\alpha t)$ is obtained. N can then be found by putting in the appropriate values for t and α. Schuster and Schramm (1958) calculated α values from the results of experiments made at two different pH values in which the deamination products were determined as a function of time. In this manner, values of 3,100 and 3,500 were obtained for N. Since there are about 6,400 nucleotides in the TMV-RNA molecule, the average value of about 3,300 for N indicates that there is about a 50 percent probability that one deamination will inactivate.

Other viruses have been inactivated with nitrous acid, among them poliovirus (Boeyé 1959). An interesting contrast was noted between polio and TMV, both of which are RNA-containing viruses, in that TMV-RNA is more rapidly inactivated by nitrous acid than whole virus, whereas with poliovirus the reverse was observed. Evidence that the polio protein is involved in the inactivation process (presumably by chemical changes in the protein that reduce significantly the ability of the viral RNA to be released) can be inferred from the decreasing slope of the whole virus inactivation curve. A similar result was observed with the spheroidal southern bean mosaic virus (Sehgal 1973) where it was found that the intact virus was inactivated twice as fast as the isolated RNA even though nucleic acid bases were deaminated to the same extent.

The cases of poliovirus and southern bean mosaic virus just cited suggest involvement of the protein in the inactivation, but a peculiar instance has been reported for tobacco rattle virus. Like the other two examples, the rate of inactivation of whole virus was substantially greater than that of isolated nucleic acid but there was no indication that the viral protein was involved (Robinson 1973). There appears simply to be an enhanced inactivation of RNA *in situ* by an unknown mechanism.

Deamination studies with phages have helped to evaluate the relative significance of the amino-purine and amino-pyrimidine bases in inactivation and mutation induced by nitrous acid. Vielmetter and Schuster (1960) treated T2 phage with nitrous acid for up to 60 hr and isolated and analyzed the DNA from aliquots at various times. As shown in Table 37, there is a great difference in the rate of deamination, depending upon pH, but at both

Table 37. Rates of Deamination of Bases in Coliphage T2 DNA by Treatment with Nitrous Acid at pH 4.2 and 4.5 Together with Rates of Inactivation and Production of Plaque Type Mutants.[a]

	Deaminations per Minute ($\times 10^{-6}$)	
	pH 4.2	pH 5.0
Adenine	18.5	<0.2
Guanine	297	8.6
5-Hydroxymethylcytosine	117	1.3
Lethal hits per minute	9	0.3
$\frac{\%\ r^- \text{ mutants per minute}}{\text{survivors}}$	2.9	3.3×10^{-2}

[a] Adapted from Vielmetter and Schuster 1960.

pH 4.2 and 5.0, guanine in T2-DNA was deaminated much more rapidly and adenine much less quickly than 5-hydroxymethylcytosine. Schuster (1960b) observed this same relationship for the deamination of the bases for DNA from a variety of sources. However, which bases are involved in deamination can apparently be dependent on conformation and structural interactions of the nucleic acid with its environment. This was illustrated by the results of studies on TMV in which it was shown that treatment of isolated viral RNA with nitrous acid results in deamination of adenine, guanine, and cytosine but that a similar treatment of whole virus deaminates two of the three bases, guanine escaping attack (Schuster and Wilhelm 1963). A corollary is the finding that at the same level of survival, more mutants are obtained by treating whole virus with nitrous acid than by the reaction with free nucleic acid (Sehgal and Krause 1968).

The hydrogen bonding between complementary strands of DNA involves amino groups of purines and pyrimidines (see Figure 24) and greatly reduces the capacity of such groups to react with nitrous acid, formaldehyde, and other reagents. However, at the pH values used in the reaction with nitrous acid, it appears that incipient separation of strands occurs, permitting a limited reaction with small molecules (Singer and Fraenkel-Conrat 1969). It has also been suggested that the numerous folds required to pack viral DNA into a phage head (T4 DNA is about 600 times longer than the T4 phage head) may cause distortions of structure that expose some of the reactive groups of the purine and pyrimidine bases (Freese and Strack 1962).

Apparently, inactivation can also be caused by nitrous acid and other reagents through formation of covalent cross-links between complementary strands of nucleic acid or even between loops of single-stranded nucleic acid (Becker et al. 1964).

Whatever the specific chemical change in viral nucleic acid produced by treatment with nitrous acid, if inactivation occurs it can be assumed that one or more of the vital functions of the nucleic acid is being blocked: (1) template for the replication of the nucleic acid, (2) transcription, or (3) translation. This same principle applies of course to changes in the nucleic acid brought about by a wide variety of agents.

d. Formaldehyde and Other Aldehydes

The inactivation of viruses by formaldehyde has been known for years, and is of particular importance because of its wide application in the production of vaccines containing noninfectious but antigenically active viruses (Potash 1968). It was long supposed that formalin inactivation primarily involved reaction of amino groups of the viral protein with formaldehyde to yield mono- or dimethylol derivatives, which can be illustrated as follows with an amino acid:

$$\underset{\underset{NH_2}{|}}{R{-}CH{-}COO^-} + HCHO \rightleftharpoons \underset{\underset{\substack{HN{-}CH_2OH \\ \text{Methylol Derivative}}}{|}}{R{-}CH{-}COO^-} + HCHO \rightleftharpoons \underset{\underset{\substack{HOCH_2{-}N{-}CH_2OH \\ \text{Dimethylol Derivative}}}{|}}{R{-}CH{-}COO^-}$$

This reaction is the basis of the classical Henriques-Sorenson (1910) formol titration method. However, the reaction of proteins with formaldehyde is more complex than the above, and it was shown, for example (Fraenkel-Conrat and Olcott 1948), that at room temperature and over a wide pH range cross-linking methylene bridges can be formed between amino-methylol groups, on one hand, and amide or guanidyl groups, on the other. Thus, after the initial fast reaction of a primary amino group with formaldehyde to form a methylol compound, a slower cross-linking reaction may occur as follows:

$$\underset{\substack{\text{Methylol} \\ \text{Compound}}}{R{-}NH{-}CH_2OH} + \underset{\substack{R \\ \text{Amide}}}{H{-}\underset{|}{N}{-}\overset{\overset{O}{\|}}{C}{-}R} \rightarrow \underset{\substack{R \\ \text{Methylene Bridge} \\ \text{Compound}}}{R{-}NH{-}CH_2{-}\underset{|}{N}{-}\overset{\overset{O}{\|}}{C}{-}R} + H_2O$$

The formation of methylol compounds is, as indicated, a reversible reaction, whereas the cross-linking reaction is much less so and is now regarded as contributing greatly to the tanning or hardening action of formaldehyde on proteins.

These reactions doubtless occur to a variable extent with viruses treated with formaldehyde, but a new interpretation of the formaldehyde reaction is suggested by the knowledge that nucleic acid is the seat of viral infectivity and that nucleic acid reacts with formaldehyde (Fraenkel-Conrat 1954). It now seems that inactivation of viruses by formaldehyde involves both protein and nucleic acid. The primary inactivating effect depends on the reaction of formaldehyde with the nucleic acid, but sufficient alteration of the protein coat by the cross-linking reaction described above may prevent release of nucleic acid in the infected cell, and hence may also contribute to the inactivation. In this connection it is conceivable that the protein may be significantly altered before the nucleic acid is rendered noninfectious.

Quantitative studies on the reaction of TMV-RNA with formaldehyde and other aldehydes have been made by Staehelin (1958, 1959, 1960), using ^{14}C-labeled compounds. It was expected that formaldehyde would react only with bases containing amino groups (adenine, guanine, and cytosine) and this seems to be the case. At least the reaction with formaldehyde stops at about the level of 60–70 moles of formaldehyde per 100 moles nucleotide, which is about the proportion of adenine, guanine, and cytosine (the amino bases) to total nucleotides present. The extent of reaction was found to be diminished by changing the phosphate buffer concentration from 0.001 M to 0.1 M or by adding divalent cations, presumably because these conditions favor formation of secondary structure and thus reduce the availability of the amino groups.

In contrast to isolated RNA, intact TMV was found to be inactivated only when about 100 times greater concentrations of formaldehyde and longer times of treatment were employed. This suggests that the amino groups of the RNA are hydrogen bonded with protein in the intact virus and therefore are less accessible, or that the formaldehyde has some difficulty in penetrating the protein coat to reach the deeply embedded RNA. Since the inactivation of TMV by aldehydes was observed to proceed less readily with larger aldehydes than with formaldehyde, whereas this was not so in the inactivation of free RNA, Staehelin (1960) suggests that penetration difficulties probably explain the difference in reactivity between RNA and intact virus.

A detailed review of the reaction of formaldehyde with nucleic acids and their constituents was written by Feldman (1973).

The reaction of TMV-RNA with glyoxal derivatives was found to differ strikingly from its reaction with formaldehyde as judged by effect on ultraviolet absorption (Staehelin 1959). In contrast to the shift of the absorption maximum to a higher wavelength (3–5 nm) and an increase in absorption with time (as much as 30 percent) with formaldehyde, treatment with glyoxal

$$\left(\mathrm{H{-}\overset{\overset{\displaystyle O}{\|}}{C}{-}\overset{\overset{\displaystyle O}{\|}}{C}{-}H}\right)$$

or kethoxal

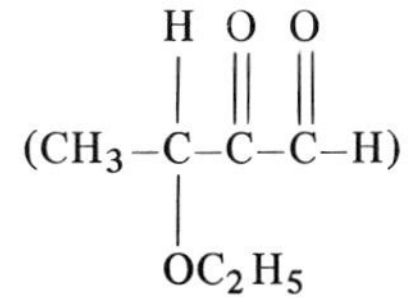

$$\mathrm{(CH_3{-}\underset{\underset{OC_2H_5}{|}}{\overset{\overset{H}{|}}{C}}{-}\overset{\overset{O}{\|}}{C}{-}\overset{\overset{O}{\|}}{C}{-}H)}$$

caused a very small increase in absorption and only a very slight shift in wavelength of the maximum absorption, and this toward the smaller wavelengths. By using radioactive kethoxal, it was shown that guanylic acid is the only one of the four RNA nucleotides that reacts with glyoxal derivatives:

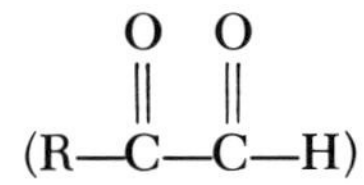

$$\mathrm{(R{-}\overset{\overset{O}{\|}}{C}{-}\overset{\overset{O}{\|}}{C}{-}H)}$$

It was suggested that the reaction of nucleic acid or its guanine derivatives with compounds related to glyoxal might involve formation of stable five-membered ring structures as shown here:

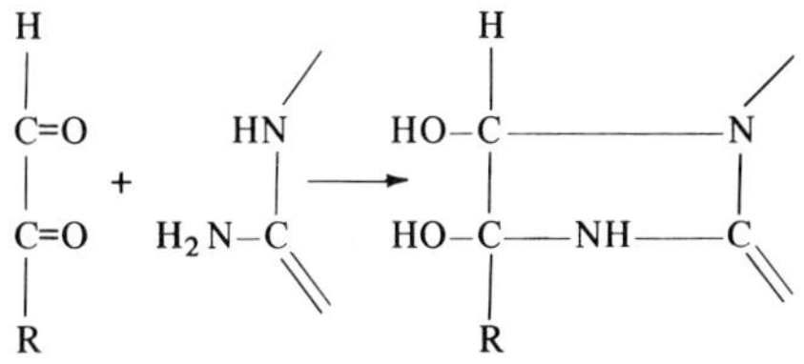

$$\begin{array}{c} H \\ | \\ C{=}O \\ | \\ C{=}O \\ | \\ R \end{array} \; + \; \begin{array}{c} HN \\ | \\ H_2N{-}C \end{array} \longrightarrow \begin{array}{c} H \\ | \\ HO{-}C{-}{-}{-}{-}{-}N \\ | \qquad\quad | \\ HO{-}C{-}NH{-}C \\ | \\ R \end{array}$$

Reaction between TMV-RNA and stable glyoxal derivatives such as kethoxal (glyoxal itself is unstable) takes place more readily and much faster than with formaldehyde, as shown in Table 38.

Kethoxal has been applied to several animal viruses with the following results (Renis 1970: (1) three picornaviruses–polio, coxsackie, and encephalomyocarditis viruses–were not activated by kethoxal although the isolated RNA of coxsackie virus was; (2) reovirus-3 and pseudorabies virus were moderately inactivated; (3) herpes, influenza, parainfluenza-3, Newcastle disease, Sindbis, and vesicular stomatitis viruses were strongly inactivated.

e. Hydroxylamine

The inactivation of bacterial, animal, and plant viruses by hydroxylamine has been reported (Kozloff et al. 1957; Freese et al. 1961a; Franklin and Wecker 1959; Holland et al. 1960; Schuster 1961). Some of the outstanding features of the reaction of viral nucleic acids with hydroxylamine are as follows:

Table 38. Inactivation of TMV-RNA by Aldehydes.[a,b]

Aldehyde	Concentration %	Activity %
Formaldehyde	0.1	0
	0.05	4
	0.025	34
Glyoxal	0.1	6
	0.05	23
	0.025	81
	0.0125	100
Kethoxal	0.02	1
	0.01	8
	0.005	32
	0.0025	73

[a]From Staehelin 1959.

[b]TMV-RNA at 1 mg/ml was treated with the aldehydes indicated for 30 minutes at about 23° in 0.001 M phosphate buffer at pH 6.8. After the reaction, the nucleic acid was precipitated 6 times with alcohol to remove unreacted aldehyde and then assayed.

1. No decrease in molecular weight of the nucleic acid occurs under conditions that lead to complete loss of infectivity. Hence, a reaction with the nucleic acid bases is indicated.
2. Hydroxylamine seems to react only with pyrimidines and not with purines (Schuster 1961; Freese et al. 1961b).
3. Among the pyrimidines the reaction rate varies with pH and the nature of substituents present in the pyrimidine ring. The reaction can be detected by change in ultraviolet absorption and also chemically, in due time, by loss of pyrimidine and gain in certain reaction products (see below).

In RNA, cytosine was found to react with hydroxylamine at pH 6 at least 30 times faster than uracil, whereas at pH 9, uracil reacts at least 8 times faster than cytosine (Schuster 1961). The reaction proceeds best (Schuster 1961) if carbon atoms 5 and 6 (or C atoms 4 and 5 by Fischer system) are unsubstituted, and pyrimidines with substituents on these atoms react only slightly or require more vigorous treatment (higher concentration of hydroxylamine and higher temperature). This point is illustrated by the data in Table 39 which show the effect on ultraviolet absorption of treatment with hydroxylamine at a pH of 7.5, a value intermediate to the optima for cytosine and uracil.

The reaction of hydroxylamine with uracil appears to destroy the pyrimidine ring yielding an isoxazolone fragment and a ribosyl urea residue. The primary reaction proposed by Schuster (1961) is as follows:

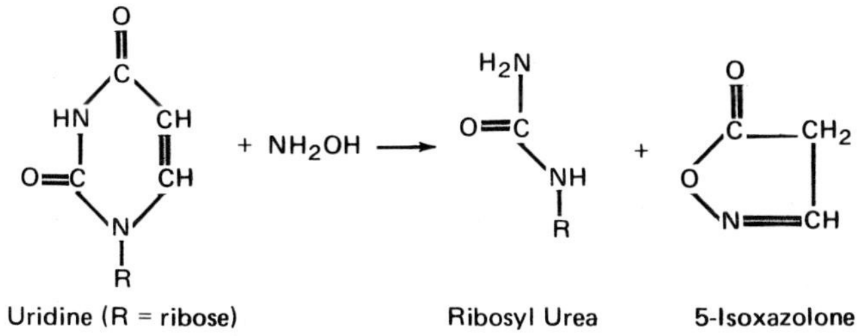

Uridine (R = ribose) + NH_2OH → Ribosyl Urea + 5-Isoxazolone

The reactions of hydroxylamine with cytosine residues in nucleic acids (Phillips and Brown 1967) may be represented as follows:

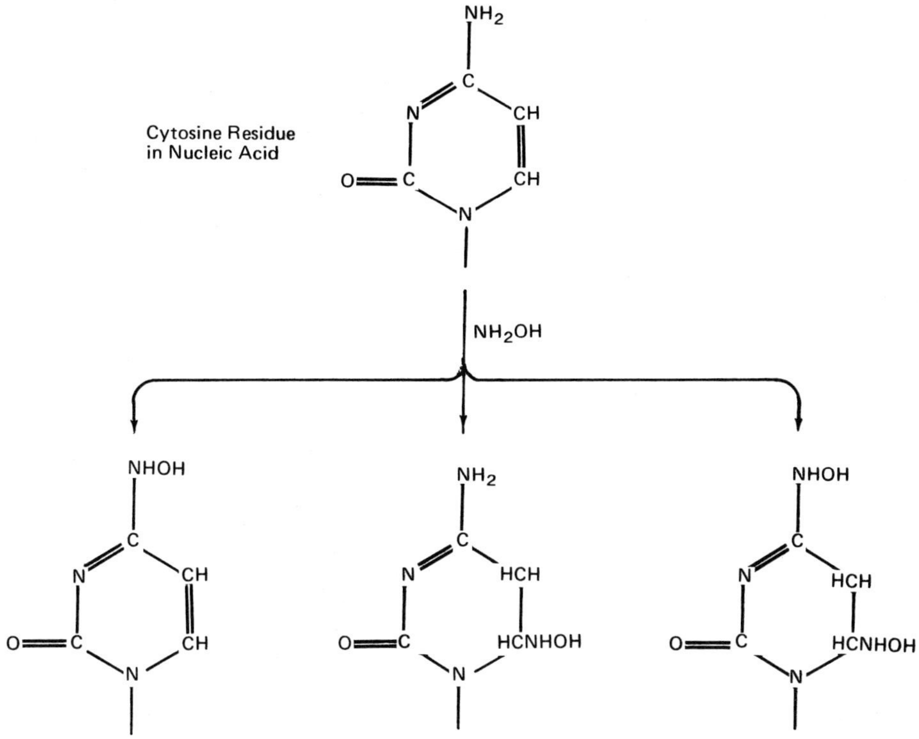

Cytosine Residue in Nucleic Acid

NH_2OH

It is not yet clear which products of the action of hydroxylamine are inactivating and which are mutagenic, although the reaction with uracil and those with cytosine in which hydroxylamine is added to the 5-6 double bonds to produce the two structures shown on the right above are considered the most likely inactivating events (Singer and Fraenkel-Conrat 1969). In addition it has been proposed that inactivation by hydroxylamine under

Table 39. Decrease of Ultraviolet Absorption of TMV-RNA and of Pyrimidines and Derivatives upon Treatment with Hydroxylamine.[a]

Compound	Wavelength, nm	Percent Decrease in Absorption
Adenylic acid	260	0
Guanylic acid	250	0
6-Aminocytosine	260	1
Thymidylic acid	270	1
TMV-RNA	260	2
5-Methylcytosine	270	3
Orotic acid	280	4
Cytidine	270	12
Uracil	260	19
Cytosine	265	20
Uridine	260	27
Thymidine	265	15[b]

[a]Adapted from Schuster 1961.
[b]Thymidine was treated with 2 M NH_2OH at pH 7.5 for 15 hr at 45°; all other reactions were with 1 M NH_2OH at pH 7.5 for 15 hr at 20°.

aerobic conditions may involve the action of free radicals (Freese and Freese 1965). The reaction of hydroxylamine with the amino group on C4 of cytosine to produce the product shown on the left above (and its tautomeric form) is considered a likely mutagenic event.

f. Alkylating Agents

Alkylating agents are chemicals that can introduce alkyl groups into proteins and nucleic acids, the current emphasis being on the latter. They may be monofunctional, di-, or polyfunctional depending on the number of reactive groups they contain. A review (Lawley 1966) and a book (Loveless 1966) provide many details about the chemistry and genetic effects of alkylating agents.

The first chemical mutagen to be reported was mustard gas, an alkylating agent (Auerbach and Robson 1947). Ethyl methane sulfonate was apparently the first alkylating agent applied to a virus, which happened to be a bacteriophage (Loveless 1958); and a rather extensive study of the effect of alkylating agents on TMV was made a few years later (Fraenkel-Conrat 1960).

Several alkylating agents were used on TMV and TMV-RNA (Fraenkel-Conrat 1960), and the use of such agents tagged with radioisotopes enabled the sites of reaction to be determined in several cases. As indicated in Table 40, all of the reagents seemed to react principally with guanine, although iodoacetate alkylated adenine as well. Where it could be

Table 40. Effect of Some Alkylating Agents on TMV-RNA.[a]

Reagent	Group Introduced	Site of Reaction	Inactivation of RNA (groups per mole RNA to reduce infectivity to 37%)	Mutational Effect
Iodoacetate	$-CH_2-C(=O)-O^-$ (carboxymethyl)	A, G	1–3	–
Ethylene oxide	$-CH_2-CH_2-OH$ (hydroxyethyl)	G	1–3	–
Propylene oxide	$CH_3-CH(-)-CH_2OH$ (hydroxyisopropyl)	?	?	–
Mustard gas (bis-β-chloroethyl sulfide)	$-CH_2-CH_2-S-CH_2-CH_2-OH$ (β-hydroxyethylthioethyl)	G	?	–
Dimethyl sulfate	$-CH_3$ (methyl)	G	?	+

[a]Data from Fraenkel-Conrat 1960.

measured, it appeared that introduction of one to three alkyl residues into the RNA caused a reduction in infectivity to the 37 percent level. Only one of the reagents, dimethyl sulfate, appeared to be consistently mutagenic. From RNA allowed to react with dimethyl sulfate much longer than required for inactivation, it was possible to isolate 7-methyl guanine:

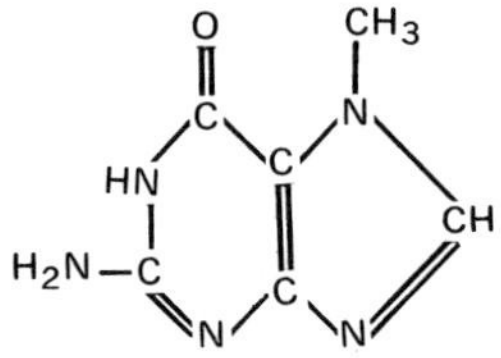

Studies of alkylations of other viruses and nucleic acids have confirmed and extended these observations (see Drake 1970; Freese 1971; Singer and Fraenkel-Conrat 1969). The following general conclusions can be stated. Alkylating agents attack principally the ring nitrogens of the purine and pyrimidine bases, predominantly N7 of guanine, and to variable smaller extents, depending on type of reagent and of nucleic acid, the N1, N3, and N7 positions of adenine and N3 of pyrimidines.

Inactivation is expected to occur when alkylation affects base pairing atoms such as N1 of adenine and N3 of pyrimidines. When the alkylating agent is polyfunctional like the sulfur or nitrogen mustards, there is a possibility of cross-linking between complementary strands or segments of nucleic acids. Such cross-linking is probably inactivating. Likewise, depurination, which is a possible side reaction, is inactivating. Mutational possibilities following alkylation will be considered later.

g. Other Inactivating Chemicals

Strong acids and bases inactivate viruses because they are able to damage both protein and nucleic acid. Proteins vary considerably in their resistance to acid denaturation, those of poliovirus being quite resistant (Loring and Schwerdt 1944), whereas those of myxoviruses and rhinoviruses are very susceptible (Miller 1944; Dimmock and Tyrrell 1962). Most viruses are disrupted by alkali with or without denaturation of protein constituents (Stanley 1938). RNA is degraded by alkali while DNA is resistant; both types of nucleic acid are attacked by acids, loss of purine bases (depurination) often resulting and degradation of the polynucleotide chains occurring to an extent determined by conditions (Loring 1955).

It has been recognized for years that oxidizing agents inactivate viruses (Stanley 1938). Examples of oxidizing agents tested on viruses include peroxide, iodine, other halogens (especially hypohalites), and, in the case of the most sensitive viruses, just air. A major structural effect of hydrogen peroxide on T-even phages is the inducing of these phages to eject their DNA (Kellenberger and Arber 1955). The oxidative deamination by nitrous acid was considered earlier.

Halogenation is generally inactivating with nucleic acid as a prime target (Brammer 1963). In experiments with TMV-RNA, adenine and uracil were resistant to halogenation but *N*-bromosuccinimide reacted with both guanine and cytosine at pH 7 and pH 9, although reaction with guanine was more favored at pH 9. Bromine gave a more specific reaction, reacting almost exclusively with cytosine at pH 7 and with guanine at pH 9. Mutagenic action has been reported for both *N*-bromosuccinimide and bromine (Singer and Fraenkel-Conrat 1969).

It can be reiterated here that any chemical that irreversibly denatures protein or that alters nucleic acid in such a way as to interfere with its use as a template will either inactivate or mutate.

h. *In Vivo* Inactivators of Viruses

The inactivating agents discussed above were considered primarily on the basis of their application to viruses in the laboratory and the results of subsequent tests in appropriate hosts. Much has been learned from such studies, but there is a growing concern about what kinds of inactivating and mutating influences occur or can be induced in cells. A few kinds of intracellular inactivations of viruses will be considered here.

1. Base analogs. It was natural when the primacy of nucleic acid in the reproduction of viruses became apparent that metabolic antagonists should be tested for effect on virus growth. Consequently, numerous purine and pyrimidine analogs have been added to viral systems and the effects noted (see Matthews and Smith 1955). Among the analogs studied are the following:

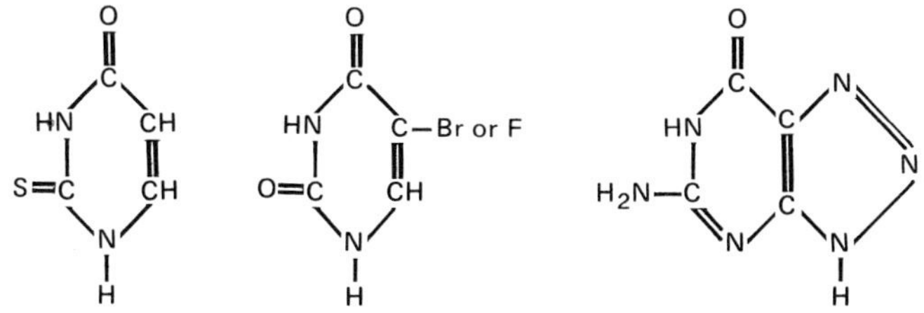

2-Thiouracil

5-Bromouracil and 5-Fluorouracil

8-Azaguanine

Some analogs show strong inhibitory effects on virus multiplication and others not. Some of the compounds effective with one virus are ineffective with another. Certain viruses such as T2 phage, not directly inactivated by incorporation of a pyrimidine analog, may be substantially more sensitive than normal phage to subsequent exposure to x-rays, ultraviolet light, or even visible light (Stahl et al. 1961). Among explanations offered for this effect are the greater absorption of radiation by bromouracil as opposed to thymine and the possibility that radiation damage to the analog may not be as readily reversed by repair enzymes as damage to thymine.

Some analogs have been shown to be incorporated into viral nucleic acids by subsequently isolating the analog-containing nucleotides. For example, although 5-fluorouracil (or better yet, 5-fluorouridine) inhibits the multiplication of TMV, the virus does multiply in the presence of this analog and replaces as much as 30–50 percent of the uracil with the 5-fluorocompound (Gordon and Staehelin 1959). This is equivalent to about 500–800 molecules of 5-fluorouracil per mole of RNA. Nevertheless, such altered virus was found to be fully infective on Xanthi nc tobacco. On the other hand, 2-thiouracil and 8-azaguanine, both of which are incorporated into viral RNA, inhibit the multiplication of TMV and other plant viruses (Matthews and Smith 1955; Loebenstein 1972).

The base analogs shown to inhibit multiplication of animal viruses are primarily the 5-halogenated uracils, and inhibition seems most effective when these are applied as the uridine derivatives; however, other compounds such as 2-thiouridine and 2-aminopurine have also been used in animal virus systems (Prusoff et al. 1965; Kaplan and Ben-Porat 1970; Sambrook et al. 1966; Slechta and Hunter 1970).

Animal viruses whose multiplication has been blocked by base analogs include rabbitpox, pseudorabies, polyoma, herpes simplex, and coxsackie viruses. Two mechanisms of action are considered most likely to explain the antiviral effects observed: (1) incorporation of analog causes lethal mistakes in some protein coded by the viral nucleic acid (that is, an amino acid is replaced by one that renders the protein nonfunctional); (2) the analogs react with and inactivate some enzyme needed for virus replication.

Base analogs are also mutagenic but this will be considered in the next section.

2. *Radiophosphorus.* It is possible to incorporate a self-destruction mechanism (so-called "^{32}P suicide") into viral nucleic acid by supplying virus-infected cells with ^{32}P phosphate of high specific activity. Such radioactive phosphorus is incorporated into nascent nucleic acid as the virus replicates. Since all nucleic acids are characterized by a linear structure in which hundreds of nucleotides are linked by phosphodiester bonds (see small segment of nucleic acid illustrated in Figure 17), decay of a ^{32}P atom to ^{32}S (with release of a beta particle) may result in rupture of the nucleic acid strand at the site of decay. The break apparently reflects the inability of the sulfodiester linkage that replaces the phosphodiester bonding to withstand the energy exchanges accompanying the transmutation.

Inactivation by ^{32}P decay is a one-hit phenomenon—whether observed with double-stranded DNA of T2 coliphage (Hershey et al. 1951; Stent and Fuerst 1955, 1960), polyoma virus (Benjamin 1965), with single-stranded DNA of coliphage ØX174 (Sinsheimer 1959b), or single-stranded RNA of poliovirus (Henry and Youngner 1963). The fraction of survivors (S) is proportional to the number of incorporated ^{32}P atoms that have decayed by the time of assay for infectivity according to the formula $S = e^{-\alpha N f}$, where α is the efficiency of inactivation per ^{32}P disintegration, N the average number of ^{32}P atoms per nucleic acid molecule, and f the fraction of ^{32}P atoms decayed by a given time. A plot illustrating the survival of ^{32}P-labeled phage T2 with time is given in Figure 41. From information available, α, the efficiency of inactivation, can be calculated. Such calcula-

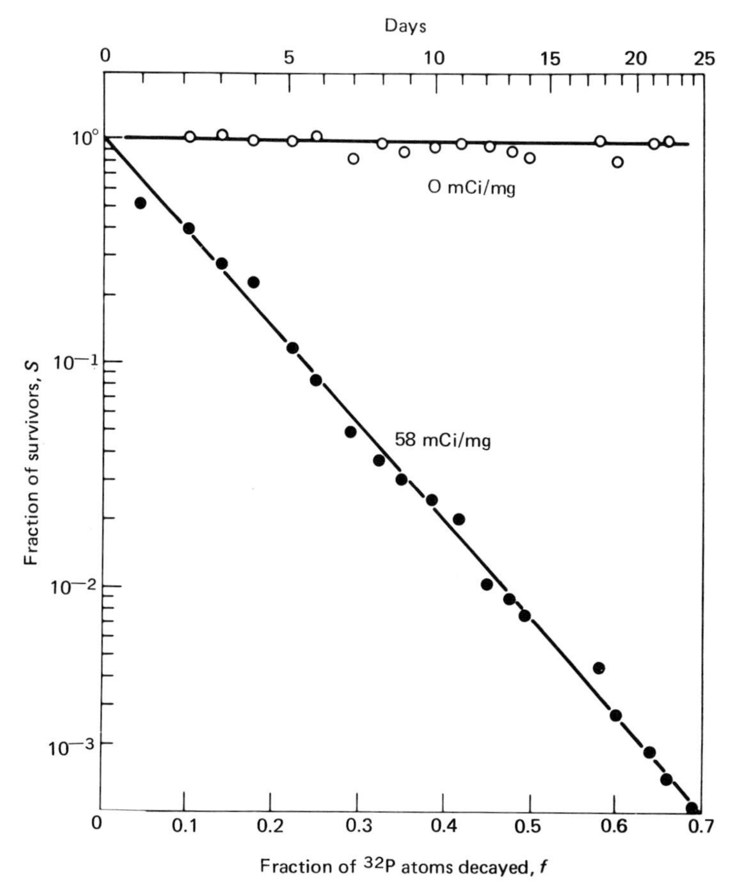

Fig. 41. Survival of ^{32}P-labeled and nonlabeled stocks of T2 bacteriophage stored at 4°C, as a function of the fraction of the ^{32}P atoms that have decayed by the day of assay. (Adapted from Stent and Fuerst 1960.)

tions indicate that only about one in ten disintegrations of P result in inactivation of T2 phage, whereas ØX174 DNA or polivirus RNA are inactivated with an efficiency near 1. This difference seems related to the necessity to have disintegrations in the same vicinity on complementary strands of double-stranded nucleic acid in order to break the molecule, whereas a break anywhere does it for single-stranded nucleic acid.

The situation is actually more complex than this (Harriman and Stent 1964); for example, a break in one strand of a double-stranded molecule may be readily repaired thus restoring full function since the molecule remains together with a nick in it. Conversely, a break in only one strand may be lethal if by chance it is in a gene coding for a needed repair enzyme. Nevertheless, there is a distinctive difference in the efficiency of inactivation of single-stranded and double-stranded nucleic acids by ^{32}P decay.

3. *Antibodies and interferons.* Vertebrates have at least two mechanisms for inactivating viruses not present in the cells of protists or plants. These are antibodies and interferons.

Antibodies are specific proteins (immunoglobulins) produced by special cells (lymphocytes) in response to foreign substances introduced into the animal. In general, viruses are good antigens (antigens are substances that elicit the formation of antibodies and react specifically with them). Consequently, recovery from a virus disease often leaves an animal immune for some time to further attack from the same or closely related strains of virus, and such immunity can usually be correlated with the presence of specific immunoglobulins (antibodies) in blood, lymph, and tissues.

This situation is exploited in the production and use of prophylactic vaccines. In some cases a "live" vaccine is employed that consists of a mild strain of virus that gives a localized infection; antibodies are induced to the mild strain that also neutralize more virulent strains. Carefully inactivated ("killed") virus is injected in a second type of vaccination. Virions that are not infectious but still antigenic can be obtained by proper treatment of virus with such agents as formaldehyde, ethylene oxide, β-propiolactone, or ultraviolet light. In some cases, for example, in some influenza vaccines (Ruben and Jackson 1972), the virions are dissociated by a surface active agent to release antigenic components of the virus, which are separated from other materials and concentrated by column chromatography for use in the vaccine. Detailed descriptions of various methods for vaccine preparation are given by Potash (1968). An excellent exposition on immunoglobulins by many specialists in various aspects of the topic is available (Kochwa and Kunkel 1971).

Viruses can be inactivated by their reaction with antibodies according to two mechanisms: (1) With some viruses the virus-antibody complex cannot attach and penetrate cells and thus such virus fails to get an infection started; (2) the complex is taken up by macrophages (scavenger cells present in various tissues) by a phagocytotic process, or in some cases, in contrast to above, the complex is engulfed by cells normally susceptible to free virus. Within macrophages or in phagocytic vesicles of other cells, the virus is degraded enzymatically. Virus-antibody complex appears more readily degraded than free virus, which can also be engulfed by macrophages.

It should be noted that inactivating ("neutralizing") antibody does not irreversibly inactivate virus simply by forming a complex with it; virus-antibody complexes can be dissociated by dilution, treatment with salts, dilute alkali, and so on with release of infectious virus.

The immune reactions of viruses with tissues and cells represent a complex area with gaps in understanding, but many aspects are dealt with in a review by Mims (1964).

Interferons are substances whose name is derived from the observation first made by Isaacs and Lindenmann (1957) that treatment of chick cells with heat-inactivated influenza virus causes release of a soluble factor that interferes with the infection of other cells. Interferon seems to be a glycoprotein of about 30,000 daltons that is heat resistant and stable over a wide pH range (Colby 1971; Kleinschmidt 1972).

Interferon is species specific but not virus specific. This means that interferon produced in a given type of cell such as in chick cells is active against a wide variety of viruses presented to chick cells but has little effect against the same viruses in mouse or monkey cells. This and other facts suggest the nature of interferon, namely, that it is a gene product of the host rather than of the virus. Hence there are many interferons each characteristic of the species in which it is produced.

Interferon does not inactivate viruses when mixed with them nor does it interfere with the attachment, penetration, or uncoating of infecting viruses. Interferon appears to be an inducer of another host-coded protein, "virus inhibitory" or "antiviral" protein. This antiviral protein is thought to interfere with translation of viral mRNA but not with host mRNA; however, there is also evidence that it may react with virus-specific RNA polymerase to prevent its function.

It might seem that interferons and/or antiviral proteins would be ideal antiviral agents either administered directly or induced in the desired host. However, at present it is impractical to get enough homologous interferon, or less yet, of antiviral protein to inject into people or other animals; similarly it is not yet feasible to induce interferon or antiviral protein in the cells where it is required. A good discussion of techniques for the study of interferons and related matters has been published (Wagner et al. 1968).

4. Miscellaneous antiviral substances. Prodigious efforts have been made to find antiviral (chemoprophylactic) drugs that might equal for viruses the success of antibiotics against bacteria. Limited success has been achieved in this objective, although the rationale for production of antiviral drugs is clear: disrupt the synthesis or function of viral nucleic acid or protein(s) without serious disturbance of host cell metabolism.

Many chemicals, belonging however to a restricted number of types of compounds, have been found to inhibit the multiplication of different animal viruses in cell cultures [see Whipple (1965) and Herrmann and Stinebring (1970) for an extensive review of antiviral substances]. These include nucleoside derivatives of many kinds, thiosemicarbazones, amantadine derivatives, thiazolidines, guanidine, and isoquinolines. However, when it comes to clinical trials with the whole animal, many chemicals prove to be ineffective virus inhibitors, produce unpleasant or even toxic side effects, or induce both resistant and dependent mutants. Only two types of antiviral compounds have shown impressive efficacy without significant toxicity or side effects. They are halogenated deoxyribosides (but see below) and amantadine derivatives (see their basic structures illustrated in Figure 42).

Halogenated nucleosides, especially deoxyribosides such as 5-iodouracildeoxyriboside (IUDR or idoxuridine), are effective inhibitors of several viruses containing DNA. A halogenated nucleoside such as IUDR is converted in the cell to the nucleoside triphosphate and incorporated into DNA in place of thymine as the viral DNA replicates. This may cause mispairing in further replication and faulty transcription of the DNA strand containing halogenated pyrimidine. Thus the replication and function of

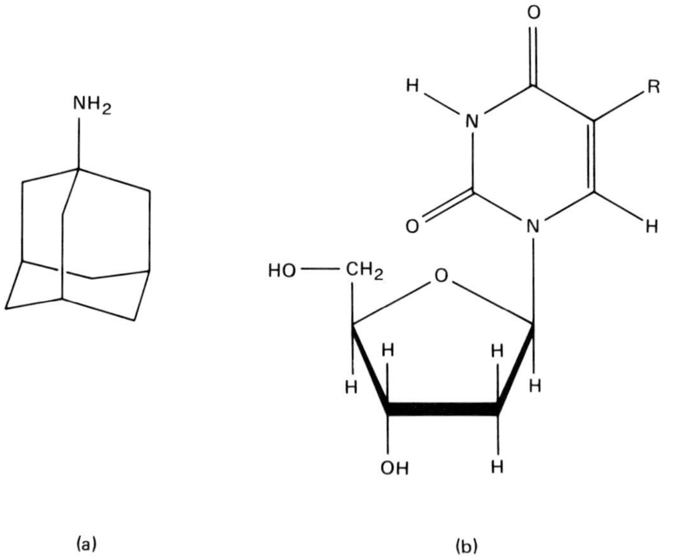

(a) (b)

Fig. 42. Basic structural formulas for two types of antiviral drug. *a*. Amantadine (1-adamantanamine); *b*. halogen-substituted deoxyuridine where R is Br, F, or I.

viral DNA are both impaired. These are not specific effects on the virus but affect cellular DNA replication and function as well. For example, after intravenous infusion of IUDR for 6 days at a total dosage of 520 mg/kg of body weight, an adult patient showed a reduction in leucocytes and platelets, abnormal liver function, and bloody diarrhea (Tomlinson and Mac Callum 1970). Obviously, halogenated nucleosides can have marked toxic effects. However, IUDR is effective and nontoxic in topical applications to the eye infected with herpes simplex or vaccinia viruses (Kaufman 1965).

Undoubtedly, the most successful antiviral agent in clinical use at present is amantadine and its derivatives [the synthesis and testing of 37 derivatives have been described by May and Peteri (1973)]. They are used as chemoprophylactic agents against influenza A and certain strains of parainfluenza (see Galbraith et al. 1970). Amantadine and its derivatives are not viricidal but appear to interfere with viral penetration into host cells and possibly uncoating after entry.

B. Mutation

A virus mutation may be defined as a heritable change in the nucleic acid (also called the genome or genotype) of the virus. The product of mutation is called a mutant, or variant, or, especially with plant viruses, a strain.

Mutants of all types of viruses are recognized by distinctive heritable features, that is, by changes in the phenotype. There are many such characteristics, but most commonly mutants are recognized by differences in (1) host or tissue specificity, (2) disease symptoms, (3) serological specificity, and (4) protein structure. In some cases it is possible to demonstrate directly mutational changes in the genotype, that is, in the structure of the nucleic acid itself. It should be noted that the consequences of mutation are not always readily detected, and that some mutants are viable and hence detectable only under a special set of conditions.

For example, a mutant may multiply in one host (permissive host) but not in another (nonpermissive or restrictive host), whereas the parental virus multiplies in both. Likewise a temperature-sensitive (ts) mutant may replicate at a certain temperature and not at a higher temperature, while the parental virus multiplies at both temperatures. Since the host-restricted and ts mutants cannot multiply under conditions in which the common or wild type virus can, they are often referred to as conditional lethal mutants; a lethal mutant is the product of mutation not able to replicate under any known conditions. While mutation is essentially a discontinuous process, occurring in discrete steps, repeated mutations can occur. Consequently, mutants are often observed that display accumulated changes in characteristics.

It is probably safe to say that all viruses are capable of mutation and do mutate spontaneously in nature. The frequency of mutation (mutation rate) is difficult to ascertain with certainty owing to the difficulty of detecting and scoring accurately the great number of variations that can occur. However, fairly accurate estimates can be made of the frequency with which specific types of mutants arise (see chap. 5 in Drake 1970; Stent 1971). For example spontaneous mutation of T2 coliphage may produce plaque type (r) mutants at a frequency of about 1 in 10^4 duplications of the phage, whereas mutants of the host-range type (h) may appear with a frequency of only 1 in 10^8 duplications of the phage. These results can be expressed respectively as probabilities of 10^{-4} and 10^{-8} mutations per phage per duplication (Stent 1971).

A similar range in frequency of spontaneous mutations has been observed with several RNA-containing animal viruses (Dulbecco and Vogt 1958; Takemori and Nomura 1960; Carp 1963; Breeze and Subak-Sharpe 1967; Medill-Brown and Briody 1955). Reliable calculations of spontaneous mutation rates seem not to have been made for DNA-containing animal viruses. However, some DNA animal viruses such as rabbitpox virus yield pock type mutants with a frequency as high as 1 percent (Gemmell and Fenner 1960). One interpretation for this high frequency is that individual mutations may occur at a frequency of only 10^{-5} but at a thousand separate genetic sites.

Only crude estimates of spontaneous mutation rates have been made for plant viruses. However, some cases seem to be high like the rabbitpox virus cited above. For example, Kunkel (1940) found about 0.5 percent mutants in the tobacco mosaic virus present in sap of plants that had been inoculated five to six weeks earlier with a pure culture of common or wild type virus.

The mutation rates just discussed are for forward mutation. When a mutant replicates it can experience other forward mutations, or mutations can occur that restore the previous phenotype. The latter events are called back mutations, or reversions. It is a true reversion if back mutation occurs at the original site, but if the lost or altered characteristic is restored by mutation at a site separable from the original one by recombination, it is called a suppressor mutation. The frequency of back mutations has not been determined accurately in many cases but those observed with coliphage T4 (Drake and McGuire 1967) vary over a tremendous range (at least 10^{-4}–10^{-11}). In both forward and back mutations the observed frequencies must reflect the molecular mechanisms involved in the mutational event. Some of these mechanisms will be considered shortly.

Various procedures are used experimentally for isolating mutants from a population of viruses. Occasionally, selection of a mutant can be accomplished by passage in a host that is different from the one favoring the wild type. More commonly, use of techniques that yield isolated colonies (clones) of virus are employed. Thus pure cultures of mutants as well as of wild type are obtained from plaques formed by phages on lawns of bacteria, from plaques of tissue culture cells infected with animal viruses, or from local lesions or other distinctive spots on leaves of virus-infected plants. Such virus colonies are illustrated in Figure 3.

1. *Molecular Mechanisms of Mutation*

Viral nucleic acids exhibit considerable plasticity in that they can withstand many changes without loss of function, and can undergo some alterations that permit function under special conditions (conditional lethal mutants). Many changes are, of course, inactivating, that is, lethal.

Two types of alterations of viral nucleic acids appear to be the major causes of mutations: (1) nucleotide substitutions (also called base substitutions since a mutational change in a nucleotide is usually accomplished by a change in the purine or pyrimidine base of the nucleotide); (2) addition or deletion of one or more nucleotides. In addition, there is some evidence for rearrangement of large polynucleotide segments through inversion (see Drake 1970); and, although not commonly considered as such, the genetic hybrids resulting from recombination between the nucleic acids of two viruses, or between the nucleic acids of a temperate virus and its host, satisfy major criteria for mutants.

The main events of the central dogma scheme (replication, transcription, and translation of nucleic acid) all depend on the phenomenon of base pairing illustrated in Figure 24. Hydrogen bonding of H with contiguous N or O atoms is the key to specific base pairing. Thus adenine pairs with thymine in the replication process and with uracil in transcription, while guanine pairs with cytosine; and this specific pairing generally occurs with great fidelity. However, the structures of nucleic acid bases permit migration of hydrogen atoms from one atom to an adjacent one (tautomeric shift) so that different pairing arrangements can plausibly occur. For example, Figure 43 illustrates how the shift of a proton from N-3 in thymine to C-4 changes the keto group at C-4 to an enol group and alters the base pairing so that thymine can pair with guanine instead of adenine. Likewise, a keto-enol shift can occur in guanine, N-1 hydrogen moving to C-6 oxygen (this also permits guanine to pair with thymine but only when the latter is in the keto form). Similar shifts of hydrogen to and from nitrogen atoms (amino-imino shift) can occur, creating various other mispairing arrangements. However, such tautomeric shifts, as well as the ionizations of protons at positions 1 and 3 of guanine and thymine, respectively, occur to such a small extent under usual physiologic conditions that mispairing is infrequent. Nevertheless, such rare structures provide a strong theoretic basis for nucleotide (base) substitution and mutation.

Obvious possibilities in base substitution include the substitution of

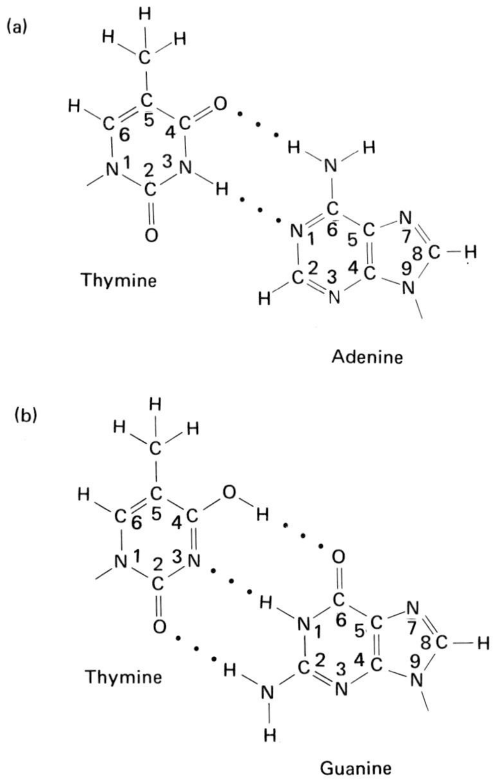

(a)

Thymine

Adenine

(b)

Thymine

Guanine

Fig. 43. Normal and abnormal base-pairing of thymine.

one purine for another or one pyrimidine for another. Such substitutions are called transitions and these may be contrasted with the possibility of purines substituting for pyrimidines or vice versa, which are called transversions. Each may be summarized in terms of the common purines and pyrimidines of DNA and RNA as follows:

Transitions: $C \rightleftharpoons T$ or U, and $A \rightleftharpoons G$
Transversions: $A \rightleftharpoons C \rightleftharpoons G \rightleftharpoons U$ or T

Spontaneous mutations of viruses can be explained on the basis of one or another of the above molecular mechanisms (see Drake 1970; Stent 1971; Freese 1971). However, evidence that a particular mechanism actually operates is more convincing for some of the chemically induced mutations than it is for the spontaneous ones.

Transitions appear to be caused by most of the major classes of mutagenic chemicals: nitrous acid, base analogs, hydroxylamine, and alkylating agents. Additions and deletions, resulting in frame-shift mutations, are induced by intercalating chemicals such as acridine and various amino acridines. It appears that some alkylating agents, notably ethylethane sulfonate, may cause transversion. Transitions and transversions affecting as they do one nucleotide are said to cause point mutations.

The type of mutational alteration in a viral nucleic acid can tentatively be identified by genetic criteria briefly summarized as follows: Transitions are induced to revert with base analogs; frame-shift mutations are generally induced to revert by treatment with acridines; and mutants not induced to revert with either type of agent but that do revert spontaneously are likely to be transversions.

Before considering individual mutagens some distinction should be made between *in vitro* and *in vivo* mutagenic action. In some cases, a viral nucleic acid, either isolated or in the virion, can be altered by chemical treatment in such a way as to represent a mutation or to induce a mutation during subsequent replication. This is *in vitro* mutagenesis; some chemicals causing mutation in this manner are nitrous acid, hydroxylamine, and alkylating agents. *In vivo* mutagenesis occurs when the mutagenic agent is present during replication of the viral nucleic acid; base analogs and intercalating chemicals can act mutagenically at this stage of virus replication.

a. Nitrous Acid

Nitrous acid is one of the most effective mutagenic chemicals. As indicated in Sect. VIA, 3c, it oxidatively deaminates primary amino groups of purines and pyrimidines in viral nucleic acids. Thus adenine (A) is converted to hypoxanthine (H), cytosine (C) to uracil (U), guanine (G) to xanthine (X). The C to U transition represents a direct substitution of one common nucleotide for another; the other two nitrous acid-induced transitions yield natural products, but ones seldom, if ever, found in nucleic acids. Upon further replication the transition of cytosine to uracil is perpetuated while the hypoxanthine resulting from deamination of adenine pairs with cytosine, which, in turn, pairs with guanine so that ultimately an adenine is replaced by guanine. Both these transitions are potentially mutagenic (they are also potentially inactivating as previously noted). Xanthine, produced by the deamination of guanine can pair with cytosine, which in the next replicative step will pair with guanine thus restoring the original sequence. However, xanthine appears to ionize more at pH values around neutrality than other bases, and this interferes with its base-pairing function producing a lethal effect. This lethality is demonstrated in a practical manner by the observation that deamination of guanine-containing polynucleotides renders them inactive as templates (Basilio et al. 1962). The direct and indirect effects of nitrous acid may be sketched for RNA and DNA showing just the relevant base or bases in the polynucleotide chain as follows:

Replicative events after oxidative deamination of bases in single-stranded RNA:

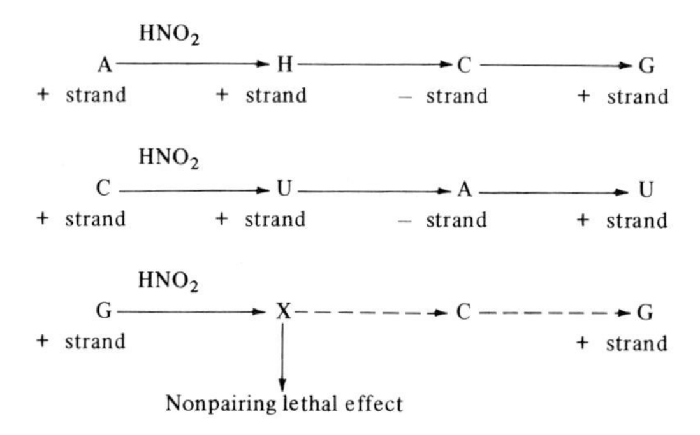

The same scheme applies to the appropriate bases in double-stranded RNA. Replicative events after oxidative deamination of bases in double-stranded DNA:

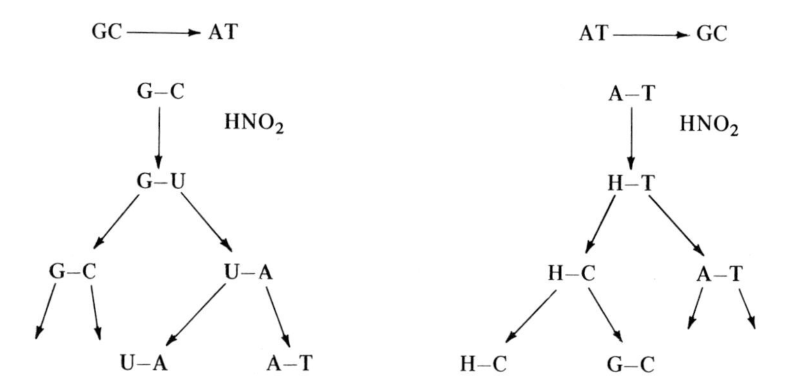

The guanine to xanthine transition in DNA proceeds as indicated for RNA except that complementary bases are carried along in the scheme.

Nitrous acid has been the most effective mutagen for RNA but relatively poor for DNA, presumably because amino groups are virtually unavailable in double-stranded nucleic acid molecules except in zones of incipient denaturation (strand separation). Instead of deamination, several inactivating side reactions tend to occur with DNA as indicated in Sec. VIA, 3c.

The interaction between proteins and nucleic acid can affect the mutagenicity of nitrous acid as shown by difference in reactivity of free and virus-incorporated RNA of TMV. Under conditions that result in considerable deamination of adenine, guanine, and cytosine in isolated TMV-RNA, essentially no deamination of guanine occurs in whole virus while 20 and

30 percent of the adenine and cytosine, respectively, are deaminated (Schuster and Wilhelm 1963). A consequence of this difference in reactivity is that the inactivating effect of nitrous acid on guanine is reduced in relationship to the mutagenic action on cytosine and adenine (Singer and Fraenkel-Conrat 1969).

Some relationships of nitrous acid mutagenesis to the genetic code will be considered in the next section.

b. Hydroxylamine

Hydroxylamine is considered one of the most specific *in vitro* mutagens since it and its derivatives (for example, methoxyamine and *N*-methylhydroxylamine) react only with pyrimidines; the reaction with cytosine (or 5-hydroxymethylcytosine in T-even phages) is very probably mutagenic while that with uracil is inactivating (Drake 1970). As indicated earlier, the reaction with uracil (and presumably with thymine also) occurs mainly at pH 9–10, whereas the reaction with cytosine takes place optimally around pH 6. The reaction shown at the left in Sec. VIA, 3e, involving the 4-amino group of cytosine, is considered to be mutagenic (Singer and Fraenkel-Conrat 1969; Phillips and Brown 1967). The base pairing of cytosine is altered by formation of the hydroxylamine derivative so that it can pair with adenine in place of the usual guanine; upon further replication (when DNA is involved) this results in an A-T pair in place of G-C, thymine having taken the place occupied by cytosine in the sequence of the original polynucleotide chain (Freese 1971).

In vivo mutagenic action on the single-stranded DNA phages S13 and ØX174 by hydroxylamine has been reported (Tessman et al. 1965). Transitions affecting all of the bases appeared to be effected when bacterial cells were treated with hydroxylamine prior to infection. The precise mechanism of this nonspecific *in vivo* induction of mutants is unclear.

c. Alkylating Agents

Alkylation is effected by treatment of viruses or their nucleic acids with nitrogen and sulfur mustards, alkyl sulfates, alkyl sulfonates, ethylene oxide, propylene oxide, and so on. The alkyl group is introduced at the 7 position of guanine and to some extent elsewhere such as the N-1 of adenine (the latter may be the main reaction with denatured or single-stranded DNA, while the former seems to be the major reaction with double-stranded DNA). The alkylation of guanine causes a tautomeric shift and a change in base pairing, resulting mainly in transitions from G-C to A-T pairs in DNA (Freese 1971).

d. Base Analogs

Base analogs are *in vivo* mutagens that cause mutations during replication of DNA, as was first demonstrated by the classic study of Litman and

Pardee (1956) with T2 coliphage and 5-bromouracil. While there is little doubt that some base analogs (for example, 2-thiouracil and 8-azaguanine) inactivate plant viruses when incorporated into the viral RNA, it is uncertain that any of them is mutagenic. Not many base analogs are mutagenic with DNA either, the halogenated uracils (5-bromo, 5-chloro, and 5-iodouracil) and 2-amino purine being most efficient in this regard, especially when introduced to the cells in the form of their respective deoxynucleosides (Freese 1971). The halogenated uracils are apparently incorporated into DNA in place of thymine, bromouracil (BU) working especially well presumably because the 5-bromo group occupies about the same space as the 5-methyl group of thymine. However, the electronic characteristics of Br and CH_3 are quite different so that a certain tautomeric shift occurs more frequently with BU than with thymine. Thus, BU, like thymine, is normally in the keto form (C=O group at position 4) in which it, like thymine, pairs with adenine; a tautomeric shift of H from N-3 to the O of the keto group at position 4 converts the keto into an enol group (C-OH) and promotes pairing with guanine instead of with adenine.

This mispairing can cause a transition in either of two ways: If BU is in the rare enol form during synthesis of a DNA strand, it can be incorporated opposite guanine in place of cytosine (error due to substrate enol, also called "mistake in incorporation" or "substrate transition"). When BU has been incorporated into a DNA strand and that strand serves as a template for replication, adenine is normally paired with BU (keto form), but if BU is in the rare enol form, it will pair with guanine instead (error due to template enol, also called "mistake in replication" or "template transition"). Thus BU causes *in vivo* transitions: GC to AT and AT to GC. The GC to AT transition results as shown in Figure 44a when BU is in the enol form at the time of incorporation (step 2 in the figure) while the AT to GC transition occurs during replication when BU, having been incorporated while in the common keto form (Figure 44b, step 2) shifts to the enol form and pairs with guanine (Figure 44b, step 3).

Comparison of a and b in Figure 44 suggests that BU-induced transitions should be reversible by treatment with BU; such reversions have been observed experimentally (Freese 1971).

e. Intercalating Chemicals

Certain acridine dyes and the structurally similar phenanthridines can intercalate between adjacent purine and pyrimidine base pairs of the DNA double helix (Lerman 1964; Cairns 1962). Examples of these two classes of intercalating molecules are shown in Figure 45. Intercalation of a molecule such as proflavin into a phage DNA in infected bacterial cells can cause *in vivo* mutations that appear based on additions or deletions usually of one or a few nucleotides (Lerman 1964; Drake 1970). Since transcription of the

(a) Error due to substrate enol

1 2 3 4

C–G G–BU BU–A A–T

(b) Error due to template enol

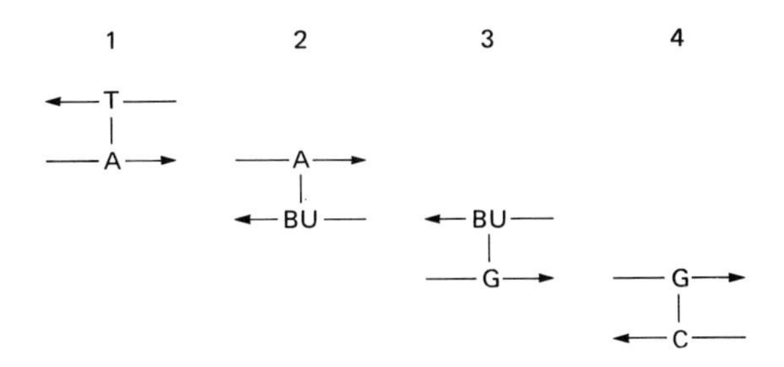

Fig. 44. Transitions caused by bromouracil.

genetic code appears to begin at a fixed point and to move along a linear series of nonoverlapping, nucleotide triplets, addition or deletion of even a single nucleotide can have a profound effect on the transcription of a DNA strand. The transcribing is shifted so that every codon (nucleotide triplet) to the right of the addition or deletion is altered. Mutants resulting from such altered transcription are called frame-shift or phase-shift mutants because the frame of reference for transcribing the nucleotides as triplets has been shifted; in other words, the transcription has been shifted out of phase. Conceptually at least, the addition or deletion of nucleotides in the form of one or more codons might be far less drastic than insertion or deletion of a single nucleotide because frame shifts would be avoided.

However, the key question of how intercalating compounds accomplish their mutagenic action is not yet clear. It is thought that single-strand breaks ("nicks") occur quite frequently in DNAs, and Streisinger et al. (1966) surmised that intercalated molecules stabilize the mispairing that may occur between complementary strands of DNA that have undergone breakage and partial strand separation. Mispairing is accompanied by looping out of one of the strands of the duplex, with or without some excision by

Proflavin (2,8-diaminoacridine)

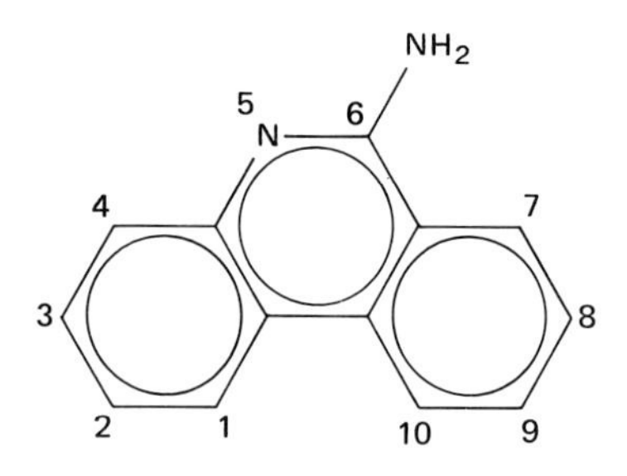

6-aminophenanthridine

Fig. 45. Basic formulas for two classes of intercalating chemicals that induce frame shifts in DNA.

nuclease, and is followed by repair, which leaves one strand with its original composition while the other is either lengthened or shortened. These steps are illustrated in Figure 46.

The reversion of mutations caused by acridines and similar compounds, which is also induced by treatment with acridines, presumably results from restoration of the proper number of base pairs by addition or deletion. [If hundreds of nucleotides have been deleted, as occasionally occurs (Benzer 1961), restoration would not be expected.] The reverting addition or deletion need not be at the site of the original alteration but can be at a closely linked site. On the basis of revertibility by acridines, it appears that the majority of spontaneous mutants of T4 phage are frame shift (Drake 1970).

The mutagenicity of acridines can be complicated by such factors as the different permeabilities of cells to various acridine derivatives and the presence or absence of visible light since acridines can also induce mutations photodynamically. The latter type of mutation appears to involve nucleotide substitution rather than addition-deletion. It has been observed with phage T4 (Ritchie 1964, 1965).

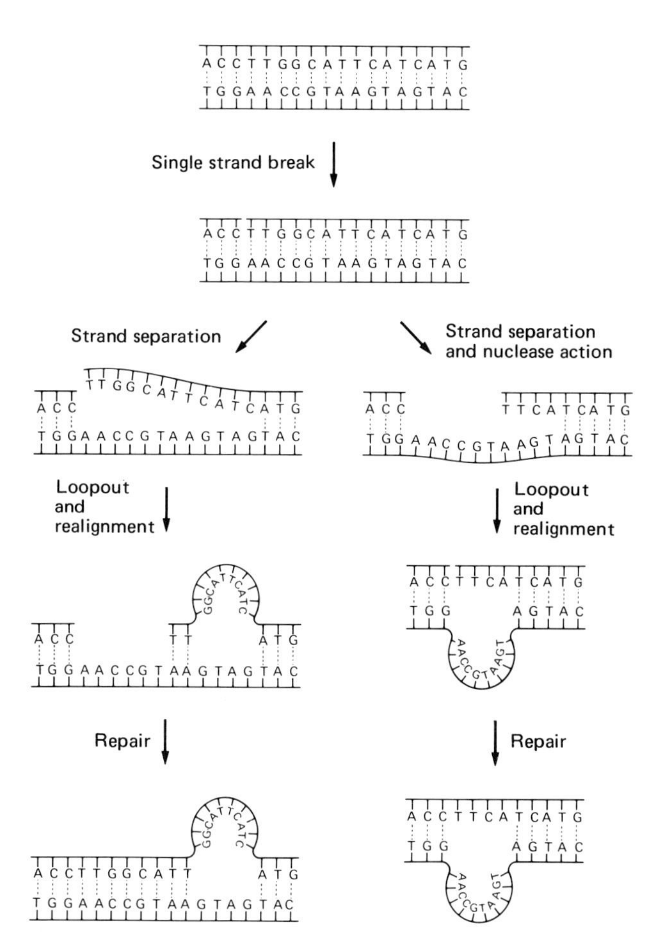

Fig. 46. Hypothetical mechanism for the production of frame-shift mutants according to the Streisinger model. As shown, after a single break has been introduced in one of the strands of DNA, strand separation occurs over a small distance, either followed by nuclease action on the open strand (*right*) or not (*left*). In both cases, loopout of a strand occurs next but in different places, and is followed by repair that closes the interrupted strand. In the next round of replication, it is evident that the top strand of the product on the left will give rise to a duplex containing an addition, whereas the top strand of the product on the right will yield a duplex with a deletion. (Adaptation Knight 1974 from Streisinger et al. 1966.)

2. *Effect of Mutations on Viral Proteins*

Mutation is based on significant changes in the composition of nucleic acids, but such alterations are directly demonstrable with great difficulty if at all by current analytic techniques. Even in a case where the nature of the chemical change induced by a mutagenic chemical is well established, such as the oxidative deamination of cytosine to uracil, the number of cytosines that must be converted to uracil in order to be detectable by analysis is many times larger than theoretically required for mutagenesis, although small changes might conceivably be detected by sequencing techniques described earlier. Consequently, mutation often has been analyzed by genetic methods which, while very valuable, are by nature indirect and highly deductive. However, since mutations are generally reflected in the structure of some protein (the sequence of amino acids is determined by the sequence of nucleotides in the appropriate gene), the consequences and mechanisms of mutation can also be evaluated by analysis and sequencing of proteins. This is technically feasible in at least some cases. A few examples will be cited here to illustrate the consequences of spontaneous mutation and of the mutagenic procedures sketched in the previous section.

Strains of tobacco mosaic virus are exceedingly numerous and varied in their properties. Furthermore, most strains are easily grown in tobacco plants and can be isolated and readily purified in sufficient quantities for analysis.

Analyses of the coat proteins of some spontaneous mutants of TMV were made many years ago (Knight and Stanley 1941; Knight 1942, 1943, 1947b). The complete analyses of strain proteins (Knight 1947b) were made by microbiological assay at a time when the only other complete analysis of any protein was that of Brand et al. (1945) of beta-lactoglobulin. Now, complete amino acid analyses of proteins are made routinely and with greater accuracy with automatic analyzers employing ion exchange columns (see Sec. IIIA, 2a). Scores of proteins have been analyzed (Dayhoff 1972, 1973), including numerous viral proteins. Nevertheless, the following conclusions drawn from the microbiological assays of the TMV strain proteins have been confirmed and strengthened by subsequent more refined analyses: (1) Strains often differ from one another in protein composition. These differences usually appear in proportions of the constituent amino acids. However, in some cases, certain amino acids (notably histidine and methionine) may be present in one strain which are completely lacking in another. (2) The closely related strains, which presumably have arisen by one or few mutational steps, are very similar in protein composition. In fact, there seem to be no differences at all in the proteins of common TMV and M. (3) Differences in strain proteins seem to involve almost any of the constituent amino acids.

These conclusions are consistent with current genetic theory, pertinent portions of which are as follows: The genetic material of any virus is its nucleic acid whose linear sequence of nucleotides comprise a few to many genes (also called cistrons), depending mainly on the size of the nucleic acid. One of these genes determines the composition and sequences of amino acids of a viral coat protein. Spontaneous mutation is essentially a random process that may or may not affect the coat protein gene, and if it does might be expected to alter randomly any one of the amino acids in the coat protein. Closely related strains would be expected to have fewer differences in coat protein than distantly related strains.

These principles are illustrated especially well by comparing the sequences of amino acids in viral proteins of related groups of viruses. A graphic way to express the results is not to present the detailed sequential information but rather to express the results in terms of the percent difference between sequences of the viral proteins. This is done in Tables 42 and 43 for the coat proteins of some strains of tobacco mosaic virus and of some small RNA phages, respectively. The viruses in each of these two groups are considered related because they exhibit common physical, chemical, and serological characteristics, as well as other similarities, such as host specificity, mode of infection, and so forth. The values shown in Tables 42 and 43 illustrate the point made earlier, that viral coat proteins of different strains of a virus may not differ in any respect (hence their coat protein genes must be identical) or they may differ somewhat or considerably, again reflecting various degrees of similarity in their coat protein genes.

The determination of the genetic code has made it possible to relate viral protein analyses to nucleic acid sequences (that is, to relate a gene

Table 42. Comparison of Amino Acid Sequences of Some Spontaneous Mutants of Tobacco Mosaic Virus.[a]

	TMV	OM	Dahl.	ORS	U2	HR
Strain	% Difference Between Sequences					
TMV (common tobacco strain)	0	2	18	26	27	55
OM (tobacco strain)	2	0	18	26	27	56
Dahlemense (tobacco strain)	18	18	0	27	30	54
ORS (orchid strain)	26	26	27	0	30	53
U2 (tobacco strain)	27	27	30	30	0	55
HR (ribgrass strain)	55	56	54	53	55	0

[a]Adapted from Dayhoff 1972.

Table 43. Comparison of Amino Acid Sequences of Some Related RNA-Containing Bacteriophages.[a]

	f2	MS2, R17	ZR	fr	Qβ
Bacteriophage	% difference between sequences				
F2	0	1	1	17	75
MS2, R17	1	0	0	16	75
ZR	1	0	0	16	75
fr	17	16	16	0	76
Qβ	75	75	75	76	0

[a]Adapted from Dayhoff 1972 but not corrected to reflect later analytical data showing three differences between the coat proteins of MS2 and R17 (Min Jou et al. 1972), which would also alter slightly the other values in the table.

product to its gene on the molecular level). In fact, during early stages of development of the genetic code, analyses of nitrous acid mutants of TMV were helpful in confirming codons as well as in testing the universality of the code (Ochoa 1963). Additional confirmation of the code has come from later extensive studies of the coat proteins of the f2-MS2-R17 group of RNA phages (see Min Jou et al. 1972).

The genetic code presented in Table 44 has developed over the course of about ten years. It was proposed by Crick et al. (1961), on the basis of several lines of evidence, that the genetic code is a triplet, degenerate (degenerate means that more than one triplet codes for a given amino acid), nonoverlapping code which is read from a fixed starting point. These premises have been validated over the years. The 64 triplets of the genetic code are shown in Table 44 using A, G, C, and U as abbreviations for the ribonucleotides: adenylic acid, guanylic acid, cytidylic acid, and uridylic acid, respectively.

The earliest assignments of nucleotides to the codons (codons are the coding triplets) were derived from data obtained in cell-free syntheses of polypeptides from radioactive amino acids using synthetic polyribonucleotides of differing compositions as messengers (Matthaei and Nirenberg 1961; Nirenberg and Matthaei 1961; Nirenberg et al. 1962). The essential components of this system were washed *E. coli* ribosomes and tRNAs, an ATP-generating system, labeled amino acid, and polynucleotides containing different proportions of nucleotides as test messengers. By noting the amount of incorporation of a given ^{14}C-labeled amino acid in response to enzymatically synthesized messengers containing different proportions of nucleotides, it was possible to deduce more than a third of the codons (Martin et al. 1962; Lengyel et al. 1961; Speyer et al. 1962a, 1962b; Lengyel et al. 1962; Matthaei et al. 1962). Later, synthetic messengers whose nucleotide sequences were more precisely known were prepared and used by

Table 44. The Genetic Code.

Ala	GCA	Gly	GGA	Pro	CCA
	GCC		GGC		CCC
	GCG		GGG		CCG
	GCU		GGU		CCU
Arg	AGA	His	CAC	Ser	AGC
	AGG		CAU		AGU
	CGA				UCA
	CGC	Ile	AUA		UCC
	CGG		AUC		UCG
	CGU		AUU		UCU
Asp	GAC	Leu	CUA	Thr	ACA
	GAU		CUC		ACC
			CUG		ACG
Asn	AAC		CUU		ACU
	AAU		UUA		
			UUG	Trp	UGG
Cys	UGA				
	UGC				
	UGU	Lys	AAA	Tyr	UAC
			AAG		UAU
Glu	GAA				
	GAG	Met	AUG	Val	GUA
		FMet			GUC
Gln	CAA	Phe	UUC		GUG
	CAG		UUU		GUU

Initiation codons: AUG; GUG.
Termination codons: amber UAG, ochre UAA, umber (opal) UGA.

Khorana and associates (1966) to test the specificities of various codons in cell-free syntheses. Still another ingenious method for testing codons was developed by Nirenberg and associates (1966). This technique was based on the synthesis of all 64 triplets and the demonstration that these would bind to ribosomes where in turn specific ^{14}C-amino acid tRNAs would attach while the nonspecific ones would not; the codon-ribosome-amino acid-tRNA complex was retained on a fine cellulose nitrate filter and the quantity could be ascertained by radioactivity measurements in a scintillation counter. If the codon was not right, the amino acid-tRNA would pass through the filter and subsequent radioactivity counts would be low.

All of the procedures sketched above plus other lines of evidence have contributed to the codon catalog presented in Table 44. Each of the procedures used to identify the codons showed some anomalies, but together generally support the listing given.

The arrangement used in Table 44 was developed from the data of P.

Leder in Sober (1970) and presented in Knight (1974). Some advantages of this unconventional arrangement are (1) the amino acids are listed in alphabetical order for quick reference; (2) all of the codons associated with a given amino acid are present in one place and are in alphabetical order; (3) some basic features of the genetic code are immediately apparent; for example, the general identity of the first two nucleotides and the variability of the third in the codons for a particular amino acid.

With the genetic code in mind and considering molecular mechanisms of mutation, the question could be raised concerning the origin of spontaneous mutants. This is a question that cannot yet be answered definitively. However, largely on the basis of what mutagenic agents can cause reversions in the T-even bacteriophages, it appears that spontaneous mutations of these viruses may involve mainly frame shifts rather than base substitutions (Brenner et al. 1961). In the case of plant viruses a comparison of the sequences of amino acids in the coat proteins of the common and HR strains of tobacco mosaic virus (Funatsu and Funatsu 1968; Hennig and Wittmann 1972) suggests that if HR evolved from TMV a rich variety of mutational events must have occurred, including transitions, transversions, frame shifts, and deletions. Further study is needed to define better the mutational mechanisms operative in nature.

The laboratory-produced mutants of the tobacco mosaic virus series have illuminated a number of relationships between genes and gene products. Some of the findings will be briefly reviewed here.

Nitrous acid has proved to be the most efficient mutagen for strains of TMV. The coat proteins of scores of these mutants were analyzed (Funatsu and Fraenkel-Conrat 1964; Henning and Wittmann 1972). In a large number of cases, the coat protein gene of the virus was unaffected as shown by the lack of any differences in coat protein. Where changes in coat protein were found in the nitrous acid mutants, most of them proved to be transitions of the A to G and C to U types expected in nitrous acid mutagenesis. This is illustrated by some data compiled in Table 45, which provide support for operation of the genetic code of Table 44 and are consistent with the idea of a universal genetic code. Although not indicated in Table 45, most of the nitrous acid mutants showing any amino acid exchange showed only one. This can be taken as evidence for a non-overlapping genetic code, since one change in an overlapping code would often have brought about two exchanges of amino acids next to each other in the polypeptide chain. No such contiguous exchanges were observed.

The nature of the code as presented in Table 44 suggests several potential effects of mutagenesis on the gene products of a viral nucleic acid:

1. If a base substitution occurs in the first or second nucleotide of a triplet, a different amino acid will almost always result from translation of the codon (arginine and leucine are partial exceptions); when the substitu-

Table 45. Amino Acid Exchanges Found and Corresponding Codon Changes Associated with Coat-Protein Mutants Obtained by Treatment of Tobacco Mosaic Virus with Nitrous Acid.[a]

No. of Mutants	Amino Acid Exchange	Corresponding Codon Change	Nature of Change
1	Arg → Cys	CGU → UGU	C → U
4	Arg → Gly	AGG → GGG	A → G
1	Arg → Lys	AGA → AAA	A → G
4	Asn → Ser	AAU → AGU	A → G
1	Asp → Gly	GAU → GGU	A → G
1	Gln → Arg	CAG → CGG	A → G
1	Glu → Asp	GAA → GAU	A → U
2	Glu → Gly	GAA → GGA	A → G
1	Ile → Met	AUA → AUG	A → G
5	Ile → Val	AUU → GUU	A → G
4	Pro → Leu	CCU → CUU	C → U
2	Pro → Ser	CCC → UCC	C → U
1	Ser → His	UCU → CAU	UC → CA
3	Ser → Leu	UCG → UUG	C → U
5	Ser → Phe	UCU → UUU	C → U
2	Thr → Ala	ACU → GCU	A → G
6	Thr → Ile	ACU → AUU	C → U
2	Thr → Met	ACG → AUG	C → U
1	Tyr → Cys	UAC → UGC	A → G
1	Val → Met	GUG → AUG	G → A

[a]Compiled from Hennig and Wittmann 1972.

tion occurs in the third nucleotide, no amino acid change will result if the amino acid concerned is one coded for by four triplets. But there may or may not be a change if the amino acid has less than four codons, depending on the base substituted. In addition there is a special case: a third-nucleotide transition in the Trp codon produces the umber peptide chain termination codon.

2. Termination codons can be generated by single transitions of Gln and Trp codons or single transversions of Glu and Lys codons.

3. Base substitution in a single nucleotide of a codon can lead to any of the following effects on a protein: polypeptide chain termination in a few cases (see point 2); no change in amino acid and hence no effect (many cases, see point 1); a change in amino acid either with or without detectable effect on protein function.

4. A frame shift has potentially the most drastic effect on the genetic function of nucleic acid because it can produce any of the consequences mentioned above, and in generous measure, since by nature it almost surely affects more than one codon.

Mutations that cause chain termination and hence loss of function of some protein are invaluable in genetic studies (Hayes 1968), especially when combined with the observation that a permissive host may be available that can respond not by chain termination but rather by translation of the termination codon as an amino acid codon. The latter response is known as suppressor effect and is based on a suppressor gene in the permissive host that appears to code for a minor species of amino acid-tRNA whose anticodon (triplet complementary to a codon) is complementary to the termination codon (Garen 1968). Thus a functional polypeptide is produced in the presence of a suppressor; for example serine may be inserted in the permissive host in response to the amber codon (Tooze and Weber 1967).

Terminations resulting from the amber codon (UAG) have been detected frequently among phage mutants and are termed amber (am) mutants (Sarabhai et al. 1964; Hayes 1968, p. 484). Amber mutants are distinguished from ochre or umber mutants by their responses to different suppressors, whose specificities were established by mutagenesis and protein analytic studies on alkaline phosphatase of *E. coli* (Garen 1968).

A series of amber mutants of the small RNA phages, f2, R17, and MS2, were produced mostly by treatment with nitrous acid but also by hydroxylamine and fluorouracil mutagenesis (Zinder and Cooper 1964; Webster et al. 1967; Gussin 1966; Tooze and Weber 1967; Min Jou et al. 1972). Some of these mutations were in the coat protein gene and were located by amino acid sequence analyses. All of them involved glutamine, except one, which affected tryptophan. These changes provide chemical confirmation for the amber nature of the mutation since the amber codon (UAG) could predictably arise by the treatments employed from the glutamine and tryptophan codons, CAG and UGG, by simple transitions.

A neat demonstration of the colinearity of gene nucleic acid and gene product (protein) was made with amber mutants of coliphage T4 by Sarabhai et al. (1964). The experiment devised was based on the concept that if there is colinearity between gene and protein, each of several different amber mutants should yield after transcription and translation a polypeptide chain whose length would depend on the distance of mutation from one end of the gene. By infecting separate cultures of nonpermissive bacteria with different amber mutants (grown up in the permissive host) and adding radioactive amino acids late in the infectious process when phage head protein is predominately synthesized, it was possible to identify a series of proteins that by comparison with wild type protein did represent polypeptides of different lengths.

As suggested earlier, mutation can change the function of a protein—whether it be plant virus coat protein, influenza hemagglutinin, or phage lysozyme. There are numerous cases of such effects but a few examples will illustrate some of the possibilities.

Some of the nitrous acid mutants of tobacco mosaic virus are difficult to pass in series, which was found to be related in certain cases to the defective nature of the coat protein of the mutant virus (Siegel and Zaitlin 1965). Such protein may be unable to assemble around the viral RNA; consequently the RNA is readily digested by plant ribonucleases when attempts are made to isolate or transfer the virus. Analysis of the protein of one of these mutants, PM2, revealed only two differences between it and wild type TMV protein: threonine in position 28 had been replaced by isoleucine, and glutamic acid in position 95 had been replaced by aspartic acid (see Table 13 for sequence of amino acids in TMV protein). The pronounced effect of these changes on the ability of the protein to aggregate into tight, rodlike (tubular) structures is illustrated in the electron micrographs of Figure 47 in which polymerized wild type protein and polymerized PM2 protein are contrasted. It is not known whether both of the amino acid exchanges noted above contribute to the abnormal aggregation of PM2 protein, but it has been suggested that the crucial exchange might be the one substituting leucine for threonine since this amounts to

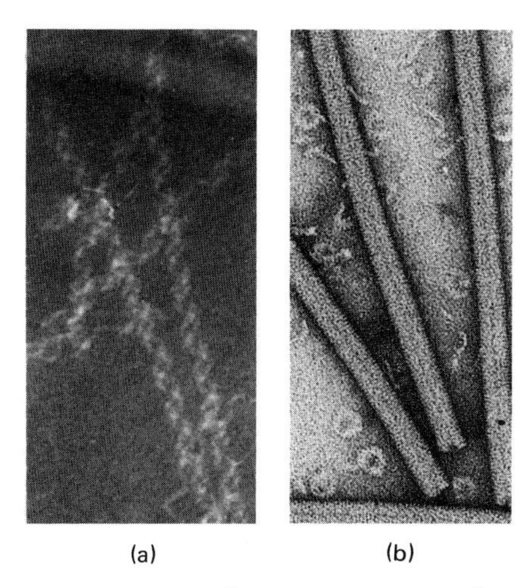

Fig. 47. Electron micrographs, prepared by the negative staining technique, of elongated structures formed in the absence of nucleic acid by aggregation of tobacco mosaic virus coat protein. *a*. Open helical structures formed by aggregation of protein of TMV nitrous acid mutant PM2. *b*. Regular close-packed helical structures formed by aggregation of TMV protein. (*a*, courtesy M. Zaitlin; *b*, courtesy R. C. Williams.)

replacing an amino acid residue having a hydrophilic side chain (Thr) with one of hydrophobic character (Leu), whereas the amino acids in the other exchange are essentially equivalent.

Two nitrous acid mutants of TMV isolated by Sengbusch and Wittmann (1965) presented an unusual opportunity to evaluate the effect on serological specificity of single amino acid exchanges in a viral coat protein. Differences in serological specificity had previously been demonstrated between TMV and TMV from all of whose protein subunits the C-terminal threonine residues had been removed (Knight 1961), but the unique feature of the two nitrous acid mutants (Ni 118 and Ni 1927) was that they exhibit the same amino acid exchange, proline to leucine, but in different positions in the polypeptide chain. The amino acid exchange occurred in position 20 (near the N-terminal) in Ni 118 and in position 156 (near the C-terminal; see Table 13) in Ni 1927. In serologic tests it was found that the mutant with the exchange at position 156 could be distinguished from TMV but the other mutant could not. From earlier studies it was known that the C-terminal of the TMV protein subunits are located near the surface of the virus particle, whereas the N-terminal is folded in toward the interior of the particle. Hence, the single exchange located near the surface of the particle in Ni 1927 registered serologically while the same exchange internally oriented in Ni 118 was serologically undetectable.

One of the virus-specific proteins of the T-even phages is the enzyme, lysozyme. This enzyme is a product of the e gene (endolysin gene) of phage T4 and can be readily purified and analyzed. It is possible to produce mutants of T4 in which the phage lysozyme is affected, including mutants resulting from *in vivo* mutagenesis with acridines (Streisinger et al. 1966). It will be recalled from the earlier discussion of frame-shift mutants that acridines such as proflavin cause deletions or additions of nucleotides (often one nucleotide) resulting in gross changes in the affected gene. However, if the deletion of a nucleotide is followed by addition of one beyond the region of mutation (or the reverse, addition followed by deletion), the damage can sometimes be corrected. Terzaghi and associates (1966), working with proflavin mutants of T4 phage, obtained a double mutant whose lysozyme differed from that of wild type in that it was partially rather than fully active and possessed a different sequence of five amino acids in an eight amino acid sequence:

Wild type: -Thr-Lys-Ser-Pro-Ser-Leu-Asn-Ala-
Mutant : -Thr-Lys-Val-His-His-Leu-Met-Ala-

The appropriate codons of the mRNA for the eight-amino acid segment of wild type lysozyme can be shown as follows:

1	4	7	10	13	16	19	22
A C A -	A A A -	A G U -	C C A -	U C A -	C U U -	A A U -	G C U -

Deletion-addition events explaining the observed amino acid sequence of the double mutant lysozyme are as follows: Delete nucleotide 3 (A), shift nucleotides beyond 3 one place to the left to form new triplets; add G between nucleotides 22 and 23. The resulting mRNA segment is

1	4	7	10	13	16	19	22
A C A -	A A A -	G U C -	C A U -	C A C -	U U A -	A U G -	G C U -

In addition to confirming postulates about the nature of frame-shift mutations, the results of these studies also confirmed the concept that mRNA is translated in the 5′ to 3′ direction.

In relating a gene to its product, the ultimate achievement on the molecular level is to make a direct comparison of the relevant mRNA nucleotide sequences and the amino acid sequences of the protein specified. Such a comparison has been made with coliphage MS2 (Min Jou et al. 1972). This virus is a small RNA phage with only three genes—one for an RNA replicase (also called RNA polymerase or synthetase), one for the main coat protein, and a third coding for a special coat protein called A protein or maturation protein. The main coat protein of this phage, a 129-amino acid polypeptide, can be readily isolated and its amino acid sequence had been determined, although with three errors according to the gene analysis referred to here. The coat protein gene was isolated and analyzed as follows.

^{32}P-labeled MS2 RNA was subjected to limited digestion with ribonuclease T_1 at low temperatures which yielded fragments that could be separated on neutral polyacrylamide gels and were small enough for direct sequence analysis by procedures described in Sec. IIIB, 2. Knowledge of the amino acid sequences of MS2 coat protein, coupled with the genetic code, enabled screening of the oligonucleotides obtained from ribonuclease T_1 digestion for the ones related to coat protein gene. This resulted in isolation and sequencing of five fragments spread over the gene (Min Jou et al. 1971). These and many other analyses were combined by taking advantage of overlapping oligonucleotides and knowledge of what the order of nucleotides should be from the amino acid sequence of the protein. The nucleotide sequence obtained is shown in Figure 48. It includes not only the coat protein gene with initiation and termination codons but also the ribosomal binding site for the coat protein gene and one for the RNA polymerase together with the initiation codon (AUG) and the first six codons for the polymerase.

The nucleotide triplets which were found to code for the 129 amino acids of the MS2 coat protein are indicated in Table 46, the figures in parentheses indicating the number of times a particular codon is used in the gene. Forty-nine different codons are employed and there is no clear basis yet for explaining the choice between degenerate codons. However, it will be noted that all of the codons found agree with standard ones listed in the codon catalog.

. . . (G)· AUA· GAG· CCC· UCA· ACC· GGA· GUU· UGA· AGC· AUG·

GCU· UCU· AAC· UUU· ACU· CAG· UUC· GUU· CUC· GUC· GAC· AAU· GGC· GGA· ACU· GGC· GAC· GUG· ACU· GUC· GCC· CCA· AGC· AAC· UUC·
Ala Ser Asn Phe Thr Gln Phe Val Leu Val Asp Asn Gly Gly Thr Gly Asp Val Thr Val Ala Pro Ser Asn Phe
1 5 10 15 20 25

GCU· AAC· GGG· GUC· GCU· GAA· UGG· AUC· AGC· UCU· AAC· UCG· CGU· UCA· CAG· GCU· UAC· AAA· GUA· ACC· UGU· AGC· GUU· CGU· CAG·
Ala Asn Gly Val Ala Glu Trp Ile Ser Ser Asn Ser Arg Ser Gln Ala Tyr Lys Val Thr Cys Ser Val Arg Gln
30 35 40 45 50

AGC· UCU· GCG· CAG· AAU· CGC· AAA· UAC· ACC· AUC· AAA· GUC· GAG· GUG· CCU· AAA· GUG· GCA· ACC· CAG· ACU· GUU· GGU· GGU· GUA·
Ser Ser Ala Gln Asn Arg Lys Tyr Thr Ile Lys Val Glu Val Pro Lys Val Ala Thr Gln Thr Val Gly Gly Val
55 60 65 70 75

GAG· CUU· CCU· GUA· GCC· GCA· UGG· CGU· UCG· UAC· UUA· AAU· AUG· GAA· CUA· ACC· AUU· CCA· AUU· UUC· GCU· ACG· AAU· UCC· GAC·
Glu Leu Pro Val Ala Ala Trp Arg Ser Tyr Leu Asn Met Glu Leu Thr Ile Pro Ile Phe Ala Thr Asn Ser Asp
80 85 90 95 100

UGC· GAG· CUU· AUU· GUU· AAG· GCA· AUG· CAA· GGU· CUC· CUA· AAA· GAU· GGA· AAC· CCG· AUU· CCC· UCA· GCA· AUC· GCA· GCA· AAC·
Cys Glu Leu Ile Val Lys Ala Met Gln Gly Leu Leu Lys Asp Gly Asn Pro Ile Pro Ser Ala Ile Ala Ala Asn
105 110 115 120 125

UCC· GGC· AUC· UAC· UAA· UAG· ACG· CCG· GCC· AUU· CAA· ACA· UGA· GGA· UUA· CCC· AUG· UCG· AAG· ACA· ACA· AAG· AAG· (U)
Ser Gly Ile Tyr Ser Lys Thr Thr Lys Lys
129 1 5

Fig. 48. Nucleotide sequence of the coat protein gene of coliphage MS2 together with the amino acid sequences specified by this gene. The gene is preceded and followed by nongenic regions; note the initiating codon (AUG) preceding the first codon of the coat protein and the double terminating codons (UAA and UAG) at the end of the gene. The numbers refer to the positions of amino acid residues in the coat protein (1–129) and the first six in the replicase gene. (From Min Jou et al. 1972.)

Table 46. Codons Used in Phage MS2 Coat Protein Gene.[a]

Ala	GCA (6)[b]	Gly	GGA (2)	Pro	CCA (2)
	GCC (2)		GGC (3)		CCC (1)
	GCG (1)		GGG (1)		CCG (1)
	GCU (5)		GGU (3)		CCU (2)
Arg	AGA	His	CAC	Ser	AGC (4)
	AGG		CAU		AGU
	CGA				UCA (2)
	CGC (1)	Ile	AUA		UCC (2)
	CGG		AUC (4)		UCG (2)
	CGU (3)		AUU (4)		UCU (3)
Asp	GAC (3)	Leu	CUA (2)	Thr	ACA
	GAU (1)		CUC (2)		ACC (4)
			CUG		ACG (1)
Asn	AAC (6)		CUU (2)		ACU (4)
	AAU (4)		UUA (1)		
			UUG	Trp	UGG (2)
Cys	UGC (1)				
	UGU (1)	Lys	AAA (5)	Tyr	UAC (4)
			AAG (1)		UAU
Glu	GAA (2)				
	GAG (3)	Met	AUG (2)	Val	GUA (3)
					GUC (4)
Gln	CAA (1)	Phe	UUC (1)		GUG (3)
	CAG (5)		UUU (3)		GUU (4)

[a]Adapted from Min Jou et al. 1972.

[b]Numbers in parentheses refer to number of times codon is used in the gene.

It should be pointed out here that, although no examples have been given to illustrate the production and consequences of mutations of animal viruses, there has been extensive activity in the area of animal virus genetics involving a wide variety of animal viruses and numerous mutagenic agents. A comprehensive review of this subject is that of Ghendon (1972).

3. *Gene Location*

All viruses contain in their nucleic acids linear sequences of nucleotides that function through the processes of transcription and/or translation to yield specific proteins. These protein-generating segments of nucleic acid are called genes (or, by some, cistrons; see Benzer 1957). As indicated earlier the number of genes to be expected is proportional to the size of the nucleic acid, but all viruses are polygenic, meaning that they have enough nucleic acid to code for several proteins. The simplest RNA phages, such as MS2, R17, f2, and others like them, have only three genes, as described in the previous section, whereas the large T phages and the pox viruses have enough nucleic acid to provide for 200 to 300 genes.

It has been shown for MS2 and R17 phages that the viral RNA, which serves both as genome and mRNA, does not consist entirely of genes. Some untranslated portions of the RNA serve other functions such as ribosome binding sites (Steitz 1969). There are doubtless similar untranscribed segments of double-stranded nucleic acids; in addition it appears that only one of the two strands of DNA is transcribed into messenger RNAs and hence translated into proteins. T4 and lambda coliphages are partial exceptions, one strand of the DNA being transcribed for certain genes and the other strand for the remaining ones (Guha and Szybalski 1968; Cohen and Hurwitz 1968).

A crude but rapid estimate of the number of genes in a viral nucleic acid may be obtained by the relation:

$$\text{Number of genes in single-stranded nucleic acid} = \frac{\text{Molecular weight of nucleic acid}}{0.3 \times 10^6}$$

For viruses with double-stranded nucleic acid, the molecular weight of the nucleic acid is divided by 2 in order to reflect the assumption that gene function is associated with only one strand. The assumptions behind this very approximate calculation are that most of the coding strand of nucleic acid is involved in gene function, that the average molecular weight of a nucleotide is 300 and of an amino acid 100, that the proteins coded for have a molecular weight of 35,000, and that the code is a triplet nonoverlapping one.

The sizes of viral genes vary considerably in accordance with the gene concerned. Thus the gene for the coat protein of phage fd has only 147 nucleotides (Asbeck et al. 1969), whereas genes for some of the larger viral proteins must consist of more than a thousand nucleotides.

Whatever their size or number, several methods have been used to determine the location of genes in viral nucleic acids: (1) mating and mapping; (2) hybridization and electron microscopy; (3) selective mutagenesis; and (4) comparison of amino acid and nucleotide sequences.

a. Mating and Mapping

The mating and mapping procedure has been used to locate genes in phage nucleic acids (Hayes 1968). The procedure depends on genetic recombination between strains of a virus when they infect the same cells, and mapping is based on the probability principle that the number of recombinants obtained will be proportional to the distance separating the genes that are recombined. Consequently, if virus strains are available that have distinctive hereditary characters that can be used as markers, crosses

Table 47. Recombination from Mating of Some Coliphages.

Phages Mated	Progeny: Number and Types of Plaques				% Recombinants[a]
Rc × Rm	5162 Rc (large, clear)	6,510 (minute, turbid)	311 RR (large, turbid)	341 cm (minute, clear)	5.3
Rs × Rc	7,101 Rs (small, turbid)	5,851 Rc (large, clear)	145 RR (large, turbid)	169 cs (small, clear)	2.4
Rs × Rm	647 Rs (small, turbid)	502 Rm (minute, turbid)	65 RR (large, turbid)	56 sm (small turbid with halo)	9.5

[a]Adapted from Kaiser (1955). Mutants designated here as Rc, Rs, and Rm were obtained by ultraviolet irradiation of a reference type of phage lambda designated RR. Percent recombinants = $\frac{\text{Sum of two recombinants}}{\text{Sum of all types}} \times 100$.

(mating) can be made between pairs of strains. From the results of the mating, the genes associated with the marker characteristics can be mapped. In a typical cross between two strains, each of which has one distinctive hereditary characteristic, four types of progeny are observed: two are the two parental types and two are new types that combine characteristics of the parental types (these are the recombinants). The results obtained in successive matings of pairs of plaque type mutants of a lambda phage variant are indicated in Table 47. From these data a linkage map can be constructed showing the relative positions of three genes that affect plaque type, as follows. Let each percent recombinants be equal to one map unit. Then the c and m genes are separated by 5.3 map units, while the c and s genes are 2.4 map units apart. This information is not enough to decide between c--s--m and s--c-----m, but after making the cross Rs × Rm, which indicated that s and m were separated by 9.5 map units, it is clear that the correct order of the genes must be s--c-----m. This is a simplified version of the simplest application of mating-mapping. In practice, allowances must often be made for such technical factors as differences in attachment efficiencies, repeated rounds of matings, and so on. Moreover, mating is often coupled with complementation tests in order to distinguish between mutants of the same gene and of different genes concerned with the same function; also, three-factor crosses, and deletion-mapping techniques are employed (see Hayes 1968).

b. Hybridization and Electron Microscopy

Another technique of gene location in DNA is by hybridization and electron microscopy (Davis and Davidson 1968; Westmoreland et al. 1969; Davis et al. 1971). The principle of this method is that the complementary strands of DNA can be separated (this is denaturation) by heat or other treatments and under optimal conditions brought back together (annealed) to form a structure that appears uniformly fibrous when viewed in the electron microscope. However, if two different DNAs are denatured and complementary strands from each are subjected to annealing, the double-stranded structure obtained (hybrid duplex) will show more or less uniformity depending on the degree of homology between the annealed strands. Nonpairing segments of nucleic acid form single-stranded loops that are seen as such in the electron microscope, or under some conditions the single-stranded loops collapse into small branched clumps called "bushes." If formamide is used in mounting DNA molecules for electron microscopy (Westmoreland et al. 1969; Davis et al. 1971), single-stranded regions of DNA do not collapse but appear in loops as filaments with measurable contour length. This latter technique, coupled with other labeling methods, has enabled the location of genes, such as those for tRNAs, in the DNA of coliphages (Wu and Davidson 1973); the formamide technique was used also to locate a bit of bacterial DNA incorporated into the DNA of the transducing phage Ø80. This is illustrated in Figure 49, which shows the heteroduplex formed between a strand of Ø80 DNA and a complementary strand in which a segment of bacterial DNA containing genes for two tyrosine tRNAs has been substituted for a segment of phage DNA. The loopout section in Figure 49 represents the area of nonhomology between strands of DNA and permits the precise location of the tyrosine tRNA genes.

Morrow and Berg (1972) also used the hybridization-electron microscopy technique to demonstrate the presence and location of a small segment of simian virus 40 (SV40) genome in the hybrid species containing both adenovirus and SV40 DNAs. First, annealing of adenovirus-SV40 hybrid DNA with just adenovirus DNA produced a loopout of the SV40 segment, thus indicating its position in the hybrid DNA. Next, the specific

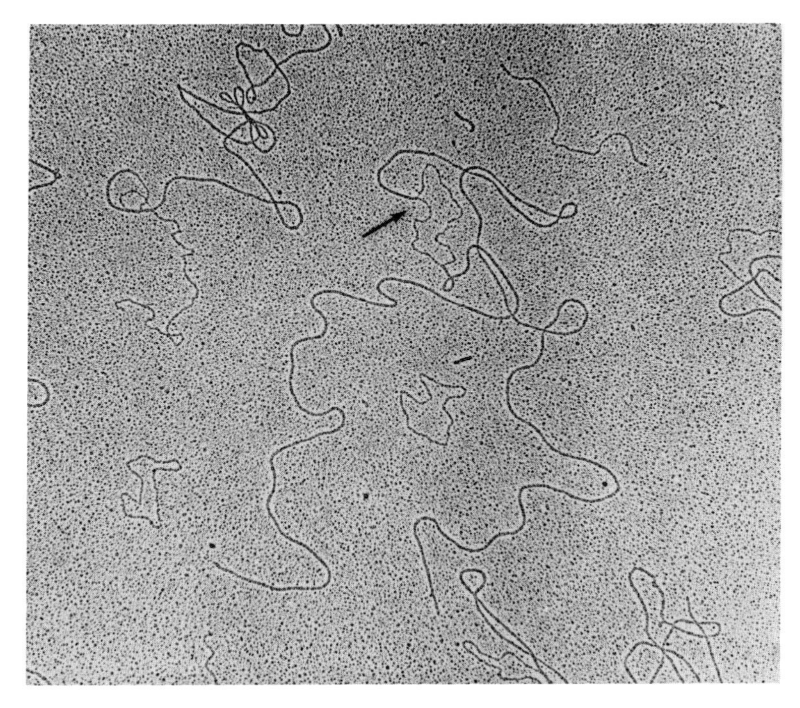

Fig. 49. Heteroduplex formed between a strand of coliphage Ø80 DNA and a complementary strand of Ø80 DNA in which a segment of bacterial DNA has been substituted for a segment of phage DNA; since there is no homology between these segments, they appear as single-stranded components of a loop in the otherwise uniform duplex. The arrow points to the junction of phage DNA and bacterial DNA, the longer part of the loop being bacterial DNA (about 3,100 nucleotides) and the shorter part, Ø80 DNA (about 2,100 nucleotides). The entire heteroduplex DNA molecule is shown; two single-stranded circular DNA molecules of phage ØX174 put in as length standards can also be seen. (From Wu and Davidson 1973.)

segment of SV40 DNA incorporated in the hybrid was identified by hybridizing the loopout heteroduplex with SV40 DNA. Hybridization produced a bit of triple heteroduplex by base-pairing between the loopout section and the complementary region on the added SV40 DNA. By inspection of the electron micrographs of this structure and by measurement of distances, it was possible to identify the specific region in the SV40 DNA from which the segment had come that was present in the adenovirus-SV40 hybrid. From other tests it was known that genes associated the SV40 DNA segment include one for a specific SV40 antigen and another enabling the hybrid to multiply in monkey kidney cells (adenovirus alone cannot multiply in these cells).

c. Selective Mutagenesis

There is little evidence that genetic recombination occurs with plant viruses, except for tomato spotted wilt virus (Best 1968). However, even with this virus, it is not yet possible to obtain the quantitative data on recombinants needed in order to locate genes by the mating and mapping procedures. Nor has the hybridization-electron microscopy method proved applicable to plant viruses thus far. However, a highly specialized procedure of selective mutagenesis, so far used only for tobacco mosaic virus, has been devised. The rationale of this procedure is based on the observation that the protein subunits spiralling around the tubular TMV particle can be stripped off from the RNA in a polar fashion starting from the end of the particle in which the 3′-terminal of the viral RNA is located (May and Knight 1965). This stripping is achieved by treating the virus with sodium dodecyl sulfate at 37°C for about 3 hr.

Large amounts of variously stripped particles obtained in this manner are separated into classes of different lengths by sedimentation of the reaction mixture on a sucrose density gradient. Such classes of particles are then treated with nitrous acid and tested for the production of a specific type of mutant (the nitrous acid can react mutagenically with viral RNA that is still ensheathed with protein but only at a fraction of the rate shown by the exposed RNA). A gene concerned with the viral capacity to produce brown necrotic spots on *Nicotiana sylvestris* tobacco in which the wild type causes a systemic green mottling response was located by this technique (Kado and Knight 1966). When the RNA segment containing the sought-for gene was exposed, the number of mutants increased significantly over background spontaneous mutants as shown in Figure 50.

d. Comparison of Amino Acid and Nucleotide Sequences

The most direct means of gene location is the sequencing of nucleic acid and identification of unique sequences with the amino acid sequences of specific proteins. This has been done with the MS2 phage as described in the previous section (Min Jou et al. 1972).

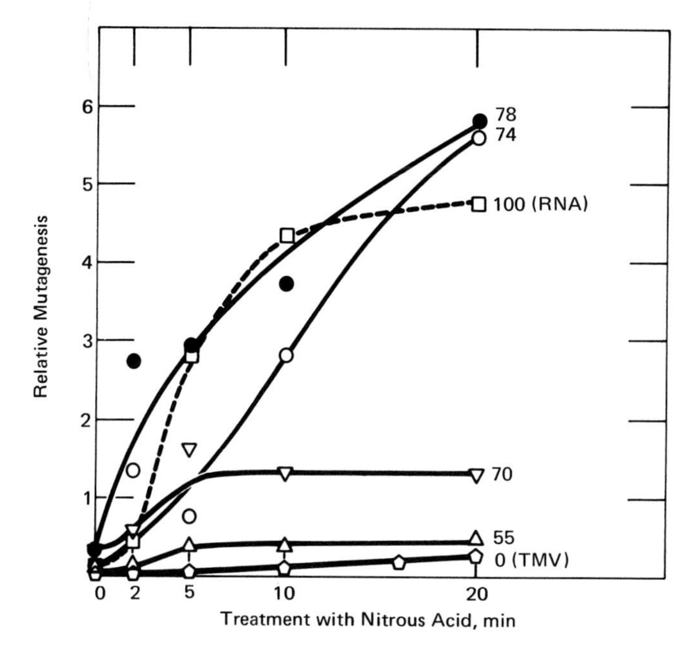

Fig. 50. Nitrous acid-induced mutagenesis of tobacco mosaic virus as a function of the fraction of the protein subunits that have been stripped from the viral RNA. Relative mutagenesis is the number of lesions on *Nicotiana sylvestris* tobacco (a quantitative measure of mutants only) per 1,000 lesions on *N. tabacum* L. var. *Xanthi nc* (a quantitative measure of total virus present). The figures at the ends of the curves refer to percent of stripping of protein from TMV particles prior to treatment with nitrous acid. (From Kado and Knight 1966.)

A somewhat less direct but related procedure was used with phage R17 RNA, which was employed as mRNA in cell-free synthesis of identifiable proteins (Jeppesen et al. 1970). As previously noted, there are three genes in the RNAs of the R17 series of phages that code for A protein (maturation protein), the major coat protein, and an RNA synthetase. Either whole phage RNA or parts of it can direct synthesis of these phage proteins *in vitro*. Two large segments of phage RNA were obtained by treatment with *E. coli* ribonuclease IV. One part, comprising 40 percent of the RNA included the 5′-end of the nucleic acid, while the remaining 60 percent contained the 3′-end. In a cell-free protein-synthesizing system with *E. coli* ribosomes it was found that the 60 percent fragment (3′-end) was translated to yield synthetase protein, while the 40 percent segment was not translated. However, when ribosomes from *Bacillus stearothermophilus* were used in the synthesis, only A protein was produced and this by the 40 percent fragment (5′-end). Thus it appears that the A protein gene is at the 5′-end and the synthetase gene is at the 3′-end; by difference, the gene for the major coat is in the middle of the viral RNA. This was confirmed by some other experiments involving sequencing of small segments of phage RNA representing initiation regions and fragments of adjacent genes. It was possible to relate these gene fragments to one or another of the three phage proteins.

VI.

Reproduction of Viruses and Viral Constituents

In order to continue to exist, viruses must multiply. This they can do only in living cells, which provide the synthetic and energy-yielding mechanisms that viruses lack. The main features of virus reproduction (also called infectious process) are all known, although reams of information are still being published on detailed and esoteric aspects of the subject. There is only one major exception to this conclusion, the viroids, whose nature and mode of interaction with cells have yet to be clarified. Viroids will be treated in Sec. 1A.

A. Virus Reproduction in Cells

No form of life seems surely exempt from virus infection. Therefore a study of the infectious process naturally focuses on the interaction of a particular virus with some kind of cell—algal, bacterial, fungal, plant, vertebrate, or invertebrate. Presumably the ability of a given virus to infect a particular type of cell (specificity) has resulted from evolutionary events involving both cell and virus. Stages at which specificity might be exerted will be apparent as the infectious process is considered.

It is convenient to view the process of infection in terms of a sequence of events: attachment, penetration, multiplication, assembly, and release.

1. Simple Infection

Simple infection ensues when susceptible cells are infected by one or a few virus particles of the same kind. In complex or mixed infection, which will be treated in the next section, the consequences of simultaneous infection with active and inactive virus particles, with two or more mutants of a virus or with two or more unrelated viruses, will be briefly considered.

The systematic investigation of various steps in virus infection has advanced mainly where it is possible to study synchronous infection of cultures of bacterial, animal, and plant cells. In such systems, some complications of whole organisms can be avoided and quantitation of events is greatly enhanced. Cell cultures also lend themselves to application of radioisotopes, which are invaluable as markers of the various chemical constituents of viruses or of enzymes and other substances involved in virus multiplication.

The process of infection can be generalized for all viruses as follows.

Virus attaches to a susceptible cell at one of several specific sites where there is a structural and electrostatic complementarity between cell surface and virus particle. Some viruses appear to have attachment organs and some not (possible attachment organs include tail fibers or knobs of certain phages, penton fibers of adenoviruses, spikes of enveloped viruses, and so on). Specific attachment of a single virus particle is sufficient to initiate an infection; conversely, inability of a virus to attach specifically to the cell surface may block infection.

Following attachment, either whole virus or viral nucleic acid penetrates the interior of the cell. Penetration of bacterial and plant cells is potentially more difficult than invasion of animal cells since they have a cell wall as well as a plasma membrane to breach. Commonly it is whole virus that penetrates in infection by animal or plant viruses. Some animal viruses such as poliovirus can pass directly through the animal cell plasma membrane even when smaller particles are excluded; the molecular mechanism for this is as yet unknown. More commonly, virus particles are engulfed by a phagocytic type of reaction sometimes called viropexis. Quite often a virus may be partly degraded by cellular enzymes in the process of penetration either at the cell membrane or within phagocytic vesicles. Bacterial viruses and some algal viruses usually inject their nucleic acid into cells to which they have attached in a specific manner. The injection process is probably facilitated by such structures as contractile tail sheaths that some phages and algal viruses have. However, nucleic acid is injected by many bacterial viruses that have no such organ. The mechanism for this remains to be clarified.

Multiplication of viruses takes place in either the cytoplasm or in the nucleus or in both. In multiplication, the virus provides most essentially its nucleic acid which has the genetic information needed in order to produce more virus particles just like the one that initiated the infection. In addition, some viruses may bring in one or more enzymes needed in some step of the replication of viral nucleic acid. Beyond that, the cell provides energy-generating systems, enzymes, and synthetic machinery (especially protein-synthesizing equipment such as ribosomes, transfer RNAs, and so on) and raw materials from which virus constituents are made (amino acids for proteins, nucleotides for nucleic acid, and so on). Viral nucleic acid functions in virus multiplication in much the same way that it does in normal cellular function as prescribed by the central dogma of molecular biology:

DNA ——Transcription——→ Messenger RNA ——Translation——→ Protein
Replication ↓↑
DNA

In the case of RNA viruses in which the virion-contained RNA can act directly as mRNA, the production of virus-specific proteins may occur in two different ways: (1) translation of viral RNA may occur discontinuously in a manner that produces a series of discrete proteins, one for each gene (for example, R17 coliphage); (2) translation may produce a large polypeptide chain that is then cleaved enzymatically ("processed") into individual proteins with various functions (for example, poliovirus). If the viral nucleic acid is single-stranded RNA, it may function as a template for its own replication, as well as serve as messenger RNA (mRNA) that is translated on the cellular ribosomes to provide virus-specific enzymes and structural proteins. Some single-stranded RNAs have the wrong "polarity" and cannot serve directly as mRNA but must first be transcribed into a complementary strand. Likewise, double-stranded RNA, which some viruses have, substitutes for DNA in the scheme outlined above. In still another instance (for example, RNA tumor viruses) single-stranded RNA may be transcribed into DNA by a viral enzyme ("reverse transcriptase") and then the sequence shown in the central dogma scheme starting with DNA can apply. In any case, viral nucleic acid and protein are produced in abundance and may accumulate in pools until a critical concentration occurs, after which assembly of progeny virus particles begins. In other cases assembly may begin before measurable pools of virus parts accumulate. The process of virus multiplication is subject to controls at either the transcriptional or translational level. Many details of these controls are still to be elucidated.

Assembly of new virus particles can be largely or wholly a spontaneous process since such assembly has been observed with a few viruses in the test tube. However, in some cases, assembly may require specific enzymes to catalyze certain steps of the process. As has already been noted, some viruses mature at cell membranes of one sort or another where they appear to acquire both viral and host materials.

Mature virus particles are finally released from infected cells ready to begin a new cycle of infection. Release can occur in one or more of three ways: (1) the infected cell may lyse releasing virus particles and cellular constituents into the surrounding area; (2) the virus may pass out through tubular structures that develop in the cell (or which connect plant cells and are there called plasmodesmata); or (3) the virus particles may mature at the plasma membrane from which they bud in a process that is the reverse in some respects of the phagocytic entrance mechanism. The yields of virus obtained from infected cells vary considerably with virus and host but range from less than 100 per cell to more than 1,000 for bacterial viruses, and usually run around 10^5–10^6 particles per cell for animal and plant viruses.

There are known instances in which the multiplication of bacterial or animal viruses is postponed after initiation of infection. The deferred production of bacterial viruses is called lysogeny; a more general term that includes animal viruses (or others if the phenomenon is eventually observed with them) is virogeny. The essence of virogeny is that the viral nucleic acid appears to be integrated into host nucleic acid where it multiplies only as the cellular DNA replicates and where its genes are usually considerably restricted in activity. In the case of lysogenized bacteria, for instance, only the gene or genes concerned with production of a repressor for the other viral genes is active. In virogenic conditions involving some DNA tumor viruses of animals, some viral genes that code for certain viral antigens remain active while viral structural genes cease functioning. The mechanism of repression of these latter genes is still obscure.

Virogenic cells can be changed into a productive state ("induced") by radiation, especially ultraviolet, or by treatment with certain chemicals, such as nitrogen mustards, mitomycin C, iodouracildeoxyriboside (IUDR) and others. Also, if the integrated genome is brought into a different environment, induction may occur and normal production of virus ensue. This happens with lysogenic bacteria if the bacterial chromosome containing integrated virus genome is transferred from one bacterial cell to another in the process of conjugation. This is called zygotic induction. Likewise, if a nonpermissive cell (a cell which may integrate viral nucleic acid but not support production of whole virus) containing integrated SV40 is fused with a permissive cell, infectious virus is produced by the fused cell (called a heterokaryon).

When viral DNA is released from integration with cellular DNA, the release may be clean, or part of the viral genome may be left behind and a piece of the cell genome brought out in a virus particle. If the piece of cellular DNA thus acquired contains a gene for a function (for example, an enzyme) lacking in another cell, the latter cell may serve to detect this, for after infection it will show a functional capacity it previously lacked. This process is called transduction. When specific genes only are transduced, the process is called specialized transduction, and when a variety of genes is transduced, the extreme being when the phage particle contains no phage DNA but only bacterial DNA, the phenomenon is termed general transduction.

2. *Complex or Mixed Infection*

Interesting molecular genetic phenomena occur when, instead of infection by a single type of active virus particle, mixed infection occurs with virus mutants, unrelated viruses, or various mixtures of active and inactive virus particles. Mating, marker rescue, and multiplicity reactivation are three examples of phenomena occurring during mixed infection, all of which appear to depend on the same process, that is, genetic recombination.

Mating has been reported for bacterial, animal, and plant viruses. The process is essentially the same for all of them. Mixed infection is made with, for example, two mutants of a virus that possess distinctive characteristics (genetic markers). The observed consequence of mixed infection is that four types of viral progeny issue: two of them are exactly the same as the two parental types, but the other two have combinations of characteristics of each of the parental types. This might be illustrated with two mutants of phage T2 whose plaque types are indicated by r and r^+ and whose host ranges are specified by h^+ and h:

$$T2rh^+ \times T2r^+h \rightarrow T2rh^+ + T2r^+h + T2rh + T2r^+h^+$$

In marker rescue mixed infection with a fully active mutant and an inactivated mutant can result in appearance ("rescue") of a genetic trait from the inactivated virus in an active virus. For example, Kilbourne and associates (1967) infected chick embryos with active A_2 influenza virus and heat-inactivated A_0 influenza virus. They obtained among the progeny of this mixed infection a viral strain having the hemagglutinin and internal (NP) antigen of A_0 parent but the neuraminidase of the A_2 strain. Thus the internal antigen and the hemagglutinin of the A_0 strain were rescued.

Multiplicity reactivation is operative when a single type of virus greatly inactivated, for instance, by ultraviolet light shows little infectivity when tested at a low multiplicity of infection (few virus particles per cell) but a disproportionately high infectivity when tested at a high multiplicity of infection.

All three of these phenomena are thought to be explained by genetic recombination between the participating molecular species. While details of the process are yet unclear (see Clark 1971; Davern 1971; Radding 1974), genetic recombination appears to involve breakage and reunion of nucleic acid molecules in such a way that there is a reciprocal exchange of parts of the nucleic acids. This is illustrated graphically in Figure 51 for the mating between phages outlined above; by changing designations of the markers the scheme will apply as well to marker rescue and multiplicity reactivation.

A special type of genetic recombination that does not require the breaking-and-reunion recombinational mechanism cited above has been observed with influenza virus and could probably occur with any virus having a segmented genome. Since the nucleic acid of influenza virus occurs in several segments, the mixed infection of a cell with two mutants of this virus can presumably lead to the production of progeny segments that can be randomly assorted at the time of assembly into virus particles. This process is perhaps properly called genotypic mixing, and is thought to be responsible for the high rate of apparent recombination observed with influenza viruses.

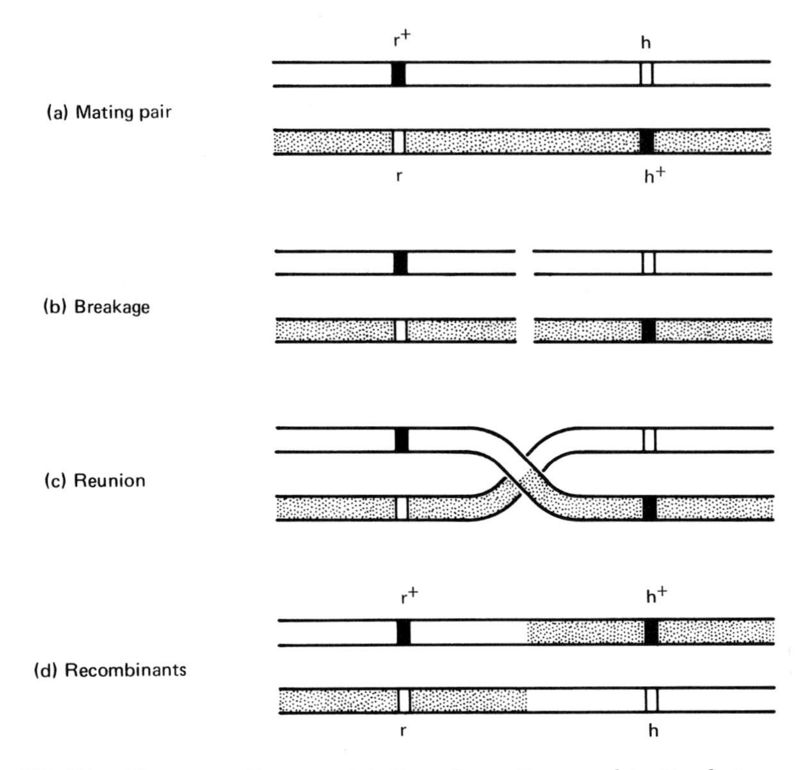

Fig. 51. Diagrammatic representation of genetic recombination between two phages by a mechanism involving breakage and reunion of their DNAs.

Some other phenomena of mixed infection that, however, do not involve genetic recombination include phenotypic mixing, complementation, and interference.

When the infectious process involves accumulation of viral parts in the metabolic pool before assembly begins, it is particularly feasible for some particles to be assembled from components of different viruses. This is phenotypic mixing. For example, mixed infection of *E. coli* bacteria with phages T2 and T4 might result in the production of some particles having all the usual constituents of T2 but the tail fibers of T4. Such a particle, upon infecting a fresh *E. coli* cell, will yield only T2 progeny since the DNA of the mixed particle is that of T2 phage.

When two viruses having different defects or deficiencies infect the same cell, they can sometimes supply each other with the missing function and thus make it possible for each to multiply whereas alone they could not. This is called complementation and has been observed many times with different viruses. Complementation may also involve defective (satellites) and nondefective viruses (Kassanis 1968; Barrett et al. 1973). A kind of complementation that seems to be very common among certain plant viruses is that which occurs between functionally different particles (Sänger 1968; Lister 1969). An example of this is tobacco rattle virus, preparations of which characteristically display short and long rods (see Figure 35). The RNA of the short rods apparently codes for coat protein for both short and long particles, while the RNA of the long rods codes for the RNA polymerase needed for both types of particles. Thus the short rods are noninfectious by themselves; the long rods are infectious but induce the production of viral RNA only.

Some mixed infections result in competition and interference between different viruses presented to a cell. The mechanisms for interference are not always clear but as a rule dissimilar phages tend to interfere mutually in the infectious process; conversely, similar plant viruses (strains or variants) tend to interfere with each other. In some instances, it appears that the interfering viruses may compete for attachment sites; in other cases one virus may induce the production of a nuclease enzyme that destroys the nucleic acid of the second virus; still another possibility is that the nucleic acid of one virus may compete better for cellular ribosomes than the other, and so forth.

Detailed accounts of the infectious process, either simple or mixed, for a variety of viruses can be found in various works. See, for example, Knight 1974; Fenner et al. 1974; Dalton and Haguenau 1973; Matthews 1970; Mathews 1971; Hayes 1968.

3. *Viroids*

Several infectious diseases of plants closely resemble virus infections except in one important regard: no virus particles can be isolated from the diseased plants. The best-studied cases are potato spindle tuber disease and citrus exocortis, although chrysanthemum stunt is also an agent of this type. In all cases the agent appears to be a single-stranded RNA about 75,000–100,000 in molecular weight (Diener 1972; 1973; Semancik et al. 1973; Diener and Smith 1973.) Diener proposed the term viroid for such agents, thus indicating their viruslike properties but suggesting basic differences.

Viroids pose two puzzling major questions: (1) How does such a small nucleic acid get replicated? (2) How does such a small nucleic acid exert its pathological effects?

A standard concept, definitely supported by evidence in some cases, of how virus nucleic acids get duplicated in cells they infect is that they code for a polymerase (replicase) enzyme that can use the invading nucleic acid as a template for synthesis of more viral nucleic acid. If all the viroid RNA is used as mRNA, a replicase of molecular weight of only about 8,000–10,000 could be made. This seems much too small in comparison with known plant RNA polymerases (Astier-Manifacier and Cornuet 1971; Peden et al. 1972; Hariharasubramanian et al. 1973) even if such a protein serves to modify an existing replicase. It is possible that the viroid RNA is replicated by a host enzyme but no such enzyme has yet been identified. If the viroid RNA acts as mRNA for a nonreplicase protein, could that protein account for the pathologic activity of viroids? The answer is unknown. Perhaps the viroid RNA causes pathologic effects by exerting some kind of abnormal regulatory control of cell processes, perhaps in the nucleus. If so this action has yet to be defined. Thus viroids represent mysterious virus-like entities whose structure, mode of replication, and pathogenic mechanism remain unknown.

B. Extracellular Reproduction of Viruses and Viral Constituents

Viruses are obligate parasites and as such do not replicate independently outside living cells. However, in the laboratory it has been possible to obtain virus constituents and to put them together in such a way as to reproduce the morphology and biological activity of certain viruses. This is called reconstitution. It has also been possible to induce cell-free synthesis of some viral proteins and nucleic acids in the laboratory.

1. Reconstitution

Reconstitution (sometimes referred to as self-assembly) is defined as the bringing together of protein and nucleic acid components of a virus in such a way that they combine to yield characteristic virus particles possessing most, if not all, of the properties (including, in the most successful cases, infectivity) of the mature virus as it is produced in natural infections. The component parts for reconstitution are obtained by degrading the virus, although sometimes the protein component is available as a natural consequence of infection and can be isolated and used in the reconstitution reaction. Reconstitution was first accomplished with TMV and some of its strains by Fraenkel-Conrat and Williams (1955) building on the results of experiments of Schramm (1947a, 1947b) and especially those of Takahashi and Ishii (1952a, 1952b, 1953). Takahashi and Ishii had demonstrated that the TMV protein found in infected plants could be aggregated to form rods that in the electron microscope looked very much like TMV but which were devoid of infectivity since they lacked RNA.

While there are several ways of obtaining protein and nucleic acid for demonstrating and studying the reconstitution of TMV, the RNA is now commonly obtained by the phenol extraction method (see Sec. IIIB, 1) and the protein by the cold 67 percent acetic acid procedure (see Sec. IIIA, 1).

Protein prepared by the latter procedure occurs at around neutral pH in the form of quasistable aggregates composed of three subunits and generally called A protein. The reconstitution reaction can be summarized for TMV and its strains as follows:

A protein	+	RNA	$\xrightarrow[\text{pH 7.3, Pyrophosphate}]{\text{1 hr at 30° C}}$	Reconstituted virus
Molecular Weight 53,000		Molecular Weight 2×10^6		Molecular Weight 40×10^6
Noninfectious		Slightly infectious		Highly infectious 18×300 nm rods

As indicated, the protein used in reconstitution has no infectivity and the infectivity of the RNA usually amounts to about 0.05 percent that of the same amount of RNA in a virus particle. In contrast, the reconstituted virus exhibits infectivities ranging from 30 to 100 percent of those in the virus from which parts were obtained for the reconstitution. The structure of reconstituted virus as revealed by electron microscopy (see Figure 52 for reconstitution sequence), the ultraviolet absorbance (Figure 53), and the stability shown to heat and various pH values by reconstituted virus are virtually identical to those exhibited by the virus obtained from infected tobacco plants.

Much is known about the details of the TMV reconstitution. The factors involved in association of the protein subunits, especially the role of water, have been exquisitely analyzed by Lauffer and Stevens (1968). They cite evidence to support the idea that the polymerization of TMV A protein is an endothermic aggregation reaction and that the observed increase in entropy is associated with the release of bound water between the polymerizing units. Thus polymerization of TMV A protein can be thought of as a transfer of organic surface from an aqueous environment to an organic environment and the importance of hydrophobic interactions in holding the superstructure of TMV rods together is stressed. In reconstitution of whole TMV there are, of course, interactions (also noncovalent) between protein and nucleic acid as well as between protein and protein (see Caspar 1963). These are important as evidenced by the fact that TMV protein rods readily disaggregate above pH 8, whereas rods containing the viral RNA are stable up to pH 10.

The composition, size, and charge of the viral protein influence the process of viral reconstitution, as can be demonstrated in three ways:

1. Mixed reconstitution can be done by using the RNA of one strain and the protein of another but the reconstitution may not go so well in this heterologous reaction especially when it involves strains with grossly different proteins (Holoubek 1962); another instance emphasizing the impor-

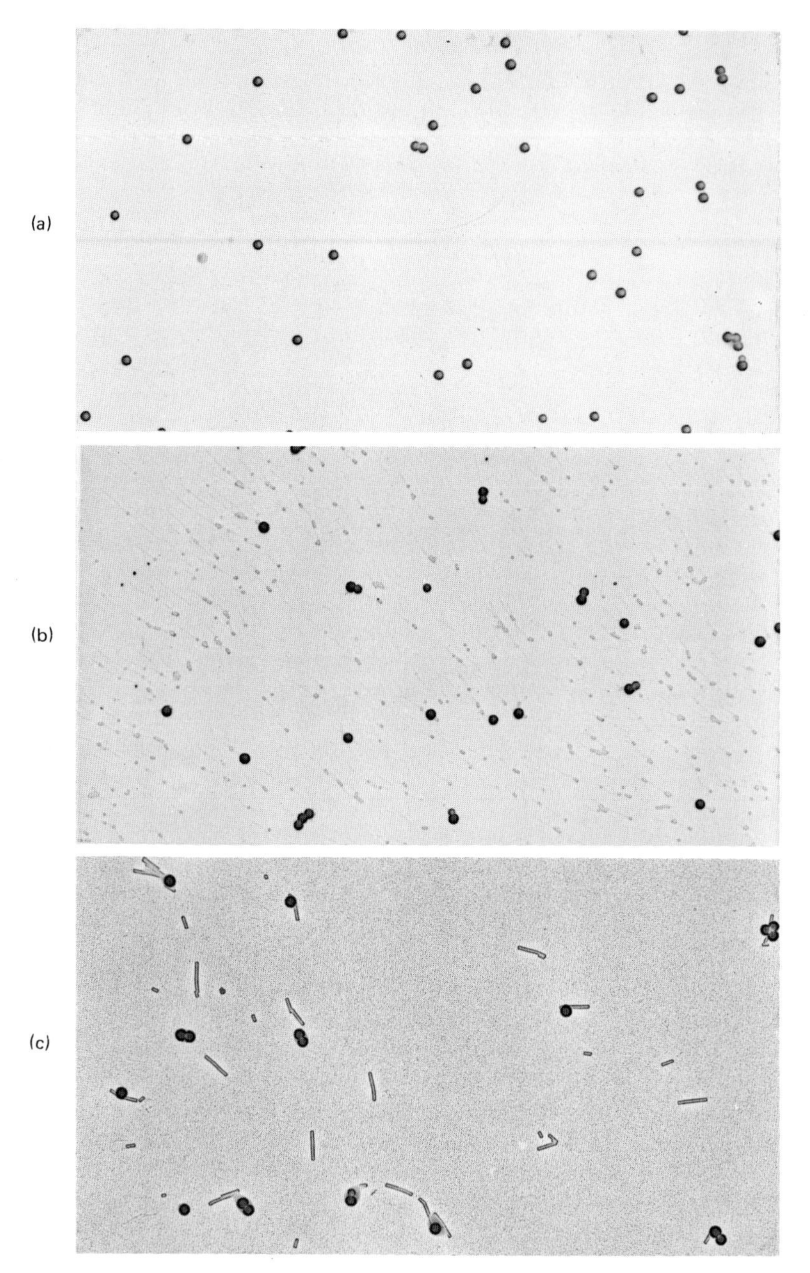

Fig. 52. Electron micrographs of a reconstitution mixture of tobacco mosaic virus A protein and tobacco mosaic virus RNA at various times after mixing. *a*. Immediately after mixing: only polystyrene latex reference spheres are visible; *b*. 2 min after mixing: many fibrous nucleic acid particles are visible, each of which has some protein assembled around one end; *c*. 6 hr after mixing: many full-size TMV particles are present. (Courtesy K. Richards.)

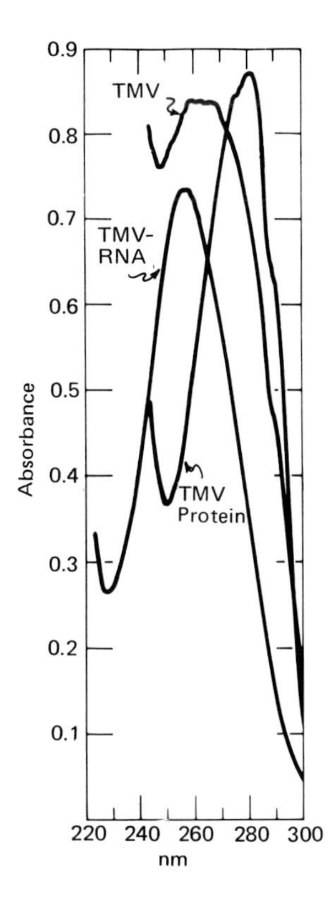

Fig. 53. Ultraviolet absorption spectra of tobacco mosaic virus, its coat protein, and its RNA.

tance of composition of the viral protein in reconstitution is the observation by Siegel and Zaitlin (1965) that the PM 2 strain of TMV obtained after nitrous acid mutagenesis of common TMV has a protein differing from that of common TMV by only two amino acid replacements but that this protein will not aggregate around viral RNA in the reconstitution reaction and aggregates by itself to form a bizarre open-helical structure.

2. If three C-terminal residues are removed from each protein subunit, reconstitution of rods is unimpaired but if 15 to 17 amino acid residues of the 158 per subunit are removed, the ability to form rodlike particles is removed (Sengbusch and Wittmann 1965).

3. The protein of common TMV has two lysine residues per subunit (at positions 53 and 68, see Table 13) and the epsilon amino groups of these residues are normally protonated; these charges can be selectively eliminated by trifluoroacetylation, which renders the protein incapable of aggregation to rodlike structures, but this capacity is restored by removal of the trifluoroacetyl groups (Perham and Richards 1968).

From the point of view of the mechanics of reconstitution of TMV, it appears from Richards and Williams' (1972) analysis that an aggregate of 34 protein subunits in the form of a disk (which arises spontaneously when A protein concentration and pH are suitable) is essential for rod initiation; addition of individual units of A protein then results in rod elongation. The process seems to be polar, the addition of protein proceeding from the 5′-end of the viral RNA toward the 3′-end (Dzhavakhia et al. 1970; Hirth 1971).

Several other viral RNAs as well as synthetic polynucleotides have been reconstituted with TMV protein in place of TMV-RNA (Fraenkel-Conrat 1970). Rodlike particles were generally obtained whose average length varied with the nucleic acid employed. Slight infectivity in Chinese cabbage was observed for the reconstituted virus containing turnip yellow mosaic virus RNA (Chinese cabbage is a common host for the turnip virus).

Another rodlike virus, tobacco rattle virus has also been successfully reconstituted, although the conditions favoring reconstitution are somewhat different from those used with TMV (Semancik and Reynolds 1969; Morris and Semancik 1973). Protein was extracted from the virus with acetic acid and RNA was obtained by a phenol extraction procedure. The two components were mixed in the proportions of 10 parts of protein to 1 of RNA and dialyzed against 0.25 M glycine buffer at pH 8 at 9°C for 12 hr. Nucleoprotein sedimenting at the same rate as native virus and resembling the latter in appearance in the electron microscope was obtained. This material exhibited infectivity.

The first demonstration of reconstitution of spheroidal viruses was not achieved until some years after success with TMV. This may be due to the apparently greater susceptibility to denaturation of proteins from spheroidal viruses; at least the earlier subjects for investigation such as tomato bushy stunt and turnip yellow mosaic viruses are in the category characterized by coat proteins that are readily and often irreversibly denatured.

However, Hiebert et al. (1968) (see also Bancroft 1970) found that spheroidal plant viruses of the brome mosaic group (for example, brome mosaic, cowpea chlorotic mottle, and broad bean mottle viruses) could be readily disaggregated and reconstituted. These viruses are serologically unrelated but share certain chemical and physical properties (particles are about 25 nm in diameter and contain about 22 percent RNA and 78 percent protein) and also have some common hosts. The RNA can be obtained from them by the phenol extraction procedure and the protein by dialyzing virus against a mixture containing 1 M NaCl, 0.02 M tris buffer (pH 7.4) and 10^{-3}M dithiothreitol (Cleland's reagent) at 4°C for 24 hr. The reaction mixture is centrifuged to separate virus protein from undegraded virus and RNA. Reconstitution is achieved by dialyzing 1 part of viral RNA and 3 parts of protein for 1 hr at 4°C against a tris-salts mixture (0.01 M tris at pH 7.4, 0.01 M KCl, 5×10^{-3} M $MgCl_2$, and 10^{-3} M dithiothreitol).

In addition to homologous reconstitution, hybrids can be formed that have the protein coat of one virus and the RNA of another. All are infectious, and the hybrids show a host range specificity characteristic of the virus from which the RNA was obtained. It was also shown that the protein from cowpea chlorotic mottle virus could reconstitute with RNAs of certain other viruses and phages, and even with a phage DNA as well as with soybean ribosomal RNA and yeast sRNA. In all cases spheroidal particles were obtained with a size similar to that of the cowpea virus. Thus, as with the TMV series, the protein constituent tends to determine the size and shape of reconstituted particles, with the reservation that the length of stable rodlike particles may depend on the length of the RNA used in the reconstitution.

Other plant viruses have also been dissociated and reconstituted but with somewhat less certain recovery (or preservation) of infectivity (see Fraenkel-Conrat 1970).

Reconstitution has also been achieved with the spheroidal RNA phages of the f2 group (for example, f2, fr, MS2, R17) and with the similar $Q\beta$ phage (see Hohn and Hohn 1970). These phages are characterized by molecules of single-stranded RNA (molecular weight 1×10^6) in a protein shell whose major protein constituent is 180 subunits of protein, each about 13,750 in molecular weight. In addition, one molecule called A protein or maturation protein (molecular weight 35,000) must be present for an infectious particle. A difficulty in testing reconstitution of these viruses is that their proteins are difficultly soluble by themselves, the A protein being especially insoluble. However, if the RNA is obtained by phenol extraction and the protein by treatment with ice-cold dilute acetic acid or guanidine hydrochloride, these components can be mixed and dialyzed against renaturing buffers (usually tris-salt mixtures similar to those previously described) to yield particles similar to native virus in size and shape. If reconstitution is done in the absence of A protein, no infectivity is found with the reconstituted particles; but it has been possible to get particles with low but definite infectivity in the presence of A protein.

Among the animal viruses, it has been reported that poliovirus can be reconstituted if a rather special procedure is employed (Drzeniek and Bilello 1972). Poliovirus can be dissociated into its protein (four species) and nucleic acid (single-stranded RNA) components by holding it at 25°C for 60 min in 10 M urea and 0.1 M mercaptoethanol. Infectious poliovirus can be reconstituted from its parts by diluting the urea dissociation mixture in five steps in cold phosphate-buffered saline at pH 7.2. Attempts to reconstitute by a one-step dilution or by dialysis have been unsuccessful.

Some of the structurally complex, tailed coliphages have been partly reconstituted by mixing phage parts under appropriate conditions. This has been done with the coliphages T4 and lambda (Wood and Edgar 1967; Edgar and Lielausis 1968; Weigle 1966; Casjens 1971) and with the *Salmonella* phage P22 (Israel et al. 1967). Appropriate conditions for assembly of these phages are simply incubation in dilute salt solutions at 30°–37°C for 1–2 hr.

For example, a magnesium-containing phosphate buffer at pH 7.4 of the following composition was used in some of the T4 reconstitutions: 0.0039 M Na_2HPO_4, 0.0022 M KH_2PO_4, 0.007 M NaCl, and 0.02 M $MgSO_4$. Advantage is taken here of conditional lethal mutants which, under restrictive conditions (such as elevated temperature), are unable to produce all of the phage parts required for infectious particles. By matching mutants whose parts are complementary, complete phage can be assembled, providing that the complementary parts will unite spontaneously. For example, a certain mutant of T4 when grown at an elevated temperature produces phage particles that are complete except for tail fibers. Since tail fibers are the attachment organs for T4, the particles lacking these are not infectious. Such defective particles can be isolated from artificial lysates (infected cells caused to burst by treatment with chloroform). Similarly, another mutant defective with respect to phage head formation produces many tail fibers that can be isolated from artificial lysates of the infected cells. These fibers, incubated with the tail fiberless particles, self-assemble to form complete, infectious particles. Since such assembly is often tested with extracts containing the complementary phage parts, the reconstitution procedure has been termed extract complementation. Aside from its use in studying phage assembly, extract complementation can sometimes be employed to determine whether the same or different morphogenetic genes have been affected in a series of mutants.

While it is clear that some portions of phage assembly can occur spontaneously, there is good evidence that some steps in the process with the structurally complex phages may require specific enzymes coded for by the virus.

A partial reconstitution of lambda phage involves the coupling of phage heads and tails. This assembly resembles the situation with the tailless RNA phages in the sense that a protein molecule (molecular weight 17,000) called the F product (that is, F gene product) is required. Thus a mixture of lambda heads and tails is no more infectious than either alone, but if F protein is included, the infectious titer rises several orders of magnitude (Casjens 1971). The lambda F protein, like the A protein of R17 phage, is thought to be essential in completing the phage head assembly. Lambda tails apparently do not join effectively to heads lacking F protein.

2. *Cell-free Synthesis of Viral Proteins*

In general, the proteins of viruses are so large that their synthesis in the laboratory by chemical techniques employed in peptide syntheses would be a major if not impossible undertaking. However, much has been learned about the basic mechanism of protein synthesis by studies with cell-free systems (see Lucas-Lenard and Lipmann 1971), and such systems have been employed in the laboratory synthesis of small amounts of some viral proteins. One difficulty of such procedures is that all viral nucleic acids are polygenic (polycistronic) so that many mRNAs are made in an infected cell from a viral DNA, or if an RNA virus is used, the RNA also codes for several different proteins. This makes *in vitro* synthesis of a specific viral protein difficult. However, this problem is minimized with the small RNA phages, which have just three genes: A or maturation protein, main coat protein, and RNA synthetase or replicase.

Cell-free synthesis of viral proteins of such phages can be summarized as follows. The RNA is isolated from purified phage by the phenol extraction procedure. This serves as mRNA. Ribosomes are obtained from *E. coli* cells by extraction with dilute salt and centrifugal clarification. Such preparations of ribosomes usually contain tRNAs (although in some experiments these are prepared separately and added to the system) and various other factors needed such as those required for polypeptide chain initiation, elongation, termination, and so on. ATP and GTP are added, as well as an ATP-generating system such as phosphoenolpyruvate and pyruvate kinase. Amino acids are usually added, including one or more radiolabeled ones (amino acids are present in the ribosomal extract as well). Mg^{++}, KCl, and tris buffer (pH 7.4–7.8) are added to complete the mixture. The mixture is incubated at 25°–35°C for 30 min and then treated with trichloroacetic acid (TCA) which precipitates the protein that has been synthesized along with that in the ribosomal extract. After suitable washing, this precipitate is analyzed by one or more procedures to identify viral protein that may have been synthesized. Since this kind of synthesis is rather inefficient and hence involves only microgram amounts, cold (nonradioactive) viral proteins are sometimes added prior to analysis. Proteins from the reaction mixture can be separated on DEAE cellulose or Sephadex columns or, very commonly now, by electrophoresis on polyacrylamide gels. Further characterization of isolated protein can be done by digesting with trypsin and comparing the resulting peptides with those obtained from protein isolated from purified virus. Such comparisons are conveniently made by a peptide mapping procedure such as was illustrated in Figure 14.

Synthesis of one or more of the virus-specific proteins of the RNA phages MS2, f2, R17 has been demonstrated in cell-free systems such as that sketched above (Eggen et al. 1967; Nathans et al. 1962; Capecchi 1966). Likewise, the cell-free synthesis of T4 lysozyme and of the major head protein of T4 coliphage have been reported (Coolsma and Haselkorn 1969; Klagsbrun and Rich 1970).

Attempts to synthesize plant viral proteins in the *E. coli* system using tobacco necrosis satellite RNA and alfalfa mosaic virus RNA as messengers yielded proteins that gave tryptic peptide patterns similar to those of authentic viral proteins (Clark et al. 1965; van Ravenswaay-Claasen et al.

1967). TCA-precipitable material was also produced in the *E. coli* system when TMV-RNA was used as messenger, but the product could not be identified (Aach et al. 1964).

In contrast, both necrosis satellite RNA and TMV-RNA appear to be translated in cell-free extracts of wheat germ in such a way that resultant coat proteins can be identified (Klein et al. 1972; Roberts and Paterson 1973). Shih and Kaesberg (1973) made a particularly favorable application of the wheat germ system with RNAs isolated from the four different species of brome mosaic virus particles, one of which appears to be monogenic and codes for the viral coat protein. Looking at it in reverse, application of *in vitro* synthesis enabled the establishment of function of the brome mosaic RNA called RNA 4. In addition, it was found that coat protein exerted an inhibitory effect on *in vitro* translation of other viral RNA messages which may indicate a regulatory function in the *in vivo* protein synthesis induced by this virus. Such inhibition by coat protein had been noted earlier in an RNA phage system (Sugiyama and Nakada 1968).

The wheat embryo system has also been demonstrated to support *in vitro* protein synthesis by a phage (Qβ) mRNA and an animal virus (vesicular stomatitis virus) mRNA (Davies and Kaesberg 1973; Morrison et al. 1974).

Several animal virus mRNAs have been translated in *in vitro* systems using cell-free extracts from a variety of cells (rabbit reticulocytes, Krebs II mouse ascites, Chinese hamster ovary, mouse L fibroblasts, HeLa, and so on). Some recent examples include vesicular stomatitis virus (Morrison et al. 1974), reovirus (McDowell et al. 1972), and Sendai virus (Kingsbury 1973). In all of these cases the objective has usually been to elucidate some feature of the viral replication process rather than to demonstrate cell-free synthesis of viral protein although the latter is a significant development of the past decade.

An important aspect of *in vitro* synthesis research with respect to the genetic code and mechanisms of protein synthesis is the demonstration that *E. coli*, rabbit reticulocyte, and wheat germ cell-free systems can translate messenger RNAs from various types of viruses. For example, the wheat germ system has been used with mRNAs from plant viruses, animal viruses, and phages. This emphasizes the universality of the genetic code and the similarity in fundamental mechanisms through which the code functions.

3. *Cell-free Synthesis of Viral Nucleic Acids*

Cell-free synthesis of viral RNAs or portions thereof have been accomplished for some years starting perhaps with the isolation of an RNA polymerase (also called synthetase, or replicase) from cells infected with the RNA coliphage MS 2 (Haruna et al. 1963). Research soon shifted to a stabler enzyme produced in *E. coli* cells infected with Qβ phage. A review of these developments is given by Spiegelman et al. (1968), and a recent survey has been made by August et al. (1973).

A summary of the cell-free synthesis of phage viral RNA is as follows. The viral RNA is required as a template. This can be isolated from the purified virus by the phenol extraction procedure. The replicase enzyme is extracted from infected cells and used in a purified form obtained from extracts by density-gradient centrifugation and chromatography. To the template RNA and replicase enzyme in a tris buffer at about pH 7.4 containing magnesium ions are added four nucleoside-5′-triphosphates (ATP, GTP, CTP, and UTP), one or more of which is usually radioactively labeled. The mixture is incubated at about 30°C for 30–60 min. [The *in vitro* synthesis may require addition of one or more host factors (August et al. 1973), although if manganese ions are provided as well as Mg^{++} these may be unnecessary (Palmenberg and Kaesberg 1974).] The product of *in vitro* synthesis may be tested directly for biological activity or isolated as an acid-insoluble product by treating the reaction mixture with trichloroacetic acid. The quantity of product in the acid-washed precipitate can be estimated by radioactivity measurements.

Haruna and Spiegelman (1965) used such experimental conditions to effect the first *in vitro* synthesis of infectious nucleic acid. Since RNA phages have a penetration mechanism that their nucleic acid lacks, this infectivity was demonstrated by use of bacterial protoplasts called spheroplasts. These are bacterial cells whose walls have been removed by treatment with lysozyme.

The specificity of *in vitro* synthesis of Qβ RNA is usually quite high, occurring mainly with either plus or minus strands of Qβ RNA or RNA of its mutants as templates; however, by use of both Mn^{++} and Mg^{++} in the reaction mixture, some synthesis of coliphage R17 RNA and of brome mosaic and tobacco mosaic RNAs can be achieved (Palmenberg and Kaesberg 1974).

Several groups of RNA-containing animal viruses appear to incorporate virus-specific RNA polymerases in their virions. Such virions can then be used directly (or in some cases after treatment with nonionic detergents or proteolytic enzymes) in *in vitro* synthesis of viral RNA using conditions similar to those described above. The products observed in such syntheses thus far have been somewhat heterogeneous and variable in the completeness of synthesis; in contrast to the phage situation, no instance of production of infectious nucleic acid has been observed, but the animal virus RNAs are all much larger in size, which makes the task more difficult with respect to maintaining the integrity of template and product during the incubation period of the *in vitro* synthesis.

Some illustrative examples of the *in vitro* action of viral RNA-dependent RNA polymerases include those of myxoviruses, for example, influenza virus (Skehel 1971; Penhoet et al. 1971); diplornaviruses, such as reovirus; cytoplasmic polyhedrosis virus of the silkworm and wound tumor virus of sweetclover (Borsa and Graham 1968; Lewandowski et al. 1969; Black and Knight 1970); and rhabdoviruses, for example, vesicular stomatitis virus (Baltimore et al. 1970).

A stable, soluble poliovirus RNA-polymerase complex has been isolated from infected HeLa cells and shown to be active in the *in vitro* synthesis of poliovirus RNA (Ehrenfeld et al. 1970). The poliovirus polymerase complex differs from the examples above in that it is a subviral structure active in infected cells but not appearing in the mature virion, which is devoid of polymerase activity. Some plant virus RNA-dependent RNA polymerases have also been isolated in partially purified form from plant tissues infected with turnip yellow mosaic, cucumber mosaic, tobacco ringspot, and brome mosaic viruses but the *in vitro* synthesis products have yet to be more than superficially characterized (Astier-Manifacier and Cornuet 1971; May and Symons 1971; Peden et al. 1972; Hadidi and Fraenkel-Conrat 1973).

Many enzymes are known that are active in some phase of DNA synthesis, and valuable *in vitro* applications have been made of this knowledge.[1] However, few attempts at *in vitro* synthesis of viral DNAs have been made, probably because these DNAs are so large and the enzymes needed for the chore are scarce. Nevertheless, an outstanding *in vitro* synthesis of infectious DNA, paralleling the similar feat with RNA performed earlier by Haruna and Spiegelman (1965), was accomplished by Goulian, Kornberg, and Sinsheimer (1967). Infectious coliphage ØX174 DNA was produced in this *in vitro* synthesis summarized in Figure 54. The enzymes used were isolated from *E. coli*.

The virions of RNA tumor viruses have several enzymes associated with them, but the one most intensively studied so far is an RNA-dependent DNA polymerase called reverse transcriptase (Temin and Mizutani 1970; Baltimore 1970). For a while, this enzyme seemed to threaten the central dogma of molecular biology, that information flows from DNA to RNA to protein; reverse transcriptase was so named because information flow goes back from RNA to DNA and then subsequently forward in the usual direction. However, Crick, an originator of the central dogma, judged this anomaly to be within the basic concept (1970). In any case, *in vitro* synthesis of radioactive DNA segments using tumor virus RNA as template and viral reverse transcriptase provides valuable material with which to probe for tumor virus nucleic acids by hybridization reactions (see Tooze 1973).

[1]See *DNA Synthesis in Vitro*, R. D. Wells and R. B. Inman, eds., Baltimore: University Park Press (1973).

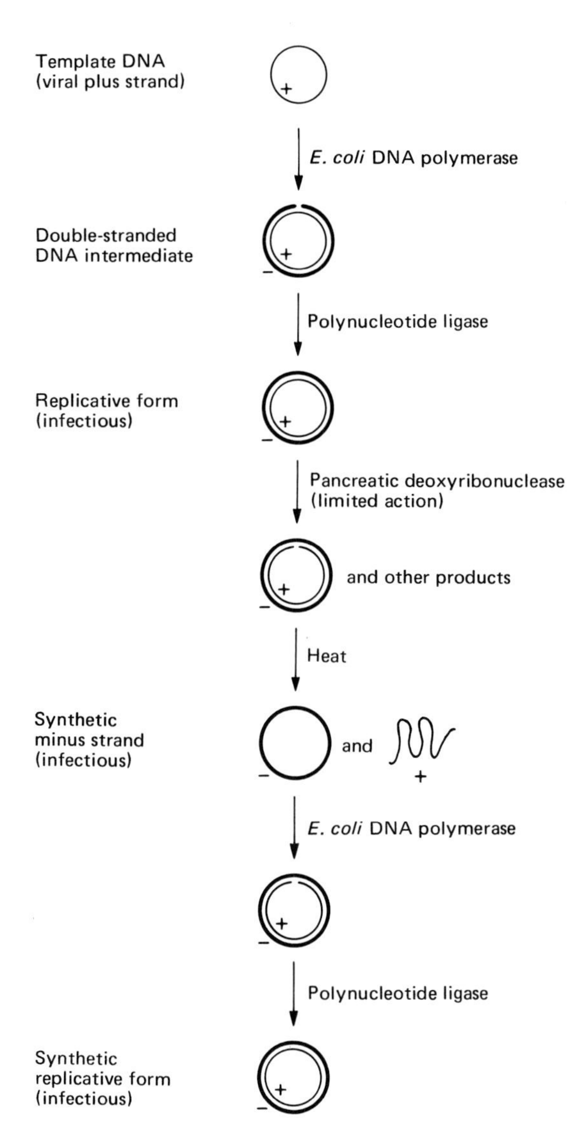

Fig. 54. Schematic representation of the enzymatic synthesis of infectious phage ØX174 DNA. (From Knight 1974 as adapted from Goulian et al. 1967 by omitting important but distracting details such as radioactive and density labeling and some of the products.)

C. Origin of Viruses

The origin of viruses and the broader question of the origin of life are matters for speculation. At least three ideas can be advanced for the origin of viruses: (1) viruses originated in early days of the planet by chemical evolution; (2) viruses arose from more complex microorganisms by a process of retrograde evolution; (3) viruses arose from the genetic material of cells.

Reconstruction experiments, starting with the raw materials and kinds of energy postulated to abound around primitive planets, have yielded amino acids, peptides, sugars, lipids, purine and pyrimidine bases, and many other organic compounds (see reviews by Fox et al. 1970; Fox 1971; Ponnamperuma and Gabel 1968). From the results of these experiments it is conceivable that nucleoproteins were among the products of chemical evolution. If it is imagined that such nucleoproteins had the ability to organize duplicate molecules from their surroundings, they might be thought of as primitive viruses. Furthermore, mutation of some of these primitive viruses may have led to more complex forms until the first cell-like assemblies were developed from which biological evolution may have proceeded. There is little to substantiate these conjectures at present.

Likewise, the notion that viruses may represent the products of retrograde evolution of higher organisms (for example, bacteria) has little to support it. There is an obvious progression in complexity from the simplest virus to the larger, more complex ones, from these to the simplest bacteria (for example, Bedsonia, Rickettsiae, and so on), and from there to the most complex bacteria and other protists. If this progression is reversed, one could imagine viruses as the ultimate product in morphologic and metabolic regression toward parasitism. However, such a series has not been induced experimentally and this idea remains speculative.

The most plausible concept of the origin of viruses seems to be that they represent detached segments of cellular DNA, or transcripts thereof in the case of RNA viruses, which achieved some measure of autonomy and evolved parallel with their hosts. Acquisition of protein coats and other morphologic structures exhibited by current complex viruses could be viewed as evidence of the operation of mutation and selection in the evolutionary process. Perhaps the most suggestive current model in support of the origin of viruses as escaped genetic elements is the bacterial plasmid. Plasmids are supernumerary DNAs of about 20×10^6 daltons that exhibit a variety of functions and accordingly are called sex factors, bacteriocinogenic factors, and resistance transfer factors. In a sense, they are all sex factors in that they share the property of promoting conjugation of and genetic transfer between bacterial cells. Those cells that have a sex factor transfer material during conjugation to other cells that do not have sex factor; the receptor cells thus acquire the capacity to be donors and in this sense have changed their sex. Resistance transfer plasmids are characterized by genes that enable them to provide to their host cells resistance, for example, to drugs; such resistance may depend on production of an enzyme coded for by the plasmid that destroys a drug, for example, penicillinase which destroys penicillin. Bacteriocinogenic bacteria produce substances toxic to other bacteria, for example, colicins produced by plasmids of certain strains of *E. coli*. In any case, the bacterial plasmid can exist free in the cytoplasm of the cell where it reproduces autonomously, or it is occasionally inserted into the bacterial chromosome where it replicates only when the latter does.

A striking similarity between a plasmid and the coliphage lambda has been noted by Hayes (1968, Chap. 24). Plasmids and phage lambda have DNA genomes of about the same size, and both genomes are independent of the bacterial chromosome in the sense that they can replicate autonomously and if lost from the cell can be reacquired only by infection. Infection by lambda is by means of the injection mechanism commonly employed by phages in initiating infection; infection by plasmids is by conjugation which they often promote. Both, under appropriate conditions, can become integrated into host cell chromosome where they replicate coordinately with the host DNA. Both can mutate and both can undergo genetic recombination. The resemblance is strong enough so that Hayes thinks it logical to regard plasmids as viruses. The missing link, then, with respect to the origin of viruses is the demonstration that any plasmid originated from a bacterial chromosome. This proof is missing but the hypothesis is supported by the apparent homology between plasmid and chromosomal DNAs. This is presumably the basis for integration of some plasmids and the degree of homology observed is consistent with expectations if the plasmids had evolved through many generations from a segment of bacterial chromosome.

Some Reference Books in Virology

General References

Multivolume Works

Advances in Virus Research: Smith, Lauffer, and Bang, eds. Academic Press, New York, 1953 on (annual).

Comprehensive Virology: Fraenkel-Conrat and Wagner, eds. Plenum, New York, 1974 on (22 volumes projected).

Methods in Virology: Maramorosch and Koprowski, eds. Academic Press, New York, 1967, 1968, 1971 (5 volumes).

Perspectives in Virology: Pollard, ed. Wiley, Hoeber, Academic Press, New York, 1959 on (annual).

Progress in Medical Virology: Melnick, ed. Karger, New York, 1958 on (annual).

Virology Monographs: various authors. Springer-Verlag, New York, 1968 on (12 volumes, more in preparation).

Single Volumes

The Biochemistry of Viruses: Levy, ed. Dekker, New York, 1969.

The Chemistry and Biology of Viruses: Fraenkel-Conrat. Academic Press, New York, 1969.

Comparative Virology: Maramorosch and Kurstak, eds. Academic Press, New York, 1971.

Animal Viruses

Basic Mechanisms in Animal Virus Biology. Cold Spring Harbor Symposia on Quantitative Biology, Vol. 27, Cold Spring Harbor, N. Y. Laboratory, 1962.

Biohazards in Biological Research: Hellman et al., eds. Cold Spring Harbor, N. Y. Laboratory, 1973.

The Biology of Animal Viruses: Fenner et al. Academic Press, New York, 1974.

The Biology of Large RNA Viruses: Barry and Mahy, eds. Academic Press, New York, 1970.
The Cytoplasmic Polyhedrosis Virus of the Silkworm: Aruga and Tanada, eds. University of Tokyo Press, Tokyo, 1971.
Insect Virology: Smith. Academic Press, New York, 1967.
Medical Virology: Fenner and White. Academic Press, New York, 1970.
The Molecular Biology of Tumor Viruses: Tooze, ed. Cold Spring Harbor, N. Y. Laboratory, 1973.
RNA Viruses and Host Genome in Oncogenesis: Emmelot and Bentvelzen, eds. American Elsevier, New York, 1972.
Slow Virus Diseases: Zeman and Lennette, eds. Williams and Wilkins, Baltimore, 1974.
An Electron Micrographic Atlas of Viruses: Williams and Fisher. Thomas, Springfield, Ill., 1974.
Fundamental Techniques in Virology: Habel and Salzman, eds. Academic Press, New York, 1969.
General Virology: Luria and Darnell. Wiley, New York, 1967.
An Introduction to Virology: Goodheart. Saunders, Philadelphia, 1969.
Molecular Basis of Virology: Fraenkel-Conrat, ed. Van Nostrand Reinhold, New York, 1968.
The Molecular Biology of Viruses: Colter and Paranchych, eds. Academic Press, New York, 1967.
Molecular Virology: Knight. McGraw-Hill, New York, 1974.
Strategy of the Viral Genome: Wolstenholme and O'Connor, eds. Churchill Livingstone, Edinburgh, 1971.
Techniques in Experimental Virology: Harris, ed. Academic Press, London, 1964.
Ultrastructure of Animal Viruses and Bacteriophages. An Atlas: Dalton and Haguenau, eds. Academic Press, New York, 1973.
Viruses Affecting Man and Animals: Sanders and Schaeffer, eds. Green, St. Louis, 1971
Viral and Rickettsial Infections of Animals: Betts and York. Academic Press, New York, 1967.
Viral and Rickettsial Infections of Man (4th ed.): Horsfall and Tamm, eds. Lippincott, Philadelphia, 1965.
Virus-Cell Interactions and Viral Antimetabolites: Shugar, ed. Academic Press, New York, 1972.
Virus Morphology: Madeley. Williams and Wilkins, Baltimore, 1973.
Viruses of Vertebrates (3rd ed.): Andrewes and Pereira. Williams and Wilkins, Baltimore, 1972.

Bacterial Viruses

Bacteriophage Biochemistry: Mathews. Van Nostrand Reinhold, New York, 1971.

The Bacteriophage Lambda: Hershey, ed. Cold Spring Harbor, N. Y. Laboratory, 1971.

Episomes: Campbell. Harper & Row, New York, 1969.

The Genetics of Bacteria and Their Viruses (2nd ed.): Hayes. Wiley, New York, 1968.

Molecular Biology of Bacterial Viruses: Stent. Freeman, San Francisco, 1963.

Papers on Bacterial Viruses (2nd ed.): Stent, ed. Little, Brown, Boston, 1965.

Phage and the Origins of Molecular Biology: Cairns et al., eds. Cold Spring Harbor, N. Y. Laboratory, 1966.

Virus-Induced Enzymes: Cohen. Columbia University Press, New York, 1968.

Plant Viruses

The Biochemistry and Physiology of Infectious Plant Diseases: Goodman et al. Van Nostrand, Princeton, 1967.

Descriptions of Plant Viruses: Gibbs et al. Commonwealth Agriculture Bureaux, Farnham Royal, Slough SL 2 3 BN, England, 1970 on.

Index of Plant Virus Diseases: Thornberry. Agriculture Handbook No. 307, U. S. Department of Agriculture, Washington, D. C., 1966.

Plant Virology: Corbett and Sisler, eds. University of Florida Press, Gainesville, 1964.

Plant Virology: Matthews. Academic Press, New York, 1970.

Plant Viruses and Virus Diseases (4th ed.): Bawden. Ronald Press, New York, 1964.

Principles and Techniques in Plant Virology: Kado and Agrawal, eds. Van Nostrand Reinhold, New York, 1972.

Symptoms of Virus Diseases in Plants: Bos, Vada, Wageningen, The Netherlands, 1963.

A Textbook of Plant Virus Diseases (3rd ed.): Smith. Academic Press, New York, 1973.

Viruses of Plants: Beemster and Dijkstra, eds. Wiley, New York, 1966.

Viruses in Plant Hosts: Form, Distribution, and Pathologic Effects: Esau. University of Wisconsin Press, Madison, 1968.

Viruses, Vectors, and Vegetation: Maramorosch, ed. Wiley, New York, 1969.

References

Aach, H. G., Funatsu, G., Nirenberg, M. W., and Fraenkel-Conrat, H. 1964. Further attempts to characterize products of TMV-RNA-directed protein synthesis. *Biochemistry* **3**, 1362–1366.

Ada, G. L., and Gottschalk, A. 1956. Component sugars of influenza virus. *Biochem. J.* **62**, 686–689.

Ada, G. L., and Perry, B. T. 1954. The nucleic acid content of influenza virus. *Australian J. Exptl. Biol. Med. Sci.* **32**, 453–468.

Adolph, K. W., and Haselkorn, R. 1971. Isolation and characterization of a virus infecting the blue-green alga *Nostoc muscorum. Virology* **46**, 200–208.

Ahmed, M. E., Black, L. M., Perkins, E. G., Walker, B. L., and Kummerow, F. A. 1964. Lipid in potato yellow dwarf virus. *Biochem. Biophys. Res. Commun.* **17**, 103–107.

Akabori, S., Ohno, K., Ikenaka, T., Okada, Y., Hanafusa, H., Haruna, I., Tsugita, A., Sugae, K., and Matsushima, T. 1956. Hydrazinolysis of peptides and proteins. II. Fundamental studies on the determination of the carboxyl-ends of proteins. *Bull. Chem. Soc. Japan* **29**, 507–518.

Albertsson, P. 1960. *Partition of Cell Particles and Macromolecules.* Chap. 7. Wiley, New York.

Albertsson, P. 1971. *Partition of Cell Particles and Macromolecules.* 2nd edit. Wiley, New York.

Alberty, R. A. 1953. Electrochemical properties of the proteins and amino acids. *In* H. Neurath and K. Bailey, eds., *The Proteins.* pp. 461–548. Academic Press, New York.

Alexander, H. E., Koch, G., Mountain, I. M., and Van Damme, O. 1958. Infectivity of ribonucleic acid from poliovirus in human cell monolayers. *J. Exptl. Med.* **108**, 493–506.

Allison, A. C., and Burke, D. C. 1962. The nucleic acid contents of viruses. *J. Gen. Microbiol.* **27**, 181–194.

Ambler, R. P. 1972. Enzymatic hydrolysis with carboxypeptidases. *In* C. H. W. Hirs and S. N. Timasheff, eds., *Methods in Enzymology.* Vol. 25, pp. 143–154. Academic Press, New York.

Ames, B. N., and Dubin, D. T. 1960. The role of polyamines in the neutralization of bacteriophage deoxyribonucleic acid. *J. Biol. Chem.* **235**, 769–775.

Ames, B. N., Dubin, D. T., and Rosenthal, S. M. 1958. Presence of polyamines in certain bacterial viruses. *Science* **127**, 814–816.

Anderer, F. A. 1959a. Das Molekulargewicht der Peptideinheit im Protein des Tabakmosaikvirus. *Z. Naturforsch.* **14b**, 24–28.

Anderer, F. A. 1959b. Reversible Denaturierung des Proteins aus Tabakmosaikvirus. *Z. Naturforsch.* **14b**, 642–647.

Anderer, F. A., Uhlig, H., Weber, E., and Schramm, G. 1960. Primary structure of the protein of tobacco mosaic virus. *Nature* **186**, 922–925.

Anderer, F. A., Wittmann-Liebold, B., and Wittmann, H. G. 1965. Weitere Untersuchungen zur Aminosaüresequenz des Proteins im Tabakmosaikvirus. *Z. Naturforsch.* **20b**, 1203–1213.

Anderson, N. G. 1955. Studies on isolated cell components. VIII. High resolution gradient differential centrifugation. *Exptl. Cell Res.* **9**, 446–459.

Anderson, N. G., and Cline, G. B. 1967. New centrifugal methods for virus isolation. *In* K. Maramorosch and H. Koprowski, eds., *Methods in Virology*. Vol. II, pp. 137–178. Academic Press, New York.

Anson, M. L. 1942. Some factors which influence the oxidation of sulfhydryl groups. *J. Gen. Physiol.* **25**, 355–367.

Appleyard, G. 1967. The photosensitivity of Semliki Forest and other viruses. *J. Gen. Virol.* **1**, 143–152.

Appleyard, G., Oram, J. D., and Stanley, J. L. 1970. Dissociation of Semliki Forest virus into biologically active components. *J. Gen. Virol.* **9**, 179–189.

Arimura, H. 1973. Effect of lysozyme from human placenta on ectromelia virus. *Acta Virol.* **17**, 130–137.

Asbeck, F., Beyreuther, K., Kohler, H., von Wettstein, G., and Braunitzer, G. 1969. Virus proteins. IV. The constitution of the coat protein of the fd phage. *Hoppe-Seyler's Z. Physiol. Chem.* **350**, 1047–1066.

Ashwell, G. 1957. Colorimetric analysis of sugars. *In* S. P. Colowick and N. O. Kaplan, eds., *Methods in Enzymology*. Vol. 3, pp. 73–105. Academic Press, New York.

Astier-Manifacier, S., and Cornuet, P. 1971. RNA-dependent RNA polymerase in Chinese cabbage. *Biochim. Biophys. Acta* **232**, 484–493.

Auerbach, C., and Robson, J. M. 1947. The production of mutations by chemical substances. *Proc. Roy. Soc. Edinburgh (B)* **62**, 271–283.

August, J. T., Eoyang, L., Franze de Fernandez, M. T., Hayward, W. S., Kuo, C. H., and Silverman, P. M. 1973. Host proteins in the replication of bacteriophage RNA. *In* F. T. Kenney, B. A. Hamkalo, G. Favelukes, and J. T. August, eds., *Gene Expression and Its Regulation*. pp. 29–41. Plenum Press, New York.

Avery, O. T., MacLeod, C. M., and McCarty, M. 1944. Studies on the chemical nature of the substance inducing transformation of pneumococcal types. Induction of transformation by a deoxyribonucleic acid fraction isolated from pneumococcus type III. *J. Exptl. Med.* **79**, 137–157.

Bachrach, H. L. 1960. Ribonucleic acid of foot-and-mouth disease virus: Its preparation, stability, and plating efficiency on bovine-kidney cultures. *Virology* **12**, 258–271.

Bachrach, H. L., and Schwerdt, C. E. 1952. Purification studies of Lansing poliomyelitis virus: pH stability, CNS extraction, and butanol purification experiments. *J. Immunol.* **69**, 551–561.

Bachrach, H. L., and Schwerdt, C. E. 1954. Purification studies on Lansing poliomyelitis virus. II. Analytical electron microscopic identification of the infectious particle in preparations of high specific infectivity. *J. Immunol.* **72**, 30–38.

Backus, R. C., and Williams, R. C. 1953. Centrifugation in field-aligning capsules: Preparative microcentrifugation. *Science* **117**, 221–223.

Bagi, G., Gulyas, A., and Solymosy, F. 1970. A differential extraction method for the isolation of host nucleic acids from TMV-infected tobacco plants. *Virology* **42**, 662–667.

Bailey, J. L. 1967. *Techniques in Protein Chemistry*. pp. 189–190. Elsevier, Amsterdam.

Baltimore, D. 1970. RNA-dependent DNA polymerases in virions of RNA tumor viruses. *Nature* **226**, 1209–1211.

Baltimore, D., Huang, A. S., and Stampfer, M. 1970. Ribonucleic acid synthesis of vesicular stomatitis virus. II. An RNA polymerase in the virion. *Proc. Natl. Acad. Sci. U.S.* **66**, 572–576.

Baluda, M. A., and Nayak, D. P. 1969. Incorporation of precursors into ribonucleic acid, protein, glycoprotein, and lipoprotein of avian myeloblastosis virions. *J. Virol.* **4**, 554–566.

Bambara, R., Padmanabhan, R., and Wu, R. 1973. Nucleotide sequence analysis of DNA. X. Complete nucleotide sequence of the cohesive ends of bacteriophage ϕ80 DNA. *J. Molecular Biol.* **75**, 741–744.

Bancroft, J. B. 1970. The self-assembly of spherical plant viruses. *Advan. Virus Res.* **16**, 99–134.

Barrett, K., and Calendar, R. 1974. Personal communication.

Barrett, K., Calendar, R., Gibbs, W., Goldstein, R. N., Lindqvist, B., and Six, E. 1973. Helper-dependent bacteriophage P4: A model satellite virus and its implications for animal virology. *Progr. Med. Virol.* **15**, 309–330.

Barry, R. D. 1960. Equilibrium sedimentation of influenza virus in cesium chloride density gradients. *Australian J. Exptl. Biol.* **38**, 499–508.

Barry, R. D., Bromley, P. A., and Davies, P. 1970. The characterization of influenza virus RNA. *In* R. D. Barry and B. W. J. Mahy, eds., *The Biology of Large RNA Viruses*. pp. 279–300. Academic Press, New York.

Basilio, C., Wahba, A. J., Lengyel, P., Speyer, J. F., and Ochoa, S. 1962. Synthetic polynucleotides and the amino acid code. V. *Proc. Natl. Acad. Sci. U.S.* **48**, 613–616.

Bather, R. 1957. The nucleic acid of partially purified Rous No. 1 Sarcoma virus. *Brit. J. Cancer* **11**, 611–619.

Bawden, F. C. 1950. *Plant Viruses and Virus Diseases.* 3rd edit. Chronica Botanica, Waltham, Mass.
Bawden, F. C. 1958. Reversible changes in strains of tobacco mosaic virus from leguminous plants. *J. Gen. Microbiol.* **18**, 751–766.
Bawden, F. C., and Pirie, N. W. 1937. The isolation and some properties of liquid crystalline substances from solanaceous plants infected with three strains of tobacco mosaic virus. *Proc. Roy. Soc.* **B123**, 274–320.
Bawden, F. C., and Pirie, N. W. 1938. Liquid crystalline preparations of potato virus X. *Brit. J. Exptl. Pathol.* **19**, 66–82.
Bawden, F. C., Pirie, N. W., Bernal, J. D., and Fankuchen, I. 1936. Liquid crystalline substances from virus-infected plants. *Nature* **138**, 1051.
Bawden, F. C., and Kleczkowski, A. 1959. Photoreactivation of nucleic acid from tobacco mosaic virus. *Nature* **183**, 503–504.
Beard, J. W. 1948. The chemical, physical, and morphological properties of animal viruses. *Physiol. Rev.* **28**, 349–367.
Beard, J. W., and Wyckoff, R. W. G. 1938. The pH stability of the papilloma virus protein. *J. Biol. Chem* **123**, 461–470.
Beard, J. W., Finklestein, H., and Wyckoff, R. W. G. 1938. The pH stability of the elementary bodies of vaccinia. *J. Immunol.* **35**, 415–425.
Becker, E. F., Jr., Zimmerman, B. K., and Geiduschek, E. P. 1964. Structure and function of cross-linked DNA. I. Reversible denaturation and *Bacillus subtilis* transformation. *J. Molecular Biol.* **8**, 377–391.
Beer, S. V., and Kosuge, T. 1970. Spermidine and spermine—polyamine components of turnip yellow mosaic virus. *Virology* **40**, 930–938.
Beijerinck, M. W. 1898a. Over een Contagium vivum fluidum als oorsaak van de Vlekziekte der Tabaksbladen. *Verslag. Kon. Akad. V. Wetenschap. Amsterdam* **7**, 229–235.
Beijerinck, M. W. 1898b. Über ein Contagium vivum fluidum als Ursache der Fleckenkrankheit der Tabaksblätter, *Verhandl. Kon. Akad. Wetenschap. Amsterdam* **6**, 1–24.
Bellett, A. J. D. 1967. Numerical classification of some viruses, bacteria and animals according to nearest neighbor base sequence frequencies. *J. Molecular Biol.* **27**, 107–112.
Bellett, A. J. D. 1968. The iridescent virus group. *Advan. Virus Res.* **13**, 225–246.
Benjamin, T. L. 1965. Relative target sizes for the inactivation of the transforming and reproductive abilities of polyoma virus. *Proc. Natl. Acad. Sci. U.S.* **54**, 121–124.
Ben-Porat, T., and Kaplan, A. S. 1962. The chemical composition of herpes simplex and pseudorabies viruses. *Virology* **16**, 261–266.
Benzer, S. 1957. The elementary units of heredity. *In* W. D. McElroy and B. Glass, eds., *The Chemical Basis of Heredity.* pp. 70–93. Johns Hopkins Press, Baltimore.
Benzer, S. 1961. Genetic fine structure. In *Harvey Lectures, Series* **56**, pp. 1–21. Academic Press, New York.

Bercks, R., Koenig, R., and Querfurth, G. 1972. Plant virus serology. *In* C. I. Kado and H. O. Agrawal, eds., *Principles and Techniques in Plant Virology*. pp. 466–490. Van Nostrand-Reinhold, New York.

Bergoin, M., and Dales, S. 1971. Comparative observations on poxviruses of invertebrates and vertebrates. *In* K. Maramorosch and E. Kurstak, eds., *Comparative Virology*. pp. 169–205. Academic Press, New York.

Bergold, G. H. 1953. Insect viruses. *Advan. Virus Res.* **1**, 91–139.

Bergold, G. H., and Wellington, E. F. 1954. Isolation and chemical composition of the membranes of an insect virus and their relation to the virus and polyhedral bodies. *J. Bacteriol.* **67**, 210–216.

Bernal, J. D., and Fankuchen, I. 1941. X-ray and crystallographic studies of plant virus preparations. *J. Gen. Physiol.* **25**, 111–165.

Best, R. J. 1968. Tomato spotted wilt virus. *Advan. Virus Res.* **13**, 65–146.

Bikel, I., and Knight, C. A. 1972. Differential action of aspergillus glycosidases on the hemagglutinating and neuraminidase activities of influenza and Newcastle disease viruses. *Virology* **49**, 326–332.

Billeter, M. A.., Dahlberg, J. E., Goodman, H. M., Hindley, J., and Weissmann, C. 1969. Sequence of the first 175 nucleotides from the 5′ terminus of Qβ RNA synthesized *in vitro*. *Nature* **224**, 1083–1086.

Bishop, D. H. L., Claybrook, J. R., and Spiegelman, S. 1967. Electrophoretic separation of viral nucleic acids on polyacrylamide gels. *J. Molecular Biol.* **26**, 373–387.

Bishop, D. H. L., Mills, D. R., and Spiegelman, S. 1968. The sequence at the 5′ terminus of a self-replicating variant of viral Qβ ribonucleic acid. *Biochemistry* **7**, 3744–3753.

Black, D. R., and Knight, C. A. 1970. Ribonucleic acid transcriptase activity in purified wound tumor virus. *J. Virol.* **6**, 194–198.

Blair, C. D., and Duesberg, P. H. 1970. Myxovirus ribonucleic acids. *Ann. Rev. Microbiol.* **24**, 539–573.

Bockstahler, L. E., and Kaesberg, P. 1961. Bromegrass mosaic virus: A virus containing an unusually small ribonucleic acid. *Nature* **190**, 192–193.

Boedtker, H. 1959. Some physical properties of infective ribose nucleic acid isolated from tobacco mosaic virus. *Biochim. Biophys. Acta* **32**, 519–531.

Boeyé, A. 1959. Induction of a mutation in poliovirus by nitrous acid. *Virology* **9**, 691–700.

Bonar, R. A., and Beard, J. W. 1959. Virus of avian myeloblastosis. XII. Chemical constitution. *J. Natl. Cancer Inst.* **23**, 183–197.

Bonhoeffer, F., and Schachman, H. K. 1960. Studies on the organization of nucleic acids within nucleoproteins. *Biochem. Biophys. Res. Commun.* **2**, 366–371.

Borsa, J., and Graham, A. F. 1968. RNA polymerase activity in purified virions. *Biochem. Biophys. Res. Commun.* **33**, 895–901.

Bøvre, K., Lozeron, H. A., and Szybalski, W. 1971. Techniques of RNA-DNA hybridization in solution for the study of viral transcription. *In* K.

Maramorosch and H. Koprowski, eds., *Methods in Virology*. Vol. 5, pp. 271–292. Academic Press, New York.

Bradlcy, D. E. 1971. A comparative study of the structure and biological properties of bacteriophages. *In* K. Maramorosch and E. Kurstak, eds., *Comparative Virology*. pp. 207–253. Academic Press, New York.

Bradley, D. E., and Kay, D. 1960. The fine structure of bacteriophages. *J. Gen. Microbiol.* **23**, 553–563.

Brakke, M. K. 1955. Zone electrophoresis of dyes, proteins, and viruses in density-gradient columns of sucrose solutions. *Arch. Biochem. Biophys.* **55**, 175–190.

Brakke, M. K. 1960. Density gradient centrifugation and its application to plant viruses. *Advan. Virus Res.* **7**, 193–224.

Brakke, M. K. 1967. Density-gradient centrifugation. *In* K. Maramorosch and H. Koprowski, eds., *Methods in Virology*. Vol. II, pp. 93–118. Academic Press, New York.

Brammer, K. W. 1963. Chemical modification of viral nucleic acid. II. Bromination and iodination. *Biochim. Biophys. Acta* **72**, 217–229.

Brand, E., Saidell, L. J., Goldwater, W. H., Kassell, B., and Ryan, F. J. 1945. The empirical formula of beta-lactoglobulin. *J. Am. Chem. Soc.* **67**, 1524–1532.

Braunitzer, G. 1954. Über den Nachweis der Carboxyl-Endgruppe im Tabakmosaikvirus durch Spaltung mit Hydrazin. *Z. Naturforsch.* **9b**, 675–678.

Breeze, D. C., and Subak-Sharpe, H. 1967. The mutability of small-plaque-forming encephalomyocarditis virus. *J. Gen. Virol.* **1**, 81–88.

Brenner, S., and Horne, R. W. 1959. A negative staining method for high resolution electron microscopy of viruses. *Biochim. Biophys. Acta* **34**, 103–110.

Brenner, S., Barnett, L., Crick, F. H. C., and Orgel, A. 1961. The theory of mutagenesis. *J. Molecular Biol.* **3**, 121–124.

Brinton, C. C., Jr., and Lauffer, M. A. 1959. The electrophoresis of viruses, bacteria and cells, and the microscope method of electrophoresis. *In* M. Bier, ed., *Electrophoresis, Theory, Methods, and Applications*. pp. 427–492. Academic Press, New York.

Brown, R. M., Jr. 1972. Algal viruses. *Advan. Virus Res.* **17**, 243–277.

Brown, F., and Cartwright, B. 1960. Purification of the virus of foot-and-mouth disease by fluorocarbon treatament and its dissociation from neutralizing antibody. *J. Immunol.* **85**, 309–313.

Brownlee, G. G. 1972. *Determination of Sequences in RNA*. American Elsevier, New York.

Burge, B. W., and Huang, A. S. 1970. Comparison of membrane protein glycopeptides of Sindbis virus and vesicular stomatitis virus. *J. Virol.* **6**, 176–182.

Burgess, A. B., and Denhardt, D. T. 1969. Studies on ØX174 proteins. I.

Phage-specific proteins synthesized after infection of *Escherichia coli. J. Molecular Biol.* **44**, 377–386.

Burton, K. 1968. Determination of DNA concentration with diphenylamine. *In* L. Grossmann and K. Moldave, eds., *Methods in Enzymology*. Vol. 12, Pt. B, pp. 163–166. Academic Press, New York.

Cairns, J. 1962. The application of autoradiography to the study of DNA viruses. *Cold Spring Harbor Symp. Quant. Biol.* **27**, 311–318.

Capecchi, M. R. 1966. Cell-free protein synthesis programmed with R17 RNA: Identification of two phage proteins. *J. Molecular Biol.* **21**, 173–193.

Carp, R. I. 1963. A study of the mutation rates of several poliovirus strains to the reproductive capacity temperature/ 40+ and guanidine marker characteristics. *Virology* **21**, 373–382.

Carpenter, F. H., and Chramback, A. 1962. On the amide content of insulin fractions isolated by partition column chromatography and countercurrent distribution. *J. Biol. Chem.* **237**, 404–408.

Cartwright, B., Smale, C. J., and Brown, F. 1970. Structural and biological relations in vesicular stomatitis virus. *In* R. D. Barry and B. W. J. Mahy, eds., *The Biology of Large RNA Viruses*. pp. 115–132. Academic Press, New York.

Casals, J. 1967. Immunological techniques for animal viruses. *In* K. Maramorosch and H. Koprowski, eds., *Methods in Virology*. pp. 113–198. Academic Press, New York.

Casjens, S. 1971. The morphogenesis of the phage lambda head: The step controlled by gene F. *In* A. D. Hershey, ed., *The Bacteriophage Lambda*. pp. 725–732. Cold Spring Harbor Laboratory.

Caspar, D. L. D. 1956. Radial density distribution in the tobacco mosaic virus particle. *Nature* **177**, 928.

Caspar, D. L. D. 1963. Assembly and stability of the tobacco mosaic virus particle. *Advan. Protein Chem.* **18**, 37–121.

Caspar, D. L. D., and Klug, A. 1962. Physical principles in the construction of regular viruses. *Cold Spring Harbor Symp. Quant. Biol.* **27**,1–24.

Caspar, D. L. D., Dulbecco, R., Klug, A., Lwoff, A., Stoker, M. G. P., Tournier, P., and Wildy, P. 1962. Proposals. *Cold Spring Harbor Symp. Quant. Biol.* **27**, 49.

Chargaff, E. 1955. Isolation and composition of the deoxypentose nucleic acids and of the corresponding nucleoproteins. *In* E. Chargaff and J. N. Davidson, eds., *The Nucleic Acids*. Vol. 1, pp. 307–371. Academic Press, New York.

Charney, J. 1957. Discussion. *Spec. Publ. N. Y. Acad. Sci. No.* 5, 171–172.

Charney, J., Machlowitz, R., Tytell, A. A., Sagin, J. F. and Spicer, D. S. 1961. The concentration and purification of poliomyelitis virus by the use of nucleic acid precipitation. *Virology* **15**, 269–280.

Chen, C., Compans, R. W., and Choppin, P. W. 1971. Parainfluenza virus surface projections: Glycoproteins with hemagglutinin and neuraminidase activities. *J. Gen. Virol.* **11**, 53–58.

Chidlow, J., and Tremaine, J. H. 1971. Limited hydrolysis of cowpea chlorotic mottle virus by trypsin and chymotrypsin. Virology **43**, 267–278.

Clamp, J. R., Bhatti, T., and Chambers, R. E. 1972. The examination of carbohydrate in glycoproteins by gas-liquid chromatography. *In* A. Gottschalk, ed., *Glycoproteins. Their Composition, Structure and Function.* 2nd edit., Pt. A, pp. 300–321. Elsevier, New York.

Clark, A. J. 1971. Toward a metabolic interpretation of genetic recombination of *E. coli* and its phages. *Ann. Rev. Microbiol.* **25**, 437–464.

Clark, J. M., Chang, A. Y., Spiegelman, S., and Reichmann, M. E. 1965. The *in vitro* translation of a monocistronic message. *Proc. Natl. Acad. Sci. U.S.* **54**, 1193–1197.

Cochran, G. W., and Chidester, J. L. 1957. Infectious nucleic acid in plants with tobacco mosaic. *Virology* **4**, 390–391.

Cochran, G. W., Chidester, J. L., and Stocks, D. L. 1957. Chromatography of TMV on a cellulose cation exchange adsorbent. *Nature* **180**, 1281–1282.

Cohen, S. N., and Hurwitz, J. 1968. Genetic transcription in bacteriophage λ: Studies of λ mRNA synthesis *in vivo*. *J. Mol. Biol.* **37**, 387–406.

Cohen, S. S. 1971. *Introduction to the Polyamines.* Prentice Hall, Englewood Cliffs, N. J.

Cohen, S. S., and Arbogast, R. 1950. Chemical studies in host-virus interactions. VI. Immunochemical studies on the purity of concentrates of various bacterial viruses prepared by differential centrifugation procedures. With an appendix summarizing data on the desoxyribonucleic acid content of these viruses. *J. Exp. Med.* **91**, 607–618.

Cohen, S. S., and Stanley, W. M. 1942. The molecular size and shape of the nucleic acid of tobacco mosaic virus. *J. Biol. Chem.* **144**, 589–598.

Cohen, S. S., and Dion, A. 1971. Polyamines and the multiplication of T-even phages. *In* D. W. Ribbons, J. F. Woessner and J. Schultz, eds., *Nucleic Acid-Protein Interactions.* pp. 266–287. North-Holland, Amsterdam.

Colby, C., Jr. 1971. The induction of interferon by natural and synthetic polynucleotides. *Progr. Nucleic Acid Res. Molecular Biol.* **11**, 1–32.

Commoner, B., Lippincott, J. A., Shearer, G. B., and Richman, E. E. 1956. Reconstitution of tobacco mosaic virus components. *Nature* **178**, 767–771.

Compans, R. W., Klenk, H. D., Caliguiri, L. A., and Choppin, P. W. 1970. Influenza virus proteins. I. Analysis of polypeptides of the virion and identification of spike glycoproteins. *Virology* **42**, 880–889.

Compans, R. W., and Choppin, P. W. 1971. The structure and assembly of influenza and parainfluenza viruses. *In* K. Maramorosch and E. Kurstak, eds., *Comparative Virology.* pp. 407–432. Academic Press, New York.

Compans, R. W., and Choppin, P. W. 1973. Orthomyxoviruses. *In* A. J. Dalton and F. Haguenau, eds., *Ultrastructure of Animal Viruses and Bacteriophages.* pp. 213–237. Academic Press, New York.

Coolsma, J., and Haselkorn, R. 1969. *In vitro* synthesis of T4 proteins. *Biochem. Biophys. Res. Commun.* **34**, 253–259.

Corbett, M. K. 1961. Purification of potato virus X without aggregation. Virology **15**, 8–15.

Cox, H. R., Van der Scheer, J., Aiston, S., and Bohnel, E. 1947. The purification and concentration of influenza virus by means of alcohol precipitation. *J. Immunol.* **56**, 149–166.

Crawford, L. 1968. The DNA viruses of the adeno, papilloma, and polyoma groups. *In* H. Fraenkel-Conrat, ed., *Molecular Basis of Virology.* pp. 393–434. Van Nostrand-Reinhold, New York.

Crawford, L. V., and Crawford, E. M. 1961. The properties of Rous sarcoma virus purified by density gradient centrifugation. *Virology* **13**, 227–232.

Crestfield, A. M., and Allen, F. W. 1955a. Improved apparatus for zone electrophoresis. *Ann. Chem.* **27**, 422–423.

Crestfield, A. M. and Allen, F. W. 1955b. Resolution of ribonucleotides by zone electrophoresis. *Ann. Chem.* **27**, 424–425.

Crick, F. H. C. 1970. Central dogma of molecular biology. *Nature* **227**, 561–563.

Crick, F. H. C., Barnett, L., Brenner, S., and Watts-Tobin, R. J. 1961. General nature of the genetic code for proteins. *Nature* **192**, 1227–1232.

Dahlberg, J. E. 1968. Terminal sequences of bacteriophage RNAs. *Nature* **220**, 548–552.

Dales, S. 1973. The structure and replication of poxviruses as exemplified by vaccinia. *In* A. J. Dalton and F. Haguenau, eds., *Ultrastructure of Animal Viruses and Bacteriophages.* pp. 109–129. Academic Press, New York.

Dales, S., and Gomatos, P. J. 1965. The uptake and development of reovirus in strain L cells followed with labeled viral ribonucleic acid and ferritin-antibody conjugates. *Virology* **25**, 193–211.

Dalton, A. J., and Haguenau, F., eds. 1973. *Ultrastructure of Animal Viruses and Bacteriophages: An Atlas.* Academic Press, New York.

Davern, C. I. 1971. Molecular aspects of genetic recombination. *Progr. Nucleic Acid Res. Molecular Biol.* **11**, 229–258.

Davies, J. W., and Kaesberg, P. 1973. Translation of virus mRNA: Synthesis of bacteriophage Qβ proteins in a cell-free extract from wheat embryo. *J. Virol.* **12**, 1434–1441.

Davis, R. W., and Davidson, N. 1968. Electron-microscopic visualization of deletion mutations. *Proc. Natl. Acad. Sci. U.S.* **60**, 243–250.

Davis, R. W., Simon, M., and Davidson, N. 1971. Electron-microscope heteroduplex methods for mapping regions of base sequence homology in nucleic acids. *In* L. Grossman and K. Moldave, eds., *Methods in Enzymology*. Vol. 21, pp. 413–428. Academic Press, New York.

Davison, P. F., Freifelder, D., Hede, R., and Levinthal, C. 1961. The structural unity of the DNA of T2 bacteriophage. *Proc. Natl. Acad. Sci. U.S.* **47**, 1123–1129.

Dayhoff, M. O., ed. 1972. *Atlas of Protein Sequence and Structure*. Vol. 5. National Biomedical Research Foundation, Washington, D. C.

Dayhoff, M. O., ed. 1973. *Atlas of Protein Sequence and Structure*. Vol. 5, Suppl. I. National Biomedical Research Foundation, Washington, D.C.

DeFremery, D., and Knight, C. A. 1955. A chemical comparison of three strains of tomato bushy stunt virus. *J. Biol. Chem.* **214**, 559–566.

Desjardins, P. R., Senseney, C. A., and Hess, G. E. 1953. Further studies on the electron microscopy of purified tobacco ringspot virus. *Phytopathology* **43**, 687–690.

Dewey, D. L. 1972. Mechanism of phage inactivation by radiation. *Israel J. Chem.* **10**, 1213–1228.

Diener, T. O. 1971. Potato spindle tuber "virus." IV. A replicating, low molecular weight RNA. *Virology* **45**, 411–428.

Diener, T. O. 1972. Viroids. *Advan. Virus Res.* **17**, 295–313.

Diener, T. O. 1973. Potato spindle tuber viroid: A novel type of pathogen. *Perspectives Virol.* **8**, 7–30.

Diener, T. O., and Smith, D. R. 1973. Potato spindle tuber viroid IX. Molecular weight determination by gel electrophoresis of formylated RNA. *Virology* **53**, 359–365.

Dimmock, N. J., and Tyrrel, D. A. J. 1962. Physicochemical properties of some viruses isolated from common colds (rhinoviruses). *Lancet* **2**, 536–537.

Dorner, R. W., and Knight, C. A. 1953. The preparation and properties of some plant virus nucleic acids. *J. Biol. Chem.* **205**, 959–967.

Doty, P. 1961. Polynucleotides and nucleic acids. *J. Polymer Sci.* **55**, 1–24.

Doty, P., Boedtker, H., Fresco, J. R., Haselkorn, R., and Litt, M. 1959. Secondary structure in ribonucleic acids. *Proc. Natl. Acad. Sci. U.S.* **45**, 482–499.

Drake, J. W. 1970. *The Molecular Basis of Mutation*. Holden-Day, San Francisco.

Drake, J. W., and McGuire, J. 1967. Characteristics of mutations appearing spontaneously in extracellular particles of bacteriophage T4. *Genetics* **55**, 387–398.

Drzeniek, R. 1972. Viral and bacterial neuraminidases. *Current Topics Microbiol. Immunol.* **59**, 35–74.

Drzeniek, R., and Bilello, P. 1972. Reconstitution of poliovirus. *Biochem. Biophys. Res. Commun.* **46**, 719–724.

Duesberg, P. H. 1968. The RNA's of influenza virus. *Proc. Natl. Acad. Sci. U.S.* **59**, 930–937.

Dulbecco, R. 1950. Experiments on photoreactivation of bacteriophages inactivated with ultraviolet radiation. *J. Bacteriol.* **59**, 329–347.

Dulbecco, R., and Vogt, M. 1958. Study of the mutability of d lines of polioviruses. *Virology* **5**, 220–235.

Dunker, A. K., and Rueckert, R. R. 1969. Observations on molecular weight determinations on polyacrylamide gel. *J. Biol. Chem.* **244**, 5074–5080.

Dzhavakhia, V. G., Rodionova, N. P., and Atabekov, J. G. 1970. Coat protein recognizing site in TMV RNA. *Proc. 5th Ann. Meeting Lab. Bioorganic Chem.* Lomonosov State Univ., Moscow, pp. 97–99.

Edgar, R. S., and Lielausis, I. 1968. Some steps in the assembly of bacteriophage T4. *J. Molecular Biol.* **32**, 263–276.

Editorial. 1932. *J. Am. Med. Assoc.* **99**, 656.

Edman, P. 1950a. Preparation of phenyl thiohydantoins from some natural amino acids. *Acta Chem. Scand.* **4**, 277–282.

Edman, P. 1950b. Method for determination of the amino acid sequence in peptides. *Acta Chem. Scand.* **4**, 283–293.

Edman, P. 1956. On the mechanism of the phenyl isothiocyanate degradation of peptides. *Acta Chem. Scand.* **10**, 761–768.

Eggen, K., Oeschger, M. P., and Nathans, D. 1967. Cell-free protein synthesis is directed by coliphage MS2 RNA: Sequential synthesis of specific phage proteins. *Biochem. Biophys. Res. Commun.* **28**, 587–597.

Ehrenfeld, E., Maizel, J. V., and Summers, D. F. 1970. Soluble RNA polymerase complex from poliovirus-infected HeLa cells. *Virology* **40**, 840–846.

Epstein, H. T., and Lauffer, M. A. 1952. Ultracentrifugal identification of infectious entities with characteristic particles. *Arch. Biochem. Biophys.* **36**, 371–382.

Eriksson-Quensel, I., and Svedberg, T. 1936. Sedimentation and electrophoresis of the tobacco mosaic virus protein. *J. Am. Chem. Soc.* **58**, 1863–1867.

Espejo, R. T., and Canelo, E. S. 1968. Properties of bacteriophage PM2: A lipid-containing bacterial virus. *Virology* **34**, 738–747.

Faulkner, P. Martin, E. M., Sved, R. C., Valentine, R. C., and Work, T. S. 1961. Studies on protein and nucleic acid metabolism in virus-infected mammalian cells. 2. The isolation, crystallization and chemical characterization of mouse encephalomyocarditis virus. *Biochem. J.* **80**, 597–605.

Feldman, M. Ya. 1973. Reactions of nucleic acids and nucleoproteins with formaldehyde. *Progr. Nucleic Acid Res. Molecular Biol.* **13**, 1–49.

Fenner, F. 1968. *The Biology of Animal Viruses.* Vols. I, II. Academic Press, New York.

Fenner, F., McAuslan, B. R., Mims, C. A., Sambrook, J., and White, D. O. 1974. *The Biology of Animal Viruses.* 2nd edit. Academic Press, New York.

Fiers, W., and Sinsheimer, R. L. 1962a. The structure of the DNA of bacteriophage ØX 174. I. The action of exopolynucleotidases. *J. Molecular Biol.* **5**, 408–419.

Fiers, W., and Sinsheimer, R. L. 1962b. The structure of the DNA of bacteriophage ØX 174. III. Ultracentrifugal evidence for a ring structure. *J. Molecular Biol.* **5**, 424–434.

Finch, J. T., and Klug, A. 1967. Structure of broad bean mottle virus. I. Analysis of electron micrographs and comparison with turnip yellow mosaic virus and its top component. *J. Molecular Biol.* **24**, 289–302.

Fink, K., and Adams, W. S. 1966. Paper chromatographic data for purines, pyrimidines and derivatives in a variety of solvents. *J. Chromatog.* **22**, 118–129.

Fish, W. W., Mann, K. G., and Tanford, C. 1969. The estimation of polypeptide chain molecular weights by gel filtration in 6 M guanidine hydrochloride. *J. Biol. Chem.* **244**, 4989–4994.

Fox, C. F., Robinson, W. S., Haselkorn, R., and Weiss, S. B. 1964. Enzymatic synthesis of ribonucleic acid. III. The ribonucleic acid-primed synthesis of ribonucleic acid with *Mircrococcus lysodeikticus* ribonucleic acid polymerase. *J. Biol. Chem.* **239**, 186–195.

Fox, S. W. 1971. Chemical origin of cells. 2. *Chem. Eng. News* **49**, 46–53.

Fox, S. W., Harada, K., Krampitz, G., and Mueller, G. 1970. Chemical origin of cells. *Chem. Eng. News* **48**, 80–94.

Fraenkel-Conrat, H. 1954. Reaction of nucleic acid with formaldehyde. *Biochim. Biophys. Acta* **15**, 308–309.

Fraenkel-Conrat, H. 1956. The role of the nucleic acid in the reconstitution of active tobacco mosaic virus. *J. Am. Chem. Soc.* **78**, 882.

Fraenkel-Conrat, H. 1957. Degradation of tobacco mosaic virus with acetic acid. *Virology* **4**, 1–4.

Fraenkel-Conrat, H. 1960. Chemical modification of viral ribonucleic acid. I. Alkylating agents. *Biochim. Biophys. Acta* **49**, 169–180.

Fraenkel-Conrat, H. 1966. Preparation and testing of tobacco mosaic virus-RNA. *In* G. L. Cantoni and D. R. Davies, eds., *Procedures in Nucleic Acid Research.* pp. 480–487. Harper and Row, New York.

Fraenkel-Conrat, H. 1970. Reconstitution of viruses. *Ann. Rev. Microbiol.* **24**, 463–478.

Fraenkel-Conrat, H., and Olcott, H. S. 1948. The reaction of formaldehyde with proteins. V. Crosslinking between amino and primary amide or guanidyl groups. *J. Am. Chem. Soc.* **70**, 2673–2684.

Fraenkel-Conrat, H., Harris, J. I., and Levy, A. L. 1955. Recent developments in techniques for terminal and sequence studies in peptides and proteins. *Methods Biochem. Anal.* **2**, 359–425.

Fraenkel-Conrat, H., and Fowlks, E. 1972. Variability at the 5′ ends of two plant viruses. *Biochemistry* **11**, 1733–1736.

Fraenkel-Conrat, H., and Ramachandran, L. K. 1959. Structural aspects of tobacco mosaic virus. *Advan. Protein Chem.* **14**, 175–229.

Fraenkel-Conrat, H., and Singer, B. 1954. The peptide chains of tobacco mosaic virus. *J. Am. Chem. Soc.* **76**, 180–183.

Fraenkel-Conrat, H., and Singer, B. 1957. Virus reconstitution. II. Combination of protein and nucleic acid from different strains. *Biochim. Biophys. Acta* **24**, 540–548.

Fraenkel-Conrat, H., and Singer, B. 1962. The absence of phosphorylated chain ends in TMV-RNA. *Biochemistry* **1**, 120–128.

Fraenkel-Conrat, H., Singer, B., and Tsugita, A. 1961. Purification of viral RNA by means of bentonite. *Virology* **14**, 54–58.

Fraenkel-Conrat, H., Singer, B., and Williams, R. C. 1957. Infectivity of viral nucleic acid. *Biochim. Biophys. Acta* **25**, 87–96.

Fraenkel-Conrat, H., and Williams, R. C. 1955. Reconstitution of active tobacco mosaic virus from its inactive protein and nucleic acid components. *Proc. Natl. Acad. Sci. U.S.* **41**, 690–698.

Francki, R. I. B. 1968. Inactivation of cucumber mosaic virus (Q strain) nucleoprotein by pancreatic ribonuclease. *Virology* **34**, 694–700.

Franklin, R. E., Caspar, D. L. D., and Klug, A. 1959. The structure of viruses as determined by x-ray diffraction. *In* C. S. Holton et al., eds. *Plant Pathology Problems and Progress 1908–1958*. pp. 447–461. University of Wisconsin Press, Madison.

Franklin, R. E., Klug, A., and Holmes, K. C. 1957. X-ray diffraction studies of the structure and morphology of tobacco mosaic virus. *Ciba Foundation Symposium on The Nature of Viruses*. pp. 39–52. Churchill, London.

Franklin, R. M. 1962. The significance of lipids in animal viruses. An essay on virus multiplication. *Progr. Med. Virol.* **4**, 1–53.

Franklin, R. M. and Wecker, E. 1959. Inactivation of some animal viruses by hydroxylamine and the structure of ribonucleic acid. *Nature* **184**, 343–345.

Freese, E. 1971. Molecular mechanisms of mutation. *In* A. Hollaender, ed., *Chemical Mutagens. Principles and Methods for Their Detection*. Vol. I, pp. 1–56. Plenum Press, New York.

Freese, E., and Freese, E. B. 1965. The oxygen effect on deoxyribonucleic acid inactivation by hydroxylamine. *Biochemistry* **4**, 2419–2433.

Freese, E., and Strack, H. B. 1962. Induction of mutations in transforming DNA by hydroxylamine. *Proc. Natl. Acad. Sci. U.S.* **48**, 1796–1803.

Freese, E., Bautz-Freese, E., and Bautz, E. 1961a. Hydroxylamine as a mutagenic and inactivating agent. *J. Molecular Biol.* **3**, 133–143.

Freese, E., Bautz, E., and Bautz-Freese, E. 1961b. The chemical and mutagenic specificity of hydroxylamine. *Proc. Natl. Acad. Sci. U.S.* **47**, 845–855.

Freifelder, D. 1965. Mechanism of inactivation of coliphage T7 by x-rays. *Proc. Natl. Acad. Sci. U.S.* **54**, 128–134.

Freifelder, D. 1966. DNA strand breakage by X-irradiation. *Radiation Res.* **29**, 329–338.

Fresco, J. R., Alberts, B. M., and Doty, P. 1960. Some molecular details of the secondary structure of ribonucleic acid. *Nature* **188**, 98–101.

Friesen, B. S., and Sinsheimer, R. L. 1959. Partition cell analysis of infective tobacco mosaic virus nucleic acid. *J. Molecular Biol.* **1**, 321–328.

Frisch-Niggemeyer, W., and Steere, R. L. 1961. Chemical composition of partially purified alfalfa mosaic virus. *Virology* **14**, 83–87.

Frommhagen, L. H., and Knight, C. A. 1959. Column purification of influenza virus. *Virology* **8**, 198–208.

Frommhagen, L. H., Knight, C. A., and Freeman, N. K. 1959. The ribonucleic acid, lipid, and polysaccharide constituents of influenza virus preparations. *Virology* **8**, 176–197.

Fulton, R. W. 1967a. Purification and serology of rose mosaic virus. *Phytopathology* **57**, 1197–1201.

Fulton, R. W. 1967b. Purification and some properties of tobacco streak and Tulare apple mosaic viruses. *Virology* **32**, 153–162.

Funatsu, G., and Fraenkel-Conrat, H. 1964. Location of amino acid exchanges in chemically evoked mutants of tobacco mosaic virus. *Biochemistry* **3**, 1356–1362.

Funatsu, G., and Funatsu, M. 1968. Chemical studies on proteins from two tobacco mosaic virus strains. *Proc. 1967 Intern. Symp. Biochem. Regulation in Diseased Plants or Injury.* Tokyo, Japan, pp. 1–9.

Funatsu, G., Tsugita, A., and Fraenkel-Conrat, H. 1964. Studies on the amino acid sequence of tobacco mosaic virus protein. V. Amino acid sequences of two peptides from tryptic digests and location of amide group. *Arch. Biochem. Biophys.* **105**, 25–41.

Galbraith, A. W., Oxford, J. S., Schild, G. C., and Watson, G. I. 1970. Protective effect of aminoadamantane on influenza A2 infections in the family environment. *Ann. N.Y. Acad. Sci.* **173**, 29–43.

Gard, S. 1957. Chemical inactivation of viruses. *In* G. E. W. Wolstenholme and E. C. P. Millar, eds., *The Nature of Viruses.* pp. 123–146. Churchill, London.

Garen, A. 1968. Sense and nonsense in the genetic code. *Science* **160**, 149–159.

Gemmell, A., and Fenner, F. 1960. Genetic studies with mammalian poxviruses. III. White (μ) mutants of rabbit poxvirus. *Virology* **11**, 219–235.

Gessler, A. E., Bender, C. E., and Parkinson, M. C. 1956a. A new and rapid method for isolating viruses by selective fluorocarbon deproteinization. *Trans. N. Y. Acad. Sci.* **18**, 701–703.

Gessler, A. E., Bender, C. E., and Parkinson, M. C. 1956b. Animal viruses isolated by fluorocarbon emulsification. *Trans. N. Y. Acad. Sci.* **18**, 707–716.

Ghabrial, S. A., Shepherd, R. J., and Grogan, R. G. 1967. Chemical properties of three strains of southern bean mosaic virus. *Virology* **33**, 17–25.

Ghendon, Y. Z. 1972. Conditional-lethal mutants of animal viruses. *Progr. Med. Virol.* **14**, 68–122.

Gibbs, A. J., Nixon, H. L., and Woods, R. D. 1963. Properties of purified preparations of lucerne mosaic virus. *Virology* **19**, 441–449.

Gibbs, A. J., Harrison, B. D., and Murant, A. F., eds. 1970. *Descriptions of Plant Viruses* (with two sets of looseleaf sheets per year). Commonwealth Mycological Inst. Ferry Lane, Kew, Surrey, England.

Gibson, W., and Roizman, B. 1971. Compartmentalization of spermine and spermidine in the herpes simplex virion. *Proc. Natl. Acad. Sci. U. S.* **68**, 2818–2821.

Gierer, A. 1957. Structure and biological function of ribonucleic acid from tobacco mosaic virus. *Nature* **179**, 1297–1299.

Gierer, A. 1958a. Grösse und Struktur der Ribosenukleinsäure des Tabakmosaikvirus. *Z. Naturforsch.* **13b**, 477–484.

Gierer, A. 1958b. Die Grösse der biologisch-aktiven Einheit der Ribosenukleinsäure des Tabakmosaikvirus. *Z. Naturforsch.* **13b**, 485–488.

Gierer, A. 1958c. Vergleichende Untersuchungen an hochmolekularer Ribosenukleinsäure. *Z. Naturforsch.* **13b**, 788–792.

Gierer, A. 1959. Die Eigenschaften der infektiösen Einheit des Tabakmosaikvirus. In *Biochemistry of Viruses* (Fourth International Congress of Biochemistry) Vol. VII, pp. 58–61. Pergamon Press, London.

Gierer, A. 1960. Recent investigations on tobacco mosaic virus. *Progr. Biophys.* **10**, 299–342.

Gierer, A., and Schramm, G. 1956a. Infectivity of ribonucleic acid from tobacco mosaic virus. *Nature* **177**, 702–703.

Gierer, A., and Schramm, G. 1956b. Die Infektiosität der Ribosenukleinsäure des Tabakmosaikvirus. *Z. Naturforsch.* **11b**, 138–142.

Gillespie, D., and Spiegelman, S. 1965. A quantitative assay for DNA-RNA hybrids with DNA immobilized on a membrane. *J. Molecular Biol.* **12**, 829–842.

Ginoza, W. 1968. Inactivation of viruses by ionizing radiation and by heat. *In* K. Maramorosch and H. Koprowski, eds., *Methods in Virology*. Vol. 4, pp. 139–209. Academic Press, New York.

Ginoza, W., Atkinson, D. E., and Wildman, S. G. 1954. A differential ability of strains of tobacco mosaic virus to bind host-cell nucleoprotein. *Science* **119**, 269–271.

Gish, D. T. 1960. Studies on the amino acid sequence of tobacco mosaic virus protein. III. The amino acid sequence of a pentadecapeptide from a tryptic digest. *J. Am. Chem. Soc.* **82**, 6329–6335.

Glitz, D. G. 1968. The nucleotide sequence at the 3′-linked end of bacteriophage MS2 ribonucleic acid. *Biochemistry* **7**, 927–932.

Glitz, D. G., Bradley, A., and Fraenkel-Conrat, H. 1968. Nucleotide sequences at the 5′-linked ends of viral ribonucleic acids. *Biochim. Biophys. Acta* **161**, 1–12.

Gold, A. H. 1961. Antihost serum improves plant virus purification. *Phytopathology* **51**, 561–565.

Goldstein, D. A., Bendet, I. J., Lauffer, M. A., and Smith, K. M. 1967. Some biological and physicochemical properties of blue-green algal virus LPP-1. *Virology* **32**, 601–613.

Good, N. E., Winget, G. D., Winter, W., Connolly, T. N., Izawa, S., and Singh, R. M. M. 1966. Hydrogen ion buffers for biological research. *Biochemistry* **5**, 467–477.

Gordon, M. P., and Staehelin, M. 1959. Studies on the incorporation of 5-fluorouracil into a virus nucleic acid. *Biochim. Biophys. Acta* **36**, 351–361.

Gordon, M. P., Singer, B., and Fraenkel-Conrat, H. 1960. The terminal phosphate groups of tobacco mosaic virus. *J. Biol. Chem.* **235**, 1014–1018.

Gordon, M. P., and Huff, J. W. 1962. Heat inactivation of tobacco mosaic virus ribonucleic acid as studied by end group analysis. *Biochemistry* **1**, 481–484.

Goulian, M., Kornberg, A., and Sinsheimer, R. L. 1967. Enzymatic synthesis of infectious phage ØX 174 DNA. *Proc. Natl. Acad. Sci. U.S.* **58**, 2321–2328.

Gray, W. R. 1972. End-group analysis using dansylchloride. *In* C. H. W. Hirs and S. N. Timasheff, eds., *Methods in Enzymology*. Vol. 25, pp. 121–138. Academic Press, New York.

Green, M. 1969. Chemical composition of animal viruses. *In* H. B. Levy, ed., *The Biochemistry of Viruses*. pp. 1–54. Dekker, New York.

Guha, A., and Szybalski, W. 1968. Fractionation of the complementary strands of coliphage T4 DNA based on the asymmetric distribution of the poly U and poly U,G binding sites. *Virology* **34**, 608–616.

Gussin, G. N. 1966. Three complementation groups in bacteriophage R17. *J. Molecular Biol.* **21**, 435–453.

Guthrie, G. D., and Sinsheimer, R. L. 1960. Infection of protoplasts of *Escherichia coli* by subviral particles of bacteriophage ØX174 *J. Molecular Biol.* **2**, 297–305.

Habel, K., and Salzman, N. P., eds. 1969. *Fundamental Techniques in Virology*. Academic Press, New York.

Hadidi, A., and Fraenkel-Conrat, H. 1973. Characterization and specificity of soluble RNA polymerase of brome mosaic virus. Virology **52**, 363–372.

Hall, B. D., and Spiegelman, S. 1961. Sequence complementarity of T2 DNA and T2-specific RNA. *Proc. Natl. Acad. Sci. U.S.* **47**, 137–146.

Hall, R. H. 1971. *The Modified Nucleosides in Nucleic Acids.* Columbia University Press, New York.

Hariharasubramanian. V., Hadidi, A., Singer, B., and Fraenkel-Conrat, H. 1973. Possible identification of a protein in brome mosaic virus infected barley as a component of viral RNA polymerase. *Virology* **54**, 190–198.

Harriman, P. D., and Stent, G. S. 1964. The effect of radiophosphorus decay on cistron function in bacteriophage T4. I. Long-range and short-range hits. *J. Molecular Biol.* **10**, 488–507.

Harrington, W. G., and Schachman, H. K. 1956. Studies on the alkaline degradation of tobacco mosaic virus. I. Ultracentrifugal analysis. *Arch. Biochem. Biophys.* **65**, 278–295.

Harris, J. I., and Knight, C. A. 1955. Studies on the action of carboxypeptidase on tobacco mosaic virus. *J. Biol. Chem.* **214**, 215–230.

Harris, J. I., and Hindley, J. 1961. The protein subunit of turnip yellow mosaic virus. *J. Molecular Biol.* **3**, 117–120.

Harris, J. I., and Hindley, J. 1965. The protein subunit of turnip yellow mosaic virus. *J. Molecular Biol.* **13**, 894–913.

Harrison, B. D., and Nixon, H. L. 1959a. Separation and properties of particles of tobacco rattle virus with different lengths. *J. Gen. Microbiol.* **21**, 569–581.

Harrison, B. D., and Nixon, H. L. 1959b. Some properties of infective preparations made by disrupting tobacco rattle virus with phenol. *J. Gen. Microbiol.* **21**, 591–599.

Harrison, B. D., Finch, J. T., Gibbs, A. J., Hollings, M., Shepherd, R. J., Valenta, V., and Wetter, C. 1971. Sixteen groups of plant viruses. *Virology* **45**, 356–363.

Hart, R. G. 1958. The nucleic acid fiber of the tobacco mosaic virus particle. *Biochim. Biophys. Acta* **28**, 457–464.

Hartman, R. E., and Lauffer, M. A. 1953. An application of electrophoresis to the identification of biological activity with characteristic particles. *J. Am. Chem. Soc.* **75**, 6205–6209.

Haruna, I., Yaoi, H., Kono, R., and Watanabe, I. 1961. Separation of adenovirus by chromatography on DEAE-cellulose. *Virology* **13**, 264–267.

Haruna, I., Nozu, K., Ohtaka, Y., and Spiegelman, S. 1963. An RNA replicase induced by and selective for a viral RNA: Isolation and properties. *Proc. Natl. Acad. Sci. U.S.* **50**, 905–911.

Haruna, I., and Spiegelman, S. 1965. The autocatalytic synthesis of a viral RNA *in vitro*. *Science* **150**, 884–886.

Haschemeyer, R., Singer, B., and Fraenkel-Conrat, H. 1959. Two configurations of tobacco mosaic virus-ribonucleic acid. *Proc. Natl. Acad. Sci. U.S.* **45**, 313–319.

Hatcher, D. W., and Goldstein, G. 1969. Improved methods for determination of RNA and DNA. *Ann. Biochem.* **31**, 42–50.

Hayashi, R., Moore, S., and Stein, W. H. 1973. Carboxypeptidase from yeast. Large scale preparation and the application to COOH-terminal analysis of peptides and proteins. *J. Biol. Chem.* **248**, 2296–2302.

Hayes, W. 1968. *The Genetics of Bacteria and Their Viruses.* 2nd edit., Wiley, New York.

Hennig, B., and Wittmann, H. G. 1972. Tobacco mosaic virus: Mutants and strains. *In* C. I. Kado and H. O. Agrawal, eds., *Principles and Techniques in Plant Virology.* pp. 546–594. Van Nostrand-Reinhold, New York.

Henriques, V., and Sorensen, S. P. L. 1910. Über die quantitative Bestimmung der Aminosäuren, Polypeptide, und der Hippursäure im Harne durch Formoltitration. *Z. Physiol. Chem.* **64**, 120–145.

Henry, C., and Youngner, J. S. 1963. Studies on the structure and replication of the nucleic acid of poliovirus. *Virology* **21**, 162–173.

Herriott, R. M., and Barlow, J. L. 1952. Preparation, purification, and properties of *E. coli* virus T2. *J. Gen. Physiol.* **36**, 17–28.

Herrmann, E. C., Jr., and Stinebring, W. R., eds. 1970. Second Conference on Antiviral Substances. *Ann. N. Y. Acad. Sci.* **173**, 1–844.

Herrmann, J. E., and Cliver, D. O. 1973. Degradation of coxsackievirus type A9 by proteolytic enzymes. *Infect. Immunol.* **7**, 513–517.

Hersh, R. T., and Schachman, H. K. 1958. On the size of the protein subunits in bushy stunt virus. *Virology* **6**, 234–243.

Hershey, A. D. 1957. Some minor components of bacteriophage T2 particles. *Virology* **4**, 237–264.

Hershey, A. D., Kamen, M. D., Kennedy, J. W., and Gest, H. 1951. The mortality of bacteriophage containing assimilated radioactive phosphorus. *J. Gen. Physiol.* **34**, 305–319.

Hershey, A. D., and Chase, M. 1952. Independent functions of viral protein and nucleic acid in growth of bacteriophage. *J. Gen. Physiol.* **36**, 39–56.

Hershey, A.D., Burgi, E., and Ingraham, L. 1962. Sedimentation coefficient and fragility under hydro-dynamic shear as measures of molecular weight of the DNA of phage T5. *Biophys. J.* **2**, 423–431.

Hershey, A. D., Goldberg, E., Burgi, E., and Ingraham, L. 1963. Local denaturation of DNA by shearing forces and by heat. *J. Molecular Biol.* **6**, 230–243.

Hiebert, E., Bancroft, J. B., and Bracker, C. E. 1968. The assembly *in vitro* of some small spherical viruses, hybrid viruses, and other nucleoproteins. *Virology* **34**, 492–508.

Hierholzer, J. E., Palmer, E. L., Whitfield, S. G., Kaye, H. S., and Dowdle,

W. R. 1972. Protein composition of coronavirus OC43. Virology **48**, 516–527.

Hirs, C. H. W. 1967. Automatic computation of amino acid analyzer data. *In* C. H. W. Hirs, ed., *Methods in Enzymology*. Academic Press, New York, Vol. 11, pp. 27–31.

Hirst, G. K. 1959. Virus-host cell relation. *In* T. M. Rivers and F. L. Horsfall, Jr., eds., *Viral and Rickettsial Infections of Man*. 3rd edit., pp. 96–144. Lippincott, Philadelphia.

Hirt, B., and Gesteland, R. F. 1971. Charakterisierung der Proteine von Polyoma Virus und von SV40. *Experientia* **27**, 736.

Hirth, L. 1971. Comparative properties of rod-shaped viruses. *In* K. Maramorosch and E. Kurstak, eds., *Comparative Virology*. pp. 335–360. Academic Press, New York.

Hoffmann-Berling, H., Kaerner, H. C., and Knippers, R. 1966. Small bacteriophages. *Advan. Virus Res.* **12**, 329–370.

Hohn, T., and Hohn, B. 1970. Structure and assembly of simple RNA bacteriophages. *Advan. Virus Res.* **16**, 43–98.

Holland, J. J. 1964. Enterovirus entrance into specific host cells and subsequent alterations of cell protein and nucleic acid synthesis. *Bacteriol. Rev.* **28**, 3–13.

Holland, J. J., McLaren, L. C., Hoyer, B. H., and Syverton, J. T. 1960. Enteroviral ribonucleic acid. II. Biological, physical and chemical studies. *J. Exptl. Med.* **112**, 841–864.

Holmes, F. O. 1929. Local lesions in tobacco mosaic. *Bot. Gaz.* **87**, 39–55.

Holoubek, V. 1962. Mixed reconstitution between protein from common TMV strain and RNA from different TMV strains. *Virology* **18**, 401–404.

Honess, R. W., and Roizman, B. 1973. Proteins specified by herpes simplex virus. XI. Identification and relative molar rates of synthesis of structural and nonstructural herpes virus polypeptides in the infected cell. *J. Virology* **12**, 1347–1365.

Hopkins, G. R., and Sinsheimer, R. L. 1955. Visible and ultraviolet light scattering by tobacco mosaic virus nucleic acid. *Biochim. Biophys. Acta* **17**, 476–484.

Horne, R. W. 1962. The examination of small particles. *In* D. Kay, ed., *Techniques in Electron Microscopy*. pp. 150–166. Charles C.Thomas, Springfield, Ill.

Horne, R. W. 1967. Electron microscopy of isolated virus particles and their components. *In* K. Maramorosch and H. Koprowski, eds., *Methods in Virology*. Vol. 3, pp. 521–574. Academic Press, New York.

Horne, R. W. 1973. The structure of adenovirus particles and their components. *In* A. J. Dalton and F. Haguenau, eds., *Ultrastructure of Animal Viruses and Bacteriophages*. pp. 67–81. Academic Press, New York.

Horne, R. W., and Wildy, P. 1961. Symmetry in virus architecture. *Virology* **15**, 348–373.

Hotchkiss, R. D. 1957. Criteria for quantitative genetic transformation of bacteria. *In* W. D. McElroy and B. Glass, eds., *The Chemical Basis of Heredity*. pp. 321–335. Johns Hopkins Press, Baltimore.

Hotchkiss, R. D., and Gabor, M. 1970. Bacterial transformation with special reference to recombination process. *Ann. Rev. Gen.* **4**, 193–224.

Howard-Flanders, P. 1973. DNA repair and recombination. *Brit. Med. Bull.* **29**, 226–235.

Hull, R., Rees, M. W., and Short, M. N. 1969. Studies on alfalfa mosaic virus. I. The protein and nucleic acid. *Virology* **37**, 404–415.

Hummeler, K. 1971. Bullet-shaped viruses. *In* K. Maramorosch and E. Kurstak, eds., *Comparative Virology*. pp. 361–386. Academic Press, New York.

Huppert, J., and Rebeyrotte, N. 1960. Extraction d'un acide desoxyribonucleique transformant par le dodecyl sulfate de sodium et le phenol. *Biochim. Biophys. Acta* **45**, 189–191.

Huxley, H. E. 1957. Some observations on the structure of tobacco mosaic virus. *Electron Microscopy Proc. Stockholm Conf. 1956*, pp. 260–261.

Huxley, H. E., and Klug, A. 1971. *New Developments in Electron Microscopy*. The Royal Society, London.

Hyde, J. M., Gafford, L. G., and Randall, C. C. 1967. Molecular weight determination of fowlpox virus DNA by electron microscopy. *Virology* **33**, 112–120.

Isaacs, A., and Lindenmann, J. 1957. Virus interference. I. The interferon. II. Some properties of interferon. *Proc. Royal Soc. London* **147B**, 258–267, 268–273.

Israel, J. V., Anderson, T. F., and Levine, M. 1967. *In vitro* morphogenesis of phage P22 from head and baseplate parts. *Proc. Natl. Acad. Sci. U.S.* **57**, 284–291.

Ivanovski, D. 1899. Über die Mosaikkrankheit der Tabakspflanze. *Cb1. Bakteriol. 2E*, **5**, 250–254.

Jacobson, M. F., Asso, J., and Baltimore, D. 1970. Further evidence on the formation of poliovirus proteins. *J. Molecular Biol.* **49**, 657–669.

Jeppeson, P. G. N. 1971. The nucleotide sequences of some large ribonuclease T1 products from bacteriophage R17 ribonucleic acid. *Biochem. J.* **124**, 357–366.

Jeppeson, P. G. N., Steitz, J. A., Gesteland, R. F., and Spahr, P. F. 1970. Gene order in the bacteriophage R17 RNA: 5′ A protein-coat protein-synthetase-3′. *Nature* **226**, 230–237.

Jesaitis, M. A. 1956. Differences in the chemical composition of the phage nucleic acids. *Nature* **178**, 637.

Johnson, M. W. 1964. The binding of metal ions by turnip yellow mosaic virus. *Virology* **24**, 26–35.

Johnson, M. W., and Markham, R. 1962. Nature of the polyamine in plant viruses. *Virology* **17**, 276–281.

Joklik, W. K. 1966. The poxviruses. *Bacteriol. Rev.* **30**, 33–66.

Joklik, W. K. 1970. The molecular biology of reovirus. *J. Cell Physiol.* **76**, 289–302.

Joklik, W. K., and Smith, D. T., eds. 1972. *Zinsser Microbiology*. pp. 713–846. Appleton-Century Crofts, New York.

Josse, J., Kaiser, A. D., and Kornberg, A. 1961. Enzymatic synthesis of deoxyribonucleic acid. VIII. Frequencies of nearest neighbor base sequences in deoxyribonucleic acid. *J. Biol. Chem.* **236**, 864–875.

Kabat, E. A. 1943. Immunochemistry of the proteins. *J. Immunol.* **47**, 513–587.

Kado, C. I. 1967. Biological and biochemical characterization of sowbane mosaic virus. *Virology* **31**, 217–229.

Kado, C. I., and Agrawal, H. O., eds. 1972. *Principles and Techniques in Plant Virology*. Van Nostrand-Reinhold, New York.

Kado, C. I., and Knight, C. A. 1966. Location of a local lesion gene in tobacco mosaic virus RNA. *Proc. Natl. Acad. Sci. U.S.* **55**, 1276–1283.

Kaesberg, P. 1967. Structure and function of RNA from small phages. *In* J. S. Colter and W. Paranchych, eds., *The Molecular Biology of Viruses*. pp. 241–250. Academic Press, New York.

Kahn, R. P., Desjardins, P. R., and Senseney, C. A. 1955. Biophysical characteristics of the tomato ringspot virus. *Phytopathology* **45**, 334–337.

Kaiser, A. D. 1955. A genetic study of the temperate coliphage λ. *Virology* **1**, 424–443.

Kajioka, R., Siminovitch, L., and Dales, S. 1964. The cycle of multiplication of vaccinia virus in Earle's strain of L cells. II. Initiation of DNA synthesis and morphogenesis. *Virology* **24**, 295–309.

Kalmakoff, J., Lewandowski, L. J., and Black, D. R. 1969. Comparison of the ribonucleic acid subunits of reovirus, cytoplasmic polyhedrosis virus, and wound tumor virus. *J. Virol.* **4**, 851–856.

Kaper, J. M., and Steere, R. L. 1959a. Infectivity of tobacco ringspot virus nucleic acid preparations. *Virology* **7**, 127–139.

Kaper, J. M., and Steere, R. L. 1959b. Isolation and preliminary studies of soluble protein and infectious nucleic acid from turnip yellow mosaic virus. *Virology* **8**, 527–530.

Kaplan, A. S., and Ben-Porat, T. 1970. Nucleoside analogues as chemotherapeutic antiviral agents. *Ann. N. Y. Acad. Sci.* **173**, 346–361.

Kass, S. J. 1970. Chemical studies on polyoma and Shope papilloma viruses. *J. Virol.* **5**, 381–387.

Kass, S. J., and Knight, C. A. 1965. Purification and chemical analysis of Shope papilloma virus. *Virology* **27**, 273–281.

Kassanis, B. 1968. Satellitism and related phenomena in plant and animal viruses. *Advan. Virus Res.* **13**, 147–180.

Kassanis, B. 1970. Tobacco necrosis virus. *C.M.I./A.A.B. Description of Plant Viruses No. 14.*

Kates, M., Allison, A. C., Tyrrell, D. A. J., and James, A. T. 1961. Lipids of influenza virus and their relation to those of the host cell. *Biochim. Biophys. Acta* **52**, 455–466.

Kaufman, H. E. 1965. *In vivo* studies with antiviral agents. *Ann. N. Y. Acad. Sci.* **130**, 168–180.

Kay, D. 1959. The inhibition of bacteriophage multiplication by proflavine and its reversal by certain polyamines. *Biochem. J.* **73**, 149–154.

Kay, D., ed. 1961. *Techniques for Electron Microscopy.* Charles C. Thomas. Springfield, Ill.

Kellenberger, E., and Arber, W. 1955. Die Struktur des Schwanzes der Phagen T2 und T4 und der Mechanismus der irreversiblen Adsorption. *Z. Naturforsch.* **10b**, 698–704.

Kelley, J. M., Emerson, S. V., and Wagner, R. R. 1972. The glycoprotein of vesicular stomatitis virus is the antigen that gives rise to and reacts with neutralizing antibody. *J. Virol.* **10**, 1231–1235.

Kelley, J. J., and Kaesberg, P. 1962. Preparation of protein subunits from alfalfa mosaic virus under mild conditions. *Biochim. Biophys. Acta* **55**, 236–237.

Kelner, A. 1949. Effect of visible light on the recovery of *Streptomyces griseus conidia* from ultraviolet irradiation injury. *Proc. Natl. Acad. Sci. U.S.* **35**, 73–79.

Kendrew, J. C. 1959. Structure and function in myoglobin and other proteins. *Federation Proc.* **18**, 740–751.

Khorana, H. G., Büchi, H., Ghosh, H., Gupta, N., Jacob, T. M., Kössel, H., Morgan, R., Narang, S. A., Ohtsuka, E., and Wells, R. D. 1966. Polynucleotide synthesis and the genetic code. *Cold Spring Harbor Symp. Quant. Biol.* **31**, 39–49.

Kilbourne, E. D., Lief, F. S., Schulman, J. L., Jahiel, R. I., and Laver, W. G. 1967. Antigenic hybrids of influenza viruses and their implications. *Perspectives Virol.* **5**, 87–106.

Kingsbury, D. W. 1972. Paramyxovirus replication. *Current Topics Microbiol. Immunol.* **59**, 1–33.

Kingsbury, D. W. 1973. Cell-free translation of paramyxovirus messenger RNA. *J. Virol.* **12**, 1020–1027.

Klagsbrun, M., and Rich, A. 1970. *In vitro* synthesis of T4 bacteriophage head protein using polysomes of T4-infected *Escherichia coli. J. Molecular Biol.* **48**, 421–436.

Kleczkowski, A. 1968. Methods of inactivation by ultraviolet radiation. *In* K. Maramorosch and H. Koprowski, eds., *Methods in Virology,* Vol. 4, pp. 93–138. Academic Press, New York.

Klein, W. H., Nolan, C., Lazar, J. M., and Clark, J. M. 1972. Translation of satellite tobacco necrosis ribonucleic acid. I. Characterization of *in vitro* procaryotic and eukaryotic translation products. *Biochemistry* **11**, 2009–2014.

Kleinschmidt, A. K. 1968. Monolayer techniques in electron microscopy of nucleic acid molecules. *In* L. Grossman and K. Moldave, eds., *Methods in Enzymology*. 12B, pp. 361–377. Academic Press, New York.

Kleinschmidt, W. J. 1972. Biochemistry of interferon and its inducers. *Ann. Rev. Biochem.* **41**, 517–542.

Klenk, H.-D., and Choppin, P. W. 1969a. Chemical composition of the parainfluenza virus SV5. *Virology* **37**, 155–157.

Klenk, H.-D., and Choppin, P. W. 1969b. Lipids of plasma membranes of monkey and hamster kidney cells and of parainfluenza virions grown in these cells. *Virology* **38**, 255–268.

Klenk, H.-D., and Choppin, P. W. 1970. Plasma membrane lipids and parainfluenza virus assembly. *Virology* **40**, 939–947.

Klug, A., and Caspar, D. L. D. 1960. The structure of small viruses. *Advan. Virus Res.* **7**, 225–325.

Klug, A., Longley, W., and Leberman, R. 1966. Arrangement of protein subunits and the distribution of nucleic acid in turnip yellow mosaic virus. I. X-ray diffraction studies. *J. Molecular Biol.* **15**, 315–343.

Knight, C. A. 1942. Basic amino acids in strains of tobacco mosaic virus. *J. Am. Chem. Soc.* **64**, 2734–2736.

Knight, C. A. 1943. The sulfur distribution in the rib-grass strain of tobacco mosaic virus. *J. Biol. Chem.* **147**, 663–666.

Knight, C. A. 1944. A sedimentable component of allantoic fluid and its relationship to influenza viruses. *J. Exptl. Med.* **80**, 83–100.

Knight, C. A. 1946a. Precipitin reactions of highly purified influenza viruses and related materials. *J. Exptl. Med.* **83**, 281–294.

Knight, C. A. 1946b. The preparation of highly purified P R8 influenza virus from infected mouse lungs. *J. Exptl. Med.* **83**, 11–24.

Knight, C. A. 1947a. The nature of some of the chemical differences among strains of tobacco mosaic virus. *J. Biol. Chem.* **171**, 297–308.

Knight, C. A. 1947b. The nucleic acid and carbohydrate of influenza virus. *J. Exptl. Med.* **85**, 99–116.

Knight, C. A. 1954. The chemical constitution of viruses. *Advan. Virus Res.* **2**, 153–182.

Knight, C. A. 1955. The action of carboxypeptidase on strains of tobacco mosaic virus. *J. Biol. Chem.* **214**, 231–237.

Knight, C. A. 1957. Preparation of ribonucleic acid from plant viruses. *In* S. P. Colowick and N. O. Kaplan. eds., *Methods in Enzymology*. Vol. III, pp. 684–687. Academic Press, New York.

Knight, C. A. 1961. Some immunochemical aspects of tobacco mosaic virus. *In* M. Heidelberger and O. J. Plescia, eds. *Immunochemical Ap-*

proaches to Problems in Microbiology. pp. 161–170. Rutgers University Press, New Brunswick, N.J.

Knight, C. A. 1964. Preparation and properties of plant virus proteins. *In* R. J. C. Harris, ed., *Techniques in Experimental Virology*. pp. 1–48. Academic Press, New York.

Knight, C. A. 1974. Molecular Virology. McGraw-Hill, New York.

Knight, C. A., and Stanley, W. M. 1941. Aromatic amino acids in strains of tobacco mosaic virus and in the related cucumber viruses 3 and 4. *J. Biol. Chem.* **141**, 39–49.

Knight, C. A., and Woody, B. R. 1958. Phosphorus content of tobacco mosaic virus. *Arch. Biochem. Biophys.* **78**, 460–467.

Knudson, D. L. 1973. Rhabdoviruses. *J. Gen. Virol.* **20**, 105–130.

Knudson, D. L., and MacLeod, R. 1972. The proteins of potato yellow dwarf virus. *Virology* **47**, 285–295.

Kochwa, S., and Kunkel, H. G., eds. 1971. Immunoglobulins. *Ann. N. Y. Acad. Sci.* **190**, 1–584.

Konigsberg, W., Maita, T., Katze, J., and Weber, K. 1970. Amino acid sequence of the Qβ coat protein. *Nature* **227**, 271–273.

Kozloff, L. M. 1968. Biochemistry of the T-even bacteriophages of *Escherichia coli*. *In* H. Fraenkel-Conrat, ed., *Molecular Basis of Virology*. pp. 435–525. Van Nostrand-Reinhold, New York.

Kozloff, L. M., Lute, M., and Henderson, K. 1957. Viral invasion. I. Rupture of thiol ester bonds in the bacteriophage tail. *J. Biol. Chem.* **228**, 511–528.

Krieg, A. 1956. Über die Nucleinsäuren der Polyeder-Viren. *Naturwiss.* **43**, 537.

Krieg, A. 1959. Die Infektiosität der Ribonucleinsäure aus Smithiavirus pudibundae. *Naturwiss.* **46**, 603.

Kritchevsky, D., and Shapiro, I. L. 1967. Analysis of lipid components of viruses. *In* K. Maramorosch and H. Koprowski, eds., *Methods in Virology*. Vol. 3, pp. 77–98. Academic Press, New York.

Kruseman, J., Kraal, B., Jaspars, E. M. J., Bol, J. F., Brederode, F. Th., and Veldstra, H. 1971. Molecular weight of the coat protein of alfalfa mosaic virus. *Biochemistry* **10**, 447–455.

Kunkel, L. O. 1940. Genetics of viruses pathogenic to plants. *Publ. Am. Assoc. Advan. Sci. No. 12*, 22–27.

Kuno, S., and Lehman, I. R. 1962. Gentiobiose, a constituent of deoxyribonucleic acid from coliphage T6. *J. Biol. Chem.* **237**, 1266–1270.

Kurstak, E. 1972. Small DNA densonucleosis virus (DNV). *Advan. Virus Res.* **17**, 207–241.

Lacy, S., and Green, M. 1967. The mechanism of viral carcinogenesis by DNA mammalian viruses: DNA-DNA homology relationships among "weakly" oncogenic human adenoviruses. *J. Gen. Virol.* **1**, 413–418.

Laemmli, U. K. 1970. Cleavage of structural proteins during the assembly of the head of bacteriophage T4. *Nature* **227**, 680–685.

Lai, M. M. C., and Duesberg, P. H. 1972. Differences between the envelope glycoproteins and glycopeptides of avian tumor viruses released from transformed and from nontransformed cells. *Virology* **50**, 359–372.

Lauffer, M. A. 1951. Purity determination with the electron microscope. *J. Am. Chem. Soc.* **73**, 2370–2371.

Lauffer, M. A. 1952. Form and function: A problem in biophysics. *Sci. Monthly* **75**, 79–83.

Lauffer, M. A., and Ross, A. F. 1940. Physical properties of alfalfa mosaic virus. *J. Am. Chem. Soc.* **62**, 3296–3300.

Lauffer, M. A., and Stanley, W. M. 1943. The denaturation of tobacco mosaic virus by urea. I. Biochemical aspects. *Arch. Biochem.* **2**, 413–424.

Lauffer, M. A., and Price, W. C. 1947. Electrophoretic purification of southern bean mosaic virus. *Arch. Biochem.* **15**, 115–124.

Lauffer, M. A., Trkula, D., and Buzzell, A. 1956. Mechanism of inactivation of tobacco mosaic virus by x-rays. *Nature* **177**, 890.

Lauffer, M. A., and Stevens, C. L. 1968. Structure of the tobacco mosaic virus particle. Polymerization of tobacco mosaic virus protein. *Advan. Virus Res.* **13**, 1–63.

Laver, W. G. 1961. Purification, N-terminal amino acid analysis, and disruption of an influenza virus. *Virology* **14**, 499–502.

Lawley, P. D. 1966. Effects of some chemical mutagens and carcinogens on nucleic acids. *Progr. Nucleic Acid Res. Molecular Biol.* **5**, 89–131.

LeBouvier, G. L., Schwerdt, C. E., and Schaffer, F. L. 1957. Specific precipitates in agar with purified poliovirus. *Virology* **4**, 590–593.

Lehman, I. R., and Pratt, E. A. 1960. On the structure of glucosylated hydroxymethylcytosine nucleotides of coliphages T2, T4 and T6. *J. Biol. Chem.* **235**, 3254–3259.

Lengyel, J. A., Goldstein, R. N., Marsh, M., Sunshine, M. G., and Calendar, R. 1973. Bacteriophage P2 head morphogenesis: Cleavage of the major capsid protein. *Virology* **53**, 1–23.

Lengyel, J. A., Goldstein, R. N., Marsh, M., and Calendar, R. 1974. Structure of the bacteriophage P2 tail. *Virology*, in press.

Lengyel, P., Speyer, J. F., and Ochoa, S. 1961. Synthetic polynucleotides and the amino acid code. *Proc. Natl. Acad. Sci. U.S.* **47**, 1936–1942.

Lengyel, P., Speyer, J. F., Basilio, C., and Ochoa, S. 1962. Synthetic polynucleotides and the amino acid code. III. *Proc. Natl. Acad. Sci. U.S.* **48**, 282–284.

Leppla, S. H., Bjoraker, B., and Bock, R. M. 1968. Borohydride reduction of periodate-oxidized chain ends. *In* L. Grossman and K. Moldave, eds.,

Methods in Enzymology. Vol. 12, pp. 236–240. Academic Press, New York.

Lerman, L. S. 1964. Acridine mutagens and DNA structure. *J. Cell Comp. Physiol.* **64**, Suppl. **1**, 1–18.

Lesnaw, J. A., and Reichmann, M. E. 1969. The structure of tobacco necrosis virus. I. The protein subunit and the nature of the nucleic acid. *Virology* **39**, 729–737.

Levin, O. 1958. Chromatography of tobacco mosaic virus and potato virus X. *Arch. Biochem. Biophys.* **78**, 33–45.

Lewandowski, L. J., Kalmakoff, J., and Tanada, Y. 1969. Characterization of a ribonucleic acid polymerase activity associated with purified cytoplasmic polyhedrosis virus of the silkworm *Bombyx mori*. *J. Virol.* **4**, 857–865.

Lewandowski, L. J., Content, J., and Leppla, S. H. 1971. Characterization of the subunit structure of the ribonucleic acid genome of influenza virus. *J. Virol.* **8**, 701–707.

Lewandowski, L. J., and Millward, S. 1971. Characterization of the genome of cytoplasmic polyhedrosis virus. *J. Virol.* **7**, 434–437.

Lewandowski, L. J., and Leppla, S. H. 1972. Comparison of the 3′termini of discrete segments of the double-stranded ribonucleic acid genomes of cytoplasmic polyhedrosis virus, wound tumor virus, and reovirus. *J. Virol.* **10**, 965–968.

Lewandowski, L. J., and Traynor, B. L. 1972. Comparison of the structure and polypeptide composition of three double-stranded ribonucleic acid-containing viruses (diplornaviruses): Cytoplasmic polyhedrosis virus, wound tumor virus, and reovirus. *J. Virol.* **10**, 1053–1070.

Lichtenstein, J., and Cohen, S. S. 1960. Nucleotides derived from enzymatic digests of nucleic acids of T2, T4, and T6 bacteriophages. *J. Biol. Chem.* **235**, 1134–1141.

Lin, J.-Y., Tsung, C. M., and Fraenkel-Conrat, H. 1967. The coat protein of the RNA bacteriophage MS2. *J. Molecular Biol.* **24**, 1–14.

Lin, T. H., and Maes, R. F. 1967. Methods of degrading nucleic acids and separating the components. *In* K. Maramorosch and H. Koprowski, eds., *Methods in Virology*. Vol. II, pp. 547–606. Academic Press, New York.

Lippincott, J. A. 1961. Properties of infectious ribonucleic acid preparations obtained from tobacco mosaic virus by a heat method. *Virology* **13**, 348–362.

Lister, R. M. 1969. Tobacco rattle, NETU, viruses in relation to functional heterogeneity in plant viruses. *Federation Proc.* **28**, 1875–1889.

Litman, R. M., and Pardee, A. B. 1956. Production of bacteriophage mutants by a disturbance of deoxyribonucleic acid metabolism. *Nature* **178**, 529–532.

Liu, T.-Y. 1972. Determination of tryptophan. *In* C. H. W. Hirs and S. N.

Timasheff, eds., *Methods in Enzymology*. Vol. 25, pp. 44–55. Academic Press, New York.

Liu, T.-Y., and Inglis, A. S. 1972. Determination of cystine and cysteine as S-sulfocysteine. *In* C. H. W. Hirs and S. N. Timasheff, eds., *Methods in Enzymology*. Vol. 25, pp. 55–60. Academic Press, New York.

Loebenstein, G. 1972. Inhibition, interference and acquired resistance during infection. *In* C. I. Kado and H. O. Agrawal, eds., *Principles and Techniques in Plant Virology*. pp. 32–61. Van Nostrand-Reinhold Co. New York.

Loeffler, F., and Frosch, P. 1898. Berichte der Kommission zur Erforschung der Maul und Klauenseuche. *Z. Bakteriol. Parasitenk* **23**, 371–391. For English translation of a relevant part of this paper, *see* T. D. Brock. 1961. *Milestones in Microbiology*. pp. 149–153. Prentice-Hall, Englewood Cliffs, N. J.

Lo Grippo, G. A. 1950. Partial purification of viruses with an anion exchange resin. *Proc. Soc. Exptl. Biol. Med.* **74**, 208–211.

Loring, H. S. 1940. Solubility studies on purified tobacco mosaic virus. *J. Gen. Physiol.* **23**, 719–728.

Loring, H. S. 1942. The reversible inactivation of tobacco mosaic virus by crystalline ribonuclease. *J. Gen. Physiol.* **25**, 497–503.

Loring, H. S. 1955. Hydrolysis of nucleic acids and procedures for the direct estimation of purine and pyrimidine fractions by absorption spectrophotometry. *In* E. Chargaff and J. N. Davidson, eds., *The Nucleic Acids*. Vol I, pp. 191–209. Academic Press, New York.

Loring, H. S., and Schwerdt, C. E. 1944. Studies on purification of poliomyelitis virus. 2. pH stability range of MVA strain. *Proc. Soc. Exptl. Biol. Med.* **57**, 173–175.

Loring, H. S., and McT. Ploeser, J. 1949. The deamination of cytidine in acid solution and the preparation of uridine and cytidine by acid hydrolysis of yeast nucleic acid. *J. Biol. Chem.* **178**, 439–449.

Loring, H. S., Al-Rawi, S. A., and Fujimoto, Y. 1958. The iron content of tobacco mosaic virus and some properties of its infectious nucleic acid. *J. Biol. Chem.* **233**, 1415–1420.

Loveless, A. 1958. Increased rate of plaque-type and host-range mutation following treatment of bacteriophage *in vitro* with ethyl methane sulphonate. *Nature* **181**, 1212–1213.

Loveless, A. 1966. *Genetic and Allied Effects of Alkylating Agents*. Pennsylvania State University Press, University Park.

Lucas-Lenard, J., and Lipman, F. 1971. Protein biosynthesis. *Ann. Rev. Biochem.* **40**, 409–448.

Luria, S. E., and Exner, F. M. 1941. The inactivation of bacteriophages by x-rays. Influence of the medium. *Proc. Natl. Acad. Sci. U.S.* **27**, 370–375.

Luria, S. E., Williams, R. C., and Backus, R. C. 1951. Electron micrographic counts of bacteriophage particles. *J. Bacteriol.* **61**, 179–188.

Luria, S. E., and Darnell, J. E., Jr. 1967. *General Virology.* 2nd edit. Wiley, New York.

Lwoff, A., and Tournier, P. 1966. The classification of viruses. *Ann. Rev. Microbiol.* **20**, 45–74.

McCarthy, B. J., and Bolton, E. T. 1965. An approach to the measurement of genetic relatedness among organisms. *Proc. Natl. Acad. Sci. U.S.* **50**, 156–164.

McCarthy, B. J., and Church, R. B. 1970. The specificity of molecular hybridization reactions. *Ann Rev. Biochem.* **39**, 131–150.

MacDonald, E., Price, W. C., and Lauffer, M. A. 1949. A yellow variant of southern bean mosaic virus. The isoelectric points of yellow and of regular southern bean mosaic virus proteins. *Arch. Biochem.* **24**, 114–118.

McDowell, M., Villa-Komaroff, L., Joklik, W., and Lodish, H. F. 1972. Translation of reovirus messenger RNAs synthesized *in vitro* into reovirus polypeptides by several mammalian cell-free extracts. *Proc. Natl. Acad. Sci. U.S.* **69**, 2649–2653.

MacFarlane, A. S., and Kekwick, R. A. 1938. Physical properties of bushy stunt virus protein. *Biochem. J.* **32**, 1607–1613.

MacKenzie, J. J., and Haselkorn, R. 1972. Physical properties of blue-green algal virus SM-1 and its DNA. *Virology 49*, 497–504.

McLaren, A. D., and Shugar, D. 1964. *Photochemistry of Proteins and Nucleic Acids.* Macmillan, New York.

McSharry, J. J., and Wagner, R. R. 1971. Carbohydrate composition of vesicular stomatitis virus. *J. Virol.* **7**, 412–415.

Maizel, J.V., Jr., 1971. Polyacrylamide gel electrophoresis of viral proteins. *In* K. Maramorosch and H. Koprowski, eds., *Methods in Virology.* Vol. V, pp. 179–246. Academic Press. New York.

Maizel, J. V., Jr., White, D. O., and Scharff, M. D. 1968a. The polypeptides of adenovirus. I. Evidence for multiple protein components in the virion and a comparison of types 2, 7A and 12. *Virology* **36**, 115–125.

Maizel, J. V., Jr., White, D. O., and Scharff, M. D. 1968b. The polypeptides of adenovirus. II. Soluble proteins, cores, top components and the structure of the virion. *Virology* **36**, 126–136.

Mandel, M., and Marmur, J. 1968. Use of ultraviolet absorbance-temperature profile for determining the guanine plus cytosine content of DNA. *In* L. Grossman and K. Moldave, eds., *Methods in Enzymology.* Vol. XII, pp. 195–206. Academic Press, New York.

Mandel, M., Schildkraut, C. L., and Marmur, J. 1968. Use of CsCl density gradient analysis for determining the guanine plus cytosine content of DNA. *In* L. Grossman and K. Moldave, eds., *Methods in Enzymology.* Vol. XII, pp. 184–195. Academic Press, New York.

Maramorosch, K., and Koprowski, H., eds. 1967. *Methods in Virology*. Vol. II. Academic Press, New York.

Markham, R. 1953. Nucleic acids in virus multiplication. *In* P. Fildes and W. E. Van Heyningen, eds., *The Nature of Virus Multiplication*, pp. 89–95. Cambridge University Press, Cambridge.

Markham, R. 1959. The biochemistry of plant viruses. *In* F. M. Burnet and W. M. Stanley, eds., *The Viruses*, Vol. 2, pp. 33–125. Academic Press, New York.

Markham, R. 1967. The ultracentrifuge. *In* K. Maramorosch and H. Koprowski, eds., *Methods in Virology*. Vol. II, pp. 1–39. Academic Press, New York.

Markham, R., Matthews, R. E. F., and Smith, K. M. 1948. Specific crystalline protein and nucleoprotein from a plant virus having insect vectors. *Nature* **162**, 88.

Markham. R., and Smith, J. D. 1951. Chromatographic studies of nucleic acids. 4. The nucleic acid of the turnip yellow mosaic virus, including a note on the nucleic acid of the tomato bushy stunt virus. *Biochem. J.* **49**, 401–407.

Markham, R., and Smith, K. M. 1949. Studies on the virus of turnip yellow mosaic. *Parasitology* **39**, 330–342.

Marmur, J., and Doty, P. 1961. Thermal renaturation of deoxyribonucleic acids. *J. Molecular Biol.* **3**, 585–594.

Marmur, J., and Doty, P. 1962. Determination of the base composition of deoxyribonucleic acid from its thermal denaturation temperature. *J. Molecular Biol.* **5**, 109–118.

Marshall, R. D., and Neuberger, A. 1972. Qualitative and quantitative analysis of the component sugars. *In* A. Gottschalk, ed., *Glycoproteins. Their Composition, Structure and Function.* 2nd edit., Pt. A, pp. 224–299. Elsevier, New York.

Martin, R. G., Matthaei, J. H., Jones, O. W., and Nirenberg, M. W. 1962. Ribonucleotide composition of the genetic code. *Biochem. Biophys. Res. Commun.* **6**, 410–414.

Matheka, H. D., and Armbruster, O. 1956a. Die Anwendung von Ionenaustauschern zur Viruspräparation. 1. Mitt.: Der Einfluss der Austauschereigenschaften auf die Adsorption von Influenza-Virus. *Z. Naturforsch.* **11b**, 187–193.

Matheka, H. D., and Armbruster, O. 1956b. Die Anwendung von Ionenaustauschern zur Viruspräparation. 2. Mitt.: Fraktionierung von Influenza-Virus. *Z. Naturforsch.* **11b**, 193–199.

Mathews, C. K. 1971. *Bacteriophage Biochemistry*. Van Nostrand-Reinhold, New York.

Matthaei, J. H., Jones, O. W., Martin, R. G., and Nirenberg, M. W. 1962. Characteristics and composition of RNA coding units. *Proc. Natl. Acad. Sci. U.S.* **48**, 666–677.

Matthaei, J. H., and Nirenberg, M. W. 1961. The dependence of cell-free protein synthesis in *E. coli* upon RNA prepared from ribosomes. *Biochem. Biophys. Res. Commun.* **6**, 410–414.

Matthews, R. E. F. 1967. Serological techniques for plant viruses. *In* K. Maramorosch and H. Koprowski, eds., *Methods in Virology.* pp. 199–241. Academic Press, New York.

Matthews, R. E. F. 1970. Turnip yellow mosaic virus. *C.M.I./A.A.B. Description of Plant Viruses No. 2.*

Matthews, R. E. F., and Smith, J. D. 1955. The chemotherapy of viruses. Advan. Virus Res. **3**, 49–148.

May, D. S., and Knight, C. A. 1965. Polar stripping of protein subunits from tobacco mosaic virus. *Virology* **25**, 502–507.

May, G., and Peteri, D. 1973. Synthese und Prüfung von Adamantan-Abkömmlingen als Virustatika. *Arzneim. Forsch.* **23**, 718–721.

May, J. T., and Symons, R. H. 1971. Specificity of the cucumber mosaic virus-induced RNA polymerase for RNA and polynucleotide templates. *Virology* **44**, 517–526.

Mayer, M. M., Rapp, H. J., Roizman, B., Klein, S. W., Cowan, K. M., Lukens, D., Schwerdt, C. E., Schaffer, F. L., and Charney, J. 1957. The purification of poliomyelitis virus as studied by complement fixation. *J. Immunol.* **78**, 435–455.

Medill-Brown, M., and Broidy, B. A. 1955. Mutation and selection pressure during adaption of influenza virus to mice. *Virology* **1**, 301–312.

Melnick, J. L. 1971. Classification and nomenclature of animal viruses. 1971. *Progr. Med. Virol.* **13**, 462–484.

Melnick, J. L. 1972. Classification and nomenclature of viruses, 1972. *Progr. Med. Virol.* **14**, 321–332.

Meselson, M., Stahl, F. W., and Vinograd, J. 1957. Equilibrium sedimentation of macromolecules in density gradients. *Proc. Natl. Acad. Sci. U.S.* **43**, 581–588.

Miki, T., and Knight, C. A. 1965. Preparation of broad bean mottle virus protein. *Virology* **25**, 478–481.

Miki, T., and Knight, C. A. 1967. Some chemical studies on a strain of white clover mosaic virus. *Virology* **31**, 55–63.

Miki, T., and Knight, C. A. 1968. The protein subunit of potato virus X. *Virology* **36**, 168–173.

Miller, G. L. 1944. Influence of pH and of certain other conditions on the stability of the infectivity and red cell agglutinating activity of influenza virus. *J. Exptl. Med.* **80**, 507–520.

Miller, G. L., Lauffer, M. A., and Stanley W. M. 1944. Electrophoretic studies on PR 8 influenza virus. *J. Exptl. Med.* **80**, 549–559.

Miller, G. L., and Price, W. C. 1946. Physical and chemical studies on southern bean mosaic virus. I. Size, shape, hydration and elementary composition. *Arch. Biochem.* **10**, 467–477.

Miller, H. K., and Schlesinger, R. W. 1955. Differentiation and purification of influenza viruses by adsorption on aluminum phosphate. *J. Immunol.* **75**, 155–175.

Millward, S., and Graham, A. F. 1970. Structural studies on reovirus: Discontinuities in the genome. *Proc. Natl. Acad. Sci. U.S.* **65**, 422–429.

Milne, R. G. 1972. Electron microscopy of viruses. *In* C. I. Kado and H. O. Agrawal, eds., *Principles and Techniques in Plant Virology.* pp. 76–128. Van Nostrand-Reinhold, New York.

Mims, C. A. 1964. Aspects of the pathogenesis of virus diseases. *Bacteriol. Rev.* **28**, 30.

Min Jou, W., Haegeman, G., and Fiers, W. 1971. Studies on the bacteriophage MS2. Nucleotide fragments from the coat protein cistron. *FEBS Letters* **13**, 105–109.

Min Jou, W., Haegemen, G., Ysebaert, M., and Fiers, W. 1972. Nucleotide sequence of the gene coding for the bacteriophage MS2 coat protein. *Nature* **237**, 82–88.

Miura, K., and Egami, F. 1960. Distribution of guanylic acid in RNA of yeast and tobacco mosaic virus. *Biochim. Biophys. Acta* **44**, 378–379.

Morris, T. J., and Semancik, J. S. 1973. *In vitro* protein polymerization and nucleoprotein reconstitution of tobacco rattle virus. *Virology* **53**, 215–224.

Morrison, T., Stampfer, M., Baltimore, D., and Lodish, H. F. 1974. Translation of vesicular stomatitis messenger RNA by extracts from mammalian and plant cells. *J. Virol.* **13**, 62–72.

Morrow, J. F., and Berg, P. 1972. Cleavage of simian virus 40 DNA at a unique site by a bacterial restriction enzyme. *Proc. Natl. Acad. Sci. U.S.* **69**, 3365–3369.

Mountcastle, W. E., Compans, R. W., and Choppin, P. W. 1971. Proteins and glycoproteins of paramyxoviruses: A comparison of simian virus 5, Newcastle disease virus, and Sendai virus. *J. Virol.* **7**, 47–52.

Muller, R. H. 1950. Application of ion exchange resins to the purification of certain viruses. *Proc. Soc. Exptl. Biol. Med.* **73**, 239–241.

Muller, R. H., and Rose, H. M. 1952. Concentration of influenza virus (PR8 strain) by a cation exchange resin. *Proc. Soc. Exptl. Biol. Med.* **80**, 27–29.

Munk, K., and Schäfer, W. 1951. Eigenschaften tierischer Virusarten untersucht an den Geflügelpestviren als Modell. II. Mitt: Serologische Untersuchungen über die Geflügelpestviren und über ihre Beziehungen zu einem normalen Wirtsprotein. *Z. Naturforsch.* **6b**, 372–379.

Narita, K. 1958. Isolation of acetylpeptide from enzymic digest of TMV-protein. *Biochim. Biophys. Acta* **28**, 184–191.

Narita, K. 1959. Isolation of acetylpeptide from the protein of cucumber virus 4. *Biochim. Biophys. Acta* **31**, 372–377.

Nathans, D., Notani, G., Schwartz, J. H., and Zinder, N. D. 1962. Biosynthesis of the coat protein of coliphage f2 by *E. coli* extracts. *Proc. Natl. Acad. Sci. U.S.* **48**, 1424–1431.

Nermut, M. V. 1972. Further investigation on the fine structure of influenza virus. *J. Gen. Virol.* **17**, 317–331.

Neuberger, A., Gottschalk, A., Marshall, R. D., and Spiro, R. G. 1972. Carbohydrate-peptide linkages in glycoproteins and methods for their elucidation. *In* A. Gottschalk, ed., *Glycoproteins. Their Composition, Structure and Function.* 2nd edit., Pt. A, pp. 450–490. Elsevier, New York.

Newmark, P., and Myers, R. W. 1957. Degradation of tobacco mosaic virus by alkanolamines. *Federation Proc.* **16**, 226.

Nirenberg, M., Caskey, T., Marshall, R., Brimacombe, R., Kellogg, D., Doctor, B., Hatfield, D., Levin, J., Rottman, F., Pestka, S., Wilcox, M., and Anderson, F. 1966. The RNA code and protein synthesis. *Cold Spring Harbor Symp. Quant. Biol.* **31**, 11–24.

Nirenberg, M. W., and Matthaei, J. H. 1961. The dependence of cell-free protein synthesis in *E. coli* upon naturally occurring or synthetic polyribonucleotides. *Proc. Natl. Acad. Sci. U.S.* **47**, 1588–1602.

Nirenberg, M. W., Matthaei, J. H., and Jones, O. W. 1962. An intermediate in the biosynthesis of polyphenylalanine directed by synthetic template RNA. *Proc. Natl. Acad. Sci. U.S.* **48**, 104–109.

Niu, C.-I., and Fraenkel-Conrat, H. 1955. C-terminal amino acid sequence of tobacco mosaic virus protein. *Biochim. Biophys. Acta* **16**, 597–598.

Niu, C.-I., Shore, V., and Knight, C. A. 1958. The peptide chains of some plant viruses. *Virology* **6**, 226–233.

Nonoyama, M., and Ikeda, Y. 1964. Ribonuclease resistant RNA found in cells of *Escherichia coli* infected with RNA phage. *J. Molecular Biol.* **9**, 763–771.

Northrop, J. H. 1961. Biochemists, Biologists, and William of Occam. *Ann. Rev. Biochem.* **30**, 1–10.

Nowinski, R. C., Fleissner, E., Sarkar, N. H., and Aoki, T. 1972. Chromatographic separation and antigenic analysis of proteins of the oncornaviruses. II. Mammalian leukemia-sarcoma viruses. *J. Virol.* **9**, 359–366.

Nowinski, R. C., and Sarkar, N. H. 1972. Serologic and structural studies of mouse mammary tumor virus. *J. Natl. Cancer Inst.* **12**, 1169–1174.

Ochoa, S. 1963. Discussion following presentation of H. G. Wittmann. *Perspectives Virol.* **3**, 36–39.

Offord, R. E., and Harris, J. I. 1965. The protein subunit of tobacco rattle virus. *Fed. European Biochem. Soc. 2nd Meeting.* Abstracts, pp. 216–217. Verlag der Wiener Mediz. Akad., Wien.

Onodera, K., Komano, T., Himeno, M., and Sakai, F. 1965. The nucleic acid of the nuclear-polyhedrosis virus of the silkworm. *J. Molecular Biol.* **13**, 532–539.

Orlob, G. B. 1967. Inactivation of purified plant viruses and their nucleic acids by photosensitizing dyes. *Virology* **31**, 402–413.

Oster, G., and McLaren, A. D. 1950. The ultraviolet light and photosensitized inactivation of tobacco mosaic virus. *J. Gen. Physiol.* **33**, 215–228.

Padan, E., and Shilo, M. 1973. Cyanophages—Viruses attacking blue-green algae. *Bacteriol. Rev.* **37**, 343–370.

Palmenberg, A., and Kaesberg, P. 1974. Replication of heterologous RNAs with Qβ replicase. *Proc. Natl. Acad. Sci. U.S.*, 71, 1371–1375.

Pauling, L, Corey, R. B., and Branson, H. R. 1951. The structure of proteins: Two hydrogen-bonded helical configurations of the polypeptide chain. *Proc. Natl. Acad. Sci. U.S.* **37**, 205–211.

Peden, K. W. C., May, J. T., and Symons, R. H. 1972. A comparison of two plant virus-induced RNA polymerases. *Virology* **47**, 498–501.

Penhoet, E., Miller, H., Doyle, M., and Blatti, S. 1971. RNA-dependent RNA polymerase activity in influenza virions. *Proc. Natl. Acad. Sci. U.S.* **68**, 1369–1371.

Perham, R. N., and Richards, F. M. 1968. Reactivity and structural role of protein amino groups in tobacco mosaic virus. *J. Molecular Biol.* **33**, 795–807.

Peter, R., Stehelin, D., Reinbolt, J., and Duranton, H. 1972. Coat protein —turnip yellow mosaic virus. *In* Margaret O. Dayhoff, ed., *Atlas of Protein Sequence and Structure 1972.* p. D287. National Biomedical Research Foundation, Silver Spring, Md.

Pett, D. M., Vanaman, T. C., and Joklik, W. K. 1973. Studies on the amino and carboxyl terminal amino acid sequences of reovirus capsid polypeptides. *Virology* **52**, 174–186.

Pfefferkorn, E. R., and Hunter, H. S. 1963. Purification and partial chemical analysis of Sindbis virus. *Virology* **20**, 433–445.

Philipson, L. 1967a. Chromatography and membrane separation. *In* K. Maramorosch and H. Koprowski, eds., *Methods in Virology.* Vol II, pp. 179–233. Academic Press, New York.

Philipson, L. 1967b. Water-organic solvent phase systems. *In* K. Maramorosch and H. Koprowski, eds., *Methods in Virology,* Vol. II, pp. 235–244. Academic Press, New York.

Phillips, J. H., and Brown, D. M. 1967. The mutagenic action of hydroxylamine. *Progr. Nucleic Acid Res. Molecular Biol.* **7**, 349–368.

Piña, M., and Green, M. 1965. Biochemical studies on adenovirus multiplication. IX. Chemical and base composition analysis of 28 human adenoviruses. *Proc. Natl. Acad. Sci. U.S.* **54**, 547–551.

Pirie, N. W. 1940. The criteria of purity used in the study of large molecules of biological origin. *Biol. Rev.* **15**, 377–404.

Pirie, N. W. 1945. Physical and chemical properties of tomato bushy stunt virus and the strains of tobacco mosaic virus. *Advan. Enzymol.* **5**, 1–29.

Pollard, M., Connolly, J., and Fromm, S. 1949. The precipitating effect of methanol on viruses. *Proc. Soc. Exptl. Biol. Med.* **71**, 290–293.

Polson, A., and Russell, B. 1967. Electrophoresis of viruses. *In* K. Maramorosch and H. Koprowski, eds., *Methods in Virology.* Vol. II, pp. 391–426. Academic Press, New York.

Ponnamperuma, C., and Gabel, N. W. 1968. Current status of chemical studies on the origin of life. *Space Life Sci.* **1**, 64–96.

Pons, M. W., and Hirst, G. K. 1968. Polyacrylamide gel electrophoresis of influenza virus RNA. *Virology* **34**, 386–388.

Porter, C. A. 1956. Evaluation of a fluorocarbon technique for the isolation of plant viruses. *Trans. N. Y. Acad. Sci.* **18**, 704–706.

Potash, L. 1968. Methods in human virus vaccine preparation. *In* K. Maramorosch and H. Koprowski, eds., *Methods in Virology.* Vol. 4, pp. 371–464. Academic Press, New York.

Price, W. C. 1940. Thermal inactivation rates of four plant viruses. *Arch. Ges. Virusforsch.* **1**, 373–386.

Price, W. C. 1946. Purification and crystallization of southern bean mosaic virus. *Am. J. Bot.* **33**, 45–54.

Prusoff, W. H., Bakhle, Y. S., and Sekely, L. 1965. Cellular and antiviral effects of halogenated deoxyribonucleosides. *Ann. N. Y. Acad. Sci.* **130**, 135–150.

Putnam, F. W. 1953. Protein denaturation. *In* H. Neurath and K. Bailey, eds., *The Proteins.* pp. 807–892. Academic Press, New York.

Radding, C. M. 1973. Molecular mechanisms in genetic recombination. *Ann. Rev. Genet.* **7**, 87–111.

Ralph, R. K., and Berquist, P. L. 1967. Separation of viruses into components. *In* K. Maramorosch and H. Koprowski, eds., *Methods in Virology.* Vol. II, pp. 463–545. Academic Press, New York.

Ramachandran, L. K., and Gish, D. T. 1959. Studies on the amino acid sequence of tobacco mosaic virus (TMV) protein. II. The amino acid sequences of six peptides obtained from a tryptic digest. *J. Am. Chem. Soc.* **81**, 884–890.

Randall, C. C., Gafford, L. G., Darlington, R. W., and Hyde, J. 1964. Composition of fowlpox virus and inclusion matrix. *J. Bacteriol.* **87**, 939–944.

Raskas, H. J., and Green, M. 1971. DNA-RNA and DNA-DNA hybridization in virus research. *In* K. Maramorosch and H. Koprowski, eds., *Methods in Virology.* Vol. 5, pp. 247–269. Academic Press, New York.

Ray, D. S. 1968. The small DNA-containing bacteriophages. *In* H. Fraenkel-Conrat, ed., *Molecular Basis of Virology.* pp. 222–254. Reinhold Book, New York.

Reddi, K. K. 1958. Ultraviolet absorption studies on tobacco mosaic virus nucleic acid. *Biochim. Biophys. Acta* **27**, 1–4.

Reddi, K. K. 1959. Degradation of tobacco mosaic virus nucleic acid with micrococcal phosphodiesterase. *Biochim. Biophys. Acta* **36**, 132–142.
Reddi, K. K. 1960. Distribution of guanylic acid residues in tobacco mosaic virus nucleic acid. *Nature* **188**, 60–61.
Reddy, D. V. R., and Black, L. M. 1973. Electrophoretic separation of all components of the double-stranded RNA of wound tumor virus. *Virology* **54**, 557–562.
Reichmann, M. E. 1959. Potato X virus. III. Light scattering studies. *Can. J. Chem.* **37**, 384–388.
Reichmann, M. E. 1960. Degradation of potato virus X. *J. Biol. Chem.* **235**, 2959–2963.
Reichmann, M. E. 1964. The satellite tobacco necrosis virus: A single protein and its genetic code. *Proc. Natl. Acad. Sci. U.S.* **52**, 1009–1017.
Reichmann, M. E., and Stace-Smith, R. 1959. Preparation of infectious ribonucleic acid from potato virus X by means of guanidine denaturation. *Virology* **9**, 710–712.
Reid, M. S., and Bieleski, R. L. 1968. A simple apparatus for vertical flat-sheet polyacrylamide gel electrophoresis. *Anal. Biochem.* **22**, 374–381.
Renis, H. E. 1970. Antiviral studies with kethoxal. *Ann. N. Y. Acad. Sci.* **173**, 527–535.
Rich, A., and Watson, J. D. 1954. Some relations between DNA and RNA. *Proc. Natl. Acad. Sci. U.S.* **40**, 759–764.
Richards, K. E., and Williams, R. C. 1972. Assembly of tobacco mosaic virus *in vitro:* Effect of state of polymerization of the protein component. *Proc. Natl. Acad. Sci. U.S.* **69**, 1121–1124.
Richardson, C. C. 1965. Phosphorylation of nucleic acid by an enzyme from T4 bacteriophage-infected *Escherichia coli. Proc. Natl. Acad. Sci. U.S.* **54**, 158–165.
Riley, V. 1950. Chromatographic studies on the separation of the virus from chicken tumor. I. Effect of salt concentration on adsorption, elution, and purification. *J. Nat. Cancer Inst.* **11**, 199–214.
Ritchie, D. A. 1964. Mutagenesis with light and proflavin in phage T4. *Genet. Rev.* **5**, 168–169.
Ritchie, D. A. 1965. Mutagenesis with light and proflavin in phage T4. 2. Properties of the mutants. *Genet. Rev.* **6**, 474–478.
Rivers, T. M. 1941. The infinitely small in biology. *Diplomate* **13**, 184–186.
Roberts, B. E., and Paterson, B. M. 1973. Efficient translation of tobacco mosaic virus RNA and rabbit globin 9S RNA in a cell-free system from commercial wheat germ. *Proc. Natl. Acad. Sci. U.S.* **70**, 2330–2334.
Robinson, D. J. 1973. Inactivation and mutagenesis of tobacco rattle virus by nitrous acid. *J. Gen. Virol.* **18**, 215–222.
Robinson, W. S., and Duesberg, P. H. 1967. Tumor virus RNA. Subviral

carcinogenesis. *First Intern. Symp. Tumor Viruses*, July 1967, pp. 3–17.

Roizman, B., and Roane, P. R. 1961. A physical difference between two strains of herpes simplex virus apparent on sedimentation in cesium chloride. *Virology* **15**, 75–79.

Roizman, B., and Spear, P. G. 1971. Herpesviruses: Current information on the composition and structure. *In* K. Maramorosch and E. Kurstak, eds., *Comparative Virology*. pp. 135–168. Academic Press, New York.

Rosemond, H., and Moss, B. 1973. Phosphoprotein component of vaccinia virions. *J. Virol.* **11**, 961–970.

Rosen, L. 1964. Hemagglutination. *In* R. J. C. Harris, ed., *Techniques in Experimental Virology*. pp. 257–276. Academic Press, New York.

Ross, A. F. 1941. The determination of some amino acids in tobacco mosaic virus protein. *J. Biol. Chem.* **138**, 741–749.

Ross, A. F., and Stanley, W. M. 1939. The amino acids of tobacco mosaic virus protein. *J. Biol. Chem.* **128**, lxxxiv-lxxxv.

Ruben, F., and Jackson, G. 1972. A new subunit influenza vaccine: Acceptability compared with standard vaccines and effect of dose on antigenicity. *J. Infect. Dis.* **125**, 656–664.

Rubenstein, I., Thomas, C. A., Jr., and Hershey, A. D. 1961. The molecular weights of T2 bacteriophage DNA and its first and second breakage products. *Proc. Natl. Acad. Sci. U.S.* **47**, 1113–1122.

Rubin, H. 1965. Virus without symptoms and symptoms without virus: Complementary aspects of latency as exemplified by the avian tumor viruses. *Perspectives Virol.* **4**, 164–174.

Rueckert, R. R. 1971. Picornaviral architecture. *In* K. Maramorosch and E. Kurstak, eds., *Comparative Virology*. pp. 255–306. Academic Press, New York.

Rueckert, R. R., Dunker, A. K., and Stoltzfus, C. M. 1969. The structure of mouse-Elberfeld virus: A model. *Proc. Natl. Acad. Sci. U.S.* **62**, 912–919.

Rueckert, R. R., and Schäfer, W. 1965. Studies on the structure of viruses of the Columbia SK group. 1. Purification and properties of ME-virus grown in Ehrlich ascites cell suspensions. *Virology* **26**, 333–344.

Rushizky, G. W. 1967. Mapping of oligonucleotides. *In* L. Grossman and K. Moldave, eds., *Methods in Enzymology*. Vol. 12, Pt. A, pp. 395–398. Academic Press, New York.

Rushizky, G. W., and Knight, C. A. 1959. Ribonuclease-sensitive infectious units from tomato bushy stunt virus. *Virology* **8**, 448–455.

Rushizky, G. W., and Knight, C. A. 1960a. A mapping procedure for nucleotides and oligonucleotides. *Biochem. Biophys. Res. Commun.* **2**, 66–70.

Rushizky, G. W., and Knight, C. A. 1960b. An oligonucleotide mapping procedure and its use in the study of tobacco mosaic virus nucleic acid. *Virology* **11**, 236–249.

Rushizky, G. W., and Knight, C. A. 1960c. Products obtained by digestion of the nucleic acids of some strains of tobacco mosaic virus with pancreatic ribonuclease. *Proc. Natl. Acad. Sci. U.S.* **46**, 945–952.

Rushizky, G. W., Knight, C. A., and McLaren, A. D. 1960. A comparison of the ultraviolet-light inactivation of infectious ribonucleic acid preparations from tobacco mosaic virus with those of the native and reconstituted virus. *Virology* **12**, 32–47.

Rushizky, G. W., Knight, C. A., and Sober, H. A. 1961. Studies on the preferential specificity of pancreatic ribonuclease as deduced from partial digests. *J. Biol. Chem.* **236**, 2732–2737.

Rushizky, G. W., Knight, C. A., Roberts, W. K., and Dekker, C. A. 1962a. Studies on the action of micrococcal nuclease. II. Degradation of ribonucleic acid from tobacco mosaic virus. *Biochim. Biophys. Acta* **55**, 674–682.

Rushizky, G. W., Sober, H. A., and Knight, C. A. 1962b. Products obtained by digestion of the nucleic acids of some strains of tobacco mosaic virus with ribonuclease T_1. *Biochim. Biophys. Acta* **61**, 56–61.

Russell, W. C., and Crawford, L. V. 1963. Some characteristics of the deoxyribonucleic acid from herpes simplex virus. *Virology* **21**, 353–361.

Russell, W. C., Watson, D. H., and Wildy, P. 1963. Preliminary chemical studies on herpes virus. *Biochem. J.* **87**, 26P–27P.

Ryle, A. P., Sanger, F., Smith, L. F., and Kitai, R. 1955. The disulphide bonds of insulin. *Biochem. J.* **60**, 541–556.

Sambrook, J. F., Padgett, B. L., and Tomkins, J. K. N. 1966. Conditional lethal mutants of rabbitpox virus. 1. Isolation of host cell-dependent and temperature-dependent mutants. *Virology* **28**, 592–599.

Sanger, F. 1945. The free amino groups of insulin. *Biochem. J.* **39**, 507–515.

Sanger, F. 1949. The terminal peptides of insulin. *Biochem. J.* **45**, 563–574.

Sänger, H. L. 1968. *Defective plant viruses in molecular genetics. In* H. G. Wittmann and H. Schuster, eds., *4 Wiss. Kong. Ges. Deutsche Arze,* Berlin, 1967, pp. 300–336. Springer-Verlag, Berlin.

Sarabhai, A. S., Stretton, A. O. W., Brenner, S., and Bolle, A. 1964. Co-linearity of the gene with the polypeptide chain. *Nature* **201**, 13–17.

Sarov, I., and Joklik, W. K. 1972. Studies on the nature and location of the capsid polypeptides of vaccinia virions. *Virology* **50**, 579–592.

Schachman, H. K. 1959. *Ultracentrifugation in Biochemistry.* Academic Press, New York.

Schäfer, W. 1959. The comparative chemistry of infective virus particles and of other virus-specific products: Animal viruses. *In* F. M. Burnet and W. M. Stanley, eds., *The Viruses.* Vol. 1, pp. 475–504. Academic Press, New York.

Schaffer, F. L., and Schwerdt, C. E. 1959. Purification and properties of poliovirus. *Advan. Virus Res.* **6**, 159–204.

Schaffer, F. L., and Hackett, A. J. 1963. Early events in poliovirus HeLa cell interaction: Acridine orange photosensitization and detergent extraction. *Virology* **21**, 124–126.
Scharff, M. D., Maizel, J. V., Jr., and Levinton, L. 1964. Physical and immunological properties of a soluble precursor of the poliovirus capsid. *Proc. Natl. Acad. Sci. U.S.* **51**, 329–337.
Schild, G. C. 1970. Studies with antibody to the purified haemagglutinin of an influenza Ao virus. *J. Gen. Virol.* **9**, 191–200.
Schildkraut, C. L., Marmur, J., and Doty, P. 1962. Determination of the base composition of deoxyribonucleic acid from its buoyant density in CsCl. *J. Molecular Biol.* **4**, 430–443.
Schito, G. C. 1966. A rapid procedure for the purification of bacterial viruses. *Virology* **30**, 157–159.
Schlegel, D. E., 1960. Highly infectious phenol extracts from tobacco leaves infected with cucumber mosaic virus. *Virology* **11**, 329–338.
Schlesinger, M. 1934. Zur Frage der chemischen Zusammensetzung des Bakteriophagen. *Biochem. Z.* **273**, 306–311.
Schlesinger, M. J., and Schlesinger, S. 1972. Identification of a second glycoprotein in Sindbis virus. *Virology* **47**, 539–541.
Schneider, W. C. 1957. Determination of nucleic acids in tissues by pentose analysis. *In* S. P. Colowick and N. O. Kaplan, eds., *Methods in Enzymology*. Vol. 3, pp. 680–684. Academic Press, New York.
Schramm, G. 1947a. Über die Spaltung des Tabakmosaikvirus und die Wiedervereinigung der Spaltstücke zu höhemolekularen Proteinen. I. Die Spaltungsreaktion. *Z. Naturforsch.* **2b**, 112–121.
Schramm, G. 1947b. II, Versuche zur Wiedervereinigung der Spaltstücke. *Z. Naturforsch.* **2b**, 249–257.
Schramm, G. 1954. *Die Biochemie der Viren.* Springer-Verlag, Berlin.
Schramm, G., Schumacher, G., and Zillig, W. 1955. Über die Struktur des Tabakmosaikvirus. III. Der Zerfall in alkalischer Lösung. *Z. Naturforsch.* **10b**, 481–492.
Schroeder, W. A. 1972. Hydrazinolysis. *In* C. H. W. Hirs and S. N. Timasheff, eds., *Methods in Enzymology*. Vol. 25, pp. 138–143. Academic Press, New York.
Schulze, I. T. 1972. The structure of influenza virus. II. A model based on the morphology and composition of subviral particles. *Virology* **47**, 181–196.
Schulze, I. T. 1973. Structure of the influenza virion. *Advan. Virus Res.* **18**, 1–55.
Schumaker, V. N. 1967. Zone centrifugation. *Advan. Biol. Med. Phys.* **11**, 246–334.
Schuster, H. 1960a. The ribonucleic acids of viruses. *In* E. Chargaff and J. N. Davidson, eds., *The Nucleic Acids*. Vol. 3, Chap. 34. Academic Press, New York.

Schuster, H. 1960b. Die Reaktionsweise der Desoxyribonucleinsäure mit salpetriger Säure. Z. *Naturforsch.* **15b**, 298–304.

Schuster, H. 1961. The reaction of tobacco mosaic virus ribonucleic acid with hydroxylamine. *J. Molecular Biol.* **3**, 447–457.

Schuster, H., and Schramm, G. 1958. Bestimmung der biologisch wirksamen Einheit in der Ribosenucleinsäure des Tabakmosaikvirus auf chemischen Wege. Z. *Naturforsch.* **13b**, 697–704.

Schuster, H., Schramm, G., and Zillig, W. 1956. Die Struktur der Ribosenucleinsäure des Tabakmosaikvirus. Z. *Naturforsch.* **11b**, 339–345.

Schuster, H., and Wilhelm, R. C. 1963. Reaction differences between tobacco mosaic virus and its free ribonucleic acid with nitrous acid. *Biochim. Biophys. Acta* **68**, 554–560.

Schwerdt, C. E., and Schaffer, F. L. 1955. Some physical and chemical properties of purified poliomyelitis virus preparations. *Ann. N. Y. Acad. Sci.* **61**, 740–753.

Schwerdt, C. E., and Schaffer, F. L. 1956. Purification of poliomyelitis viruses propagated in tissue culture. *Virology* **2**, 665–678.

Sehgal, O. P. 1973. Inactivation of southern bean mosaic virus and its ribonucleic acid by nitrous acid and ultraviolet light. *J. Gen. Virol.* **18**, 1–10.

Sehgal, O. P., and Krause, G. F. 1968. Efficiency of nitrous acid as an inactivating and mutagenic agent of intact tobacco mosaic virus and its isolated nucleic acid. *J. Virol.* **2**, 966–971.

Sekiguchi, M., Taketo, A., and Takagi, Y. 1960. An infective deoxyribonucleic acid from bacteriophage ØX174. *Biochim. Biophys. Acta* **45**, 199–200.

Semancik, J. S., Morris, T. J., and Weathers, L. G. 1973. Structure and conformation of low molecular weight pathogenic RNA from excortis disease. *Virology* **53**, 448–456.

Semancik, J. S., and Reynolds, D. A. 1969. Assembly of protein and nucleoprotein particles from extracted tobacco rattle virus protein and RNA. *Science* **164**, 559–560.

Sengbusch, P., von, and Wittmann, H. G. 1965. Serological and physicochemical properties of the wild strains and two mutants of tobacco mosaic virus with the same amino acid exchange in different positions of the protein chain. *Biochem. Biophys. Res. Commun.* **18**, 780–787.

Senseney, C. A., Kahn, R. P., and Desjardins, P. R. 1954. Particle size and shape of purified tomato-ringspot virus. *Science* **120**, 456–457.

Setlow, R. B. 1968. The photochemistry, photobiology and repair of polynucleotides. *Progr. Nucleic Acid Res. Molecular Biol.* **8**, 257–295.

Shainoff, J. R., and Lauffer, M. A. 1956. Chromatographic purification of southern bean mosaic virus. *Arch. Biochem. Biophys.* **64**, 315–318.

Shainoff, J. R., and Lauffer, M. A. 1957. An application of ion exchange chromatography to the identification of virus activity with characteristic particles. *Virology* **4**, 418–434.

Shapiro, A. L., Viñuela, E., and Maizel, J. V. 1967. Molecular weight estimation of polypeptide chains by electrophoresis in SDS-poly-acrylamide gels. *Biochem. Biophys. Res. Commun.* **28**, 815–820.

Shapiro, A. L., Viñuela, E., and Maizel, J. V. 1969. Molecular weight estimation of polypeptides by SDS-polyacrylamide gel electrophoresis: Further data concerning resolving power and general considerations. *Ann Biochem.* **29**, 505–514.

Sharp, D. G. 1953. Purification and properties of animal viruses. *Advan. Virus Res.* **1**, 277–313.

Sharp, D. G., Hook, A. E., Taylor, A. R., Beard, D., and Beard, J. W. 1946. Sedimentation characters and pH stability of the T2 bacteriophage of *Escherichia coli. J. Biol. Chem.* **165**, 259–270.

Sharp, D. G., Taylor, A. R., Beard, D., Beard, J. W. 1942. Electrophoresis of the rabbit papilloma virus protein. *J. Biol. Chem.* **142**, 193–202.

Shatkin, A. J., and Sipe, J. D. 1968. RNA polymerase activity in purified reoviruses. *Proc. Natl. Acad. Sci. U.S.* **61**, 1462–1469.

Shatkin, A. J., Sipe, J. D., and Loh, P. J. 1968. Separation of ten reovirus genome segments by polyacrylamide gel electrophoresis. *J. Virol.* **2**, 986–998.

Shih, D. S., and Kaesberg, P. 1973. Translation of brome mosaic viral ribonucleic acid in a cell-free system derived from wheat embryo. *Proc. Natl. Acad. Sci. U.S.* **70**, 1799–1803.

Siegel, A., and Hudson, W. 1959. Equilibrium centrifugation of two strains of tobacco mosaic virus in density gradients. *Biochim. Biophys. Acta* **34**, 245–255.

Siegel, A., and Wildman, S. G. 1954. Some natural relationships among strains of tobacco mosaic virus. *Phytopathology* **44**, 277–282.

Siegel, A., and Zaitlin, M. 1965. Defective plant viruses. *Perspectives Virol.* **4**, 113–125.

Silber, G., and Burk, L. G. 1965. Infectivity of tobacco mosaic virus stored for fifty years in extracted "unpreserved" plant juice. *Nature* **206**, 740–741.

Silbert, J. A., Salditt, M., and Franklin, R. M. 1969. Structure and synthesis of a lipid-containing bacteriophage. III. Purification of bacteriophage PM2 and some structural studies on the virion. *Virology* **39**, 666–681.

Simpson, R. W., and Hauser, R. 1966. Influence of lipids on the viral phenotype. I. Interaction of myxoviruses and their lipid constituents with phospholipases. *Virology* **30**, 684–697.

Sinclair, J. B., Geil, P. H., and Kaesberg, P. 1957. Biophysical studies of wild cucumber mosaic virus. *Phytopathology* **47**, 372–377.

Singer, B., and Fraenkel-Conrat, H. 1961. Effects of bentonite on infectivity and stability of TMV-RNA. *Virology* **14**, 59–65.

Singer, B., and Fraenkel-Conrat, H. 1966. Dye-catalyzed photoinactivation of tobacco mosaic virus ribonucleic acid. *Biochemistry* **5**, 2446–2450.

Singer, B., and Fraenkel-Conrat, H. 1969. The role of conformation in chemical mutagenesis. *Progr. Nucleic Acid Res. Molecular Biol.* **9**, 1–29.

Sinsheimer, R. L. 1959a. Purification and properties of bacteriophage ØX174. *J. Molecular Biol.* **1**, 37–42.

Sinsheimer, R. L. 1959b. A single-stranded deoxyribonucleic acid from bacteriophage ØX174. *J. Molecular Biol.* **1**, 43–53.

Sinsheimer, R. L. 1960. The nucleic acids of the bacterial viruses. *In* E. Chargaff and J. N. Davidson, eds., *The Nucleic Acids.* Vol. 3, pp. 187–244. Academic Press, New York.

Sinsheimer, R. L. 1968. Bacteriophage ØX174 and related viruses. *Progr. Nucleic Acid Res. Molecular Biol.* **8**, 115–167.

Sjöquist, J. 1953. Paper strip identification of phenyl thiohydantoins. *Acta Chem. Scand.* **7**, 447–448.

Skehel, J. J. 1971. RNA-dependent RNA polymerase activity of the influenza virus. *Virology* **45**, 793–796.

Skehel, J. J., and Schild, G. C. 1971. The polypeptide composition of influenza A viruses. *Virology* **44**, 396–408.

Slechta, L., and Hunter, J. H. 1970. Antiviral action of 2-thiouridine. *Ann. N. Y. Acad. Sci.* **173**, 708–713.

Smith, J. D. 1955. The electrophoretic separation of nucleic acid components. *In* E. Chargaff and J. N. Davidson, eds., *The Nucleic Acids.* Vol. 1, pp. 267–284. Academic Press, New York.

Smith, J. D., Freeman, G., Vogt, M., and Dulbecco, R. 1960. The nucleic acid of polyoma virus. *Virology* **12**, 185–196.

Smith, J. D., and Markham, R. 1950. Chromatographic studies on nucleic acids. 2. The quantitative analysis of ribonucleic acids. *Biochem. J.* **46**, 509–513.

Smith, K. C., and Allen, F. W. 1953. The liberation of polynucleotides by the alkaline hydrolysis of ribonucleic acid from yeast. *J. Am. Chem. Soc.* **75**, 2131–2133.

Smith, K. M. 1967. *Insect Virology.* Academic Press, New York.

Smith, K. M. 1971. The viruses causing the polyhedroses and granuloses of insects. *In* K. Maramorosch and E. Kurstak, eds., *Comparative Virology.* pp. 479–507. Academic Press, New York.

Smith, W., Belyavin, G., and Sheffield, F. W. 1955. The host-tissue component of influenza viruses. *Proc. Roy. Soc.* **B143**, 504–522.

Snell, D. T., and Offord, R. E. 1972. The amino acid sequence of the B-protein of bacteriophage ZJ-2. *Biochem. J.* **127**, 167–178.

Sober, H. A., ed. 1970. *Handbook of Biochemistry.* Chemical Rubber Company, Cleveland.

Solymosy, F., Fedorcsák, I., Gulyás, A., Farkas, G. L., and Ehrenberg, L. 1968. A new method based on the use of diethyl pyrocarbonate as a nuclease inhibitor for the extraction of undegraded nucleic acid from plant tissues. *European J. Biochem.* **5**, 520–527.

Spackman, D. H. 1967. Amino acid analysis and related procedures: Accelerated methods. *In* C. H. W. Hirs, ed., *Methods in Enzymology*. Vol. 11, pp. 3–15. Academic Press, New York.

Spackman, D. H., Stein, W. H., and Moore, S. 1958. Automatic recording apparatus for use in the chromatography of amino acids. *Anal. Chem.* **30**, 1190–1206.

Spear, P. G., and Roizman, B. 1972. Proteins specified by herpes simplex virus. V. Purification and structural proteins of the herpesvirion. *J. Virol.* **9**, 143–159.

Speyer, J. F., Lengyel, P., Basilio, C., and Ochoa, S. 1962a. Synthetic polynucleotides and the amino acid code. II. *Proc. Natl. Acad. Sci. U.S.* **48**, 63–68.

Speyer, J. F., Lengyel, P., Basilio, C., and Ochoa, S. 1962b. Synthetic polynucleotides and the amino acid code. IV. *Proc. Natl. Acad. Sci. U.S.* **48**, 441–448.

Spiegelman, S., Pace, N. R., Mills, D. R., Levisohn, R., Eikhom, T. S., Taylor, M. M., Peterson, R. L., and Bishop, D. H. L. 1968. The mechanism of RNA replication. *Cold Spring Harbor Symp. Quant. Biol.* **33**, 101–124.

Spies, J. R., and Chambers, D. C. 1949. Chemical determination of tryptophan in proteins. *Anal. Chem.* **21**, 1249–1266.

Sreenivasaya, M., and Pirie, N. W. 1938. The disintegration of tobacco mosaic virus preparations with sodium dodecyl sulphate. *Biochem. J.* **32**, 1707–1710.

Stace-Smith, R. 1970. Tobacco ringspot virus. *C.M.I./A.A.B. Descriptions of Plant Viruses No. 17.*

Staehelin, M. 1958. Reaction of tobacco mosaic virus nucleic acid with formaldehyde. *Biochim. Biophys. Acta* **29**, 410–417.

Staehelin, M. 1959. Inactivation of virus nucleic acid with glyoxal derivatives. *Biochim. Biophys. Acta* **31**, 448–454.

Staehelin, M. 1960. Chemical modifications of virus infectivity: Reactions of tobacco mosaic virus and its nucleic acid. *Experientia* **16**, 473–483.

Stahl, F. W., Crasemann, J. M., Okun, L., Fox, E., and Laird, C. 1961. Radiation sensitivity of bacteriophage containing 5-bromodeoxyuridine. *Virology* **13**, 98–104.

Stanley, W. M. 1934a. Chemical studies on the virus of tobacco mosaic. I. Some effects of trypsin. *Phytopathology* **24**, 1055–1085.

Stanley, W. M. 1934b. Chemical studies on the virus of tobacco mosaic. II. The proteolytic action of pepsin. *Phytopathology* **24**, 1269–1289.

Stanley, W. M. 1935. Isolation of a crystalline protein possessing the properties of tobacco mosaic virus. *Science* **81**, 644–645.

Stanley, W. M. 1936. Chemical studies on the virus of tobacco mosaic. VI. The isolation from diseased turkish tobacco plants of a crystalline

protein possessing the properties of tobacco mosaic virus. *Phytopathology* **26**, 305–320.
Stanley, W. M. 1937. Chemical studies on the virus of tobacco mosaic. VIII. The isolation of a crystalline protein possessing the properties of aucuba mosaic virus. *J. Biol. Chem.* **117**, 325–340.
Stanley, W. M. 1938. Biochemistry and biophysics of viruses. I. Inactivation of viruses by different agents. *In* R. Doerr and C. Hallauer, eds., *Handbuch der Virusforschung, erste Hälfte.* pp. 447–458. Julius Springer, Vienna.
Stanley, W. M. 1939. The architecture of viruses. *Physiol. Rev.* **19**, 524–556.
Stanley, W. M. 1940. Purification of tomato bushy stunt virus by differential centrifugation. *J. Biol. Chem.* **135**, 437–454.
Stanley, W. M., and Lauffer, M. A. 1939. Disintegration of tobacco mosaic virus in urea solutions. *Science* **89**, 345–347.
Stanley, W. M., Lauffer, M. A., and Williams, R. C. 1959. Chemical and physical properties of viruses. In *Viral and Rickettsial Infections of Man.* pp. 11–48. Lippincott, Philadelphia.
Stark, G. R. 1972. Use of cyanate for determining NH_2-terminal residues in protein. *In* C. H. W. Hirs and S. N. Timasheff, eds., *Methods in Enzymology.* Vol. 25, pp. 103–120. Academic Press, New York.
Stark, G. R., and Smyth, D. G. 1963. The use of cyanate for the determination of NH_2-terminal residues in proteins. *J. Biol. Chem.* **238**, 214–226.
Steere, R. L. 1956. Purification and properties of tobacco ringspot virus. *Phytopathology* **46**, 60–69.
Steere, R. L. 1959. The purification of plant viruses. *Advan. Virus. Res.* **6**, 1–73.
Steere, R. L., and Ackers, G. K. 1962. Purification and separation of tobacco mosaic virus and southern bean mosaic virus by agar gel filtration. *Nature* **194**, 114–116.
Steere, R. L., and Williams, R. C. 1953. Identification of crystalline inclusion bodies extracted intact from plant cells infected with tobacco mosaic virus. *Am J. Bot.* **40**, 81–84.
Steinschneider, A., and Fraenkel-Conrat, H. 1966. Studies of nucleotide sequences in tobacco mosaic virus ribonucleic acid. III. Periodate oxidation and semicarbazone formation. *Biochemistry* **5**, 2729–2734.
Steitz, J. A. 1969. Polypeptide chain initiation: Nucleotide sequences of the three ribosomal binding sites. *Nature* **224**, 957–964.
Stent, G. S. 1971. *Molecular Genetics. An Introductory Narrative.* Chap. 13. W. H. Freeman, San Francisco.
Stent, G. S., and Fuerst, C. R. 1955. Inactivation of bacteriophages by decay of incorporated radioactive phosphorus. *J. Gen. Physiol.* **38**, 441–458.
Stent, G. S., and Fuerst, C. R. 1960. Genetic and physiological effects of the decay of incorporated radioactive phosphorus in bacterial viruses and bacteria. *Advan. Biol. Med. Phys.* **7**, 1–75.
Stoltzfus, C. M., and Rueckert, R. R. 1972. Capsid polypeptides of mouse-

Elberfeld virus. I. Amino acid compositions and molar ratios in the virion. *J. Virol.* **10**, 347–355.

Strauss, J. H., Jr., Burge, B. W., and Darnell, J. E. 1970. Carbohydrate content of the membrane protein of Sindbis virus. *J. Molecular Biol.* **47**, 437–448.

Streisinger, G. Y., Okada, Y., Emrich, J., Newton, J., Tsugita, A., Terzaghi, E., and Inouye, M. 1966. Frameshift mutations and the genetic code. *Cold Spring Harbor Symp. Quant. Biol.* **31**, 77–84.

Stromberg, K. 1972. Surface active agents for isolation of the core component of avian myeloblastosis virus. *J. Virol.* **9**, 684–697.

Stubbs, J. D., and Kaesberg, P. 1964. A protein subunit of bromegrass mosaic virus. *J. Molecular Biol.* **8**, 314–323.

Studier, F. W. 1965. Sedimentation studies of the size and shape of DNA. *J. Molecular Biol.* **11**, 373–390.

Studier, F. W. 1973. Analysis of bacteriophage T7 early RNAs and proteins on slab gels. *J. Molecular Biol.* **79**, 237–248.

Subak-Sharpe, H., Burk, R. R., Crawford, L. V., Morrison, J. M., Hay, J., and Keir, H. M. 1966. An approach to evolutionary relationships of mammalian DNA viruses through analysis of the pattern of nearest neighbor base sequences. *Cold Spring Harbor Symp. Quant. Biol.* **31**, 737–748.

Sugiyama, T. 1962. Identification of the terminal nucleosides of TMV-RNA. Ph.D. thesis, University of California, Berkeley.

Sugiyama, T., and Fraenkel-Conrat, H. 1961a. Identification of 5′ linked adenosine as end group of TMV-RNA. *Proc. Natl. Acad. Sci. U.S.* **47**, 1393–1397.

Sugiyama, T., and Fraenkel-Conrat, H. 1961b. Terminal structure of tobacco mosaic virus ribonucleic acid. Abstracts, National Meeting of the American Chemical Society, Chicago, Sept. 3–8, 1961, p. 550.

Sugiyama, T., and Fraenkel-Conrat, H. 1963. The end groups of tobacco mosaic virus RNA. II. Nature of the 3′-linked chain end in TMV and of both ends in four strains. *Biochemistry* **2**, 332–334.

Sugiyama, T., and Nakada, D. 1968. Translational control of bacteriophage MS2 RNA cistrons by MS2 coat protein: Polyacrylamide gel electrophoretic analysis of proteins synthesized *in vitro*. *J. Molecular Biol.* **31**, 431–440.

Summers, W. C., and Szybalski, W. 1967. Gamma-irradiation of deoxyribonucleic acid in dilute solutions. II. Molecular mechanisms responsible for inactivation of phage, its transfecting DNA and of bacterial transforming activity. *J. Molecular Biol.* **26**, 227–235.

Suzuki, J., and Haselkorn, R. 1968. Studies on the 5′-terminus of turnip yellow mosaic virus RNA. *J. Molecular Biol.* **36**, 47–56.

Tabor, H., Tabor, C. W., and Rosenthal, S. M. 1961. The biochemistry of the polyamines. *Ann. Rev. Biochem.* **30**, 579–604.

Takahashi, W. N. 1948. Crystallization of squash mosaic virus. *Am. J. Bot.* **35**, 243–245.

Takahashi, W. N., and Ishii, M. 1952a. The formation of rod-shaped particles resembling tobacco mosaic virus by polymerization of a protein from mosaic-diseased tobacco leaves. *Phytopathology* **42**, 690–691.

Takahashi, W. N., and Ishii, M. 1952b. An abnormal protein associated with tobacco mosaic virus infections. *Nature* **169**, 419–420.

Takahashi, W. N., and Ishii, M. 1953. A macromolecular protein associated with tobacco mosaic virus infection: Its isolation and properties. *Am J. Bot.* **40**, 85–90.

Takemori, N., and Nomura, S. 1960. Mutation of polioviruses with respect to size of plaque. II. Reverse mutation of minute plaque mutant. *Virology* **12**, 171–184.

Takemoto, K. 1954. Use of a cation exchange resin for isolation of influenza virus. *Proc. Soc. Exptl. Biol. Med.* **85**, 670–672.

Tan, K. B., and Sokol, F. 1972. Structural proteins of simian virus 40: Phosphoproteins. *J. Virol.* **10**, 985–994.

Taussig, A., and Creaser, E. H. 1957. Chromatographic purification for T2r bacteriophage. *Biochim. Biophys. Acta* **24**, 448–449.

Taylor, A. R. 1944. Chemical analysis of the influenza viruses A (PR8) strain and B (Lee strain) and the swine influenza virus. *J. Biol. Chem.* **153**, 675–686.

Taylor, J. F. 1953. The isolation of proteins. *In* H. Neurath and K. Bailey, eds., *The Proteins.* pp. 1–85. Academic Press, New York.

Temin, H. M. 1970. RNA-dependent DNA polymerase in virions of Rous sarcoma virus. *Nature* **226**, 1211–1213.

Temin, H. M. 1972. The RNA tumor viruses—Background and foreground. *Proc. Natl. Acad. Sci. U.S.* **69**, 1016–1020.

Temin, H. M., and Mizutani, S. 1970. RNA-dependent DNA polymerase in virions of Rous sarcoma virus. *Nature* **226**, 1211–1213.

Terzaghi, E., Okada, Y., Streisinger, G., Emrich, J., Inouye, M., and Tsugita, A. 1966. Change of a sequence of amino acids in phage T4 lysozyme by acridine-induced mutations. *Proc. Natl. Acad. Sci. U.S.* **56**, 500–507.

Tessman, I., Ishiwa, H., and Kumar, S. 1965. Mutagenic effects of hydroxylamine *in vivo*. *Science* **148**, 507–508.

Thomas, C. A., Jr., and MacHattie, L. A. 1967. The anatomy of viral DNA molecules. II. *Ann. Rev. Biochem.* **36**, 485–518.

Thomas, R. S. 1961. The chemical composition and particle weight of *Tipula iridescent* virus. *Virology* **14**, 240–252.

Timasheff, S. N., Witz. J., and Luzzati, V. 1961. The structure of high molecular weight ribonucleic acid in solution. A small-angle x-ray scattering study. *Biophys. J.* **1**, 525–537.

Tiselius, A. 1954. Chromatography of proteins on calcium phosphate columns. *Arkiv Kemi* **7**, 443–449.

Tiselius, A., Hjerten, S., and Jerstedt, S. 1965. Particle-sieve electrophoresis of viruses in polyacrylamide gels, exemplified by purification of turnip yellow mosaic virus. *Arch. Ges. Virusforsch.* **17**, 512–521.

Tomlinson, A. H., and Mac Callum, F. O. 1970. The effect of iododeoxyuridine on herpes simplex virus encephalitis in animals and man. *Ann. N. Y. Acad. Sci.* **173**, 20–28.

Tooze, J., ed. 1973. *The Molecular Biology of Tumor Viruses.* Cold Spring Harbor Laboratory, Cold Spring Harbor, N. Y.

Tooze, J., and Weber, K. 1967. Isolation and characterization of amber mutants of bacteriophage R17. *J. Molecular Biol.* **28**, 311–330.

Townsley, P. M. 1959. The electrophoretic analysis of plant virus preparations. *Can. J. Biochem. Physiol.* **37**, 119–126.

Townsley, P. M. 1961. Chromatography of tobacco mosaic virus on chitin columns. *Nature* **191**, 626.

Tsugita, A., and Fraenkel-Conrat, H. 1960. The amino acid composition and C-terminal sequence of a chemically evoked mutant of TMV. *Proc. Natl. Acad. Sci. U.S.* **46**, 636–642.

Tsugita, A., and Fraenkel-Conrat, H. 1962. Contributions from TMV studies to the problem of genetic information transfer and coding. *In* J. H. Taylor, ed., *Progress in Molecular Genetics.* pp. 477–520. Academic Press, New York.

Tsugita, A., Gish, D. T., Young, J., Fraenkel-Conrat, H., Knight, C. A., and Stanley, W. M. 1960. The complete amino acid sequence of the protein of tobacco mosaic virus. *Proc. Natl. Acad. Sci. U.S.* **46**, 1463–1469.

Tsugita, A., and Hirashima, A. 1972. Isolation and properties of virus proteins. *In* C. I. Kado and H. O. Agrawal, eds., *Principles and Techniques in Plant Virology.* pp. 413–443. Van Nostrand-Reinhold, New York.

Tung, J.-S., and Knight, C. A. 1972a. The coat protein subunits of cucumber viruses 3 and 4 and a comparison of methods for determining their molecular weights. *Virology* **48**, 574–581.

Tung, J.-S., and Knight, C. A. 1972b. The coat protein subunits of potato virus X and white clover mosaic virus, a comparison of methods for determining their molecular weights and some *in situ* degradation products of potato virus X protein. *Virology* **49**, 214–223.

Tung, J.-S., and Knight, C. A. 1972c. Relative importance of some factors affecting the electrophoretic migration of proteins in sodium dodecyl sulfate-polyacrylamide gels. *An. Biochem.* **48**, 153–163.

Uyemoto, J. K., and Grogan, R. G. 1969. Chemical characterization of tobacco necrosis and satellite viruses. *Virology* **39**, 79–89.

Vago, C., and Bergoin, M. 1968. Viruses of invertebrates. *Advan. Virus Res.* **13**, 247–303.

Vaheri, A., and Hovi, T. 1972. Structural proteins and subunits of rubella virus. *J. Virol.* **9**, 10–16.

van Iterson, G., Jr., den Dooren de Jong, L. E., and Kluyver, A. J. 1940. *Martinus Willem Beijerinck—His Life and His Work.* pp. 118–121. Martinus Nijhoff, The Hague.

van Ravenswaay-Claasen, J. C., van Leeuwen, A. B. J., Duijts, G. A. H., and Bosch, L. 1967. *In vitro* translation of alfalfa mosaic virus RNA. *J. Molecular Biol.* **23**, 535–544.

van Regenmortel, M. H. V. 1964. Separation of an antigenic plant protein from preparations of plant viruses. *Phytopathology* **54**, 282–289.

van Regenmortel, M. H. V. 1966. Plant virus serology. *Advan. Virus Res.* **12**, 207–271.

van Regenmortel, M. H. V. 1967. Biochemical and biophysical properties of cucumber mosaic virus. *Virology* **31**, 391–396.

van Regenmortel, M. H. V. 1972. Electrophoresis. *In* C. I. Kado and H. O. Agrawal, eds., *Principles and Techniques in Plant Virology.* pp. 390–412. Van Nostrand-Reinhold, New York.

van Regenmortel, M. H. V., Hendry, D. A., Baltz, T. 1972. A re-examination of the molecular size of cucumber mosaic virus and its coat protein. *Virology* **49**, 647–653.

van Slyke, D. D. 1912. The quantitative determination of aliphatic amino groups. II. *J. Biol. Chem.* **12**, 275–284.

Venekamp, J. H. 1972. Chromatographic purification of plant viruses. *In* C. I. Kado and H. O. Agrawal, eds., *Principles and Techniques in Plant Virology.* pp. 369–389. Van Nostrand-Reinhold, New York.

Vidaver, A. K., Koski, R. K., and Van Etten, J. L. 1973. Bacteriophage Ø6: A lipid-containing virus of *Pseudomonas phaseolicola. J. Virol.* **11**, 799–805.

Vielmetter, W., and Schuster, H. 1960. Die Basenspezifität bei der Induktion von Mutationen durch salpetrige Säure in Phagen T2. *Z. Naturforsch.* **15b**, 304–311.

Vinograd, J., and Hearst, J. E. 1962. Equilibrium sedimentation of macromolecules and viruses in a density gradient. *Fortschr. Chem. Org. Naturstoffe* **20**, 372–422.

Vinson, C. G. 1927. Precipitation of the virus of tobacco mosaic. *Science* **66**, 357–358.

Vinson, C. G., and Petre, A. W. 1929. Mosaic disease of tobacco. *Bot. Gaz.* **87**, 14–38.

Vinson, C. G., and Petre, A. W. 1931. Mosaic disease of tobacco. II. Activity of the virus precipitated by lead acetate. *Contr. Boyce Thompson Inst.* **3**, 131–145.

Viswanatha, T., Pallansch, M. J., and Liener, I. E. 1955. The inhibition of trypsin II. The effect of synthetic anionic detergents. *J. Biol. Chem.* **212**, 301–309.

Vogt, P. K. 1965. Avian tumor viruses. *Advan. Virus Res.* **11**, 293–385.
Vogt, P. K. 1973. The genome of avian RNA tumor viruses: A discussion of four models. *In* L. Sylvestri, ed., *Possible Episomes in Eukaryotes.* North Holland Publishing Co., Amsterdam.
von Tavel, P. 1959. The purification of tobacco mosaic virus by chromatography. *Arch. Biochem. Biophys.* **85**, 491–498.
Wacker, W. E. C., Gordon, M. P., and Huff, J. W. 1963. Metal content of tobacco mosaic virus and tobacco mosaic virus RNA. *Biochemistry* **2** 716–719.
de Wachter, R., and Fiers, W. 1969. Sequences at the 5′-terminus of bacteriophage Qβ RNA. *Nature* **221**, 233–235.
Wagner, R. R., Levy, A. H., and Smith, T. J. 1968. Techniques for the study of interferons in animal virus-cell systems. *In* K. Maramorosch and H. Koprowski, eds., *Methods in Virology,* Vol. 4, pp. 1–52. Academic Press, New York.
Wahl, R., Huppert, J., and Emerique-Blum, L. 1960. Production de phages par des "protoplastes" bacteriens infectes par des preparations d'acide desoxyribonucleique. *Compt. Rend.* **250**, 4227–4229.
Watkins, J. F. 1971a. Fusion of cells for virus studies and production of cell hybrids. *In* K. Maramorosch and H. Koprowski, eds., *Methods in Virology.* Vol. 5, pp. 1–32. Academic Press, New York.
Watkins, J. F. 1971b. Cell fusion in virology. *Perspectives Virol.* **7**, 159–178.
Watson, J. D. 1950. The properties of x-ray inactivated bacteriophage. *J. Bacteriol.* **60**, 697–718.
Watson, J. D. 1954. The structure of tobacco mosaic virus. I. X-ray evidence of a helical arrangement of subunits around the longitudinal axis. *Biochim. Biophys. Acta* **13**, 10–19.
Watson, J. D., and Crick, F. H. C. 1953a. A structure for deoxyribose nucleic acid. *Nature* **171**, 737–738.
Watson, J. D., and Crick, F. H. C. 1953b. The structure of DNA. *Cold Spring Harbor Symp. Quant. Biol.* **28**, 123–131.
Watson, J. D., and Littlefield, J. W. 1960. Some properties of DNA from Shope papilloma virus. *J. Molecular Biol.* **2**, 161–165.
Weber, K. 1967. Amino acid sequence studies on the tryptic peptides of the coat protein of the bacteriophage R17. *Biochemistry* **6**, 3144–3154.
Weber K., and Konigsberg, W. 1967. Amino acid sequence of the f2 coat protein. *J. Biol. Chem.* **242**, 3563–3578.
Weber, K., and Osborn, M. 1969. The reliability of molecular weight determinations by dodecyl sulfate-polyacrylamide gel electrophoresis. *J. Biol. Chem.* **244**, 4406–4412.
Webster, R. E., Engelhardt, D. L., Zinder, N. D., and Konigsberg, W. J. 1967. Amber mutants and chain termination *in vitro. J. Molecular Biol.* **29**, 27–43.
Webster, R. G. 1970. The structure and function of the haemagglutinin and

neuraminidase of influenza virus. *In* R. D. Barry and B. W. J. Mahy, eds., *The Biology of Large RNA Viruses*. pp. 53–74. Academic Press, New York.

Webster, R. G., and Darlington, R. W. 1969. Disruption of myxoviruses with Tween 20 and isolation of biologically active hemagglutinin and neuraminidase subunits. *J. Virol.* **4**, 182–187.

Wecker, E. 1957. Die Verteilung von ^{32}P im Virus der klassischen Geflügelpest bei verschiedenen Markierungsverfahren. *Z. Naturforsch.* **12b**, 208–210.

Wecker, E. 1959. The extraction of infectious virus nucleic acid with hot phenol. *Virology* **7**, 241–243.

Weigle, J. 1966. Assembly of phage lambda *in vitro*. *Proc. Natl. Acad. Sci. U.S.* **55**, 1462–1466.

Westmoreland, B. C., Szybalski, W., and Ris, H. 1969. Mapping of deletions and substitutions in heteroduplex DNA molecules of bacteriophage lambda by electron microscopy. *Science* **163**, 1343–1348.

Westphal, H., and Dulbecco, R. 1968. Viral DNA in polyoma- and SV40-transformed cell lines. *Proc. Natl. Acad. Sci. U.S.* **59**, 1158–1165.

Westphal, O., Luderitz, O., and Bister, F. 1952. Über die Extraktion von Bakterien mit Phenolwasser. *Z. Naturforsch.* **7b**, 148–155.

Whipple, H. E., ed. 1965. Antiviral substances. *Ann. N. Y. Acad. Sci.* **130**, 1–482.

Wilkins, M. H. F., Stokes, A. R., and Wilson, H. R. 1953. Molecular structure of deoxypentose nucleic acids. *Nature* **171**, 738–740.

Williams, R. C. 1953a. A method of freeze-drying for electron microscopy. *Exptl. Cell Res.* **4**, 188–201.

Williams, R. C. 1953b. The shapes and sizes of purified viruses as determined by electron microscopy. *Cold Spring Harbor Symp. Quant. Biol.* **18**, 185–195.

Williams, R. C. 1954. Electron microscopy of viruses. *Advan. Virus Res.* **2**, 183–239.

Williams. R. C., and Fisher, H. W. 1974. *An Electron Micrographic Atlas of Viruses*. Charles C. Thomas, Springfield, Ill.

Williams, R. C., Kass, S. J., and Knight C. A. 1960. Structure of Shope papilloma virus particles. *Virology* **12**, 48–58.

Williams, R. C., and Smith, K. M. 1958. The polyhedral form of the *Tipula iridescent* virus. *Biochim. Biophys. Acta* **28**, 464–469.

Williams, R. C., and Wyckoff, R. W. G. 1945. Electron shadow-micrography of virus particles. *Proc. Soc. Exptl. Biol. Med.* **58**, 265–270.

Wimmer, E., Chang, A. Y., Clark, J. M., Jr., and Reichmann, M. E. 1968. Sequence studies of satellite tobacco necrosis virus RNA: Isolation and characterization of a 5′-terminal trinucleotide. *J. Molecular Biol.* **38**, 59–73.

Wimmer, E., and Reichmann, M. E. 1969. Two 3′-terminal sequences in satellite tobacco necrosis virus RNA. *Nature* **221**, 1122–1126.

Wittmann-Liebold, B., and Wittmann, H. G. 1967. Coat proteins of strains of two RNA viruses: Comparison of their amino acid sequences. *Molecular Gen. Genetics* **100**, 358–363.

Wood, W. B., and Edgar, R. S. 1967. Building a bacterial virus. *Sci. Am.* **217**, 60–74.

Woody, B. R., and Knight, C. A. 1959. Peptide maps obtained with tryptic digests of the proteins of some strains of tobacco mosaic virus. *Virology* **9**, 359–374.

Wu, M., and Davidson, N. 1973. A technique for mapping transfer RNA genes by electron microscopy of hybrids of ferritin-labeled transfer RNA and DNA: The Ø80 hpsu$^{+,-}_{III}$ system. *J. Molecular Biol.* **78**, 1–21.

Wyatt, G. R. 1955. Separation of nucleic acid components by chromatography on filter paper. *In* E. Chargaff and J. N. Davidson, eds., *The Nucleic Acids.* Vol. 1, Ch. 7. Academic Press, New York.

Wyatt, G. R., and Cohen, S. S. 1952. A new pyrimidine base from bacteriophage nucleic acids. *Nature* **170**, 1072.

Wyatt, G. R., and Cohen, S. S. 1953. The bases of the nucleic acids of some bacterial and animal viruses: The occurrence of 5-hydroxymethylcytosine. *Biochem. J.* **55**, 774–782.

Wyckoff, R. W. G. 1937. An ultracentrifugal study of the stability of tobacco mosaic virus protein. *J. Biol. Chem.* **122**, 239–247.

Yamazaki, H., Bancroft, J., and Kaesberg, P. 1961. Biophysical studies of broad bean mottle virus. *Proc. Natl. Acad. Sci. U.S.* **47**, 979–983.

Yamazaki, H., and Kaesberg, P. 1961. Biophysical and biochemical properties of wild cucumber mosaic virus and of two related virus-like particles. *Biochim. Biophys. Acta* **51**, 9–18.

Yamazaki, H., and Kaesberg, P. 1963. Degradation of bromegrass mosaic virus with calcium chloride and the isolation of its protein and nucleic acid. *J. Molecular Biol.* **7**, 760–762.

Yoshiike, K. 1968. Studies on DNA from low-density particles of SV40. *Virology* **34**, 391–401.

Young, R. J., and Fraenkel-Conrat, H. 1970. The 3′-linked terminus of Qβ RNA. *Biochim. Biophys. Acta* **228**, 446–455.

Zamenhof, S. 1957. Properties of the transforming principles. *In* W. D. McElroy and B. Glass, eds., *The Chemical Basis of Heredity.* pp. 351–377. Johns Hopkins Press, Baltimore.

Ziff, E. B., Sedat, J. W., Galibert, F. 1973. Determination of the nucleotide sequence of a fragment of bacteriophage ØX 174 DNA. *Nature New Biol.* **241**, 34–37.

Zinder, N. D., and Cooper, S. 1964. Host-dependent mutants of the bacteriophage f2. 1. Isolation and preliminary classification. *Virology* **23**, 152–158.

INDEX

WILLIAM IRWIN THOMPSON

A stunning synthesis of speculative fantasy and myth, ISLANDS OUT OF TIME, by acclaimed cultural historian William Irwin Thompson, is a cosmic journey of the imagination. It recounts the final days of the legendary lost continent of Atlantis, whose end is brought about not by natural disaster but because the Atlanteans seek to overreach nature.

Thompson's multifaceted fantasia is as much an archaeology of the unconscious as it is a vision of an advanced civilization in its final flowering. The Silver Age of Atlantis is an era of high technology and higher aspiration, dominated by archetypes like the High Priest, ruler of the Theocracy and sponsor of an awesome scheme to consummate human evolution in a single, daring, irrevocable leap; Bran of Eiru, whose destiny it is to escape the prophesied end of Atlantis and to lead

a chosen few into a new age, bearing the seeds of memory and myth; Viracocha, musician of genius and the unwitting instrument of this escape; Brigita, Abbess of the One Thousand of Eiru, Viracocha's lover first on the spiritual plane and later on the physical; the Archon, Director of the Temple Institute, the indelibly Satanic personification of science run amok.

ISLANDS OUT OF TIME is high adventure of the mind and spirit, woven out of mythologies ranging from Mexico and South America to Ireland and the Middle East. It is also a study of the folly of transcendence and the nature of evil in science and religion which provides new understanding of ourselves, our past, and the precarious course of our imminent future.

William Irwin Thompson is the founding director of the Lindisfarne Association, a contemplative educational community with bases in Colorado, New York City, and Bern, Switzerland. He is the author of six previous books, including *Passages About Earth, Darkness and Scattered Light,* and *At the Edge of History,* which was nominated for a National Book Award.

Islands Out of Time

A MEMOIR OF THE LAST DAYS OF ATLANTIS

Also by William Irwin Thompson

THE IMAGINATION OF AN INSURRECTION
AT THE EDGE OF HISTORY
PASSAGES ABOUT EARTH
EVIL AND WORLD ORDER
DARKNESS AND SCATTERED LIGHT
THE TIME FALLING BODIES TAKE TO LIGHT
BLUE JADE FROM THE MORNING STAR

Islands Out of Time

A MEMOIR OF THE LAST DAYS OF ATLANTIS

A Metafiction by

WILLIAM IRWIN THOMPSON

The Dial Press/Doubleday
GARDEN CITY, NEW YORK
1985

Library of Congress Cataloging in Publication Data

Thompson, William Irwin.
Islands out of time.
I. Title.
PS3570.H64518 1985 813′.54
ISBN 0-385-19571-0
Library of Congress Catalog Card Number 85-1478

Printed in the United States of America

First printing

FOR JOHN AND
DAVID SPANGLER

Foreword

Fiction appeals to us because the ego is a fiction. This conventionally maintained identity we call the self is a matter of perspective, and if we look too closely, as when we look at the dots in a newspaper photograph, the recognizable image dissolves.

If we take a good look at our minds, either through Western introspection or Eastern *zazen*, we discover that there are large gaps between each thought, and in the interval between we cannot be certain at all "where" we were, or where the "I" had gone. In *zazen* particularly there is a curious reversal of *figure* and *ground* that comes about as the gap between each thought widens. The gap becomes the *figure*, and thoughts become the background, or rather a kind of background noise that has no absolute and essential *ground*. In that groundlessness which we uncover, we no longer fill up the *structure* of consciousness with new *content:* either with sensory perceptions or psychic visions; we immerse that structure of *knowing* in *being*, like someone taking off his or her clothes to float in the sea. As we return to the condition of conventional mind, we see how it had been possible to stretch one thought to the next, to ignore the gaps and make the sequence of thoughts appear to be a continuous line with a solidly inked-in identity.

It is a simply plotted, linear fiction, this ego of ours, but one does not have to be a zen master to discover just how fictional personal identity is, for one only has to try to write an autobiography. It soon becomes obvious that our memories of childhood experiences are adult constructions that we never experienced

in that way as children. The memories are not so much descriptions of the past as they are performances of the present; they are artifacts with an independent literary life of their own. And so we discover that the ego is a fiction in the act of creating a fictional narrative about the unfoldment of its "true" identity.

If autobiography, as a *recherche du temps perdu,* is a fiction, then a memoir of a past life on a lost continent in a mythical prehistory is even more of a fiction of identity. Or, perhaps, it would be better to call it a parody of identity, a parody of autobiography, a parody of prehistory, a parody of the genre of the science fiction novel.

People who believe in egos write novels with characters. And people who believe in the solidity of materialism write realistic novels. Not surprisingly, the rise of the novel is related to the historical period of the rise of the middle class and the expansion of industrial materialism. Every epoch has its dominant elite, and its dominant literary genre, and in these epochs the descriptions of these genres are so powerful that social reality soon begins to imitate art. The duke and the duchess take Don Quixote into their palace as an honored knight errant of old because they have read Part One of the book of his adventures. Even as far back as the time of Cervantes, the media could create reality. The instant feedback of Don Quixote's madness onto himself in play makes his obsession a shared fantasy, and the play of shared fantasies is what we know as *culture.* Whether the knight errant is Don Quixote or President Reagan, "the Great Communicator," the shared fantasy in which the media create reality is the paradoxical *nature* of *culture.*

In the age of aristocratic, land-owning warriors, from Homer to Ariosto, the cultures of Europe made the epic their dominant literary genre. In the age of industrial capitalism, books and money were printed, and gold and paintings became forms of accumulated wealth stored in banks and museums that, architecturally, tended to look alike. The middle classes believed in things and egos and liked novels that had characters and were "true to life." They were also brashly confident enough to think that they knew exactly what "true to life" meant.

Now we are in the period of transition from one age to an-

other, and from one elite to another. With contemporary society being shaped by physics, microelectronics, and genetic engineering, it looks as if Thatcher, Kohl, and Reagan are the sunset-effect of industrial capitalism. Like Don Quixote on Rocinante, they ride forth in their Rolls, Mercedes, and Lincolns right at the time when the age of the automobile is in its climactic finale. The forests, orchards, and vineyards are dying in Europe, and even though the air pollution could be immediately reduced by 18 percent if the Germans would simply impose a speed limit, the middle class consumers refuse to do so, for the car, like the sword of old, is a symbol of manhood and power. For a German-speaking people, raised on the fairy tales of Grimm, the forest is the dark place where the evil witch lives; it is a place of female control; but the highway is a place of masculine assertiveness; consequently a speed limit is not simply a matter of rational environmental management. But there is also an archetypal power to whine in the European folk-soul and unconscious, and as Europe continues to lose its orchards and vineyards, the shock will be profound, and some new form of ecological management will be inevitable. It looks as if the scientists are going to become the new governing elite, but what the dadaistic and anarchic Greens of Germany do not seem to realize yet is that ecological management will require computerized, authoritarian forms of control. The law will have to dictate just what kind of fuel is tolerable, how big an engine can be, and how fast a vehicle is permitted to go. The feedback of pollution on culture will have to be electronically coded; therefore, some form of electronics is likely to become the Hammurabian Code of the future.

Artists and social critics have been aware for some time that we were living in an informational society, so they have been expressing themselves in new forms that are not the old genre of the bourgeois novel with its attendant English Department criticism, of Eng. Lit. Inc. Elsewhere I have called this new genre *Wissenskunst;*[1] other critics seem to prefer the Greek-Latin macaronic term of *metafiction,*[2] but from whatever point of view one adopts to look at our new society, it is becoming increasingly clear that realistic characters locally defined with a

sense of place may indeed map on to Faulkner's Yoknapatawpha County, but they do not embody the new electronic-aerospace culture of the American Sunbelt. What is the individual and where is the place for a person plugged into an *Apple* and networked up to "The Source" in a conference that is not occurring in real-time or real-space? *Per-sona* in Latin means to sound through, and the term refers back to the masks of the old medium of classical Graeco-Roman theater from which the actor's voice sounded through. Who is the being behind the mask? Who is the being behind the terminal? Literary man had a stable character he acted out through a lifetime, through a *story* of rags to riches; but electronic humanity does not have a stable identity, or even a stable sexuality. He-she flips channels and sexes like a Boy George with a remote-control TV switcher in its hand. Esoteric individuals in our cybernetic era may watch their minds in *zazen,* but exoteric and electronically conditioned individuals watch themselves experience identity as the shared response of the medium to entry. Both the esoteric and the exoteric persons begin to experience identity as something that the old Buddhists liked to call, not eternal essence, but "co-dependent origination."

Artists have always had a way of anticipating technological innovations, and the multi-dimensionality and interpenetrating spaces that are now so common in music video, thanks to video synthesizers and computers, were first anticipated on canvas by painters like Magritte and Escher. Similarly, James Joyce, in *Finnegans Wake,* was one of the first novelists to replace characters with patterns. As "the pattern that connects" replaced the character that holds identity, the barriers between fiction and nonfiction broke down. Nonfiction, like Tom Wolfe's *Electric Kool-Aid Acid Test,* or my own *At the Edge of History,* became more narrative than expository. At the same time that the barriers between fiction and nonfiction were broken down, the barriers between political activity and journalistic reporting dissolved. And with the dissolution of these boundaries also came a loss of orientation, as Left and Right lost their old meanings, for the gonzo journalism of Hunter Thompson and the media performances of Ronald Reagan are *structurally* both emanations

of the same electronic culture. They are both pure Southern California.

We had been warned about all of this by McLuhan in Canada, Borges in Argentina, and Escher in the Netherlands, but Americans are quintessentially a practical, industrial, and middle class people; they are not inclined either to Buddhist Emptiness or European epistemological complexity. They always look at the *content* but rarely see the *structure,* so they go to Disneyland and see the Midwestern town. They look at Reagan, our first divorced President, who never goes to church, and they see him as a small town Midwesterner who is a pillar of Southern fundamentalism.

In my earlier books and public lectures I have said that "all scholarship is disguised autobiography," so it occurred to me that a disguised autobiography might provide an interesting approach to scholarship. An imaginary autobiography of a past life on a lost continent requires no historical research and gives the writer a degree of freedom that no historical novelist or seriously responsible autobiographer could ever hope to have, so the task was compelling.

Precisely because the ego is a fiction, identity is a guise. Conventionally, we like to think of heroes versus villains, but the *characters* are not the real descriptions of identity. The narrative itself is our character, for we as human beings are a story of beauty and beast, hero and villain. That is why Blake was right when he said that when Milton wrote about God in *Paradise Lost,* his spirit lay in chains, but when he wrote about Satan, his spirit soared. I find that in this archaeology of the unconscious I too became more interested in the voices of evil in the narrative, and so I must own up to the fact that the villains speak for me as much as the stupid genius who thinks the story is about him. In this chamber work for five voices, there are *personae,* but they are not so much characters as instruments of myth.

I apologize for having had to frame this narrative, to put it into a context, but I found that some readers wished to put it into the old historical context of the realistic novel, and this is a waste of time, even with a work about lost time. To put one

historical context inside another historical context seems appropriate, especially since Atlantis is not admitted into the narratives of scientific prehistory, but that may be why it serves so well as a metaphoric description of scientific post-history.

May 10, 1985
Bern, Switzerland

Notes

1. *See the author's* Passages About Earth *(New York, Harper & Row, 1974), p. 3, for a definition of* Wissenskunst; *also his* Evil and World Order *(New York, Harper & Row, 1976), p. 77; and his* The Time Falling Bodies Take to Light *(New York, St. Martin's Press, 1981), pp. 4, 248.*
2. *William H. Gass coined the term "metafiction" in his essay "Philosophy and the Form of Fiction" in 1971. See Gass's* Fiction and the Figures of Life *(Boston, Nonpareil Books, 1978), p. 25. The term has been picked up and put into general circulation by Patricia Waugh in her recent study,* Metafiction *(New York, Methuen, 1984). Unfortunately, Dr. Waugh has added lead to the silver of Gass's brightly reflective prose. Professors of English Literature now seem so intent on protecting their employment and justifying their existence to educational management that they have retooled themselves to reappear in the marketplace as behavioral scientists. They now write as badly as educational administrators and sociologists. Consider, for example, Waugh's use of such words as "foregrounded" and "problematizes" in the following definition of metafiction:*

   In all of these what is foregrounded is the writing of the text as the most fundamentally problematic aspect of that text. Although metafiction is just one form of post-modernism, nearly all contemporary experimental writing displays some *explicit metafictional strategies. Any text that draws the reader's attention to its process of construction by frustrating his or her conventional expectations of meaning and closure problematizes more or less explicitly the ways in which narrative codes—whether "literary" or "social"—artificially construct apparently "real" and imaginary worlds in terms of particular ideologies while presenting these as transparently "natural" and "eternal." (Waugh, p. 22.)*

Both the philosophical playfulness of Gass's thinking and the artistic delight of his prose are lost in Waugh's academic approach. The kind of metafiction she seems to favor and promote is simply a professional and technical self-conscious discourse between the edges of genres and the margins of texts. Ultimately, the accumulating ironies of her favored self-indulgent authors end in a lattice of negations that permit the critic to have the last word by surrounding the artist with the containment of the dying cosmology, the careerist nihilism of the bureaucratic professor. Wissenskunst, *by contrast, plays with the narratives of science; it does not try to imitate science or chase after a scientistic professionalism. The* Wissenskünstler *lives at the edge of science the way the bard lived at the edge of kingly power. Always at an edge with socially accepted definitions of reality, the* Wissenskünstler *becomes a "Juggler of Our Lady" who found no room for his work in the corridors of the monks and so juggled alone at midnight in front of the statue. The juggler has to play the fool among experts to find a new place for the archaic in the contemporary. Metafiction need not, therefore, be merely an ironic commentary on its own composition; it can also become a shift of consciousness, a movement from* noia *to* metanoia, *from travel to homecoming, from domestic habituation to cosmic astonishment.*

Islands Out of Time

A MEMOIR OF THE LAST DAYS OF ATLANTIS

1

Mostly the mind dwells on the Tower. Even in other incarnations millennia afterward, the mind is drawn back to that serenely wrought instrument of perfected terror, and only then brought back to those four who stand out against the backdrop of the brown volcanic rocks to remain more durable than temple stones or the rubble of broken desires. Enduring into other times, other cultures, they persist like some unknown architecture of the mind. In all the civilizations that were to follow there would be legends and esoteric murmurings, or sometimes even an obscure compulsion to re-create in form what had been stricken from history. And always, they would be there: unfree, untransformed, still seeking the ultimate deliverance, still unforgiving of the angelic violation that had raped their spirits and thrust their souls screaming into animal bodies.

It is too late to blame anyone. The four—no, call it five—are an entirety, even a single entity. Time has at least accomplished that much in writing. Still, the terror, the incomprehension of Evil, remains. Cresting a hill in San Francisco, you stop mechanically at a traffic light, and then become taken over by the horror of involuntary and impersonal memory at the sight of the Transamerica Pyramid. Of course, the builders cannot admit, nor can the members of the academy accept, the pull of racial memory; and so a dumb and brutal doom compells them to act it out again: to find great seismic faults in the earth and there of all places choose to thrust momentary towers into their acrid, rotting sky. More than a mere quotation, the Transamerica Pyramid is an evocation of the Tower: you have only to set the solar crystal into its crown to conjure it all up again.

The prehistoric bay, however, was not so large or irregular in shape. The remains of the collapsed cone of an ancient volcano,

the bay was almost circular in form where the land with its crescent wings gave protection from the winds to favor its deep natural harbor. Upon these wings were the steep ledges into which had been set all the white buildings of the temple city of the High Priest. In the center of these mustard brown volcanic ledges stood the Great Tower itself, and immediately below it, the Palace of the High Priest and the great hypostyle hall. To the west of the southward facing Tower and Palace were the numerous buildings of the Temple Institute, which extended to the westernmost cliffs above the sea. And there at the extreme edge of the cliff stood the large double doors of copper that led into the Archon's Laboratory of Incarnation. Despite the spray of the salt seawinds, the doors were never allowed to cover with patina, but were polished every morning by the guards themselves. Sometimes at sunset they would glow a lurid red in the dying light, and on nights of the full moon they would glisten like the armor of soldiers hiding in the night.

On the opposite side of the Great Tower, to the east of the Palace, were the buildings and cloistered gardens of the Sacred College. The largest of these enclosed gardens separated the hypostyle hall from the residential quarters and the great room of the Convocation. To the extreme east was the small chapel with its seventy-two seats for the Members of the Sacred College.

Below the broad stone-paved street that connected the College Gate, the hypostyle hall, and the copper doors of the Institute, were the administrative buildings, and below these, the storehouses which extended to the quais at the water's edge. The storehouses, which had practically taken over the old capital, were here not allowed to extend to the center of the bay, which was reserved for the processional ramp that climbed from the moorings of the barges to the portals of the hypostyle hall itself. The boats would dock to east or west, depending on whether their affairs took them to the Institute or the College, but the center was always kept open for the processional barge that would take the High Priest to his garuda.

Lovely as the barges and sailing ships were, the greatest sight of all was to watch the great vessels of the air, the silver garudas,

set down in the middle of the harbor and then glide swanlike to their appointed rest.

The peninsular City of the Tower was the creation of the High Priest himself. As the Empire had grown, commerce had grown with it, and the causeways and canals had become littered with the debris of merchants and farmers. With the extension of the Empire to the colonies in both the Eastern and Western continents, the Military too had pressed its demands for barracks, docks, and storehouses, until at last the imperial capital was more of a cluttered bazaar than the seat of the High Priest and the Mother City of the Theocracy. It was then that the High Priest conceived the vision of the City of Knowledge to be built on the isolated peninsula to the south. Here he imagined that the Theocracy could express itself in utmost purity. Here the traditional arts and the new sciences could be lifted to great heights in the very consummation of Atlantean civilization itself. To encourage the former, the High Priest created the Sacred College, and for the latter he established the Temple Institute. With these two strong institutions as his right and left arms he believed he could restrain the disordered growth of the military and the commercial classes to hold all within the form of the ancient and sublime Theocracy. And to that end the Temple Institute served the High Priest with loyalty and skill, for from its lore of herbs and medicines had come a new knowledge of drugs and poisons that made the ingeniously catapulted darts of the Atlanteans the most feared weapons in the world.

To abandon the old capital and to build anew the center of civilization was not only visionary, but costly as well. Provinces that had before participated in trade had become colonies providing tribute; but as great as this burden was, nothing caused as much hatred as the small, but steady stream of pregnant women and children that was led in small groups through the large double doors of copper into the Institute.

On one particular day of the final summer there was more than the usual flow of traffic in the bay. A Solemn Assembly had been called, for it had been decided by the High Priest that if the Provincial Governors knew how the City of Knowledge

itself had become a vessel for the evolutionary transformation of the human race, then they could serve with renewed enthusiasm to direct the sullen peoples toward humanity's appointed destination. And so a call was sent out for all the Governors to meet with the Supreme Council of the High Priest, the Members of the Sacred College, and the faculty of the Temple Institute.

The Solemn Assembly must have been a source of deep anxiety for Bran, the Captain of the largest garuda assigned to the Institute. As a provincial himself, an islander from Eiru in the Remnants of Hyperborea, Bran's sympathies were not always imperial, but he loved to fly, and there were no other flying vessels anywhere else in the world. It was only in the last year, when he had been set to transporting the pregnant women and children that his work had grown intolerable. Always before the conspiracy of the Society of the 144 had kept his spirits up with the thought that he had been called to flying to fulfill a great destiny: to escape the end of civilization that Abaris had prophesied and to carry the chosen ones into a new age. When he had only to wait and spend his time carrying officials from one colony to another, he could be patient; but now that he and his garuda had been assigned to the Institute, waiting on prophecies had become impossible.

The Society of the 144, however, did not seem to mind the wait. They would irritate Bran with their servile meticulousness as they went over, once again, their lists of secret storehouses, their roster of the chosen technicians and horticulturalists the colony would need to survive, and their maps of the chosen location by a large lake high in the mountains of the Southern Continent. But they had only to look at their plans, they did not, like Bran, have to look into the dark obsidian eyes of the Governor of Anahuac.

Bran would dwell on that scene often in the attempts he would make later to chronicle the last days. He would try to recreate the mood of that silent crowd of people standing in the clearing as the Provincial Governor of Anahuac himself came to protest the conscription of his subjects.

The day had been unbearably hot and clouds of steam hung

just above the red flowers on top the mimosa trees. Bran had wanted to rip off his silk uniform, but it had stuck to his skin like a plaster on a festering wound. The sweat had streamed down into his eyes, blinding him with salt, until finally he had tied his ceremonial scarf around his forehead as if it were a mere rag. Whatever authority he had once felt left him as he looked down at the small man with his amethyst medallion of office on his bare chest. Once the mythic quality of being the tall blond god descending from the skies and bringing the natives the mysteries of corn and tubers had appealed to him, but that was in a time when he was giving and not taking; now that little man seemed to tower over him.

The Governor stood there looking up with those alien black eyes under a severely straight edge of dark hair that covered his entire forehead. When he spoke there was no trace of rage or passion, only the stiff formality of words that were alien to his tongue. It was as if he had become a figure in some tragedy and his face were a mask:

"You are a disciple of Abaris. Under your robe you wear the emblem of the winged serpent. How can you do this?"

"It will come to an end *soon,*" Bran said with all the emotion missing from the Governor's voice. "Everything will come to an end soon, but we need this vessel if Abaris' prophecies are to be fulfilled. If I revolt or give myself away now, the whole plan will be lost!"

"What good will your plans do for those already inside your vessel, Captain? If Abaris is willing to sacrifice them for his plan, then he is no different from your High Priest. But I did not come here to argue. I came to go back with you. I will go with you now to your Solemn Assembly and there in front of all, I will kill your High Priest. In the civil war that is sure to follow, the Empire will not have time to bother about Anahuac. But when the army tries to return here, my people will be ready. This will serve more than waiting for the end of the world."

The Governor passed by Bran in a silent contempt that wounded him more deeply than any attack. He had taken the lead, and by turning to follow him into the vessel, Bran felt that

at least he was doing something by ferrying this assassin into the beginning of civil war.

The Governor remained silent throughout the journey, nor did he speak as he left the ship to follow the pages up the processional ramp to the Palace of the Provincial Governors. As Bran watched him disappear, his words did not go away but continued to weigh upon Bran's conscience.

If civil war were to come, Bran wondered, would his garuda be taken from him? Was he about to lose the one thing that had kept him holding on to Abaris' plan? But if civil war were about to break out, that itself could mean that the final days were at hand. But why hadn't the leader appeared? Abaris had been so insistent on that point: in the final weeks immediately before the catastrophe another man who had been trained by Abaris would appear. Brigita would be the first to recognize him, but Bran was to know him from the passwords.

Then more than ever did Bran feel isolated and alone. With Abaris dead and Brigita on the other side of the sea in Eiru, there was no one who could tell him what to do. Should he report the Governor to stop the assassination and civil war that could result in the loss of his ship? Or should he move to escape with the 144 precisely at the moment of confusion?

Bran looked up at the City as if he could read his instructions from that citadel of knowledge, and then he dropped his eyes to his feet in disgust and began to walk toward the ramp that would take him up to the Institute. There was little in the City of Knowledge that he cared to look at anymore. The architectural power of the City had long since worn thin, and in the rigid symmetries of Institute and Sacred College, there was not enough diversity and life to hold him. All the ramps had been so constructed that only the top of the Tower could be seen as one climbed up from the harbor to the Broad Street, and this perspective made the Tower appear not to rise from the earth but to float above, weightless in the air. It was not until one came to the Broad Street itself that one could take in the whole expanse of the City, from the copper doors of the Institute to the west, to the Chapel of the Sacred College to the east. But Bran had seen it all before a thousand times; he knew every little vista at the

turns of the ramps that celebrated the capital in prearranged instances of self-praise.

Bran had always preferred the old capital. Its architectural styles spanned the millennia, and its carefully planned symmetries were caught in a tussle with spontaneity, like a celibate caught in bed with a woman. Everywhere that order tried to transcend into perfection, just there disorder camped at its edges. All the races and occupations of the world flowed through the dirty streets and cluttered canals of the ancient city, and Bran knew how to find his way by smell alone: from fish market to flower stalls, from the sweet thick rot of the slaughterhouse to the hilltop temples where the tropical winds from the south were subtle and far more mysterious than the heavy and obvious incense the priests used to obscure the candlelight.

This City of Knowledge was a priest's dream of perfection, order, and control, and so Bran kept his eyes to the ground and pounded his decisions down into the stones with his feet: he would not report the Governor of Anahuac; he would not give the signal to move to the 144; he would simply wait to see what the Solemn Assembly would bring.

In disgust at his own powerlessness, Bran broke out into a run as he cleared the last three steps and turned to the left to come out of the ramp and onto the Broad Street. But the ramps were designed for a slow and meditative ascent and not an athletic race, and no sooner had he turned the corner than he smashed into a dignitary and sent the man sprawling to the ground. Bran recognized at once from the pale blue robes that he had collided with one of the chosen few, one of the seventy-two Members of the Sacred College.

"You clumsy ass! You wear the uniform of a Captain of a garuda, but you don't even look where you're going. Pity the poor wretches that have to fly with you!"

That hurt. And for that remark alone, Bran wanted to strike out at him, but the Members of the Sacred College were the aristocracy of aristocracies and outranked everyone except the Supreme Council of the High Priest.

"Please forgive me, Your Grace. I'm afraid that birds that can be graceful in the air are often clumsy on earth."

The dignitary reached up to take Bran's extended hand and smiled back at him:

"I like that! That's a good image, Captain. 'The bird that soars in air, waddles on the muddy earth.' For the sake of your poetic nature, I accept your apology. You just might have the beginnings of a good song there."

"Is Your Grace a poet?" Bran said as he bowed in deference.

"No, a composer. If you finish the words to your poem, Captain, come to me and I will set it to music. I am Viracocha. I am one of the precious hothouse flowers that decorate the morning end of the City."

The dignitary pointed in the direction of the College, smiled, and with a slight nod of the head, turned and moved off down the Broad Street toward the College Gate.

At least he doesn't take himself too seriously, Bran thought. What a life he must lead in his palace of arts.

As Bran watched the man walk rapidly down the street, he began to imagine himself as a Member of the Sacred College for poetry. He had no idea of what the living quarters would be, but he saw himself sitting at a desk by a window to the sea: he had been up writing all night and now he could see the sun coming up over the ocean. As clearly as if he had just written the words down, he could see the quatrain he had been working on:

Body burdened on the ground
Startled heaven when it found
Arms it had falsely used
Were wings that force had bruised.

But the image of wings only recalled to his mind the emblem of the winged serpent and the black accusing eyes of the Governor of Anahuac. The daydream of living the life of a poet ended in the pain that reminded him only too well of who he really was and what he had to do. As the dignitary disappeared through the College Gate, Bran turned around to face in the opposite direction of the Institute.

Bran moved toward the copper doors with a prayer that this would be one of the last times that he would have to report there. He couldn't see how it could go on much longer, for either the 144 would be discovered, or he would give the signal to go and not wait for the prophesied leader to appear.

The guards at the doors saluted him as he passed through, but the Officer of the Guard inside the entrance post surprised him with an unexpected message as Bran handed over his flight report:

"Good Morning, Captain. His Excellency the Archon has requested that you stop in to see him before you go to your quarters."

An immediate spiral of fear spun around in Bran's stomach. Had they learned about the 144? Had their spies found out about the Governor? Bran took a deep breath to slow his racing heart, saluted the Officer of the Guard, and continued down the long artificially lit corridor.

Bran thought it strange that for all the people he had brought into the Institute, he had himself never gone beyond the Office of the Guards. But he did not need to be told that the double doors at the very end of the corridor led into the Laboratory of Incarnation, and he knew at once that was where the women and children would be. He did not want to go in there, and he could see from the seal of the Archon on the door to his left, that he did not have to go further toward those double doors.

Bran knocked twice and waited for the silent page to open the door to admit him to the Archon's study. But it was not a page who admitted him, but another armed guard, and this made Bran feel more ill at ease. Was this normal, he wondered, or was the City already prepared for a civil war?

The guard knocked twice on the door to the inner study, and then opened it to let Bran pass. The Archon was standing at the other end of the room with his back to Bran. A row of Mercury Tablets neatly aligned on the table gleamed in front of the robed figure and gave an aura of pink light to his silhouette. The Archon turned slowly to look at Bran, and then with a welcoming smile, he moved with the help of his cane back to his desk.

"Good Morning, Captain. Please sit down. Don't wait for me

to be seated, for, as you can see, it will take me a little longer to settle myself into my own chair."

Bran tried not to be obvious as he considered the Archon's condition. The right leg seemed stiff and the right arm was not in its sleeve. It appeared to him as if the Archon had suffered some sort of a stroke since he had last seen him.

Slowly the stricken figure adjusted himself into the chair, took a breath of resigned acceptance, and smiled again at Bran in a very grandfatherly way:

"Ironic, isn't it? Here we are at the Institute studying the nature of the blood, and I become afflicted with a blood clot."

"I am very sorry, Your Excellency," Bran said. "I hope that your own staff here can find the remedy soon."

"There is much about the blood that even we don't know, but we are coming very close to a great climax in our work, you can practically feel the excitement in the air, so, perhaps soon we shall. For the present, I must concentrate on our primary work. Which brings me to the subject: we have had some reports concerning an emotional disturbance in the Governor of Anahuac. I believe he traveled here for the Solemn Assembly aboard your vessel. Did you notice any unusual behavior?"

"He was strangely silent and withdrawn, Your Excellency. But I confess that I still don't understand what is 'usual' for these people. You look into their eyes, but it is like looking into one of their black obsidian mirrors. You can never tell what is going on inside. They never seem to signal their emotions before they express them. One minute they are peaceful, the next violent. All that I can report is that every time we prepare to ascend, the clearing is surrounded by hundreds of silent men. They do keep well beyond the range of our dart-catapults, but sooner or later, I think, they will revolt."

"I see," the Archon said as he began to toy with the handle of his cane. "And the Governor, do you see him leading such a revolt?"

"He or anyone of them could, Your Excellency. The Governor is an enigma to me. I must confess that I really don't like these people. Their moodiness disturbs me so that I always feel uneasy around them."

The old man nodded in silence as if to say that he knew exactly what Bran meant, and then in a manner that was so paternal that it bordered on the patronizing, he tried to explain the situation to Bran:

"You see, that is because their emotions are collective, not personal. But that is precisely why they are so good for our work here. The Atlantean race is too civilized, too highly individuated. We had need of a race that was more psychically malleable. You can have no idea of how important the work is that is going on here. But, soon, you shall. For now, let us say that the Institute has found a way to accelerate human evolution. These pregnant women that you have brought here are, in a way, truly the mothers of a new human race. I would like to thank you for the sensitive manner in which you have executed this mission. It is a delicate situation, but you have handled it with the appropriate delicacy. The people of Anahuac, of course, neither understand nor appreciate our work, but this Solemn Assembly should be helpful in bringing about a more sympathetic union among the peoples of the Empire. I believe that is all that I need to know for the moment, Captain. Thank you."

Bran stood up quickly to attention, bowed, backed up three paces with his head still held low, and then turned to go out the door. He passed quickly by the guards in the outer study and at the entrance and did not take a deep breath until he was beyond the copper doors by the low wall overlooking the sea.

As Bran took in a freer breath he tried to take in his whole situation. It felt to him as if the Archon had believed his report, but the remaining danger lay with the spies. Was there one of his own crew that was reporting to the Institute, and if so, who? It was clear now that the Governor of Anahuac would be arrested and that the assassination attempt would be aborted, but what about the 144? If the 144 were linked to a revolt in Anahuac, or were thought to be part of a coup by the military, then they could all be rounded up before the signal of escape was ever given. If only Abaris were still alive!

Bran turned away from the low wall, moved down the Broad Street, and then took the small alley to the left that led to the residential quarters of the Institute. He wished that there were

some way that he could get into the Palace of the Provincial Governors to give the doomed Governor of Anahuac a warning, but he doubted whether the Governor would even try to escape. The person that he really needed to see was Brigita. If Brigita would only attend this Assembly, he thought that some solution to their dilemma could be worked out. But he knew that Brigita might avoid a Solemn Assembly whose only purpose was to worship the Empire.

Bran felt trapped: he couldn't warn the Governor, he couldn't get in touch with Brigita, he couldn't, like her, simply project out of the body to find the spirit of Abaris, and he couldn't give the signal to the 144. It was then that he began to feel the panic of an animal in a trap: had that seemingly kind and solicitous old man been toying with him? Did they already know everything? Was this Solemn Assembly simply a ruse to bring everyone into the City to eliminate the military coup and the 144 all at once?

Bran fought down the feelings of panic as he entered the officers' quarters and went into his room. The one thing he was not prepared to do was the one thing that he had to do, and that was nothing.

2

The light from the clerestory dropped onto the table as somewhere beyond a cloud moved out of the way of the sun. As the sunlight flashed on the engraved metal sheets, Viracocha took a step back from the table. He had already seen enough and the light of understanding had flashed for him within seconds of looking at these proposals for the increased elevation of the Tower: the whole project was a mistake, perhaps even a very dangerous mistake.

The old man, however, seemed happily occupied in his examination of the engravings. Viracocha studied the bent figure for a moment, observed the light bouncing off his bald head, watched the long bony finger moving slowly over the calculations, and listened to the little voice laboring the obvious as he hummed tiny exclamations of agreement to himself. When the old man thought he had found a particularly challenging point, his tongue would protrude from his mouth and remain exposed until the exclamation of agreement would allow it to return to its well-deserved obscurity. Viracocha found it difficult not to become annoyed.

Viracocha knew that the old man, as Member of the Sacred College for Architecture, thought he was being generous and companionable in inviting the youngest Member to join him in a look at the new and exciting project for the City of Knowledge. And he also could sense that the old man was trying to make him feel less isolated within the College, but, in fact, Viracocha preferred his isolation and had absolutely no desire to work in groups or be part of the circle of which the Member for Architecture was the most senior.

As Viracocha watched the long skeletal finger make its slow deliberate passage among the figures, he thought of a blue

heron moving along an icy shore in winter in search of fish. Could he see himself as now this writing can from the perspective of the millennia, he might see himself as the wild horse. He was of medium height but for his broad shoulders and full chest had delicate hands. His thick shoulder-length brown hair contrasted with his red beard, but his eyes like the sea seemed to change colors with his moods: one minute they appeared to be blue, the next green. Even when he stood at rest, he raced on with an animal instinctiveness that did not think twice about its moves. Viracocha never needed to deliberate; words would rush out of him with such speed that it made thinking appear to be as natural as seeing or running. Conversation was only possible if one were to run along with him, but that rarely happened. Most people, like the old man, believed that sureness of thought and slowness of thought were the same thing.

While the Member for Architecture carried on with his official duty of examining the Institute's proposed changes in the Tower, Viracocha let his mind return to his own work. For the first time in Atlantean history, he planned to set two solo voices, a baritone and a soprano, against the chorus. He was excited and his imagination was pleased with the image of a man and a woman standing out against the massed power of the entire choir.

At last the old man came to the end of his examination. With a dramatic exhalation of self-importance and relief from his heavy responsibility, he sat down in the chair by the table and turned toward Viracocha with a priestly formality that seemed to say: "And now the Member for Architecture would like to ask a sometime architect and now Member for Theoretical Music and Composition to pronounce an opinion on the matter at hand."

"Well, Your Grace?"

As if he were thinking of more important matters, Viracocha remarked rather casually that the whole project was a mistake.

"I'm sorry, Your Grace, I'm afraid my hearing is going. Could you please repeat that?"

"Yes, of course," Viracocha replied as he closed his eyes to review his calculations. "No, it's very clear. If you change the

alteration of the Tower, you will alter its resonance with the bay in the etheric range and the whole wave pattern will be disrupted. Rather than enhancing the etheric bodies of plants and animals placed in the foundation vault of the Tower, what will happen now is that the etheric body will be separated from its connection to the physical. I'm afraid it would be dangerous for anyone to be in the foundation vault when the solar crystal was activated. Surely, Your Grace, you can see that the Institute's calculations are so consistently wrong that they suggest a pattern of conscious misrepresentation."

"I must confess, Your Grace," the old man said in disbelief and discomfort, "that I do not see what you're talking about. It's true that the site for the Tower was chosen because of the etheric properties of the rocks and the depth of the bay, which provided a convenient collector, but surely with the sea all around us, the stabilizing quality of water will not be lost?"

"It is not simply the etheric quality of water," Viracocha answered with impatience, "but the geometry of the bay as well. The underwater configuration of the bay is nearly perfectly conical, and this organizes the etheric radiance of the seawater in a way that is adjusted by both the elevation of the Tower *and* the frequency-signature of the solar crystal. Change anything, and you change everything."

"But surely you can't think that the entire faculty of the Institute, not to mention the Supreme Council of the High Priest, has made a mistake?"

"I thought that was what I just said," Viracocha replied as he looked down at the seated old man in wonder that anyone so stupid could ever be appointed Member of the Sacred College. "Does Your Grace intend to approve the designs considering, at least, the questions I have raised?"

That was going too far for the Member for Architecture. He reached out to grasp the edge of the table to hold onto solidity as he stared up at Viracocha in state of shock and disgust:

"This proposal comes from both the High Priest himself and the Archon of the Temple Institute! You do seem to have a problem of reading perspective into these engravings. How-

ever, I will ask the Archon for a clarification. I'm certain he will be able to put your mind at ease."

The old man bowed to indicate that the consultation was at an end. Viracocha returned his bow and gladly left him to his afternoon rest.

I suppose he imagined that he was doing me a favor, Viracocha thought to himself as he left the apartment and walked out into the cloister, the oldest Member giving the youngest Member a sense of belonging by taking him into the project that all his cronies were buzzing about. Viracocha could understand why the old man had let him see the plans, but what he couldn't understand was why the Institute would want to alter the whole dynamics of the Tower.

Viracocha turned away from the colonnade that led to his quarters on the eastern side of the quadrangle and decided to walk out to the end of the southern peninsula. The puzzle of the Tower began to interest him precisely because of its obvious absurdity. He began to feel that intuitive pull of growth that had taken hold of him at critical times in his life to effect the most unsuspected transformations. It was in one of those earlier transformations that he had abandoned the formulaic conventions of temple architecture to take up music. And it was through musical composition that he had been appointed to the Sacred College, to become at thirty the youngest Member since its founding.

He could feel the pull, but this time it had an almost spatial quality to it. He looked around as if his environment could give him the answer he was seeking, but he knew the surroundings only too well. Viracocha had begun to feel a sense of confinement during his three years of residence in the great City of Knowledge. As he passed out of the quadrangle of the residential quarters of the College, he followed the colonnade through an enclosed garden with a fountain in the center. And in the center of the fountain was the required obelisk to the sun. He quickly passed through and came out on to the processional walkway that wound around the entire crescent directly above the bay. It all seemed so controlled, so boringly bisymmetrical, with the Institute on the extreme western side of the proces-

sional and the College Chapel on the extreme eastern side. Directly in the center was the Broad Street, the hypostyle hall, and above it, the Tower itself. Only the Tower rose above the level of the plateau behind it, but there was nothing on that plateau except sparse grass and scrub bushes. For the most part the isolation of the City of Knowledge was unrelieved by either village or town. It was more than forty miles of poor grazing lands dotted with a few goats before one came to the ancient capital, but there were no roads on the peninsula, and all traffic between the seat of the High Priest and the old capital was either by sea or air. Once one had taken the official promenades along the processional ramps, the Broad Street, or through the labyrinthine paths in the gardens and cloisters, there were very few places left simply to go for a walk. To the extreme east, on the fields and low cliffs in front of the Sacred College there was a goat trail that led down to a small cove with a tidal spouting cave, but to the western end by the Institute there was nothing but very high cliffs that dropped from the laboratories to the surf-beaten rocks below.

Viracocha turned to his left to follow the processional walkway to the entrance to the Chapel; as he walked along the edge he looked over the low wall down into the deep blue waters of the bay. He had never seen so many garudas at rest on the waters at one time. Although it had been explained to him on several occasions by the older Members of the College, it still did not make much sense to him. A Solemn Assembly was not a decision-making body, for that was the work of the Supreme Council. A Solemn Assembly was not a forum of discussion, for that was the work of the convocations of the College or the Institute. A Solemn Assembly was an invocation of the Spirit of the Empire, and although voices of questioning were invoked in the liturgy, they were not really intended to be anything more than occasions in which to appreciate the unanimity that filled the great hypostyle hall.

The whole thing seemed something of a farce to Viracocha. First, they sent over the plans for the proposed increase in the elevation of the Tower from the Institute to the College, but they didn't really want anybody to do anything but register a

ritual approval. Next, they called a Solemn Assembly to celebrate the approval that they hadn't really received. It was more theater and politics than science or architecture, but perhaps, Viracocha thought, theater was what the politics of Empire was all about.

Viracocha liked to walk because he never had any need of plans or writings. Whether he was doing the plans for a building or the notation for a musical composition, the transcription with emerald stylus onto the thin metal sheets was always for the sake of others. He had only to look at a building to imagine its original plans, or hear a piece of music to transcribe it into the hieratic notation. So as he looked down into the deep waters of the bay and then turned to look up at the Tower, he did not need to go over the plates again to know for certain that he was right.

The copper doors of the Institute gleamed in the sunlight in response to his inquiring look across the bay, but no answers came for the questions in his mind. Only an image came out of nowhere, as if he were dreaming with his eyes open, and that was the image of some maniacal idiot pulling on a head until it was torn from the spine. In that instant he knew that he was correct in his calculations, that the etheric body would be pulled away from the physical.

Viracocha turned away from the Institute to stare at the stone wall in front of him. He had no desire to move to the left and go into the Chapel, so he reached up to the top of the wall, pulled himself up and swung himself over onto the other side without giving a thought to how ridiculous he looked scurrying over walls in his priestly pale blue robes. At first glance nothing was there but jagged brown rocks. The land did not drop away with the high dramatic cliffs of the western side of the peninsula, but simply leaned toward the sea with a collection of rubble that no architect would care to landscape. Viracocha could see why the architect had built a stone wall to mark the end of the city and the entrance to the Chapel.

There were no goat trails to follow, but the slope of the land seemed to pull him down farther to the eastern side away from all sight of the harbor. He followed the inclination and moved

down out of sight of the Tower or any of the buildings until he came upon a collection of broken basaltic columns. One of the hexagonal columns was a foot or so lower than the other three surrounding it and so it seemed nicely formed to serve as a chair.

As Viracocha sat down in the natural throne, he felt free of the psychic confinement of the city and was simply happy to be able to look out over the sea and thankful that there was nothing human to look at. He stretched his legs out, placed his hands on his chest, and closed his eyes, as if to test whether or not the place could serve for composition. But he was startled to find that his mind did not return to his composition at once. Instead the elevated Tower stood out vividly in his imagination and he saw an inner shaft of light go from the solar crystal to the center of the earth. Light met light as the brilliant shaft was met by a stream of molten lava shooting up from the depths of the earth. And then he understood. The new space of the singularity would not be in the pathway from the crystal to the center of the foundation vault; it would go all the way down to the center of the earth: the distance from the solar crystal to the center of the earth over the distance to the sun. They were trying to alter the whole geometrical dynamics of the solar system!

In the very instant that Viracocha became aware of the project, he became opposed to it. He did not think about it. It was a completely instinctive response. Once he had decided against it, thought was needed only to determine the way in which to oppose the undertaking. It was at that point that the artist in him decided to take the most dramatic and unexpected form of all: he would speak out against the project during the Solemn Assembly. When the voice of the Oracle rang out in ancient liturgical Atlantean: "Who among you could even conceive of not being moved by so great a vision?" he would rise, sing out a newly created ancient response, and proceed up the aisle to the Speaker's Stone to challenge the entire project. If theater was the politics of Empire, then, Viracocha vowed to himself: Theater we will have!

Viracocha was enormously pleased with himself as he jumped up from his basaltic throne and resolved to rush back to the

library to make certain that his responsorial chant would be composed in perfect liturgical Atlantean. The dramatic power of the challenge would be ruined if his strong tenor voice rang out in the great hypostyle hall in faulty grammar. It wouldn't matter if only the Member for Poetry recognized the error; the idea had to be perfect for Viracocha, or else the idea of perfection was lost.

Now Viracocha was truly excited in ways he had not felt before during his three years in the City. He jumped up onto his volcanic chair and leaped up into the air and began to run as fast as he could up the hill to the stone wall. The rocky ground was higher on the southern side of the wall, so Viracocha did not have to break his stride, but grasping the top of the wall with his left hand, he simply flung himself over with a great display of power, some grace, and much silliness.

He was the first to realize the silliness of it, for as he came hurtling over the wall he nearly collided with a dark blue hooded figure staring across the bay toward the copper doors of the Institute. As Viracocha landed scarcely a pace away from the figure, he put out his hand to the pavement to steady himself and looked up at the hooded face as it slowly turned to take notice of him. In the deep shadows of the hood all Viracocha saw at first were two green eyes staring at him without the slightest trace of surprise.

"Good Afternoon, Your Grace," the voice of a young woman said in a bow that by its slight inclination said that though she recognized the pale blue robes of the Sacred College, she herself was of no mean rank.

"Good Afternoon," Viracocha said as he stood up and brushed the dust from his hands. "I'm sorry to have startled you. No one is usually here at this hour, and, well, I was feeling celebrative for having made a decision."

"You didn't startle me, Viracocha. I was here to wait for you. So, you have decided to speak out against the Institute. I'm glad. It has to be stopped."

Viracocha stared at the woman first in disbelief, then in irritation:

"Whoever you are, young lady of the hooded dark, I wish you would come out into the open in *every* way."

She did not raise her arms immediately, but stood for an instant to study him, and then slowly two white hands lifted and folded the hood back over her shoulders. The skin of her face was as pale as her hands except for a faint bridge of freckles which passed over the nose and seemed to match the color of her long reddish-brown hair.

"As you can see, Your Grace, the hood is not for hiding but for protection from the sun in this southern climate."

"From the coarseness of your dark blue robe," Viracocha said as he began to examine her more closely, "I assume you come from somewhere near the ice sheets."

She was young, neither plain nor pretty, but a constantly changing relationship between the two that seemed to have as much to do with her own ambivalence as with any natural endowments. Her eyes were fiercely intense, almost threatening in a defensive way that said: "I'm not very good with people and I know it, so don't try to befriend me." She appeared to be slight and Viracocha thought she looked lost in the huge tent of her ankle-length robe. He imagined her as skinny, angular, and hard, with bony shoulders and hips that tolerated their fleshly covering the way she tolerated her ill-fitting cloak.

She looked at him with exasperation as much as annoyance and then blurted out with impatience:

"I am Brigita, the Abbess of the One Thousand of Eiru, and I don't understand how you can possibly be so ignorant of your own spiritual life not to remember me."

"Of course, I know *of* Brigita the Abbess of Eiru. Abaris told me about you, but unless I am getting feebleminded at thirty-three, I don't remember ever having had the pleasure . . ."

"Not in the body," Brigita interrupted, "*out* of the body *in* the spirit, in the *real* world."

"I'm afraid I don't work very well in your 'spiritual' world, and now I'm beginning to understand why you *seem* so good at reading minds. I'm not what you would want to call a good student of Abaris. I've never had an 'out-of-the-body' experi-

ence in my life, so although I'm flattered to have appeared as the man in your dreams, I must confess my innocence."

That annoyed Brigita just as Viracocha hoped it would. Her green eyes hissed in serpentine menace at the mere suggestion that he or any man was part of her dream life.

"Why are we wasting time with this silly banter? You are not the man of my dreams; if any man were to have that place, it would certainly be Abaris, and not you. You have had 'out-of-the-body' experiences every night for the last few years, and it is absolutely impossible for me to understand how you can forget that subtle life to focus your identity on this dumb, obvious, crude, masculine personality of yours!"

Viracocha looked over Brigita's shoulder to see two colonials coming up the processional walk to the Chapel. As he returned his glance to Brigita, she immediately turned around to consider the approach of the two men.

"I don't think we should go on talking like this here," Viracocha said. "I don't know yet what you want, but *I* want to talk to you again. We're not allowed to have women in the Sacred College, so we'll need to meet outside."

"Strange isn't it," Brigita commented with returning calmness. "You're allowed to take a concubine to your bed, even though you're all priests here, but you can't take a woman colleague into your study."

"That's because there is no such thing as a woman colleague here in this City of Knowledge," Viracocha replied. "The only women here are servants or concubines. Do you know where the spouting cave is on the eastern side of the peninsula? It's down out of sight from any of the buildings or gardens of the College. We could meet there tomorrow afternoon, after the Solemn Assembly."

"Yes, I can see the place you mean. I had hoped to be able to work some plan out with you, but, if they don't arrest you, I will meet you tomorrow."

"Why would they arrest me for a loyal observance of the most ancient ritual, me, a Member of the Sacred College? Even if you could read minds, I don't think you read the situation here in our City."

"You don't remember your spiritual life, and you can't see what's going on around you," Brigita said in disbelief. "And to think Abaris called you 'the new man,' the 'Man of the Future'! May the One help us, for I certainly can't see how I am going to be able to go through with it all with you."

Brigita did not wait for a reply, but pulling her hood over her face, she turned and went into the Chapel. Viracocha stood a moment in silent consideration of her face as it stood out so vividly in his imagination. Everything about her was at an edge. Her face was exactly at an edge and the slightest shift could make her homely or beautiful. There was not much body to her so she seemed to haunt the psychic edges of the physical like some wraith haunting a lover's bed: repelled and attracted to the physical copulation that would pull her soul into life.

In considering Brigita, Viracocha recognized that she brought him to an edge in himself, for part of him found her secretive plainness attractive, and he had erotic fantasies that if she were a concubine, he would find all sorts of unexpected and pleasant surprises in her nakedness. But another part of him found her to be an arrogant and condescending purist with a contempt for the world she could not understand, much less master. Except for the grandfatherly Abaris, she certainly did not seem to like men. The war of the sexes, Viracocha thought to himself with a smile. Well, you can't have that with a concubine, so unless they do arrest me, I will be there tomorrow at the spouting cave.

Viracocha returned the bow from the two colonial visitors to the Chapel, and then hurried down the walkway to return to his quarters in the College. He knew that he had taken up enough time with the Institute's business and that it was time to get back to his own. The High Priest had requested a performance of Viracocha's choral works for the evening's reception of the Provincial Governors. If he didn't attend the choirmaster's rehearsals, he knew he could always count on the master to work against him to make the work sound safe, conventional, and soporific.

When Viracocha returned to his rooms, he was surprised to find a large packet of drawings and a note from the Archon:

We at the Institute are pleased to learn that His Grace, the Member for Theoretical Music, has taken a keen interest in our work. We thought it fitting, therefore, that you have your own copy of the plates for study.

Nicely done, Viracocha thought to himself. He politely invites me in and tells me to be sure that I do my homework. The old man certainly did not waste any time in informing the Institute of my objections.

Viracocha did think that was curious, but he did not have time either to open the package or reflect on the Member for Architecture's motivations. He had the rehearsal, then the reception, the dinner, and then the performance. The package would have to wait.

When Viracocha returned to his rooms late that evening, he was too tired and too preoccupied with the problems of the performance to recall the packet of engravings on his desk. With his back to the table and his eyes set for infinity, he unbound the gold chain to his cape and let it drop to cover the images of the Tower like some velvet, enveloping wave. Within a few seconds, he was deeply asleep.

Only when he awoke the next morning did Viracocha remember the plates, but even then they did not seem to be central to his concerns, for he turned over the thin metal sheets in a quick and cursory examination that was only to make certain that they were simply copies of those he had seen before. Some other thought left over from sleep or dreams was trying to get at him, but he could not remember anything other than a feeling that it had to do with the Solemn Assembly. He picked up the emerald stylus and toyed with it absentmindedly for a moment, and then he moved to the side of the table where his Mercury Tablet sat.

The metallic gray liquid shimmered like the surface of a frosted mirror, but as he touched it with the stylus it transcribed his motions into gleaming figures of gold. With an elegantly artistic hand he inscribed the hieratic script for *Ome ulubi ra, se tata mak*, and then, with the head of the stylus, he activated the command for glyphic resolution. Within an instant the linear

script of gold melted into six hieroglyphs with their tonal consorts.

Viracocha hummed the tones, but did not find them to be the melody that he thought he half-remembered from his dreams. He tried reversals and inversions, but it was not until he took the occult reciprocals for the tones that he discovered the chant he wanted. It was not simply a chant, but an astonishingly haunting call to the stars.

As he chanted the sequence over and over again, he began to remember what he was looking for. It was the Solemn Assembly itself that was the art form that held out the solution to the problems that had developed in the previous performance of his choral work. The Solemn Assembly could be transformed into an oratorio in which several soloists' voices competed with the chorus of the whole civilization. The Solemn Assembly was the most archaic, most collective ritual for celebrating the tradition of Atlantean chant, that ancient, droning monody that pulled the mind down into trance.

"What a perfect occasion for contrasting the soloist singing a melody even more ancient than the plainsong of the chant!" Viracocha exclaimed to himself and he returned the Mercury Tablet to its silent silver gray. As he rose with enthusiasm to dress for the part the day was to bring, it was hard for him to contain his own excitement.

What made it difficult to sustain such enthusiasm was the unanticipated long wait of the morning. The processions were unending. First came the line of green-robed Provincial Governors, then the long line of red-robed Military Commanders, then the even longer line of the indigo-robed faculty of the Temple Institute, and finally, the line of the seventy-two pale blue-robed Members of the Sacred College. When these had taken their places in the hall, then the Supreme Council of the High Priest entered, each attired in his robe of magenta. One by one, they came onto the raised dais at the end of the hypostyle hall: the Commander of the Military, the Governor of Trade, the Archon of the Temple Institute, the Master of the Sacred College, and, finally, the Oracle herself.

Ages before, the High Priest had been a woman, not so much

a priest as a matriarch. But as trade and warfare had grown, and with them the competition to control trade through superior weaponry, the clusters of villages had grown into cities, and a new priesthood had developed with the mystery of writing that could maintain control of life with facts. Conflict followed conflict, until at last there was an open civil war between tradition and innovation. The result of that conflict was the establishment of the Theocracy. The priests took on the ancient dress of the matriarchs, and the High Priest took over the role of the Great Mother herself, but to honor the gifts of the past, woman was given the role of Oracle. In trance, she would speak for the old gods in the new assemblies of men; but that was long before in the early days of the Theocracy. Now her role was restricted to being the conductor of the ceremony of the Solemn Assembly. She no longer sat on the Supreme Council of the High Priest, but stood alone in her tower-like pulpit, above all except the High Priest, yet more of a captive than an ancient voice entoning the monodies of a cherished tradition.

After all the dignitaries had taken their places, the minor colonial officials and provincial priests were allowed to fill the back of the hall. And then there was absolute silence for the space of half an hour.

Viracocha closed his eyes to meditate and to perform the kriya he had been taught by Abaris. Silently he intoned the solar syllable as he drew up a golden stream of energy from the base of his spine. He enjoyed the subtle physical sensation of feeling the energy spread out over the back of his brain and then pour into his inner eye. The cobra stood erect: the expanded hood was his brain and his spinal column was the long body of the snake. The image of Brigita came to his mind, and he saw again the bright green eyes peering out from under the shadows of her hood. Slowly, he pronounced the lunar syllable and felt the cool silver of the stream flow back down through the spinal channel to come to rest at the bottom, like moonlight on the surface of water in a deep well.

From somewhere, very far away, he heard the sound: two wooden clapping sounds, a metallic thud, and then a high tingling sound of a small bell. The sound was coming from behind.

He listened again, as now everyone did, and musician that he was, he was able to recognize the objects from their timbre. Someone was walking on wooden shoes, punctuating the rhythm of his slow gait with a heavy metallic staff that struck the floor and set the tiny bell on top of the staff to ringing. Viracocha was hearing the approach of the High Priest. Like a great drum roll the three beats and crystalline answer continued as the High Priest slowly came from outside the back doors of the hypostyle hall and moved down the central aisle. Over and over again the four sounds repeated themselves. The High Priest was chanting with his feet and staff even before a single voice had been lifted. An enormous physical power seemed to come out of those four sounds, as if the reverberations filling the entire hall had become the incarnate body of the High Priest himself. As the sounds grew louder Viracocha found that they penetrated his body and seemed to resonate in the centers along his spine. And then he opened his eyes and saw the High Priest.

It was his newly acquired height which first amazed Viracocha. The High Priest towered above the Assembly as he struck the stones with his *cothurni.* It was not walking, it was setting the earth under him into vibration until the earth itself became the shoes on which he walked before the stars. The golden staff with its silver bell on the top was the sun and the moon, which he had taken in hand to make his journey through space. The enormous headdress of peacock feathers was broader than his shoulders, and the golden spiral of twin ram's horns over his ears suggested that he was all-seeing, all-hearing. The brocaded robes were an epiphany of the colors of the robes of the other priests, for he was their summation. He had become "the Rainbow Body of the Higher Self," and in him humanity stood out and declared its rightful patrimony.

As the High Priest approached the ramp to the pyramid that held his throne, he stopped at the dais which supported his Supreme Council; then, very slowly, he turned to face the Assembly. Viracocha found himself wondering like a little boy how the High Priest was going to ascend that pyramid to his throne. As the High Priest faced the Assembly, he struck the platform with his staff and immediately a shaft of light came

through the skylight and shone directly into the enormous emerald that was at the center of his peacock headdress. He began to turn slowly to face the ramp up to his throne, when suddenly a scream rang out through the hall: "Death to the Murderer of Anahuac!"

At first it was not fear that paralyzed everyone, but something more like awe at the inconceivable. No one in his state of religious peace had the power to jump to his feet to try to take command of a situation no one could yet understand. All saw one of the Provincial Governors break out of ranks, run into the aisle, and fire dart after dart from his catapult. But the High Priest did not fall; instead he raised his arms and his staff and commanded the guards and the Military Commanders in a voice that was much deeper than the high-pitched scream of the Governor: "Be still!"

No one moved. No one breathed. It was now a direct and elemental agon of the assassin and the High Priest. Raised upon the platform and his *cothurni,* the High Priest towered above the small man with straight black hair. They both stood motionless for an instant, and then the Governor threw down the catapult with disgust and drew out a long dagger from inside his robes. With another high-pitched scream, he started to run toward the High Priest.

A brilliant flash of flat, metallic white light burst from the staff of the High Priest, and Viracocha immediately closed his eyes in pain. It took several seconds for all the afterimages and colors to disappear so that he could see again. When he did open his eyes once more, the assassin was gone and the High Priest was seated on top of the pyramid in his throne. A single exhalation in awe of a miracle came out of the multitude, but even as they all began to draw breath, the Oracle sang out the opening words of the traditional chant and the whole Assembly answered in an enormous wave of emotion that seemed to express the very power of their entire civilization.

Over and over the hypnotic cycle of the chant turned, and as Viracocha stared at the green light of the great emerald on the High Priest's forehead it became the point where all sound and light gathered. Everywhere around him he could feel the pull

of the entire Assembly moving into trance. Nothing could be other to the single mind of the Assembly in the purity of that consecrated time of miracle. Viracocha's breath softened, his heart slowed in its beat, and the interval between each pulse widened until his mind passed through. When his heartbeat would return, it reminded him who he was, and he began to wonder where or who he had been in that rich emptiness.

He could not discern just at what point the chanting receded into stillness, or for how long the Assembly rested in that silence that was not an absence of sound but a thick presence that filled the hall. Out of that presence a voice did come. The timbre was so perfect for that sense of presence that voice became for silence what soul was for body. There was no longer an inside versus an outside, for the voice of the High Priest filled the hall and yet seemed whispered inside everyone.

"Children of the Sun, rejoice.
You are the generation of the Great Return.
For thirty-five thousand years
Our temples have labored to perfect
This animal body into which our spirits
Were cast by an errant god.
For millennia we have striven to purge
Beast and monster from our troubled blood.
Abandoned by the gods,
We did not sink down
Into the wretchedness that was expected of us,
But instead we lifted our Towers up into the Sun,
Listened with our crystals
To the melodies that came and surrounded
All living things with robes of color,
Robes of Power.
We studied how sound and color touched
The hidden organs of the body,
And we placed our bodies in the color,
in the Sound.
Then like a snake sloughing off its skin,
We dropped the hairy mantle of bestial fur

To stand out naked, reborn in light.
This has been our labor,
And now it is finished.
Now begins the new work,
Not with the body of flesh,
But with the body of light
We receive from the sun.
As we have re-created a body of flesh,
So now shall we re-create a body of fire.

First, we shall begin humbly,
Taking small lumps of clay
To knead them together
To create a larger vessel.
With our new knowledge we shall take
A dozen bodies of light, make them one,
And into this form project
Our chosen spirits.
Mothers in the Tower shall conceive
A new race of gods.
And each generation will be greater
Than the last as we ourselves
Breed ourselves out of this oblivion.
We are the last Atlantean generation.
You are the first generation
of demi-gods about to become gods.
Then we will not need vessels
To carry us back to the stars!
We will be *stars.*
Now, in the condition of our exile and
Abandonment,
We are in *space and time;*
Ah, but then, my children,
Space and Time will be in us."

There was silence again: that pure, rich, and perfect silence in which the great bronze voice of the High Priest still resounded.

Then the high sibylline voice of the Oracle sang out the liturgical call, the summons to lift the spirits in response.

Viracocha opened his eyes as if to question space, to ask what could possibly follow the poetry of the High Priest's vision. Slowly, the Archon of the Temple Institute walked with the aid of his cane to the Speaker's Stone. His first word shattered Viracocha's mood. The voice was all wrong: it was flat, avuncular, without resonance, but with a practiced charm that seemed to be a mockery of human warmth. There was posturing, but no spiritual flow to the words for Viracocha. For him they seemed extruded from his mouth to drop still-born to the floor.

The aesthetic repulsiveness of the Archon's presentation annoyed Viracocha. This was not the time or place for lectures in biology and engineering. He could hardly bring himself to listen, until he noticed the pattern. The Archon was excessively technical concerning the obvious, but when he came to the difficulties inherent in the project, he quickly passed over them with facile glibness.

Viracocha looked around to see what response the Archon's speech was calling forth from the Assembly. He could scarcely believe it, but it seemed as if half the Assembly were asleep, the other half still in trance. Why was it such a foregone conclusion, Viracocha wondered. Why did no one else seem to feel the very wrongness in the timbre of the Archon's voice?

The Archon finished his presentation, and the Member of the Sacred College for Architecture voiced his affirmation without even bothering to walk to the Speaker's Stone. The whole thing was becoming sloppy and coming apart. They had destroyed the beautiful aesthetic space created by the High Priest, and Viracocha resented the fact that this was taking the *kairos* away from his own planned response. He could see so vividly that his own speech would have been perfect if it could have come immediately after the High Priest's. The dramatic opposition would be perfect. But now he wondered whether he should go ahead with his own presentation.

The Oracle sang out the call: "Who among you could even conceive of not being moved by so great a vision." As Viracocha listened to her soprano voice, he remembered the full baritone

voice of the High Priest, and was grateful that the Oracle's voice was perfect for the moment. He could see it all so vividly: the single tenor standing alone in the central aisle of the great hypostyle hall and singing out the words that no civilization had ever heard:

"I am here. And here now do I stand against this act that would destroy the very ground on which we all stand."

Viracocha could feel the shock waves move all around him, but he could also feel how dramatically perfect the moment had become, and how right he had been to take that step out into the central aisle. He had shattered the collective trance, and now everyone's attention was on him.

Slowly, gracefully, with his arms crossed and hidden in his sleeves, he began to walk and sing the ancient-sounding chant, *Ome ulubi ra, se tata mak.* Few were there who could know its meaning of "How shall we be, if this is done?" but no one was there who could not feel the haunting cry of a melody that called to the stars to question the sad origin of the human race.

As Viracocha reached the Speaker's Stone before the dais, he bowed to the Supreme Council, to the Oracle high in her stone pulpit, and then he faced toward the High Priest and bowed very low, with his palms covering his knees. When he raised his head, he looked again into the eyes of the High Priest. There was no anger or hostility, but simply a curious fascination.

"Your Holiness, were not your spirit inspiring me, I could not stand here before you. And though it would seem that I am here to stand against this action, that is only an appearance. The moon seems to oppose the light of the sun as it fills the night, but a higher knowledge can show us how the moon takes its light from the sun. And so do I from you. But as the moon gives a different light and a different mood to the beauty of the night, it can only be true to the sun by being completely different.

"Your Holiness, alter the geometry of the Tower and you will lose the geometry of the bay. A great rift will appear in the nature of things: between the land and the sea, between humanity and the earth, between the physical body and the etheric. A person asleep or in trance in the Tower would have his etheric body completely severed from his physical form.

And the space from the solar crystal at the top to the foundation vault in the bottom would no longer be the space of the singularity. The opening would extend from the center of the solar crystal to the very center of the earth. The ground under the Tower would crack open and all the volcanic fury of the underworld would become a new darkness brought to light.

"Your Holinesss, it is loyalty to you and to our whole civilization that moves me to stand here. This proposed change touches death as it reaches out for life. Perhaps because the Temple Institute has been so taken up with war, its genius is found in destruction, and it falls to the more ancient role of music to question this misuse of vibration. But now I fear it is not the enemies of civilization that we will destroy, but civilization itself. And so it is that I must say to the call of the Oracle, *'Ome ulubi ra, se tata mak.'* "

The strong clear tenor voice of Viracocha filled the hall with the beauty of ancient liturgical Atlantean, and it did not matter that the meaning had been forgotten; all recognized the language and knew the power of tradition it summoned. No one could suspect that this archaic-sounding chant had been composed by Viracocha. No one, except perhaps for the High Priest, could appreciate that Viracocha was making his inconceivably individualistic action seem to be part of a tradition of which they all were ignorant.

As Viracocha ended his chant, he bowed low before the High Priest, and then turned to bow to the Oracle, to the Supreme Council. With his head still held low, he stepped down from the Speaker's Stone, took three paces back, and then turned round to walk up the central aisle. He could see and feel the confusion all around him. No one had any way to judge what had just happened. No one knew if one should be outraged or honored; it was so completely outside of any frame of reference, that none knew what to think or feel, and so all began to drop their eyes to their feet to avoid looking at Viracocha in the hope that someone would do or say something so they would know how to react.

Viracocha returned to his place and sat down. Once again the Oracle sang out the liturgical call, but this time the Archon and

the other members of the Supreme Council sang out the opening words of the final hymn, and relieved to know finally what to do, the entire Assembly sprang to its feet and began the cycle of the closing chant. Over and over the droning chant spun on its circularly linked syllables, and as Viracocha entered into the collective monotony, he began to think of the ways he could turn the Solemn Assembly into a new form of oratorio. He would need a stronger ending than this mumbled chant.

The trouble with the work, Viracocha thought to himself, is that there are not enough female voices; there is only the Oracle. Viracocha looked up to the pulpit and noticed that they had all left: High Priest, Oracle, and Council. The emptiness of the dais brought no ideas on how to end his work, and so Viracocha decided he would finish the choral work at hand before he took on this more ambitious project.

As the chanting continued, the Members of the Sacred College began to file out, followed by the faculty members of the Institute. As Viracocha's turn came to join the recessional, he noticed that everyone was trying to keep a good distance from him. By the time he reached the end of the hall he was completely isolated, and so it was all the more noticeable to those who came behind him when the page of the High Priest approached him and signaled that he was to follow at once.

Viracocha had expected the page to lead him outside the hall and then around to the twin staircases that led up to the balcony just before the Palace, but the page turned in the opposite direction down a narrow corridor that went along the entire length of the hypostyle hall. Since they seemed to be heading in the direction of the heart of the rocky ledge on which all the buildings stood, Viracocha began to wonder if he was being led into the foundation vault of the Tower itself.

After a few moments they passed beyond the length of the hypostyle hall, left the corridor of polished stone behind, and approached a wall of enormous megalithic masonry composed of the darker, native rock. Viracocha could see very clearly in his imagination of the path they had taken that the open spinal passageway inside the Tower had to be not very far on the other side of that wall. The page, however, was not interested in the

wall itself, but the corner where the white polished stone facade came to touch the megalithic wall. He felt with his fingers for a moment, and then when he was satisfied, he held his ring to the point until the panel of veneered stone slid to the side and revealed a stairway moving up to the left.

Viracocha followed the page, then waited while he set the panel shut with his ring, and waited for the panel to close and the dark steps to brighten like an activated Mercury Tablet. As they began to climb up the stairwell, Viracocha knew that they were now moving up to the level of the Palace, immediately above the hypostyle hall. When the page touched the wall with his ring, another panel opened, and they stepped into a hall lit with sunlight from the skylights above the broad corridor of polished white marble.

The page walked down the corridor toward the light coming from the great balcony above the harbor. Guards were stationed everywhere at intervals of a dozen paces. As they came into the great reception hall near the balcony, Viracocha began to wonder why the page had taken the long way around, and then he saw the two squads guarding the stairs at the end of the balcony, and knew the answer. A state of emergency had been declared after the assassination attempt, and now all strategic positions would have a doubled complement of guards with their catapults at hand.

As Viracocha considered the squadron of soldiers blocking the twin stairway that led up from the sides of the entrance to the hypostyle hall, he felt again the strong sense of confinement he had always felt in the City of Knowledge. A slight change of emphasis and the whole place became a true prison.

"Your Grace?" the page called to lead Viracocha out of his musings and to prepare him to enter the presence of the High Priest. They stood in the center of the hall and looked toward the double doors in front. Like the entrance to the hypostyle hall, they too were framed by stairs to left and right that led up to the High Priest's study. Viracocha knew it well, for it was in this study that the High Priest had first informed him that he was to be appointed to the Sacred College.

The Captain of the guard marched out from the squad guard-

ing the double doors, and after having satisfied himself that Viracocha's robes did not conceal catapult or dagger, he pointed to the stairs to the left, and the page proceeded to lead the way once again. Once at the top of the stairs, the page motioned for Viracocha to wait between the two guards as he moved to knock twice on the door, then open it, and with a bow and gesture of the hand, give the signal for him to pass through.

Viracocha stared for a moment at the enormous mosaic on the wall in front of him. For the uninitiated, it was simply an abstract mandala, but for the few it was an open explanation of the celestial dynamics by which the Tower affected the flow of energy from the sun into the earth.

After he had waited for the length of time required by protocol, Viracocha turned to his right to see the High Priest seated at his table and staring at him with his eyes out of specific focus. The table was a thick slab of white marble that looked more like an altar than a working desk. Absolutely nothing was on the table, not gold stylus or Mercury Tablet, nothing except for the clear quartz crystal skull. Viracocha had often wondered what purpose the crystal could have—humorous, decorative, or gnostic, but he had never seen it touched or even recognized in its presence by the High Priest.

It seemed to Viracocha that the High Priest was letting him stand there in silence for an inordinately long time. Viracocha began to look more closely to determine his situation. The High Priest had removed all his ceremonial outer robes and headdress and was wearing a simple white sleeveless robe, a gold chain about the neck, and a large pectoral emerald crest. Everything about him was as before except the eyes. The eyes kept their blank and lunatic stare, and Viracocha began to feel uncomfortable. Then suddenly the eyes snapped back into focus, the High Priest smiled warmly, and motioned to the chair of pink quartz and inlaid lapis lazuli. As he moved closer to the table, Viracocha saw the hem of magenta curling round the side of the second throne-like seat in front of the High Priest's altar-like table. He did not need to see the tip of the cane to know that the other person hidden in the chair had to be the Archon of the Institute.

"Please be seated, Your Grace. His Excellency and I were just remarking on what an extraordinary display of courage that was this morning."

As Viracocha approached the twin thrones of pink and blue that faced toward the center of the white marble table, a flash of sunlight came from the noonday sun through the tall slender portals that overlooked the bay, gleamed on the polished surface of the table, and forced him to avert his eyes to look to the floor. The light passed beyond him as he moved more closely, bowed, first to the High Priest, then to the Archon, and took his place in the chair to the left.

"Before we discuss your dramatic presentation this morning," the High Priest said with a gleam of amusement in his eyes, "let me congratulate you on last night's concert. You have come a long way since you first introduced polyphony into our court. I still remember quite vividly how amazed I was when I heard your 'Canticle to the Sun' for the first time. By the way, Your Excellency, do you love music?"

"Love?" The Archon responded with a slight tilt of the head. "That would be going too far. I do enjoy it as much as time permits, but given my age and illness, there is nothing now for me but to see the work completed. I think, Your Holiness, that music has a much larger place in your life than mine."

"That is a pity, Your Excellency, for, as I think you learned this morning, music and General Vibration Theory are not all that far apart, and an initiate in theoretical music can offer a technician a few surprises."

"I will certainly have to concede that point, Your Holiness," the Archon said as he shifted his cane and placed both hands on its top. "But, then, that is the reason you founded both Institute and College, isn't it? I must say, however, that I do think His Grace is wasted in the College. We certainly could use him in the Institute."

The Archon turned toward Viracocha and regarded him with what Viracocha imagined the Archon considered to be companionable warmth:

"After all, Your Grace does seem to be more interested in

innovation and not the conservation of tradition. What is the Institute but the very embodiment of the spirit of innovation?"

Why is it, Viracocha mused to himself, that I always get the feeling that he is playing at being human. There is something strangely abnormal about this studied normalcy. But what he said, turning from the Archon to the High Priest was less troublesome:

"I am grateful to His Holiness for the appointment to the College precisely because I am more interested in composition than technical research. At the moment I am working on a new form, not a simple chorus, as we heard last night, but a dramatic oratorio. There will be two or three soloists who will stand out against the chorus, and it is precisely *that* tension between tradition and innovation that interests me."

"I see," said the High Priest. "Then what we witnessed this morning was a dress rehearsal of sorts. Be careful, Your Grace. If you begin to confuse life with art and start to live your life as if you were *in* a work of art, you just might turn yourself into a tragic hero. It is one thing to command the elements of sound into a harmonic whole, and quite another to command living human beings into a civilization."

"If I needed to be convinced of it," Viracocha said in earnest, "this morning's miracle is certainly overwhelming proof of that. I still don't quite understand what I was seeing, or not seeing."

"And you may never know. The only one who will know is my successor, and I have not selected him yet—though the problem is beginning to interest me. But now, let's turn to the matter at hand."

"How may I serve Your Holiness?" Viracocha asked.

With a wave of his hand, the High Priest gestured to the Archon to answer Viracocha's question.

"First of all, Your Grace, you might begin by not opposing our project in public with such rapid sketches and caricatures," the Archon said, not bothering to mask his annoyance with politeness. "If you do have questions, your position certainly entitles you to be granted some answers. We regret that we could not send you copies of the plates much sooner, but we only learned of your interest on the day before the Solemn Assembly. And, as

you should have been able to envision, you with your exceptional powers of visualization, Solemn Assemblies do not take place every year, so we were rather pressed to be prepared for the arrival of all the Provincial Governors."

"All of that says nothing about the conceptual flaws in the project," Viracocha said with an immovable resolution that indicated that although he was willing to respect the High Priest's authority, the Archon had yet to prove himself. "It seems clear to me that you were purposely hiding things this morning. You were elaborate, almost tedious, on the obvious and deceptively facile on the difficulties. For example, your calculation of the singularity is dead wrong. There is absolutely no geometrical restraint to limit the field of the singularity to the foundation vault. It will extend to the center of the earth."

"That is one opinion, and is the opinion of a musician and not a scientist. Over a hundred trained specialists in the Institute would choose to differ from Your Grace's recently discovered expertise. What Your Grace does not know—and, of course, how could you?—is that the sun is coming into a new galactic alignment with a geometrical pattern that is appropriate to our needs. You have not seen the sheets for these calculations, and, considering your irresponsible and arrogant approach, we doubt that you will."

"I am being responsible to the lives of those living on this island. It is Your Excellency who is irresponsibly toying with the geometry of the solar system for some still unidentified, but I suspect, personal ambition," Viracocha said, and stared defiantly into the Archon's eyes.

The Archon turned his glance away from Viracocha and looked to the High Priest. He was quite surprised to observe that the High Priest did not seem to be so much disturbed as fascinated by Viracocha's assertiveness.

"What you seem to be overlooking, my dear Viracocha," the High Priest said with enjoyment of the scene before him, "is that neither you nor the Institute is working in a vacuum. We knew that the Governor of Anahuac was opposed to the project, but we did not know whether we were being confronted with a single act of revolt or the stirrings of a revolution in which the

military class would join with the Provinces to overthrow the Theocracy. In such a context, it was important to bring everyone into the City of Knowledge, use the intended assassination to our advantage, but not give out all the details of the project to our enemies. If I thought for a moment that your narcissistic display was part of the designs of the revolution, you would find yourself to be the permanent guest of His Excellency in the Institute."

"You mean that Your Holiness walked into that hall knowing full well that there was going to be an assassination attempt?" Viracocha said in a realization that he truly did not know what was going on around him. "But why not simply arrest him beforehand?"

"And why did you not simply go to His Excellency the Archon of the Temple Institute beforehand? Why did you choose to use the vehicle of the Solemn Assembly as a medium for personal artistic expression?"

"I really am confused," Viracocha exclaimed. "Was this morning miracle or theater?"

The High Priest smiled and leaned back in his chair as he regarded Viracocha with affection:

"I must say it is a pleasure to see you confused and not so sure of yourself. It won't do you any harm. So you will understand why I choose not to satisfy your curiosity."

Viracocha bowed his head in submission and then turned to the Archon:

"Well, whether this morning was the theater of miracle or the miracle of theater, I am not likely to know. And supposing that you are correct that the new galactic alignment restrains the singularity, I still cannot see why you want to sever the etheric body from the physical. It seems to me that the project could take us back to the Dark Ages when the soul did not penetrate into the body, but hovered above the dumb animal form."

"What you interpret as a vice," the Archon said, "is in fact a virtue. What this transformation will bring about is a condition in which the spirit will be able to inhabit a body completely."

"But that transformation is reckoned to be millennia away from us in the future!" Viracocha exclaimed in a growing sense

that his original intuition was right and that something was deeply wrong about the Archon.

"Millennia, as expressed in the oracles," the High Priest interjected, "is simply a poetic figure for a goodly length of time. Evolution moves by sudden transformative innovations and long consolidating pauses. No one can be certain as to when an opening will appear. Not even the ancient sages would have thought we could be ready in this cycle, and so they prophesied that it would be the next one. But we have moved faster than anyone would have imagined. It is going to be in this cycle, which interestingly enough, just happens to be during the period of my ascendancy. So, I have not hesitated to move with speed, a quality, I am told, Your Grace can appreciate."

"I confess that Your Holiness moves too fast for me," Viracocha said with a slight tilt of the head, and then turning again to the Archon, he went on the attack: "But just what galactic geometry are we discussing, Your Excellency? Surely, you can't mean the Lemeurian Path? That would hardly provide an excuse for ripping off the etheric body, nor would it likely produce cosmic results."

"Will you please restrain this penchant for irresponsible caricature!" the Archon said as he lifted his cane and stamped the tip on the floor. "We are not 'ripping' etheric bodies away. We are surgically removing them with sound in an act of the utmost medical precision."

The High Priest rose, but in extending the palms of his hands, he indicated that both the Archon and Viracocha were to remain seated. He walked slowly to the corner of the room and looked out of the portals at the harbor far below:

"The view is much different up here, Your Grace. Human beings are fragments to begin with. They are not yet a unity that can become integral with the universe. The universe is open, full of novelty and surprises, like your music. But human beings are only fragments of machinery, moving about in complete unconsciousness. You seem to wish to treasure and protect this individual as if it were a unity. But the individual is only a metaphor for a unity that may evolve tens, perhaps even hundreds of thousands of years from now. We are not going to wait."

The High Priest turned round from the portal, leaned against the wall, and folding his arms over his chest, looked at Viracocha as if he were considering the amount of knowledge he could impart to him.

"What you think of as the solar system is merely the seed inside a fruit. But two different suns in complex rotation nourish the growth of this fruit, and one of these suns is more stimulating for the new genetic information in the inner seed. So first we must remove the fruit, then subject the seed to the influence of the primary sun, and then the new tree will be the true unity, the unity that you mistakenly now apply to a mere human being. If you understood your own 'Canticle to the Sun' as well as I do, Your Grace, you would be more fully aware. I appointed you to the Sacred College because that piece of music, in its inner structure, is isomorphic to the new galactic geometry."

Viracocha concentrated as he had never done before in his life. The silence in the room became infinite. In his imagination he transposed his score into geometrical correlates, but not knowing enough astronomy he did not have any memory images of star charts to lay over the patterns of geometry. In frustration, he discarded the imagery and became very quiet again. Not till he was secure in that silence, did he dare to speak again:

"The subtle bodies are a recapitulation of the larger pattern of fruit and seed. They form the most natural singularity of all."

"Not quite, Your Grace," the High Priest said as he walked back to the center of the white marble slab. "The human race, yes; the individual, no. When you die, my dear Viracocha, what happens? First, your physical body begins to decay and return to its constituent elements. Then the etheric body disintegrates and returns to the universal stream of life-force that flows down from the sun. Your consciousness drifts away in its vaporous psychic body and then unites with the emotional afterimages of other lives and other beings. Then this steamy gas swirls around in the chaos of the dreaming collective mind of the human race, and there other images begin to cling to it. Bits and pieces of other times, other personalities, begin to mix with the residuum once called Viracocha. Clusters of images mass and form more

patterned, but still conglomerate images of desire. Unfulfilled longings, unsatisfied desires, and dominant appetites begin to collect, like garbage floating in a canal at the end of market day. And whatever image of desire is strongest begins to pull it down where living human beings excrete their thoughts into the collective psychic sea. The soldier thought-form comes down to haunt battlefields. The dream of sexuality is drawn to linger over copulation after copulation until finally its fascination causes it to congeal in the liquids dripping into a womb. And where, please tell me, is *Viracocha* in all of that?"

The High Priest regarded Viracocha with an affectionate but mocking smile as he slowly pulled out his chair and returned to his throne before the white altar of his desk. Viracocha did not answer but he found himself thinking that the High Priest's words would make a perfect aria for his oratorio. If he could make the denunciation of the individual strong, then the yearning of the individual to free itself from the pull of the collective could become all the stronger and much more hauntingly beautiful in its passionate longing.

"Precisely," the Archon said as he broke in on Viracocha's artistic reverie with his flat avuncular voice. "We simply take a dumb, mechanical process of *karma* and turn it into a conscious process of initiatic transformation. We surgically remove the etheric bodies from a dozen 'individuals' and then we forge them together through controlled emotions. This controlling of the emotions stops the random driftings which create these mindless clusters. The new etheric vessel is placed inside the singularity in the foundation vault of the Tower where it forms the incarnational vehicle for a more highly evolved entity. Through this more immaculate form of conception, generation by generation, the dark ocean of floating garbage will become thinner and thinner. Finally, it will disappear altogether and there will be no more sea separating Spirit and Matter, and when these two realms come together, our exile will be over. His Holiness said it better than I ever could when he said this morning that space and time will then be inside us."

"Although there is not much time," the High Priest said with a change of tone that indicated that he had made a decision and

the philosophic discussion was at an end, "there is still some time. It should not take more than a month to add a mere seventy-two feet to the Tower. You are known, Your Grace, to disappear for periods of solitude and composition. I have been informed that you have a small cabin on one of the tiny islands near the great western continent, not far from the province where you were born. So, you will 'disappear' for a month. Some will think that you are being banished for impudence, and that perception will serve me nicely. Perhaps you do need discipline, if not disciplining. Genius without wisdom or self-restraint can become a very self-destructive force, and you might give that some thought when you think about a proper aria for the hero of your oratorio.

"I confess, Your Grace, that you fascinate me, and if I were inclined to write oratorios or epics, I would make you my tragic hero. Part of your being is restless and growing unsatisfied with its sheltered environment. It has a longing to see the whole. It is that restless part of you that pushed you out of the ranks of your colleagues and set you down, alone, in front of me this morning.

"Very well, we shall begin to let you see the whole, to understand what it means to create, not simply a work of art, but an entire civilization—not a militaristic state, but an expressive civilization of genius in which works of art such as your own are valued *parts* of a greater whole. And yet, and this is the fascinating aspect to your character, I can also see that another part of your being is afraid and suspicious of the world of power. It is the infantile part of you, the part that loves its own creations and sees the whole world as simply material for its own artistic manipulations. That part of you likes to protect itself through innocence, naïveté, even ignorance. It will be interesting to watch how you will deal with this little civil war between the need not to know and the restlessness to know.

"So, I banish you, Viracocha, but only for a month. Return from your hermitage, and I shall see how it stands with you. If I approve of what I see, then I shall personally supervise the next stage of your education. But now, although I am giving you some time, I cannot take any for myself. There are a few mat-

ters concerning Anahuac to clear up. Please excuse us, Your Grace."

Viracocha stood up immediately, bowed, and took the three customary steps back before he turned to leave. This time the page escorted him down the stairs, through the reception hall, and out onto the balcony. When the page bowed and left him at the stairs leading down to the Broad Street, he stopped a moment to steady himself.

It felt as if the High Priest had reached down into the flowing stream of his soul and stirred up the bottom. All the debris and images were swirling around inside with such force that he no longer had that clear perception of himself.

Viracocha walked to the edge of the balcony above the entrance to the hypostyle hall. As he gazed out over the bay below he struggled to find a new orientation. He noticed that the point on the low wall before him marked the center of the bay, and as he lifted and extended his hands, the right hand covered the copper doors of the Institute and the left covered the wooden doors of the Chapel. Where do these points come into balance? he thought to himself as he felt the pain in his chest.

With no clear resolution in his mind, he dropped his arms and walked slowly past the guards down the stairs to the Broad Street. His thoughts sank down to his feet and as he watched one foot after another take its step, he did not pay any attention to the path. There was no path, he was not walking on firm ground but was creating a path by walking.

He passed mindlessly through the College Gate on the Broad Street, but rather than going directly back to his quarters, he turned and wandered out to the eastern slopes in front of the walls of the College. Once his feet were on the narrow goat trail that went through the low-lying scrub, he remembered. He was startled to find that both the pace of his heart and his feet were quickening, and even his imagination began to race on ahead to wonder whether or not she would still be there.

3

△△

She was. He could see her as soon as he came to the cliff's edge and looked below. She was sitting on the rocky ledge that he had always used when he fled his colleagues to come down to watch the waterspout and listen to the wind and surf thrust the darkness deeper into the cave. But it was low tide at that moment and no jet was being hurled up into the air.

Viracocha waved, but he was too high and directly above her, and to shout might call too much attention to them both; so he ran down the path as it continued along the edge of the cove to come to the fork where the second trail turned toward the sea.

There was an outcropping of rock at one point of the path that jutted overhead and blocked all sight of the ledge from the plateau above. Viracocha felt almost conspiratorial as he crouched down and made his way under the overhanging rock. When he stood up again he could see her sitting and staring into the darkness of the cave. Her hood was folded back this time and her long braided hair curved round her neck and fell from her right shoulder into her hands. She toyed with the knot at the end of the long braid, then, as she noticed him, she stood up quickly, almost in embarrassment, to face his approach.

Her hair seemed redder in the afternoon light, and the contrast of its color against the dark blue cloak with its silver clasp made her seem to belong to no land he knew: from another world and not simply from an eastern province. As Viracocha walked toward her he decided it was her large eyes that made her seem so strange. If they had been dark, and her hair black, she would have been considered beautiful. But her large green eyes were disturbing. He knew that if he came upon her in a crowded market, he would stare at her, just as he was, in an

effort to decide whether her face expressed an unearthly attraction or an unsettling distortion.

"I'm glad you've come," Brigita said as he came nearer. "I was afraid that after this morning's assassination attempt that they might not be so tolerant of your views."

"So, you were there this morning," Viracocha said. "I didn't see you."

"I was in the rear of the hall, next to one of the columns, but I could hear you very clearly. You have a beautiful voice. But how did they take such a song of rejection?"

"The Archon is angry, and I would imagine that I have made a real enemy there. But I think that the High Priest was fascinated. He banished me for a month but promised to supervise my reindoctrination personally on my return. Look, why don't we sit down. You'll be more out of the sun if you sit here on the end of the ledge."

As Brigita sat down on the rocky slab, she put her arms around her ankles and pulled her legs in tight against her chest. She seemed to huddle in her cloak as if it were a protective tent.

"You know," Viracocha said, "I still find it hard to understand how you can stand to wear that great cloak in summer."

"It's my shield. I don't really feel comfortable in this world here on the Great Island, and I won't be relaxed until I am in my vessel on my way home. I don't see how you can live and work in this place. The Institute has poisoned the whole atmosphere." Brigita looked at Viracocha questioningly, but also accusingly.

"You seem angry at *me.* What would you have me do? Storm the Institute, free the peoples of the Empire, and fly off into the dawn of a new era?"

"Now who's defensive?" Brigita responded. "No, I'm not accusing you. Obviously, I wouldn't sit here waiting for you if I thought that you were part of all this. Let's just say that I am constantly amazed by your powers of concentration and self-absorption. You can focus so intently on your work that you just shut everything out, not only the Institute, but your own psychic life as well."

"You mean these 'out-of-the-body' experiences I am supposed to have had with you?"

"Yes, I do. You see, for me what goes on in my psychic life is so much more real, more vast, more important, that it is always a shock to wake up in the morning and find myself encased in the fleshly body again. It's as if every morning I die *from* the spirit *into* the body."

"Well, then, there's your answer. Every morning for me is another birth, another call to work and creation. Though it's true that some mornings I do wake up feeling that I have just come from some place where I was listening to the most beautiful music in the universe. Then I rush to my desk to write out the melodies before they fade out, and it's always a race between my hand and my memory. But I have to write, to work with it. If I simply listened to the music, I would become entranced and float in it forever, making a kind of hell out of heaven."

"I'm not saying that I'm right and you're wrong," Brigita said as she began to relax her defenses slightly. "I'm just saying how hard it is for me to understand you and to accept Abaris' remark that we were to work together. He said that you were 'the man of the future,' that you represented what I needed to understand, I in my ancient ways. I suppose I'm being resentful. I wish merely to study with Abaris, but he keeps lecturing to me about you."

"I'm flattered. But you keep talking about Abaris as if he were alive. I thought that he died over a year ago."

"Death is irrelevant for someone like Abaris," Brigita said with a smile of love and respect that indicated just what depths of admiration she had for her teacher. "I will probably see him again tonight. . . . You know, you don't look very Atlantean. Your skin is too olive."

"My mother was a native woman from the Western Continent. Only my father was Atlantean. I wasn't born on any of the Great Islands. Brigita, excuse me, but what are we doing here? I mean, you said that you wanted to talk to me?"

Brigita looked at Viracocha as if she were thinking over her

reasons once again. She hugged her knees with her arms and looked down to the ground as she spoke to him:

"You said that you've been banished for a month. I would like to talk to you, but not here. I want to get out of here. Could you come with me to Eiru? I mean *right now,* without going back or letting anyone see where or with whom you were going?"

Viracocha looked at her in surprise as she turned her face toward him. What surprised him even more was that he liked the idea of leaving immediately.

"I haven't exactly come here prepared to leave, but then again I guess I have been wanting to leave for some time."

"Oh, don't worry," Brigita laughed at him, "we have plenty of heavy woolen cloaks for you to wear on Eiru. My house is small, it's Abaris' old house, but it's large enough for both of us."

Brigita's laugh and her flirtatious references to her house made Viracocha take a deeper look at her. She in no way was as beautiful as the concubines chosen for the College: her alien eyes, her sickly white face with its bridge of freckles across the nose, and her hair that was neither red nor brown but something in-between made her seem plain and strangely a-human at the same time. There was no way of telling what kind of woman's body lay hidden in the vast recesses of her great cloak, but he guessed from her delicate wrists that she would be bony and flat-chested. And yet, there was something very compelling about her, as if some hidden treasure were within her, something more subtle. He knew that he did want to go away with her.

"All right. For a moment I thought you were simply asking me to come to Eiru, but I'd much rather stay with you. You know, I've never lived with a woman for a whole month. It's never been more than two days . . ."

"Wait a moment," Brigita interrupted. "I think there is something I had better tell you about myself. We have no concubines, male or female, on Eiru. Celibacy is a serious vow in our religion, and I took my vows when I was sixteen, and I have *never* broken them."

"Good God! Then why on earth ask me to come to live with you? Perhaps I had better tell you a few things about myself.

Yes, I am a priest, but that is only because you cannot be a musician, not to mention a Member of the Sacred College, without being a priest. And, yes, that does mean that technically I have taken the official vows, but, Thank God, no one takes those archaisms seriously anymore. In fact, it was our High Priest himself who created the institution of the temple concubines to make certain that no one threatened the priesthood by doing something silly, like falling in love. I don't know too much about the other priests here, for I'm not exactly the kind of man who goes off with the fellows, but I do love the female body. I have been known to get lost in it and disappear for hours, sometimes even days."

"You mean that you couldn't live with me for a month," Brigita said in genuine astonishment, "having meals, going for walks, and talking late into the night, without making love?"

"For a day, perhaps, even two. For a month, I doubt it. Besides, the best heart-to-heart talking late at night is in bed."

"I don't know why you have to translate everything into sexuality. Why can't we just *be* together, simply. I mean, if you like to talk in bed, why can't we just sleep together, simply, like children?"

"Because we are not children anymore. Besides, *life,* you know *living* things, is the translation of everything into sexuality. That's how they got to *be* in the first place. You can't just *be.* Being means thinking, walking, composing, eating, sleeping, making love. I mean, why on earth would you want to sleep with someone and not make love? It sounds perverse."

"It may be *re*versed, but it's not perverse. I had a vision once . . . there is a way to make sexuality into a sacrament, a way of reversing the fall into the body, to create a sacred ritual."

"But sexuality is already sacred. It doesn't need to be *made* sacred. And I didn't fall into the body, I jumped in. And you might too if you didn't try to deny in your religion half of what the body is all about. Look, you have spent, I almost said 'wasted' all these years in virginity. Then you have a dream or vision in which you see the other side of life, how sexuality is not profane and chastity sacred, and the vision threatens your whole life as a priestess. You can't accept that, so as you try to

reach out to a man, you tell yourself that you're not. That it is 'pure' and that you're just going for a walk, or taking an innocent nap. I accept your extending yourself to me, but not on the basis of some 'out-of-the-body' fantasies, but on the basis of a real in-the-body friendship. I've never had a woman as a friend before."

Brigita's eyes grew distortedly larger in anger: "Friend! You don't want a friend, you want a concubine. Do you really think you can exert some kind of control over me to give me an ultimatum that if I don't become your concubine, you will not come to Eiru."

"Control!" Viracocha laughed in amazement. "Who's trying to control whom? Who sets herself into my path, claims to be my old out-of-the-body companion, tells me what position to take in College matters, and then wants me to escape to Eiru to spend a month in some 'Let's live together, but not love together' arrangement. Can't you see the contradictions? Can't you see that you've set this whole thing up?"

"And what you can't see in your patronizing psychology and your demeaning of my visions is that they are not the product of some sex-starved virgin. I am as familiar with the psychic world as you are with the world of music. I know the difference between dream and fantasy, projection and experience. What I saw did come out of the mind of the future. You love counterpoint so much, then why can't you see that two notes held in unresolved proximity can produce another sound?"

"The overtones of sexuality?" Viracocha became silent for a moment as he let the metaphor play about in his imagination. "I'm sorry if I seemed to be demeaning your own religious experiences. I guess I'm really trying to protect myself. I don't know whether I could take the intimacy of living with you for a month."

"Is that an insult or a compliment?" Brigita said beginning to speak with increasing speed and intensity. "But then you're not sure whether I satisfy your 'love of the female body' or not, are you? Of course you're protecting yourself. You're always protecting yourself, whether through concentration on your work in complete oblivion to what's going on around you, or through

some hedonism that you like to think of as a 'contemplative meditation on the cosmic mysteries of the female genitalia.' But while you were losing your abstract mind in all those women's bodies, I'm sure that the concubines probably gained a better understanding of you than you ever did of them!"

"We're even," Viracocha said softly. "I flippantly demeaned your visions. Now you have demeaned both my art and my sexuality. That doesn't leave much of me, does it?"

"Oh God!" Brigita exclaimed in frustration and anger at herself. "How did we get into all of this? I'm sorry. I really am. Anyway, there is a lot more to you than your art or your sexuality. I know one and don't know the other, but it wasn't on the basis of either that I asked you to come to Eiru."

Viracocha looked at Brigita in surprise. Brigita did not avert her eyes but looked all the more strongly into his, as if she were trying to make him see what she saw in him.

"Then on the basis of that," Viracocha said, "I accept your invitation. Brigita, I don't know you or what I am getting into. All I do know now is that there is nothing in me that wants to turn and go back into that closet of a Sacred College."

"Then that makes two of us who want to get out of here. My vessel is down here in the little cove around these rocks."

Viracocha followed Brigita down the path around the rocky walls of the inlet. He was just beginning to wonder how long it would take to sail from the Great Island to Eiru when he came out of the narrow canyon and saw the small silver garuda floating in the rising tide of the cove.

"But how did you ever get one of these?" Viracocha asked in surprise. "All garudas are restricted to the Empire."

"This is imperial," Brigita answered. "And the pilot is Atlantean. I am the hereditary custodian of the Treaty of Atlantis with Hyperborea. The vessel enables me to stay in official communication with the Empire, so effectively it is there whenever I need it."

"I've never seen one this small. They are actually more beautiful on this scale," Viracocha said, as he admired the grace with which the vessel responded to Brigita's signal and began to glide toward them.

When the garuda reached the edge of the rocks where they stood, it lifted silently into the air and the hull opened to extend a footbridge to them. Brigita gathered up the hem of her cloak and her white smock and stepped up. She did not turn around until she was inside, where she stood at the door and waved to Viracocha to follow.

There was almost a dream-like quality to the floating vessel and the foreign woman beckoning, and as Viracocha stepped up off the island he felt a great sense of excitement and anticipation.

As the hull closed behind them, one of the two crew members turned back to face the controls. Viracocha looked around to orient himself.

"The best view is from the observation room all the way to the back," Brigita said. "The sleeping cabin is here, the storage room here, and the toilet there. I'll join you in the observation room in a few moments after I have met with the crew. We won't arrive in Eiru until night, but the summer sun will still be up."

Viracocha began to explore the trim little ship, from the small closet of the toilet, to the storage room, and to the small sleeping cabin with its two bench-like beds on either side of the narrow aisle.

The observation room was the largest of all and had a curving oval portal that enabled one to see up as well as to the side. Two swivel chairs were mounted to the floor, but there was a low curved, padded bench all along the stern of the vessel under the window. Viracocha moved to the portal and watched the ship pass out of the cove. As he looked up at the cliffs above him, he noticed a figure standing at the edge and looking down on his departure. Even at that distance he recognized the colors of the uniform of a page of the Palace, and Viracocha knew that his own pale blue robes and Brigita's small garuda would make no mystery of where he was going and with whom. Even at the moment of his flight, his world was still maintaining its hold on him.

Viracocha stood by the window, watching his observer grow smaller as the vessel moved away from the shore. It sped along

just a few feet above the surface of the water and it was not until they were several miles from the coastline that the garuda began to rise in the air to pass through the clouds. Yet even from that distance, Viracocha could still make out the Tower as the afternoon sun struck it and displayed its great corona of light. He stared for a long time at the tip of the Tower, for he knew that the next time he saw it, the slender pinnacle that housed the solar crystal would be unstably higher in an unbalanced sky.

As Viracocha turned away from the window to step up into one of the observation chairs, he saw Brigita leaning against the entrance and studying him silently with that same diffused gaze that he had first noticed with the High Priest. She had removed her cloak and was wearing a white sleeveless linen smock that was bound at the waist with a simple hempen cord. Her hands clasped her elbows so that the upper arms framed her breasts. His attention was immediately drawn to her breasts, for he had wondered what kind of feminine body lay hidden in the folds of her great cloak.

Her breasts were small, which did not surprise him, but what he noticed with a growing sense of fascination was the outline of her nipples, which showed very clearly through the white material. The areolas were quite large, but there was no suggestion of a nipple thrusting through the smock. She appeared to have those delicate, full, virginal areolas that were not tough and leathery with a stubby teat protuberance sticking out. He had only known one concubine who had had areolas like that, and he had been amazed to discover how different the satin texture of the skin was where the breast gathered itself into the concentration of the circle. He had become so fascinated with the young concubine's girlish breasts that her areolas had become a fetish, an icon that allowed him to become lost in astonishment. Viracocha found his breath quickening in surprise and anticipation. Now he wanted to know everything about Brigita's body: would she have those long and full thighs, the very soft and straight pubic hair. Almost as if in answer to his question, Brigita's knee moved forward, and the smock fell around the form of her thighs. Viracocha could not sense through the linen smock just what the texture of her maiden hair would be, but he

raised his glance to examine her eyebrows, for he knew from experience that you can tell the texture of a woman's pubic hair from her eyebrows. But as he raised his glance from her thighs to her face, he found himself staring into the fierce, wide eyes of an angry Brigita:

"You're beyond belief!" Brigita exclaimed in exasperation. "I've seen men look at women before with lust in their minds, but you've turned lust into a religion. I mean admiring a pretty face is one thing, but you've got a shopping list of thighs, pubic hair, and nipples, as if they were all separate items you could pick up in market stalls."

"Oh God!" Viracocha exhaled. "So you really do read minds, and it wasn't a joke. Well, if I've just undressed your body, you've just undressed my mind, which just might be the greater invasion of one's privacy and personal integrity."

"Believe me," Brigita said as she reached down to grasp the hem of her smock, "it's not a talent I worked to develop, for it's *my* privacy that is lost." The last word was mumbled into the folds of her smock as she pulled it over her head and threw it to the floor in defiance and disgust. "There! Satisfy your wretched curiosity, so we can get beyond this completely absurd level."

Viracocha stared in amazement at the naked Brigita and watched her areolas shrink, shiver, and harden into defensive points. But he accepted her defiance with his own defiance and did stop to notice that, though she was slight, she was not bony, and her thighs and hips were well proportioned for her light frame. As his glance descended along the inner line of her legs, his eyes came to rest on the crumpled linen smock on the floor. With a bow that seemed as much to express honor to her body as much as a request for forgiveness, he stepped toward the white cloth, reached down to return it, and said nothing as she lifted her arms to dress. And yet, for all his surprise at the display of her naked anger, Viracocha could not stop from noticing just how her small breasts stood out as she lifted her arms above her head.

Brigita smoothed the smock over her hips, threw the cord to the side on the bench and walked over to the second observa-

tion chair. She remained silent and looked away from Viracocha toward the vanishing land.

"It's my privacy that is lost," she said without turning to face him. "I can't go for a walk without people's thoughts coming to me, like the scent of manure on the wind. I can only have peace in absolute and total privacy. But you," she said as she turned round to consider him, "your mind is a booming chorus. No wonder you're a composer. I don't know how I'm going to get through this month."

"Then why did you ask me?" Viracocha said as he moved to sit in the other observation chair.

"Because of the prophecies of Abaris, of course, and because of my own visions," Brigita said, and sat down in the chair next to him. "But I had no idea that it was going to be so, so difficult, and so irritating. I mean, why can't you accept a woman for what she *is?* Why all this fussiness over disconnected bits of anatomy?"

"I don't know. I guess I'm trying to find out what woman *is.* It's an aesthetic response, much like a response to certain sounds or colors. Who knows? Maybe I am precognitive, the way you are telepathic, and I have had this vision of you from the beginning and have been searching about among the bodies of other women for the breasts of Brigita, the thighs of Brigita, the . . ."

"Stop! Please stop this, this courting of me! Thank God, I have not had to put up with all this courtship nonsense over the years. *Please* don't court me. I loathe it. It's like watching a little boy trying to show off in front of a little girl, and all the time he makes such an ass of himself and can't see what it is the girl really likes in him."

"So, you see," Viracocha said smilingly, "you do have your own aesthetic response to what is pleasingly masculine and what is not. There are certain masculine *actions* you like, just as there are certain feminine physical *forms* that I like."

"But that's just the point," Brigita said, "these forms are objects to you. The woman isn't a person to you. She's either a collection of objects, or an idea, *Woman.*"

Viracocha laughed in enjoyment as he noticed just how much

this young woman liked being courted with the mind. Small wonder that the only man in her life had been a teacher, he remarked to himself, but said: "In other words, women are right and men are wrong. Now, please tell me what is a mere object? If one truly *sees* a flower, a mountain, a woman's breast, it is full of power and Being even at rest. It declares *presence.* It celebrates the Creation . . ."

"But—" Brigita exclaimed.

Viracocha held up his hand. "Wait! Don't interrupt me yet, my solitary hermit. We're in the Great Male-Female Debate, the cosmic dance of opposites, and you have to move according to the rules."

"And if I don't play according to the rules?" Brigita asked.

"Then you lose by default and I win."

"And if I do play according to the rules?" Brigita asked in mock flirtation.

"Then I will win and you will lose. That's what rules are for; just ask any civilization and it will tell you that."

"I thought it was something like that," Brigita said. "But go ahead, finish your point."

"I was going to say," Viracocha said with a grand flourish, "that if you look at an *action,* a bird flying, a horse running, a woman making love, the action too declares *presence.* It too is a celebration of the Creation. So, if you truly see an object saturated with Being, or if you see an act that is truly expressive, they both come from the same center, or the Center. So, I don't see, for example, those large, marvelous, virginal areolas of yours as mere objects, or disconnected bits. You know, the difference that seems to keep coming up between us is that you keep talking about 'Being' as if it were limited, almost contaminated, by bodies or physical forms . . ."

"You do promise to finish, don't you?" Brigita interjected.

"I'm nearing the dramatic end, don't go away. As I was saying," Viracocha continued, gesturing with his finger in a magisterial fashion, "I see the physical form as projected by the being, much in the way an organism will extrude a lovely and delicate shell. Yet if one looks at the geometry of the shell, the 'pure'

mathematics of the *thing*, then one sees into the Mind of God, and the *object* declares the *presence* of Creation. Finished."

"In other words, my dear Brigita," she said mockingly, "thus we can conclude that Men are right, and women are wrong."

"Precisely." Viracocha nodded in approval. "I couldn't have put it better myself. By the way, if you can read minds, why am I talking? Am I supposed to just think at you?"

"You're talking because you obviously enjoy it, so far be it from me to deny you *that* pleasure. But as for thinking at me: no, words are better. They sharpen the images and give clarity and sparkle to what would otherwise be a muddy stream of confused emotions."

"Good," Viracocha said as he began to swivel in the chair and noticed that his feet could just touch hers. "I would rather talk to you than think at you. I would also like to let our great cosmic debate drop to understand a few more simple things, basic things, like information."

"As for example?" Brigita said as she turned her chair so that her feet touched his. Then she closed her eyes, as if she were listening to a faint sound. Without opening her eyes, she pushed off her sandals and placed the tips of her toes, tentatively, against his feet. Viracocha quickly kicked his sandals off and placed the soles of his feet against hers. Two neat parallel lines appeared on her forehead as Brigita seemed to concentrate. She move slightly in her chair to adjust the position of the soles of her feet against his, and then slowly opened her eyes.

"What are you doing, Brigita?"

"I don't know, listening I suppose. Close your eyes, and listen with your feet for a change. There are more vibrations than sound, you know."

Viracocha closed his eyes and then became aware of a very high hum. He sensed that if he could only move his mind up to its level, the hum would begin to differentiate into a more melodic pulse.

"What's going on?" Viracocha asked in serious puzzlement. "Is this part of your 'Mind of the Future'?"

"I don't know," Brigita said very quietly, as if every sense in

her being was alerted. "I think we should let it be for the moment. You said you wanted to have some basic information?"

Viracocha closed his eyes again to try to hear the melodic line at the edge of his perception, but it was too far away, so he decided to relax. Perhaps, he thought, if he didn't strain to reach toward it, it might descend naturally. He opened his eyes and looked at Brigita and was amazed to notice how different she seemed. All her features were the same, but now everything that had once seemed plain or distorted became arrestingly, dynamically beautiful. Yet at the same time, she was frightening, and the energy that poured out of the soles of her feet made bands of color appear to pulse around her whole body.

"Try to relax," Brigita said with a calmness that expressed both knowledge and authority. "It's too soon to jump into analysis and interpretation. What did you want to ask me?"

"Well, for one thing," Viracocha said as he gave up trying to understand or control what was happening, "what do you mean when you say that you are the hereditary custodian of the Treaty with Hyperborea? You can't be Hyperborean, can you? I mean, they left for the Pleiades ages ago."

"The colony returned, yes. But a presence was maintained until just a few centuries ago. On my father's side, my lineage goes all the way back. On my mother's side, I am Eireanne, though I wasn't born on Eiru, but on a small island to the northeast called Iohnah."

"That accounts for the large eyes then," Viracocha said as he moved the soles of his feet away and then discovered that he could no longer hear the ethereal humming sound. Quickly, he placed his feet against hers, and almost in gratitude, Brigita pressed her feet against his. "And my second question is: How did you first meet Abaris?"

"To answer that question," Brigita said, leaning back and placing her hands behind her head and closing her eyes, "I would have to tell you something about Eiru."

Once again Brigita's forehead became creased with two tiny lines as she seemed to be listening to the sounds: "Yes, I think this is the beginning of what I saw in my vision."

Viracocha closed his eyes again. The sound was still high and undifferentiated, and yet one part of him knew that at another level it was a melodic pulsation. It seemed stronger now; he could feel it move up from the soles of his feet into the base of his spine. "What does it mean?" he asked, unable to contain his need to know just what was happening to him.

"I think it means," Brigita said as she opened her eyes to look into his, "that the subtle bodies are becoming harmonized, or perhaps even becoming connected."

"A mind is a strange thing, isn't it?" Viracocha commented as he returned her gaze. "I don't know what your explanation really means, and yet I feel better already. Perhaps, because you make it sound so, well, sexual."

"It is sexual," Brigita said as her eyes went out of focus and took on the diffuse stare he had first seen in the eyes of the High Priest. "Don't be alarmed, I'm just looking at your aura, at the edges where ours pass into one another."

"The High Priest looked at me that way," Viracocha said. "Does that mean he can see auras and read minds?"

Brigita's eyes snapped back into sharp focus. "I'm sorry to hear that. That will make him more dangerous. If he can see auras, he can pick up your feelings. It doesn't mean that he can read minds. God! I'm glad to be off that wretched island."

The mood snapped with Brigita's change of feeling, and Viracocha felt completely and suddenly disconnected as Brigita removed her feet, wrapped her arms around her knees, and huddled into a tight defensive position.

"No!" Viracocha found himself almost shouting. "I feel ripped apart. That was too sudden a move. Please, let's go back where we were."

"I'm sorry," Brigita said quietly, then placed the soles of her feet against his. "Close your eyes, then, and listen. It will take me a little while to answer all your questions about me, about Abaris, about Eiru."

Viracocha leaned back into his chair and was relieved once again to feel her naked feet, and to enter again that atmosphere of color and sound that was erotic, and yet completely alien to any sexual touch he had ever experienced before.

"Let's begin with God," Brigita almost whispered. "You Atlanteans see God in the image of the sun: powerful, life-giving, transcendent but remote. We see God in the image of earth: powerful, life-giving, close, and intimate. You see God as masculine, but for us God is the Mother, the womb from which everything comes."

The sound of the high vibration that surrounded them seemed to descend and begin to penetrate the vibrations of Brigita's voice. Her voice became physical, as if Brigita's tongue were not simply speaking but passing all over his body like the kisses of a skilled concubine. Just as Viracocha loved to stretch out, close his eyes, and drift into trance as the concubine would kiss, lightly trail her free-falling hair, or softly move her breasts over his thighs, so now did he stretch out and float into an erotic reverie as Brigita's entrancing voice moved through and over his entire body.

"Because you Atlanteans see God as a star, a concentration of light, you see society in terms of cities, star-like concentrations of power. We have no cities on Eiru. Oh, we have great fairs in the summer when thousands of many-colored tents are set up. And then we have plays, dance and song, and the recitals of the poets. But these cities of summer melt away, and people go home to prepare for the harvest and the coming winter. Because we see God in the earth, we have no temples. It is difficult to build a temple more beautiful than the earth itself. Ours is a religion of listening, but yours a religion of chanting. Your temples are always full of a hypnotic droning. You'll notice the difference between Eiru and the Great Island right away in the landscape itself. Your island is raw, rocky, elemental. Human civilization has been imposed on those volcanic rocks by an act of will, and you can feel that the spirits in the rocks are indifferent to you, almost anti-human. Since you're not listening to them, they're not listening to you, and the concerns of your civilization are irrelevant to them. We've built up a whole culture from listening to streams and rocks and trees, and the spirits like it. They reach out to us, and when they appear, they put on a half-human form to please us. So we're part of the landscape, and they're part of us."

Almost as if he too could read minds, Viracocha followed Brigita's narration, seeing the images exactly as she must have seen them. The tone of her voice, the rhythm of her speech, and the strong sense of physical connection through the soles of their feet worked upon his whole being. The stream of her words entered and moved deeply inside him. It was not what she said, but the way in which she said it that seemed to exercise such complete authority, as if she were the absent owner of his own mind who had returned at last, not to claim it, but simply to move in to make it her home. The boundary between them had shifted, and as Viracocha opened his eyes to question the edges of his own body, he looked at Brigita and it seemed as if he were standing on a mirror in which the images had become reversed: male had become female, Atlantean had become Eireanne.

In some part of his being, Viracocha knew that Brigita had put him into trance, but he did not care, anymore than he had cared when the voice of the High Priest had put him into trance. The voice of the High Priest had expressed dominion, not dominance. The voice of Brigita expressed entrance, not intrusion. It felt as if he had become a woman and now her voice was the masculine power impregnating him with visions of a life to come.

"Please don't stop," Viracocha said softly, in an effort not to break his own receptive mood. "Go on, tell me more about Eiru. I can see it as you speak."

"Because we know that the rocks are alive," Brigita said in her foreign accent that Viracocha was beginning to find not alien at all, but the right way, the most musical way to pronounce the words, "we do more than talk to them, we compose poems to them. There isn't a great rock, stream, or mountain that doesn't have five poets competing for her favors. The poet who wins the prize in the summer festival spends the year traveling around the countryside and being taken in as a guest everywhere. It's considered an honor and a blessing to have a prize-winning poet stay in your home. The competition among the poets during the summer fairs is very keen. And if there is a tie, the rock or stream is consulted, and a person like myself is called in to ask the opinion of the spirit."

Viracocha opened his eyes to look at Brigita. The passive trance began to recede as his mind returned to try to interpret just what sort of woman it was who held the naked soles of her feet against his. "And have you always been able to see these spirits in nature?" Viracocha asked.

"From childhood, yes," Brigita said in a changed tone of voice, as if she accepted his coming up from the depths to resume normal talk. "If a child is strongly favored by the spirits, the parents will bring him or her to us, and we begin the slow training. It's all fairly simple. Since we have no temples or cities, there are no taxes to levy. The people donate the little that we need, and in return we will officiate at the festivals, or at a birth, a marriage, a funeral. I lived with my parents until I was twelve, then I went to live in the forest with a priestess."

Viracocha was now fully alert and his mind began teeming with questions. He wanted to know everything about her:

"But you are 'Brigita, Abbess of the One Thousand of Eiru.' That sounds like a fairly large temple to me."

"The one thousand huts are spread all over the island. We prefer the solitude of prayer or communion with the spirits. The men and women who live apart only come together for the celebrations of the four quarters of the year. When I left my parents . . ."

"What were they like?" Viracocha interjected.

"My father was a craftsman, a goldsmith, but one who knew a great deal about the healing qualities of various gemstones. My mother was a singer and a weaver. When I left my parents I went to live with a foster-mother, a hermit who had her hut in the forest nearby. It was a good life in the forest. I had a hut to myself, next to the old woman's, and my only chores were the small domestic ones for the two of us. Then, when I turned sixteen, I started to have intense visions, nightmares, and I was beginning to spend more time out of my body than in it. That was when the old priestess realized that I had a special calling. So she sent me to Abaris. He was living in a rock house on top of a mountain by the sea, the house where we will be staying."

"But Abaris didn't come from Eiru, or did he?" Viracocha asked.

"No one knows for certain," Brigita answered. "I always assumed he came from Sham'vahlaha, the etheric city above the great lake within the Vast Continent. Abaris was always traveling, and yet he always seemed to be there when I needed him. He taught me several meditational practices and insisted that I dance with intense physical exertion every day. That helped. The nightmares went away; my lunar period came, and then as the meditational practices began to work on my body, my lunar cycle went away entirely. It was then that this sensitivity to people's thoughts and feelings developed, though I have always been able to see auras."

"You make me feel positively leaden and crude," Viracocha said. "I've never seen an aura in my life."

"Yes, I can see that. But the strange thing about you is your ability to see archangels. I can only see the angels in nature or the ones that come down closest to the human level."

"I hate to disappoint you, but I don't know what you're talking about."

"Then what do you call those ice-blue crystalline forms that dance in your imagination?" Brigita asked as she lifted her hands to hold them behind her head. She stared through him in such a peculiar way that he felt awkward and embarrassed at the thought of just how completely naked he was to her sight. Almost in a compensatory defense, his glance was again drawn to her breasts, which were lifted by the raising of her arms. He looked again in wonder at the very large circles of her areolas, but he found his mind drawn back to the original experience of wonder when he first saw those ice-blue crystals:

"I wouldn't call those archangels. I think of them more as ideas of pure form that seem to underlie the performance of nature. I only see them when I am really deep in the best kind of concentration for composing. And I wouldn't say they dance, but that their movements seem to go in and out of more than three dimensions as they rotate. I think of them as the perfect notation. I'd like to re-form musical notation to make it like them, for then musical notation would be to a concert what they are to external nature."

Viracocha closed his eyes, for he felt from the energy that

seemed to pass between them now that perhaps an old problem would find a new solution.

"That is exactly how Abaris explained archangels to me," Brigita said, lowering her hands to her lap. "He said that I was the ancient mind of humanity and could see and feel into the spirits of trees and streams. But he said that you were the new mind that stepped back and away from nature to see beneath into the architecture of being, the archangels, the ideas in the mind of God. I can see now that that explains your peculiar sexual intensity."

"What!" Viracocha exclaimed in complete astonishment. He felt as if he were a musical instrument in Brigita's hands, and that she was testing it by going from one end of its range to the other very quickly to see the chromatics of the extremes.

"Can't you see?" Brigita asked as if it were patently obvious. "You don't *express* lust; you lust after lust. You're fascinated by the female body, but you're no more in the physical body than I am. We just seem to move in different directions when we leave it. I can begin to understand now why Abaris thought we would be drawn together. It's not just the crisis brought on by the work in the Institute. It's something else."

Viracocha remained silent as he tried to understand the emotions swirling around inside him. Strangely, he felt surprised that his confusion was making him angry. This baffling woman was constantly alluding to knowledge she never explained, constantly plucking at his emotions, and constantly leading him around like a cow that was too stupid to go into the shelter before the storm. As he stopped to consider just exactly how little he knew of where they were going and what they were going to do during this magical month alone together, he began to become really annoyed. But Brigita made matters only worse by beginning to laugh at him.

"I like you!" Brigita exclaimed in delight. "When you're confused, you get angry and would slap the world that dared to toss you around. When I become confused, I close up like a sea-anemone and huddle deep inside myself. All right, I'll stop being so cryptic. I may be a virgin, but I have seen men's thoughts in all their physical vividness. You are different. You

didn't 'jump into the body,' as you claim. You were pulled in and you awoke with astonishment to find yourself within it. You're drawn to women's bodies because you're searching for the exact shore of incarnation where the ocean of Spirit touches the continent of the flesh. You stare up into a woman's vagina as if you were looking for the secret of the universe. And yet you can't, or I should say that you haven't yet been able to join that fascination with women's bodies with the vision of the archangels that takes you over. The picture I see of you as I press my feet against yours like this is of a solitary man walking along high cliffs and rocks. He knows that somewhere the immensity of the cosmic sea touches the earth in a gentler way, but he has not yet found the sandy beach where he can lie down and be *in* the sea and *on* the land at once."

Viracocha's anger disappeared as he felt just how much Brigita was alluding to the thought that everything he had been looking for in woman, he was now going to find in her. The sheer flirtation of it was tremendously exciting as he began to appreciate just how much more of everything marvelously feminine there was when a woman was not a concubine but a companion on every level. As he looked at her he felt a deep hunger for everything that she was. The energy pouring out of her body into his awakened every cell in him and as an intense erection began to rise up, he felt the erection go inward as well as out, as if his whole spinal column were part of it.

Brigita looked down at Viracocha's erection lifting up beneath his robes like some volcano coming up to alter the ground between them, and then she looked to his forehead, as if she were looking for something in his aura. With a quick movement she released the touch of the soles of her feet, swung her legs over to the side, and stood up and stretched.

"Before we deal with that kind of hunger," Brigita said as she twisted and released her body in some dancer's exercise, "I think we should start with a more basic hunger. I'm starved. I waited for you for hours, and I don't think you were exactly having lunch with the High Priest, so let me serve you, Your Grace."

Brigita bowed with the gesture of a concubine, and Viracocha

thought to himself, as he felt the intense pressure of his erection, "I really wish you would."

Brigita looked into his eyes, smiled, and then looked down at his robe:

"You're like the Great Island: a volcano hidden and waiting for release. But here, there's another way to treat this situation, something I'm sure your concubines haven't shown you."

Brigita moved closer and placed her hands on Viracocha's head. The touch was gentle, almost compassionate, and the energy was even stronger than what he felt when their naked feet had touched. Then slowly she dropped her left hand to the base of his skull and with the tip of her finger she began to move in little circles right at the point where the spine entered the base of the skull. The sense of pleasure at her touch was exquisite. Then slowly she moved her right hand down, and with thumb and forefinger she began to pinch and shake the bridge of his nose. Whereas the touch of her left hand had been gentle and entrancingly relaxing, the touch of her right hand was strong and almost harsh. Instantly, he felt a snap, then a crack of lightning inside his spine, and his eyes rolled up, almost it seemed, to the top of his head. He could feel his eyelids begin to flutter like the wings of hummingbirds, and then he went up in some brutally fast ascension into a state of complete and thoughtless energy.

"Stay up *here,*" Brigita said as she pushed hard on the point between his eyes. Then she released her hold on him. He heard the panel slide as she moved into the galley, but he remained stunned, suspended in some state of etheric shock.

Very slowly, Viracocha came down, opened his eyes, and exhaled in mental and physical exhaustion, He stood up and moved to the bench to stare out at the sky. His erection had not dropped, but it had let go of its intense claims on him, and he knew that as soon as he ate some food, it would go away.

Is this what it is going to be like for an entire month? he thought to himself, and then realized just how completely ignorant, powerless, and confused he really was. Part of him wanted to follow her lead into whatever secret states of mind she had knowledge of; the other part of him wanted nothing to do with

it, but wanted to end this frustrating flirtation and make love to her immediately, then and there in the garuda flowing through the clouds. With the energies she had aroused in him, he knew that he could make love to her again and again and again and again.

"I'm sure you could," Brigita said behind him as she returned with a tray of food. "But you must let *me* lead in this. We will become lovers all right, but not in the simple and prosaic way you imagine from your experiences with the concubines."

"You know, for a self-proclaimed virgin, you certainly seem to know a lot about sex," Viracocha said as he turned to watch her carry the tray and set it down on the cushioned bench under the rear portal. Brigita sat down on her side of the tray and then motioned for him to sit on the other.

"That's because you make the mistake of thinking of virginity as the absence of sexuality. It isn't, anymore than silence is the absence of prayer or thought. In fact, virginity, at least as I know it and practice it, is an intensification, a concentration on the mystery underneath sexuality."

Once again Brigita spoke of sexuality in a way that seemed to hold the promise of numerous undiscovered delights.

"I wish you wouldn't flirt like that," Viracocha answered. "If you can't stand to be courted, I am beginning to think that I can't stand to be flirted with, at least, by you. You get me so damned excited that I begin to feel like a mastodon in raging must."

"Yes, I can see that is going to be a problem for us," Brigita said with a lovely smile that made Viracocha realize that defenses were too late and that he was already hopelessly, stupidly infatuated with her. The grand scheme of the High Priest, to protect his geniuses from emotional stupidity, had failed. Brigita had him eating out of the palm of her hand.

With a laugh at his thoughts, Brigita reached down for a grape and placed it at Viracocha's mouth:

"Here. For heaven's sake, stop feeling sorry for yourself. If you'd only bothered to remember our out-of-the-body nonsense, as you call it, you'd know why I am here with you. Whatever vision I had of the future was a vision of us. I may be

leading in some of this, but I'm not manipulating you. You're not clay, and I'm not the potter. *We* are being called. But let's drop it for the moment and enjoy the food. Here, try this; it is a dry mead that we make on Eiru."

As she held out the clay cup to him, Viracocha looked at the strange inscription that encircled the lip:

"What does this mean? Is it Eireanne?"

"Yes." Brigita said. "It's the prayer of the maker of mead. It says: Men are flowers, gods are bees, and honey what we shall become."

"Then I will take that as the grace and drink it in gratitude," Viracocha said as he held the cup aloft. "But this is good! It doesn't taste like wine or brandy. I confess I thought it would be rough and sweet, sort of like you."

"And instead, how do you find it?" Brigita asked with a smile filling out her lower lip in a way Viracocha found to be very attractive.

"Unpredictable, an interesting combination of sharp, precise, but unidentifiable herbs, and very smooth in a sensuous way that is kind to the lips and tongue. Is it strong?" As he spoke, Viracocha returned Brigita's smile, took the cup to his lips, and looked into her eyes as he took another exploratory sip.

"Very," Brigita said in acceptance of his little game. "You had better be careful."

"Good," Viracocha said, and took a very large drink and then turned his attention to the bread and cheese.

They ate silently for a moment, but Brigita never took her eyes off Viracocha and seemed to be studying him intently as he ate and drank the entire beaker of mead. After a few moments, she looked at him more quizzically and asked:

"What is the name of that island in the middle of that blue lake? It seems to keep returning to your thoughts."

"You mean Tiahuanaco?" Viracocha answered. "Yes, for some reason the smell of the herbs and the taste of the mead started me thinking of it again. I haven't for some time. It's a small island high in the mountains of the Southern Continent. When I was very young I worked there on an agricultural mission establishing varieties of tubers that would grow well in the high

altitude. We were studying the effect of cosmic rays at high altitudes and their ability to enhance certain enzymes within these foods."

Viracocha turned to the side and stretched his feet out and became silent once again as he ate and thought back on Tiahuanaco. After a moment, he began to speak again:

"The top soil by the lake was very good. We could sink one of our metal rods into the earth until it completely disappeared. There was a little village there by the shore of this incredibly blue lake. But though the soil was good, the natives did no farming, for the marshes by the edge of the lake provided them with plenty of fish and game. I keep remembering the place because that is where I decided to go into music. One morning as I was working in the fields I kept hearing this strange music that seemed to be coming from the small island in the middle of the lake. The natives were afraid to go to the island, and only the shaman would go there once or twice a year, but I knew he was still in the village. I tried to ignore it, but it only called me the more strongly. So, finally, I threw down my tools, took one of the small reed boats, and paddled out to the island. When I came to the shore I found a stream flowing down the hill, so I followed the stream up to its source in a rocky crevice on top of a hill overlooking the bay where I had left my reed boat. I could hear the music very strongly there, as if it came out of the spring, so I sat down to listen. I seemed to drift off into some reverie as a stream of images began to flow into my mind.

"I saw a priest playing a flute—but I seemed to be behind him. We were on the top of some huge man-made mountain or temple. The priest felt me standing behind him and turned around. I asked him where I was, and he said: 'This is the land of the serpent that has learned how to fly.' Then I woke up, but it was already dark. I drank some of the water from the spring, and then curled up to sleep there until morning. I don't remember dreaming, but I heard the music again and simply floated in it. There was no sense of time passing, but when I woke up again, it was morning.

"Anyway, I forgot about it all, returned to the Great Island to study architecture, and then one day, about ten years later, I

saw a tall, gray-bearded figure alight from one of the barges in the harbor. He did not look at me, but walked past, but as he did I saw the medallion on his chest, the winged serpent. I ran after him, insisted he tell me the meaning of the medallion. And that was it. I spent the day with him; he gave me various meditation practices like kriya, and then I never saw him again."

"You mean, you never saw him again in the physical body. Actually you saw him every night. It's quite unusual that he gave you so many advanced practices so quickly. But clearly Abaris knew what he was doing, for all your centers are intensely bright."

"Well, as you know, I don't know about those sorts of things. I have absolutely no memories of any 'out-of-the-body' experiences at all, with you or Abaris. I kept up the practices because, . . . I don't really know. It was simply that there was something about Abaris, his eyes mainly, that had taken hold of me. They seemed to express a love that had been forged in hell. I had fantasies that he had confronted an Evil that was beyond me. As if he had been tortured once and found even in the torturer the desperate act of man screaming for God, a man twisting and negating, but all the time screaming out to God: 'How can you let this be? What do I have to do to make you destroy me?' As I talked with him on that single day, I had the image of a torturer wringing infants' necks with his bare hands and crying out: 'How can you let this be?' "

"That was no fantasy," Brigita said quietly. "What you saw was one of his past lives, only it was Abaris who was the torturer. We, all of us who are now his students, were once his victims. Even you, or else you would not have so quickly recalled those images."

Viracocha's eyes darted away from Brigita's glance. He looked down to his feet, then realizing that he did not want to look at anything, he closed his eyes tightly. He could feel it coming, and he worked to get control of himself: "I don't want to hear that! I don't want to hear that!" He started to pound the cushion of the bench as if he could beat back the wave of emotion that was rising up and beginning to knock aside all his controls and definitions. He did not want that wave to touch

him. He did not want to sit there with Brigita looking at him. But more than anything he did not want that wave of emotion. He jumped up, as if he could run away from it, ran toward the door, flung it aside, and saw nowhere to go but into the small sleeping room to the left. He was just able to reach the narrow cot and throw himself face down onto it before the wave struck him.

He had cried before in his life, but always from his eyes. But these tears did not fall gently from his eyes, they were torn out of his guts in great, wrenching sobs. He struggled to gain control of himself, to get up on top of that totally irrational wave of emotion that had engulfed him, but there was no longer the person there that he knew as himself. He had been violently knocked to the side by this other being of a thousand lives, a being that dwarfed and mocked the feeble definitions of self that he liked to think of as "Viracocha." There was no way that that tiny little aggregation of habits and opinions was going to control this other being. The emotions that had nothing to do with being Viracocha tore him to pieces and he sobbed into the corner of the wall as if that other being were wringing the tears out of him, twisting him and smashing him on a rock like a piece of washing in a stream. He could not stop. One spasm of tears was followed by another. There was no object to his crying, no feelings of guilt or innocence. He was both torturer and victim at once and he cried out in memory of the agony of both. Pathetically, in moments when he struggled to catch his breath, he would try to gain control of himself, but it did no good. He could not stop it; it had stopped him.

He sobbed violently for more than half an hour, and then finally the attack ceased, and he lay ravaged like a landscape through which an earthquake had passed.

When Brigita sensed that the seizure had passed, she came from the observation room into the sleeping cabin and slid the panel door closed behind her. She did not try to reach out to console him; she simply let him be and came in to be with him.

Brigita sat down on the opposite cot against the other wall and stretched her legs out on the narrow bed. Viracocha turned

his face from the corner to look at her, and then slowly he sat up.

"I'm sorry," Viracocha said.

"For what?" Brigita replied in a very subdued voice.

"I guess I'm sorry for everything, everything there is. My God! Where did that come from?"

"You're very raw and open now. You'll have to be careful. Kriya can work like that. You go on for years, thinking that nothing is happening, almost becoming bored with meditation, and then all of a sudden, everything changes. All kinds of seizures of consciousness can take you over."

"I don't think it's simply kriya. Everything seems to speed up when I'm with you. Sight, touch, smell, they all seem more intensely heightened. It's like the cactus they chew in Anahuac. Or is this all coming from the effect of the mead? Just what kind of herbs do they put in it?"

"It's not the mead," Brigita said. "Maybe it is us, for I can also feel something that is working on the two of us. I don't know what it is yet. I suppose I will have to be careful as well. It looks as if there are going to be risks in our coming together like this. Do you want to turn back? We still can you know."

"No." Viracocha said with conviction. "From the moment you placed the naked soles of your feet against mine, for me that was the point of no return. But I do think it is time for you to tell me just what you saw in this 'mind of the future.' "

"I don't know if I can," Brigita said, closing her eyes to see the vision more clearly. "It's like learning how to dance. The teacher has to move your body into place. Even if I could express it, what would the words mean? What if I said: 'Come on, let's take our sandals off, touch the soles of our feet together, and then you can go off and have a good cry.' Those words wouldn't begin to touch what you've just gone through."

"No, they certainly would not," Viracocha exclaimed as he looked down at the floor. "Wherever this vessel we're in, and I don't just mean this garuda, is taking us, I know I don't want to turn back. I couldn't. When I'm with you, my emotions begin spinning around like a compass needle gone mad. I feel myself constantly switching from one emotion to the other."

Viracocha moved from his cot to Brigita's. He lay down with his head in her lap, and then encircled her waist with his arms. The odor of her body began to excite and overwhelm him, and as if to find its source, he began to slide down to place his face between her thighs. He rested there for a moment, then he reached down to try to slide the hem of her smock up, but Brigita stopped the motion of his hands:

"Don't. Don't leap into action. Go inside yourself, not inside me." She spoke softly, compassionately, and began again to massage the point at the back of his neck where the spine entered the skull.

Viracocha released the hem of her smock and brought his hands back to encircle her hips. Her touch was so gentle and yet so precise that he shivered in pleasure as the motion of her finger expressed the power to lead him softly into trance. As he began to relax and let go of his efforts to take possession of her, Brigita shifted her body slightly and eased his face more comfortably between her thighs. He could feel his mind slowly falling like the feather of a bird in still air, and he could feel his head moving easily with his mind until it came to rest on the cushion of her hair underneath her smock. And all the time she continued the slow circling massage with the tip of her finger at the base of his head.

Her touch was more tranquilizing than exciting and he found himself slipping into a state of mind removed from the familiar actions of making love. The odor of her vulva was like a drug. He could smell an amethyst scent that he imagined was her "aura," but it blended with the less ethereal and more exciting odor that came from inside her. He inhaled deeply and felt the odor become a colored stream that passed into his lungs; his own breath seemed to be affected by it, by her presence inside him, and then he saw his breath flow out and pass into her vagina. It was as if his breath had become the penis that would enter her, and then it would reverse as his lungs became the vagina and her odor the full sexual presence that filled his chest and penetrated into his heart.

The slow sexual rhythm of breathing in and out, in and out, continued, and all the while Brigita slowly turned the tip of her

finger that pulled his eyes upward and made his eyelids flutter with flashes of light. The light grew more intense as a window seemed to open right at the point on the bridge of his nose that she had squeezed so tightly before. And at the same time he felt this high vibrating sound humming in his genitals and giving him an intense erection. He wanted to move inside her, and yet another part of him, a part that felt to be she talking inside his mind, seemed to say: "You are already inside me. What makes you *think* that touch is more real a sense than smell?"

As if he were following Brigita's suggestion, Viracocha let go of his habitually sight-interpreted world and passed back into a time before the mind had been thought of, back into a world, not of objects separated in visual space, but of mingling odors and interpenetrating presences. The light touch of her fingertip seemed to draw up the pressure and lessen the demands of the independent animal of his erection, and she made his entire spine feel phallic and alive and offered her whole body to enclose it. How wrong he had been to think of the dark as the absence of light. Space was not void. Space too had her thighs around the sun. Everything was here. Everything came from here in joy, returned here in joy. There was no need to strive toward ecstasy or seek some other sexual consummation. It was here. Words dropped away, and then images: like someone slipping off his clothes to go to bed, he slipped out of his mind and fell into dreamless, unfathomable sleep.

4

He was free. He had run too far and too fast for her ever to be able to catch up with him. He could slow down now, but he didn't want to; instead, as he came down from the mountain path on to the level plain, he picked up speed. Now it was no longer simply a matter of escaping from her, but of finding the true physical mate for the energy throbbing in his gut. The exhilaration of speed had taken hold of him and he knew he could run himself to death and never feel the need to stop. The sheer wild perfection of breath and heart, mind and muscle, put him together into a unity of terror he had never known before.

The full moon, like a conspirator in his escape from her, lit up his path and the pieces of flashing quartz in the gravel seemed to signal the way of his release. He knew he should be feeling the pain of running in his bare feet, but the fury of running stark naked in the night, with the silk robe still clutched in his hand and streaming behind him like a banner, was so elemental that he was beyond the reach of pain. Now he knew how warriors could run screaming naked into battle, picking up arrow after arrow in their pierced flesh, but continuing to run until they had come to the enemy and with one slash cut off the head to fling it high over the armored lines of astonished eyes, and then and only then to collapse in death on top the bleeding trunk. He could do that, for he saw now those parts of himself that had come from afar to be with him that night.

But what he did not know, could not understand, was why he still had an erection. Why would it not drop? Why did it feel as if it were straining to reach up and touch the throbbing point below his navel? Had she drugged him again, or was it the witch's scum that still covered it? He would not be completely free until he had washed every trace of her body off his own.

He would wash in the sea. The stream was here, but he would not wash there, for that stream had to have its source in the spring by the side of her house. No, he would wash only in the open sea.

He was out of the meadow now and past the stream. He could see the village up ahead, see the silver gleam of the garuda at the end of the quai. But first he had to be clean, to be completely free of the gelatinous coating of witch's slime.

With one last effort of driven speed he ran to the edge of the wharf, flung down his streaming pennant, and screamed, slicing the air apart, before he struck the sea. The shock of the cold salt water broke the demands of his raging body. The irritating buzzing in his lower stomach mumbled and receded as his erection dropped, releasing its hold on him. When his head broke through the surface of the water, he hovered with his arms extended and gasped for several minutes. When he had finally regained his breath, he washed himself thoroughly on the groin to cleanse himself completely, and then he began to set his mind back to strategies of escape.

Viracocha swam over to the stone steps that descended into the water and then climbed up to the street. He saw the clump of his robe lying at the edge of the dock, slipped it over his shoulders, and swept his wet hair back and away from his face. As he turned he saw them all: every villager, man, woman, and child, leaning out of their windows and staring at this strange man from Atlantis. They knew who he was. But when they had heard his high-pitched scream they must have thought that the Angel of Death had come to announce the flood that was to end the world. They were ready.

But they were not ready for this man, this man who was out of his place, and perhaps now, they must have thought, even out of his mind. One by one the shutters closed as he passed by. But that was all right with Viracocha. He didn't want them, he wanted the crew of the garuda, and he knew which house on the quai was theirs. His only strategy of escape was to commandeer that garuda, to intimidate them into returning before she could awaken from her trance to find him gone.

The pilot was there, standing in the light of his doorway.

Puzzlement blended with uneasiness as he looked into Viracocha's eyes.

"It is a matter of the highest imperial urgency that I return to the Great Island immediately," Viracocha said.

"Immediately, Your Grace?" the Captain asked. "We are prepared to take off in the morning. The vessel is ready, for we received the Supreme Council's summons to its presence this evening . . ."

"Then that's all the more reason to stop this quibbling about delays and leave *immediately.* If I am not standing in front of His Holiness by tomorrow morning, you are going to regret it in ways you can't begin to imagine."

"Very well, Your Grace." The Captain bowed and turned into the room to address his partner, who was seated at the table: "Get your things. We leave now."

"And one other thing," Viracocha commanded, "is there another sleeping cabin besides the one she uses?"

"No, Your Grace. There's only the closet behind the control room. It has a small cot that one of us uses for long flights . . ."

"That will do fine," Viracocha interrupted. "You can use her cabin. Now, let's go quickly."

Viracocha moved to the vessel and stood there glowering in his impatience, an impatience intensified to cover his anxiety that she might appear before they could leave. But he had made his point and the crew servilely rushed about to prepare the vessel for the overnight journey. As soon as the door was open, Viracocha rushed in, found the narrow closet behind the control room, and closed himself in. He stood there waiting in nervousness and dread for the moment when the vessel would begin to move. When the garuda finally began to glide away from the stone quai, he relaxed only slightly, pulled off his soggy robe, and wrapped himself in the blanket on the cot.

As he sat back on the narrow bed with his back against the wall, he could see the full moon come into the window and disappear as the ship turned to face into the southwest. Still shivering with horror and repulsion, he remembered the moment the moon came into the little window in her meditation chamber to reveal her completely. The horrible faces that

flashed across her head came back to him, each one more hideous than the last. But one face seemed to sum up them all and hold them in her true identity: those white bulbous eyes over which the eyelids fluttered like the leather wings of bats, the evil ecstatic grin, and that line of drool falling from the corner of her mouth.

How could he have not seen it before? It was all so clear now, from the drug in the mead she had given him on the flight, to the hypnotic suggestions she murmured as he fell asleep in her fetid embrace, to the day by day meditations of the last month. It was so clear now that Abaris himself must have set him up for her. All the teachings of kriya had been nothing but a hypnotic lock to which she had held the key. The so-called secret and mystical teachings were nothing but a preparation for her takeover of his mind. It was all a plot. Abaris wasn't dead. He was waiting to use him to get at the High Priest. It was all his own fault. He had no business with any of it.

As the garuda began to lift off into the air and pick up speed, Viracocha felt his fear begin to dissolve and flow away in streams of self-disgust, and he resolved to stick to his music.

But how many other young men, Viracocha began to wonder, had that witch lured into her trap, maintaining the illusion of her beauty while consuming them? He should have known from the start. She was too flawlessly embodied in all those erotic details of a woman's body that were his own private fascinations. Nothing was there to break the spell: the long and full thighs, the soft and puffy large areolas, the lightly spun and fine pubic hair. If she had had only tough leathery nipples that stuck out like pig's teats, or short squat legs, or a thick black bush of pubic hair that hung down in snarls, he would have been delivered, and the spell would have been broken long ago. But there was nothing, not even mole or momentary pimple, to break the bewitchment. She had drugged him and then simply reflected back to him everything he wanted to see.

For a month he had never left her house. For a month he had been closed in like a shaman in his cave. He would touch her, kiss her, sleep with her, but never was he permitted to enter her. It was as if she were making some witch's brew and his

body was the cauldron. Each day she would look at his aura to see how he was doing, the way a cook would check a pot. And after a few weeks of cooking over the stove of her body, he began to feel as if he were reaching the boiling point. They would kiss endlessly for hours, but always she would murmur in his ear about the need to control, to retain the semen. As she sucked on his tongue and tried to drink in his saliva, she kept mumbling about the mysteries of the seed and seemed obsessed with it, as if it were the heady brew she was boiling up inside him.

And boiling he was. In the last week he began to feel his lust turn into rage and he filled up with visions of raping her and flooding her with the semen she had dammed up inside him. But then she would smile, look at his aura, and say: "Soon. Not yet." And so she would stir the pot, smack her lips, and promise endless delights to come.

He had decided that he could not go on, that the next time she began one of those long kissing sessions, he would overpower her and put an end to the whole ordeal. Then she stood in front of him and said: "It is to be tonight, the night of the full moon."

She prepared the meditation chamber, but would not let him enter it, and, of course, now he knew why, knew why she didn't want to let him see what was hidden there in the dark. Then she came out into the bedroom, stood before him, smiled, and pulled the silk shift over her head. Always before, she had slept with him in that thin light shift; always before, except for that brief moment of anger on their journey in the garuda, she had kept that almost transparent covering on her body. Always before, he had waited eagerly for those moments when, with her back to the light, the form of her body would show through like reality through appearances. The power of illusion and suggestion that thin white shift had created had been unlimited, and now it was at her feet like a cloud on which she stood to come down tranquilly to the earth of his raging desires.

The witch's powers of deception were enormous, for she stood there like an apparition: with her long hair brushed out and falling to her waist, but not covering those little breasts that

did not weigh downward but lifted up with all the emphasis of those large areolas that then in the warm summer's swampy night did not shrivel but relaxed, waiting for his touch, and with those long thighs that framed the triangle of bright hair, copper hair, not hair at all, but, like her eyes, traces of innumerable illusions of reflected, intimately held, candled light.

She walked toward him slowly, savoring her powers and the state of his astonishment, then, kneeling down before him, she took hold of the hem of his robe and slowly began to lift it. And when the erection leaped into her face, she simply smiled, kissed it, and continued to pull the robe over his shoulders and arms. But how many men must she have known and consumed to get that precise knowledge of a man's body? She knelt down in front of him again and taking his penis between her prayerfully folded hands, she began to lecture to him, in that soft murmuring voice, about the esoteric nature of the phallus:

"Don't let the seed flow until I tell you. There are really two penises here, the outer and rather obvious one," she said with a flirtatious smile, "and the other, inner and secret one. One is the penis, but the other is the phallus. With one you give birth to the race, with the other you give birth to the Higher Self. The inner ones goes as far inside you as this one goes outside of you. But there is a point inside where this inner penis is rooted, and that is where you must place and hold your mind. I will show you precisely where this point is."

And then, putting her mouth over the head, she hummed a deep "Ommm" and he shivered as the reverberations went down the rigid shaft. As he felt the vibrations inside, she reached up behind the scrotum and with one finger placed the tip precisely where the vibrations ended:

"There is where you are, and there is where you must remain until you have transformed the seed inside you."

She cupped the testicles with her hands, as if she were blessing them, bowed with her forehead resting on his erection, said a prayer in a language that was neither hieratic Atlantean or Eireanne, and then stood up to lead him by the hand into the meditation chamber.

"I will not be able to take all of you inside me," the lying witch

had said, "since I am still a virgin. But we need only touch there for the energy to pass back and forth between us."

It was completely dark in the chamber. This time there were no candles lit on the low altar, but she knew the arrangement and position of every object in the room. She sat him down on the cushion, and then, placing her hands on his shoulders, she eased herself down and wrapped her thighs around his waist. With her left hand still on his shoulder, she reached down and placed his penis so that it gently touched but did not enter her vagina.

Instantly he felt a humming vibration and heard again that sound he had listened to when first their feet had touched. His eyes were pulled upward, and again with precise knowledge she touched the point between his eyes and spoke quietly:

"Keep your mind poised there, and at the place I touched below. Those two points of light are now far apart, but soon the point of light below will rise to meet the other. When they touch inside you, it is I coming up inside you, then, and only then, release all your semen into me."

The witch shifted her hips a little back and forth to settle into position. He should have known then, for immediately the head of his penis passed without obstruction into her vagina. But his mind was too drawn up into his eyes, seeing with astonishment a third eye open like a window in his brain, and hearing now more clearly the vibration begin to differentiate and become, not a buzzing hum, but a melody. She clasped her hands together behind his neck, exhaled deeply, warmly on his chest, and then became very still.

His breath quickly joined with hers until it had become a single lifting and falling tide. Then their breathing grew softer and softer until it stopped. Now there was no sound other than that high ethereal melody that seemed to pass into the crown of his head. But as it grew more ethereal, higher, and more rarefied, the point below began to vibrate in a denser, more physical way. The vibration in his testicles was thick and buzzing like a swarm of bees, and his penis became so distended that it felt as if it were twice as large as it had ever been before. It kept expanding and moving deeper inside her, and then when it was

fully and completely inside, he began to feel this vortex roar around it. Inch by inch the great black vortex began to pull him up inside her until it seemed as if the head of his penis had to be up inside her womb, but still the vortex did not stop but began to pull that inner, hidden penis from its roots. How could she be a virgin, he wondered, when she was so enormous that it felt as if his whole body was being sucked up inside her?

It was then that the moon came into the narrow window and caused him to open his eyes to question this dark vortex that was dragging his whole body up into her womb. He opened his eyes and in that flat, deathly white light, he saw it shine in reflection. On the low altar by her right hand, they were there, the chalice and the dagger. He pulled back to look at her face and then he watched in horror as her faces changed in rapid succession until the demonic in her lay fully revealed and he realized just for what abominable rite the witch had prepared him. He was her witch's brew, her elixir of eternal youth. She was waiting for the moment when he could no longer contain himself, then she would strike, slash, and squeeze every last drop of blood and semen into her chalice, her demented fountain of youth. Already he could see a line of drool falling from the corner of her mouth, see the white of her teeth where her lips curled up in a wild, maniacal grin.

He tried to pull himself out of her, but it was as if she were a parasite moistly fastened to his groin. Her womb had a lock on him, but by using all his strength he was able to lift her up by the elbows to pull her off. She did not awaken, but seemed intently concentrating in her ecstasy, waiting only for that moment of phallic explosion before she sprang into life with death.

He prayed to God she would not awaken as he unlocked her legs and eased her down onto the floor. In the shadows of the room there was still moonlight enough to see her sagging breasts, distended nipples, wrinkled skin, and withered thighs. As he let go of her, she stayed with her eyes all white and her lips pulled up into a leer, but her bony knees fell to either side of her gaping, devouring maw.

He backed out of the chamber, always afraid that it would awaken. He could imagine it stiffen, arch its back as the shape-

changer became a spider and that open maw a consuming mouth. Every object around the room was filled with menace as he turned to find his way out, seized his robe from the floor, and fled thoughtless in terror into the still, warm, stagnant summer night.

He did not go over it merely once in his mind, but again and again, trembling in shock and self-disgust. With his knees held tight against his chest he tried to pull himself together and clear his mind of the effects of the drugs she must have been feeding him ever since he went away with her. He was still huddled in that position, asleep, when the vessel set down in the gray light of a foggy morning and floated toward its mooring at the western quai in the City of Knowledge.

Viracocha was grateful for the fog of that cloudy morning, for more than anything he did not want to be seen as he returned, depressed, embarrassed, and an unpriestly mess of damp, wrinkled robes and stringy hair. But when he came to the gate of the quai and saw the guard he realized in anger that the pilot had docked in error at the quai of the Institute. Of all the places he did not want to be, it was certainly at the western side of the City. But even more, the guard at the gate did not want him to be standing there:

"I'm sorry, sir, but you cannot pass. May I see your permit for entry into the Temple Institute?"

"I'm not going to the Institute," Viracocha said. "The pilot of my garuda did not know the proper procedures and set me down here on the western quai. I am a Member of the Sacred College and I was expecting to be taken to the quai on the other side of the harbor."

"I'm sorry, sir, but you cannot pass," the guard said, now beginning to look at Viracocha with even more suspicion. "A guard will have to escort you to the Institute to have your identity confirmed."

"This is getting a little ridiculous," Viracocha said as he thought to himself just how much he did not want to end up in the Institute in his present condition. "If you simply send up a page to the porter of the College and say that His Grace

Viracocha requests an escort upon his return from Eiru, that will be sufficient."

"It is no longer sufficient, sir," the guard said flatly. "Since the Solemn Assembly, no one is allowed in the City of Knowledge without a permit, and the permits are administered by the Temple Institute."

The guard saluted as an officer approached with his aide.

"Good morning, Captain. How many are in your party today?" The guard spoke with a familiar warmth that contrasted sharply with his officious tone, a tone that only annoyed Viracocha all the more.

"Twelve," the tall Captain said as he looked down at the strange figure beside him.

Viracocha pretended to be deep in concentration, but managed to sneak a sideward glance at the Captain, who returned his glance and rather impolitely continued his examination of the absurdly dressed Viracocha. As the guard placed a seal on the Captain's documents and returned them to him, the Captain suddenly snapped to attention:

"I'm sorry, Your Grace, I did not recognize you immediately."

The guard looked back and forth between Viracocha and the Captain.

"Do you know this man?" the guard asked.

"Yes, of course," the Captain replied. "We are in the presence of His Grace, Viracocha, Member of the Sacred College for Music."

"Now I remember," Viracocha said, "you are the Captain of the Institute's garuda, the officer who also writes poetry."

"I will need you to sign this confirmation of identity, Captain, before His Grace can enter without the new permit."

"Certainly," the Captain said as he stepped forward to write down Viracocha's name and sign his own. "I must say, Your Grace, that I find it surprising to see you down here so early in the morning."

"It was an error made by my pilot," Viracocha said. "I have been away in Eiru for a month and they have changed the procedures in my absence."

"From Eiru?" the Captain said with quickening interest. "But

I am from Eiru. May I accompany Your Grace to the Central Stairway?"

"Yes, of course," Viracocha said. "I appreciate your assistance."

The Captain turned back and stepped over to his aide, who stood standing with catapult in hand by a small group of children:

"Go up ahead of me. I'll walk with His Grace to the Central Stairway and then join you to make our report."

The aide saluted in silence, took the sealed documents from the guard, and ordered the small group to march on ahead. Viracocha's ears picked up as he heard him speak in Anahuacan. The group moved out of the fog and passed them in silence. As they went by, Viracocha noticed that they were all about twelve years of age. He thought it strange that children would be led to the Institute under armed guard and he turned to the Captain, but as Viracocha noticed the lines of strain around his eyes, he thought better than to ask in the presence of so many.

Viracocha watched each of them as they passed into view and then disappeared into the fog ahead. The sadness seemed to be as thick a medium as the fog, but not one of the children was crying or showing any emotion at all. The girls were wearing the traditional embroidered costume of puberty that would be changed after menarche, and the boys wore the blue sash that would be exchanged for a red one at the time of their initiation.

After the children had filed passed them in silence, the Captain turned to Viracocha and with a gesture of the hand indicated that Viracocha should precede him through the gate. When they were outside and on the street alone, the Captain moved up to walk by Viracocha's side.

"My name is Bran, Your Grace. I confess that I have been working on the poem, as you suggested a month ago. What does Your Grace think of this quatrain:

Body burdened on the ground
Startled heaven when it found
Arms it had falsely used
Were wings that force had bruised.

"The meter of the last line is a little stiff, I think," Viracocha answered. "What about something more like:

It had no arms to fight
But only wings for flight.

Bran stopped in his steps and turned to Viracocha with strong interest. "May I hear that again, Your Grace?"

"Of course," Viracocha answered. "I would suggest 'It had no arms to fight/But only wings for flight.'"

Bran smiled, bowed to Viracocha, and looked up with a strong expression of approval:

"That is indeed better, Your Grace. The meter of mine was too mechanical, too martial. Yours is lighter, freer, more appropriate to the subject at hand. Your Grace, I wonder if I might be permitted to make an unusual request?"

"To make the request? Certainly," Viracocha answered. "I at least owe you that much for rescuing me from this morning's embarrassing situation. Whether I have the ability to grant it is another matter. But what would you wish?"

"I would like to request permission to visit Your Grace this evening. I have never been on the Sacred College's side of the City, and I am told that the gardens are extraordinarily beautiful. But more than that, I would simply appreciate the opportunity to speak with you. Life in the Military, I am afraid, can become narrow and, well, culturally boring."

"That seems an easy enough request to grant, Captain. The porter at the entrance to the College's residences can show you to my door. If you don't mind missing the evening prayers, why don't you come before dinner?"

"Excellent," Bran exclaimed as if his life had just taken a happy turn. "That would work out well for me. Thank you, Your Grace. Until this evening."

Viracocha returned his bow and watched Bran turn at the Central Stairway to the left and begin his climb up the series of ramps and staircases that led to the Broad Street. Viracocha watched him move away and wondered about their coming conversation. It was clear that the Captain had a fanciful image

of the life in the College and more than anything wanted to be a poet, and it was also clear from his manner and the tone to his whole being that he was not likely to become one.

Oh well, Viracocha thought to himself as he turned to the right to take the opposite stairway, an evening of polite artistic chatter would be welcome relief after the intensity of the last month.

As he climbed up the series of ramps and stairs, Viracocha felt as if he were an invading party of one, for the fog was so thick that he passed by all the posts of the pages and guards unseen and unheard. It was not until he passed into the quadrangle in the center of the residential quarter of the College that the porter saw him and greeted him:

"Good Morning, Your Grace," the old man said without registering any notice of Viracocha's unusual appearance. "It certainly is a thick morning, isn't it? Reminds me of the times I was back in the sheds at shearing with the fleece flying around in the air. There's a message here for Your Grace. I'm to send over a page to inform His Holiness of your return and you're to wait for word of when your audience is to be."

"Thank you. Instruct the page to say that I have returned and wait upon His Holiness's further instructions."

Viracocha continued under the portico of columns that surrounded the quadrangle and came to his room. He was tired and wanted little more than to go to bed and sleep for the whole day, but he suspected that his audience with the High Priest would be sooner than he wished.

As he opened the door to enter his own rooms, Viracocha stopped for a moment, and then slowly kicked the door shut without turning round. All was clean and in order, for the servants had not let the dust of a month settle on anything except the thin metal sheets that covered his desk, but it was precisely in its order that the place felt so foreign, as if he were coming to stay in an inn or guest house. The College was not home, but where then was home?

Viracocha walked through the room, trailed a finger through the dust on the metal sheets on his desk, and walked into the bathroom. He stood there for a moment before he pulled his

robe over his head and then reached down to pull out the stops to fill the pool with the water that was fed by the volcanic hot springs on the island. There was no place that was home, he realized, as he stood there completely naked; his work, like his official robes, defined him. But even his work did not belong to any place or tradition; it was so purely mathematical, more a celebration of the architecture underneath things, the sacred geometry beneath appearances, a bony skeleton of celestial anatomy.

Viracocha stepped down into the pool and screamed out as the hot sulphured water touched all the cuts and blisters on his feet, but he did not jump back out of the bath; instead he clenched his fists, closed his eyes, and forced himself to remain there standing as the water slowly moved up higher in the pool.

What has this pain to do with that skeleton under the flesh? he asked himself, and then slowly eased down into the bath. It seemed so perfectly clear to him, in the pain of that moment, how ridiculous it would be for him to try to write some oratorio that was a musical parody of the Solemn Assembly. A Solemn Assembly was a celebration of an entire civilization, and he hadn't any idea of what this or any civilization was about. It was one thing to write a "Canticle to the Sun" that cleverly played on the geometrical relationships of the inner architecture of the solar system, but a Solemn Assembly had to have real voices, real people. The opposition between himself and the High Priest was not one of innovation versus tradition, it was a competition of two vain concubines competing for favors.

As the pain in his feet began to subside, Viracocha stretched out to let the tightness in his body float away in the water. Now that the pain was gone, he began to be more aware of the strange buzzing sensation in his lower abdomen and the uncomfortable feeling of engorgement inside his testicles. He thought of releasing what would have to be a flood of semen right then and there, but the metaphor of two vain concubines set him to thinking. After a month of unrelieved excitation, he needed much more than relaxation. He had never been with two concubines at the same time, and as he stretched his arms out to float in the pool, he began to let his mind drift in imagina-

tion of what it would be like to have the lips of two women moving all over his body, trailing their long hair lightly over his skin as they moved from place to place.

The erotic fantasy began to take him over, began to take control of his imagination, but one of the concubines became the young Brigita, while the other turned into Brigita, the aged witch. Viracocha bolted upward in the bath and grabbed his head with his hands as if he could force it to envision what he wanted. A feeling of weakness overcame him, weakness filled with doubts, frustrations, anxieties. Was Brigita a witch? Or just simply a stupid woman playing with drugs? Was Abaris part of some conspiracy, or just a silly old man who thought he knew the secrets of the ancients?

The truth was that he didn't know anything: not about Brigita, Abaris, Eiru, or Atlantis. He was some sort of idiot-savant that had absolute pitch and could work in music or architecture without plans or transcriptions, but he couldn't see, know, or understand anything that was going on all around him. It was ridiculous for him to send for a concubine, he was the biggest concubine in the place. In fact, the whole Sacred College was nothing but a collection of artistic concubines kept for the pleasure of the High Priest.

Viracocha stood up in disgust, grabbed a towel with an angry jerk of his wrist, and stepped out of the bath. He had always taken some pride in his reputation as a fast thinker, but now the absurdity of that description shamed him. He realized for the first time that he knew absolutely nothing about what was going on around him. The Archon of the Institute could be joining with the Military to overthrow the High Priest for all he knew.

As Viracocha dried himself vigorously, he continued his line of attack against himself until another voice seemed to enter the tribunal. "Now you're indulging yourself in an orgy of self-abasement," it seemed to say. "Don't be silly. You know you had to position yourself in this civilization through music, and it has served you well. Now you are in position. Today is the beginning of the new pattern." And then the voice disappeared, as if any further word on the matter would only cater to Viracocha's tendency toward self-indulgence of one kind or another.

The voice inside him had a peculiar effect on Viracocha, for he felt as if he had just stepped out of some sort of psychic bath that had washed away his confusions. Although he did not *know* anything yet, he did know that his orientation to knowing had shifted. He looked back in amazement at himself to consider with what complete ignorance he had walked up the aisle to challenge the Solemn Assembly. He had acted in an arrogance so pure and unselfconscious that it amounted to a peculiar kind of innocence. But whatever innocence or stupidity had moved him that morning, he knew that it was gone now. Now he was no longer interested in acting in complete self-absorption. The High Priest had been right: he had been protecting himself with a need not to know. But now he wanted to know. He wanted to know whether Brigita was a witch, or whether she and Abaris were part of some plot against the Empire. He wanted to know why the High Priest was moving so quickly to change the elevation of the Tower. And he wanted to know just exactly what was going on inside the Institute.

Viracocha had just time enough to dry himself and put on his pale blue robes of office, when the knock came on the door. He knew at once that if the page was ignoring the sign on the door that warned visitors he was not to be disturbed at his work, then it could only mean that the page had come to summon him to an audience with the High Priest. Clearly, the little voice in his head had been right. Things were not going to wait for tomorrow. Today was the beginning of the new pattern.

And somehow Viracocha felt ready for it. He did not want to wait in his ignorance any longer. He moved quickly across his rooms at the sound of the second knock, opened the door, accepted the page's bow and silent gesture of summons, and followed him out into the courtyard. The morning fog had lifted and although the sun was shining brightly there was a change of season in the quality of light and the odor of the wind. The oranges on the trees were ripe and as they walked through the quadrangle, Viracocha reached up and pulled one off a tree to take his breakfast casually on the run. The taste of the orange was sweet, and as he deposited the heap of peelings into the hands of the startled but still silent page, he appreciated

through his own sense of well-being that it really was a beautiful morning. Soon the great rainstorms and raging winds would sweep up from the south, but now was one of the lovely moments of change when the summer was going, but the hurricanes had not yet come.

They had better be finished with the Tower by then, Viracocha thought, or else the winds will blow everyone off the scaffolding. But as he came out of the College Gate onto the Broad Street, Viracocha could see that there was little reason to be concerned about the weather, for they had already finished the increase of elevation and were now at work on the new housing for the solar crystal.

Viracocha knew that the real challenge of the project was not in the architectural engineering of the increased elevation, but in the geometrical accoustics of the foundation vault. The normal approach would be to build an icosahedron-shaped room out of iron, but if the Great Year of 25,920 years was coming to an end, and the geometry of the earth's position in the galaxy was going to be altered, then the old foundation vault would not be strong enough.

As they walked down the Broad Street to the stairs by the entrance to the hypostyle hall, Viracocha tried to imagine what configuration would be powerful enough to effect what the High Priest had described. There were only three possible frequency-signatures that could be induced into the crystal, and only the strongest would most likely be effective. If they were seeking . . . and then it burst into his imagination in completely finished detail . . . they were not simply seeking to enhance plant and animal growth by placing seeds in energizing fields aligned to the sun; they were realigning the earth to this second sun to make the earth the center of the old solar system. As if he were dreaming while he walked, Viracocha could see the new geometry needed for the foundation vault: the five Hyperborean Solids constructed out of their corresponding metals and set inside one another according to the planetary intervals. The distance from the center of the solar crystal to the center of the five Hyperborean Solids over the distance from the center of the sun to the center of the earth.

The foundation vault would recapitulate the geometry of the solar system so that the earth would receive the galactic emanation directly! They were actually going to try to take the energy from the sun to reach out into the galaxy.

But why? Viracocha almost yelled inside himself. If they were going to try to invoke that much energy, he knew that there was no way that they could contain the singularity simply inside the Tower; it would pass through to the center of the earth itself, and then anything could happen. The crust of the earth could crack open, if they were lucky, or, if they were unlucky, the sun could be pulled closer to the earth.

Now Viracocha began to get some sense of what they were doing in the Institute. They were working in haste because they were planning on skipping more than one Great Year's cycle of human evolution. They were trying to rush on ahead to the very end and consummation of human evolution to escape the earth. This kind of daring could not come from the Archon. Only the kind of man who had single-handedly rebuilt his civilization would be willing to press on with the reconstruction of human evolution itself.

As they reached the middle of the Broad Street, in front of the twin stairways that framed the entrance to the hypostyle hall, Viracocha stopped momentarily and looked down the street toward the copper doors of the Institute. The doors shown brightly in the morning light, and Viracocha thought of the children he had seen that morning in the fog. What were they doing in the Institute? How could removing the etheric body through sound affect this attempt to change the geometry of the solar system? As he turned away to follow the page up the stairs, Viracocha decided that he would ask the High Priest's permission to visit the Institute.

They moved up the stairs and across the balcony to the entrance of the Palace. The guards were everywhere, at every door and stairway, and their dart-catapults were not suspended in their leather slings but were ready at hand. Had there been another attempt on the High Priest's life? Viracocha wondered as they passed through into the entrance hall of the Palace and encountered yet another squadron of guards. He listened to his

own footsteps in the echoing hall and could feel the impact of the watchful eyes of the guards as he came to the final set of stairs.

The guard knocked twice on the door. Viracocha waited and tried to prepare himself for that diffuse glance that would tell him that the High Priest was looking at his aura. But as the door opened, it was no page who stood there, but the High Priest himself who grabbed Viracocha by the arm and commanded him to follow as he practically ran down the internal corridor of the Palace:

"Good Morning, Your Grace. Welcome back. Come on, don't stand there, come with me." The High Priest laughed at Viracocha's amazement and stopped momentarily in the hall. In his white sleeveless robe his muscular arms stood out and he seemed far more physically powerful than one would expect for a man of fifty.

"Come on, will you. I told you that I would take you personally in hand when you returned. They've finished with the elevation of the Tower this morning. Let's go up."

And without a further word the High Priest turned and continued down the corridor at such a pace that Viracocha had to run to keep up with him. They passed through the double doors that separated the study, the reception hall, and the private dining rooms from the residential quarters, and kept moving deeper into the heart of the Palace. As Viracocha considered the layout of the rooms, he realized that the High Priest had either his bedroom or a meditation chamber directly against the spinal axis of the Tower.

As they passed through the last set of double doors, Viracocha could see that there was nothing ahead except the naked megalithic stonework of the Tower itself. The High Priest reached the wall first and then turned to Viracocha:

"Hurry up, now. For 'the fast Viracocha' you can be amazingly slow. I've told the engineers to clear the Tower, so we will have it to ourselves."

"Are we going up to the solar crystal, Your Holiness?" Viracocha asked as he caught up with the High Priest.

"To the housing, yes," the High Priest said. "But the crystal

itself has not been remounted. They're restudying the facets again to determine what the new frequency-signature should be."

The High Priest turned and touched his ring to a hidden point in the stone and waited as the large trapezoidal megalith began to shift to the side.

"There are only three possible frequency-signatures, Your Holiness, and only the highest, and I might add, the most dangerous will work."

"Ah, so you still are the fast Viracocha, even if you seem to have become the timid Viracocha. Here, follow me."

The High Priest passed through the opening onto a small platform, motioned to Viracocha that he was to join him, and then pressed his ring to a second point that began to close the opening just as their platform began to lift upward. High above them Viracocha could see the light from the pyramidal apex of the Tower, but as he looked below he could see only the darkness of the empty vault. He could not take his eyes off the vault, and he saw in his imagination the thin metal models of the Five Hyperborean Solids all inside one another, and all vibrating with their appropriate frequency-signatures in resonance with the solar crystal. Finally, he turned back to the High Priest:

"No, Your Holiness, there is only one signature that will work with the Five Hyperborean Solids in the vault."

The platform reached the apex of the Tower and came to a stop. The High Priest stepped out onto the ledge that went around the four sides of the open square directly under the pyramid.

"*Five* Hyperborean Solids, you say?" the High Priest said with quickened interest. "You know, I think it's time that I sent you over to the Institute. You may just have come back in time to save them from making a mistake. Look at it, Your Grace!" the High Priest said as he gestured to the entire city and bay spread below.

And it was beautiful. On one side of the Tower to the north, all was green and natural, and no human settlement was permitted for miles; on the other all was cultural and man-made with the priestly city cut into the steep ledges of the living rock.

The whiteness of the Tower and the Palace contrasted with the darkness of the megalithic rocks and the deep blueness of the curved bay in which a dozen metallic garudas gleamed in the sun like the stones in a necklace. The air was bright and the wind carried those tropical odors that indicated that the season of storms would soon come. And yet at that moment all was peaceful, and it was hard to hold on to the sense of anxiety, unless, of course, one turned with Viracocha to look directly down into the darkness of the central channel into the foundation vault that was level with the sea.

"You look down, Your Grace, when you should look up," the High Priest said. "But I can see that you've had quite a shock. And yet, our dear Abbess seems to have awakened all your spinal centers. You're in for quite a change, Your Grace. But your sense of timing is perfect! We may be able to take this great voyage together."

Viracocha turned away from inspecting the darkness in the Tower to see the High Priest looking at him with that diffuse stare that he now understood.

"Voyage?" Viracocha questioned. "I'm afraid Your Holiness is too fast, no matter what my reputation may appear to be."

The High Priest did not take his eyes off Viracocha but continued to stare in complete concentration. And then, almost with a snap, his eyes came back into direct focus and he smiled warmly:

"Yes, perfect timing. When I sent you off into solitude I had no idea you would use your time so well. It seems you have chosen the fast and dangerous way yourself. You were not in this condition a month ago. The center in your solar plexis is much more active. Despite the shock you seem to have sustained, I think that a good dose of fear and an awakening of the perception of power are two qualities you needed to develop. You were too self-indulgent, too self-centered and impulsive before. Just what were you and our esteemed Abbess of Eiru up to in your absence?"

"I'm not certain I know, Your Holiness. I believe it's called witchcraft. Whatever it was, it certainly has an amazing power to make appearance seem reality."

"Witchcraft?" the High Priest said. "Witchcraft *can* make appearance seem reality, but it can't effect those changes inside you. I think I'll need to find out more about our esteemed Abbess of the One Thousand of Eiru, but I didn't bring you up here to discuss Eiru. Look at the Tower, Your Grace. It's finished. Tomorrow we remount the solar crystal. Just look at it! No moving parts, no noise, no sweaty slaves grunting in labor. Simply a silently vibrating crystal: pure geometry, all unheard music. There's no instrument that you can compose on, Viracocha, that can touch it. Music is a metaphor for transformation, a holding action while humanity waits out its appointed term, but this *is* the transformation, the release of spirit into the universe."

"But what of the danger, Your Holiness? If I understand the project, and I don't think I do, you are running enormous risks, you could . . ."

"Of course it's dangerous," the High Priest said as he impatiently brushed Viracocha's caution aside. "Everything of value is either dangerous or delicate. We prize things exactly to the degree that they are delicate. I can smash a harp, a vase, a baby's skull, or a civilization for that matter, in an infinitely shorter time than it took to create any of them. Does that mean we should create things that are not so delicate and susceptible to damage? No, we treasure them because of their fragility, because that fragility speaks of our condition as men. Danger is no different. We value it because it speaks of our condition and makes us more conscious, more aware of our fragility."

"And if the earth cracks under the Tower, what then Your Holiness?"

"It depends on *when* it cracks," the High Priest said with a smile. "If it cracks *before* we have made it out, then that is tragic and some future poet should write our tragedy, but if it cracks *as* we are leaving, then it's no matter, for the thirteen of us will reabsorb humanity in the Great Death. And if we fail, it's better that the civilization be destroyed at its peak than for it to degenerate as the ancient Lemeurians did. We're condemned to a world where everything dies. Very well, then; so we take this given death and use it as a vehicle of expression, just as you take

sound and use it to create music. You know, I think His Excellency the Archon was right, you're wasted in the College. It just encourages your tendency to self-indulgence. It's time you got acquainted with the work in the Institute and lent your quick genius to the real work of the spirit. How long will it take you to transcribe your notes onto plates?"

"If Your Holiness means the designs for the Five Hyperborean Solids, no more than an hour. It's all complete in my mind, I just have to copy it all out."

"Good. I'll send a page over with you to wait while you transcribe your designs. The page can take them over today, and then later this afternoon, we'll arrange for you to be given a tour of the work. In your condition now, you just may be able to effect a lifetime's growth in a day. This is going to be a very important day for you, Viracocha. You now are being given a chance to learn from me, and, who knows, perhaps someday even succeed me. But be careful with your volatile emotions, for if they get the better of you now, they will destroy you. Your stay with Brigita and your studying with Abaris have opened you up to the psychic realms, and that will be dangerous unless you can shoot through them quickly to get above the psychic. Otherwise, like a swimmer who doesn't return to the air because of the beauty of the mermaids' songs, you'll drown in that medium of illusions."

"How does Your Holiness know about Abaris?" Viracocha asked, beginning to feel a sense of fear that he did not know or understand anything or anyone, not Abaris, Brigita, or the High Priest.

"How would I *not* know?" The High Priest laughed at Viracocha's naïveté. "Of course I know all about everyone who comes and goes in and out of the City of Knowledge. Abaris especially, since he was traveling far too much to be up to much good."

"How did Abaris die?" Viracocha asked, almost not wanting to know.

"To understand his death, you have to understand his life," the High Priest said as he turned, and grasping the edge of the railing, looked out over the City and the bay. "Abaris was part of

a secret police force. They claimed to be the custodians of human evolution, but actually they were the jailers sent to keep us locked into earth. Supposedly humanity was to wait for a certain cosmic moment when one of Abaris' group would come to release us. Our project here was challenging their scheme of things, for we were advancing the time of our escape by one entire Great Year. He came here, demanding that the project be stopped, a guard misinterpreted one of his so-called secret *mudras* as a threatening gesture, and shot him with a dart. Thus ended the career of Abaris the Great, though I suppose his ghost flies at night to cohabit with his consort, our Abbess of Eiru."

The High Priest turned round and leaned against the corner of the Tower. His face was in the shadow, but the sunlight gleamed on the large emerald in the medallion on his chest. Viracocha said nothing but gazed into the medallion as if in its light he could find an end to all the darkness and confusion inside himself.

"You see," the High Priest said softly, "it's a quarrel about time, or timing. Revolts in Anahuac or attempted coups by the military, these are the pieces of the game, but not the hand that moves the pieces."

"But Abaris' teachings," Viracocha blurted out in bewilderment, "they work, they're powerful, you can feel the effects inside yourself."

"Of course they work," the High Priest said in quiet confidence, "that is, they work on their level. But all his out-of-the-body techniques disassociate the student from the physical realm in order to energize the volatile and shifting psychic body. Then all these awakened emotions are focused on the teacher. Surely, you must have noticed during your stay with our witch of Eiru that she idolizes Abaris. He is, or was, everything to her: father, lover, teacher, god. And what was Abaris doing with all this devotion from his disciples? He was traveling around the world, setting up a shadow-structure to the Empire. He was clever, for he never used the same strategy. In Anahuac, he stirred up the indigenous people. With the Military leaders, he appealed to their ambition to rule. With the powerless, he appealed to their desire to escape to some new colony in a

remote corner of the world. And with you? He simply gave you some meditational techniques to fascinate your overly abstract mind. So you block up your eyes and ears, you listen to inner sounds, you raise a stream of light up and down the inside of your spinal column, or you project out of your body to visit phantasm-schools with mysterious teachers. But all this trafficking with the dead doesn't *use* death to advance life. The disciples of Abaris keep spinning around from life to death and new life in the psychic dream world. Humanity could keep that up for millions of years, and still go nowhere. It's not escape, it's the trick of the jailer."

"I don't understand," Viracocha said quietly. "I truly don't understand. Where have I been all these years?"

"With humanity in its prison of illusions. But now you're waking up. It's time for you, just as it's time for humanity. I want to see you at the end of this day, Viracocha. You are fast, Your Grace, but today you are going to have to test the limits of that speed of growth. I look forward to seeing you tonight, for then I will know if I can take you up to the next level of initiation. So, let us say that we shall have a small dinner with the Archon and the Master of the College. But now it's time to go down to the Palace. I'll send the page back with you to wait for the Plates."

The High Priest walked from the corner of the Tower into the platform and Viracocha followed him. He touched his ring to the receptor and the platform silently began its descent. Viracocha did not speak, but kept going over all his memories and impressions of Abaris. Nothing in the way Abaris felt to him made any sense. The only thing that at all made him doubt Abaris was Brigita. She was clearly a witch, a woman who hated men and yet was insanely drawn to them in a lust for destruction.

The platform reached the level of the Palace and once again the High Priest touched his ring to the crystal receptor embedded in the stone and the trapezoidal megalith slid quietly to the side. The High Priest did not turn back until he came to the door of his private quarters, and then with a gesture to Viracocha to remain, he waited while the page opened the door.

"Go with His Grace. Wait for the metal Plates he shall give you to take to the Archon and inform His Excellency that His Grace will inspect the entire project at the hour of none. Until dinner, Your Grace. Full ceremonial dress will not be required."

The High Priest gave a last investigating glance at Viracocha, as if to say, "I wonder if you can survive the test," and then turned to disappear into his chambers. Viracocha bowed, and as he turned to leave he could hear the bells of noon begin to sound across the city; but he could not see the shadows from the sun shift quietly to the other side of the Tower.

5

△△△

It was all a cunningly designed trap. He could see that clearly now. Five Hypoborean Solids, five etherically-resonant metals, five planetary intervals, and five archetypal frequency-signatures: the systems of correspondence were certainly elegant. The ultimate correspondence was, of course, to the five human bodies: the physical, etheric, psychic, mental, and archetypal. If you took one Hyperborean Solid and resolved it from a geometrical to an algebraic expression, it gave the formula for the next Solid in the ascending sequence. The inseparability of the five human bodies came from the interlocking and overlapping patterns of vibration. The Hyperboreans had clearly known this and their Five Solids were not the teaching models for solid geometry that the Atlanteans took them to be; they were the architecture of incarnation.

Yet from another point of view, they were teaching models, and what they taught was that our incarnation was not simply *on* earth, but *in* a complete field of energy that was the solar system. Without the appropriate scale of the planetary intervals, the field of resonance in the five metals would not interact with the five bodies. The metals in the earth, like the blood in the physical body, were collectors that changed tonality under the strain induced by the changing geometry of the moving planets: and not just metals, but gems and crystals as well.

If a person were to fall asleep or go into a trance inside the geometrical lattice of the properly aligned Five Solids, the interlocking system that held all his own five bodies together would become unlocked. But what then? The being would ascend to the archetypal realm, there to cry out in joy of vision or torment of imprisonment. But he would not be able to move out of that archetypal realm, out of that meta-pattern of chains of

causation that held the manifest world together. No, Viracocha thought to himself, there was no doubt about it; you could not unlock the system from inside the system itself. They were locked in.

And in more ways than one. The distance from the solar crystal to the center of the Five Solids in the foundation vault was the intended opening in space and time, but his hunch had been correct, for the real opening would be the distance from the center of the solar crystal to the center of the earth. Then the rip would not be simply in the subtle body but in the subtle body of the earth itself. The attempt to break into heaven would only raise hell in the form of a volcano. Any attempt to escape automatically activated a system of recontainment. Was the earth some kind of galactic prison or leper's colony from which escape was impossible? Or was there a missing key, a missing crystal to open the wall? The High Priest spoke of being abandoned by the gods, but it looked much more as if humanity had been intentionally locked in. Why? What crime was humanity guilty of?

And yet there was such beauty and elegance to the pattern of containment that was our incarnation. Whoever our jailers were, whether gods or Abaris' secret police, they certainly had an appreciation of geometry and music. Why had it taken him so long to see it? The High Priest had recognized it at once when he heard "The Canticle to the Sun" and appointed him to the Sacred College. It was all there, for the "Canticle" was theme and variation to the frequency-signature for the Fifth Hyperborean Solid. Its beauty and uncanny yearning were the cry of the soul that had ascended, without any Tower at all, to the archetypal realm, there to sing out alone against the lattice of the celestial geometry that still contained it. That piece of music expressed the affinity that linked him with the High Priest, and was the reason why the High Priest had first offered to supervise his indoctrination, and now dropped suggestions that, with work, he could become the successor.

As Viracocha compared the structure of his own music with the drawings of the Solids that he had etched on the thin metal sheets, he saw something he would not have believed before.

Always before he had prided himself on his differences, his genius and individuality, which had set him apart from everyone else. Now he could see what a fiction that had been. The more he had moved into the essence of his own individuality, the more he had moved closer to the essence of his own culture. The Tower and his music were isomorphs of one another, each expressing the yearning for transcendence. What one tried to effect through science, the other tried to effect through art. The Tower was a piece of literalism, a clumsy attempt to break out that was clearly doomed to fail. But what of his own music? All human civilization, every science and art in it, now seemed so much motion on the surface of a sphere in a futile search for the center. No amount of traveling could turn one in or lift one off. They were locked in. To be ignorant meant to ignore this fact; to be intelligent only brought one to an awareness of the geometrizing demiurge that had locked them in.

It hurt to think that there was no difference between his music and their Tower. Always before he had nurtured this feeling of difference between himself and his colleagues, between the College and the Institute. The intellectual style of the faculty of the Institute had always repelled him as aesthetically distasteful. It had been so simple before: they were ugly, they with their poisonous darts, but his music was beautiful. Now he had to admit that the patterns of beauty and ugliness in a civilization were related. The feeling of difference that he so treasured was simply an emotional illusion, a membrane that enabled the cell to work more effectively within the organism.

And yet somehow that did not feel right. There had to be another crystal to unlock the system, but, clearly not within the system itself? But where then? The system of correspondences worked on the principle of homology. Could the key be in the mirror-opposite, in difference? What was the difference that made a difference?

Viracocha stood up from his desk, placed the metal sheets within a portfolio, and tied the string. As he walked across his room to the door to give the portfolio to the page waiting outside, he realized just how tired he was. The month with Brigita, the still confusing and unresolved ambiguity of his flight

from her, the revelations about Abaris and his planetary secret police, and now this collapse of the emotional foundation of his music: too much was happening too fast, even for him. He opened the door, handed the portfolio to the page in silence, and then closed it again to avoid contact with anyone.

It felt as if knowledge had hit him like one of the paralyzing darts of the Institute. He had no energy to move, to care. He was too tired to care about anything: Brigita was a witch, the High Priest was a great man, Abaris was evil; Brigita was good, Abaris was a prophet, and the High Priest was evil. He didn't know anything anymore. All he knew was that knowing didn't seem to make much of a difference. All he could feel was that he was tired, and that if he went to bed he would not wake for days. He knew that he couldn't afford deep sleep now, for he still wanted to know what was going on inside the Institute. If he could not know anything else, he wanted to know that much.

Not daring to sink down into the deep oblivion of his bed, he moved away from the door, grabbed a cushion from the chair, and collapsed on the rug on the floor. As he stuffed the pillow behind his head he told himself to listen for the bells of none, and then he let his mind drift out of control on a stream of images.

After a while the images collected into a dream. He was standing on top a high sea cliff with Brigita. He kept trying to stare up into the sun, but the light was blinding and he had to turn away. They were both naked and Brigita began to laugh and tickle him, saying, "It's not up there, silly!" And then she began to tickle him harder, pushing him closer to the edge. He tried to get her to stop, but she only laughed the harder at his elaborate and awkward movements at the edge. Then he lost his footing and began to fall, but all the while Brigita kept laughing at him. As he fell down through the air, she looked at him and laughed, and then she cupped her hands to her mouth and yelled: "Turn around, silly!" He turned over as he continued to fall and then he saw the water coming up to meet him. He stretched out his arms and tried to protect his head and neck, and then he hit the water and was surprised to find how good it felt all over his body. He wanted to see if he could make

it all the way to the bottom, for somehow he sensed that there was something down there on the bottom that he wanted. He would take hold of it, crouch down on his knees, and then spring up to the surface to hold it up in the air so that Brigita could see it, wet and glistening in the sunlight. But it was dark at the bottom and the shadows from the large rocks made it harder to see. After a moment of searching, he saw it flash and he reached out to grab it, still not knowing what it was that he had taken hold of; then, pulling his legs under him, he crouched down to spring up from the bottom. He hurled himself toward the surface, and as his head came above the water he held it aloft to Brigita as the bells began ringing and awakened him.

It took Viracocha a minute to remember where he was and why the ringing of the bells was important to him. His first impulse, when he did recall, was to run out the door to be on time, but then he decided the Archon could wait and go over the plates once more while he walked over.

Viracocha crossed the room to a small cupboard that he kept filled with food for the times he was working and did not want to leave his desk for the society of the dining hall. He reached into the jars and filled his hands with raisins and almonds, and as he stuffed his mouth he thought again of the High Priest's allusion to his becoming the successor. The absurdity of the idea was immediately apparent to Viracocha, and he wondered why it was not to the High Priest.

With a mouthful of raisins and almonds, Viracocha filled his hands again, gulped down some cider from the ceramic jug on the shelf, closed the cupboard, and turned to leave. The more he thought about the High Priest's idea, the more bizarre it seemed. As he walked out of his rooms into the quadrangle, he began to take an imaginative inventory of the place. He tried to see himself presiding over meetings, making appointments, performing secret acts of ancient mysteries in the Solemn Assemblies, and fighting back assassination attempts with his own squad of secret assassins. It was a useful exercise, for the inventory of the City of Knowledge, as he walked through the College's cloistered gardens out to the Broad Street, made him quite aware that not only did he not want to be High Priest, he

did not want to live in the City anymore. But where could he go?

Viracocha stopped for a moment inside the College Gate and looked up at the Tower. What was the difference that made a difference between his music and that instrument of cosmic vibration. If he left this City, there was no other place in the world where either his music or the hieroglyphic notation for it could be understood. The memory of listening to music in the island within the lake came back to him, but he pushed it aside at once because he knew it would be impossible for him to live his whole life in so primitive a place.

There was no place to go, except . . . But the idea was too frightening, and Viracocha knocked it aside; nevertheless, it came back. What had the High Priest meant when he spoke of the thirteen and said that "We may be able to take this great voyage together"? What would it be like to be inside the Solids within the foundation vault when the solar crystal was activated? He could see the High Priest put him into trance, as he had done before in the Solemn Assembly, he could see the volcano lift from beneath the ground to hurl their physical bodies into the air while their subtle bodies ascended to the archetypal lattice of imprisonment. He could hear the cries of millions as all the Islands began to explode and drop in fragments to the sea, but he could see no way to pass through the lattice of causation to the free world behind the stars; all that he could see was the entwined bodies of himself and the High Priest dropping into the sea where once the City of Knowledge had stood.

It was not in thought, but in the fullness of that atavistic faculty of vision that Viracocha realized why the High Priest could look upon him as a possible successor, why he could gaze at his spinal centers with a fascination equal to Brigita's. It was so clear that he wondered how he could not have seen it reflected in the ceremony in which the High Priest elevated him to the Sacred College. He should have noticed that the institution of the temple concubines was a test, an institution that expressed a contempt for women and the lesser men who were ruled by the mere biological urges of reproduction. Both the

High Priest and Brigita were obsessed with a mystery of the spiritual phallus with its magical potion that turned the flesh itself into some ritual of escape from incarnation. Both had taken the innocence of sex and twisted it into another direction in the hope that by reversing sexuality they could escape it altogether in a realm beyond male and female, generation and death.

Viracocha was walking very slowly down the Broad Street as the visions of the end flashed through his mind, but all the time he never took his eyes off the copper doors of the Institute. Behind those doors lay the Laboratory of Incarnation, but even before he had passed through he began to understand just how cursed incarnation was in everyone's imagining. Perhaps, he thought, humanity will never escape as long as it is trying to. Perhaps it is the very effort to escape that locks us in.

The guards at the copper doors saluted Viracocha and allowed him to pass unquestioned. The page at the desk of the guard inside immediately sprang to life, bowed, and motioned to Viracocha to follow him down the long corridor. At the end of the corridor, Viracocha could see another set of large double doors, but the page stopped at a door to the left, knocked twice, and then opened it to allow him to pass.

The Archon was seated in front of a table in which five Mercury Tablets were set up. The Emerald Stylus was still in his hand, and as Viracocha approached he could see the Five Hyperborean Solids shimmering in the liquid metallic gray of the Tablets. The Archon did not bother to stand or turn to greet Viracocha, but simply gestured with the stylus that he was to take the seat next to him in front of the table. Only when Viracocha was seated did the Archon turn from his study of the Solids to consider him.

"I was beginning to wonder if 'the fast Viracocha' was as slow in walking as he was fast in thinking," the Archon said as he turned with some difficulty to face him. "And then you appeared. I have, as Your Grace can see, induced your Plates into the displays."

"Good Afternoon, Your Excellency," Viracocha replied. "Yes, I can see that you have not wasted any time."

The Archon sat in silence for a few moments as he considered Viracocha. The only sound in the room was the gentle thud of the Emerald Stylus as he struck it against the cushioned arm of the chair. The silence deepened, and Viracocha decided that he would not be the one to break it. Instead, he began to study the Archon's face and noticed just how much his condition had deteriorated in the month he had been away.

Finally the Archon took in a deep breath and spoke:

"I knew Your Grace had a love for ancient knowledge and mysteries, a love that pulled him away from the Institute to the College, but I never thought that what would finally bring us together were these children's toys of the Solids."

So, the hostility is out in the open, Viracocha thought to himself, but what he said was simply:

"And does His Excellency consider the Five Hyperborean Solids to be merely toys?"

"No, not the Solids themselves," the Archon said with a stiff smile, "although I do confess to some residual anti-Hyperborean prejudice. No, it is not the Solids themselves, but your almost absurdly simple little gadget of thin metal rods with which you construct them. I mean, after all, we've been working on this project for years. The stones in the foundation vault weigh hundreds of tons, and now you come along and in a few hours suggest that pieces of induction wires and thin metal rods are all we need. Surely, even you must see the ironic nature of the proposal?"

"I think Your Excellency gives too much weight to matter and volume," Viracocha answered without animus. "As you will recall, there is more empty space in the universe than there is material, and what matter there is, if one truly understands it, is made up out of music and geometry. You know as well as I do, or else you would not have been able to move those megaliths into place so easily, that it is not size that is important, but the scale, the alignment, and the frequency-signatures you induce. You have the Mercury Tablets here to check my calculations, so what are we arguing about?"

"We are not arguing, Your Grace. We are discussing the ironies of your proposal, but it seems the irony of my remarks

escaped you. When I used the word 'toy,' I was merely saying that it was ironic that these old toys, these children's blocks with which we were taught solid geometry, should have a role to play in the successful completion of our project. Am I clear now?"

"Yes," Viracocha said, "but Your Excellency is not exactly famous for his lighthearted and joking manner, so I was not prepared for irony and wit."

"To return to the subject at hand," the Archon said, annoyed but determined to ignore Viracocha's jibe, "which of the three possible frequency-signatures do you think will be the most effective?"

"The highest one, but it will also be the most dangerous. I hope His Excellency is fully aware of the risks of this project."

"Yes, of course, we are aware that there could be an earthquake as the planet adjusts to the new galactic position, but that is a risk we shall just have to take."

"For what purpose, may I ask?" Viracocha said.

"For what purpose!? Yes, I had forgotten, you are not really aware of just what we have accomplished here. I had forgotten that this is actually your first visit to the Temple Institute. Well, that is a failure on our part that we must remedy immediately. As for the risks, His Holiness has said that the society that does not take risks degenerates. So, as you can see, we have given the matter some thought, but we are going ahead with His Holiness's encouragement."

"It's not my responsibility to make decisions," Viracocha conceded, "but it is my responsibility to pass on *all* that I know."

"Quite. And you have done that. I must say, Your Grace, that your knowledge is particularly significant for me because of all the time we shall save. I had not thought of the connection between the Hyperborean Solids and the planetary intervals, not to mention the quantum-steps of the subtle-bodies. We have ample stores of the five metals in question, and since the whole model is only a mere thirty-six feet in diameter, we can have the whole thing in place for you, oh, even by as early as tomorrow afternoon. As you can see from my worsening condition, this saving of time is critical."

Viracocha tilted his head forward in a not very deep bow of acknowledgment and then thought to ask in a very serious tone:

"Who goes into the center of the lattice inside the foundation vault?"

"I'm sorry. Yes, of course, you don't know yet about the biology of the project. Come, it's time we showed you around. I think what will be of particularly interest to Your Grace is our study of the physiological effects of music."

The Archon rose, leaning heavily on his cane. When he was standing securely balanced on his two feet, he extended his cane over a receptor-crystal on the table to summon the page. "I think that with your knowledge of architecture and music that the Laboratory of Incarnation will be especially fascinating to you, for it is precisely with sound that we have come to understand the architecture of the body."

The Archon passed by Viracocha as the page entered and bowed before them; he walked slowly, taking a step with his left foot and then dragging his right side forward with the help of his cane.

The page held the door open for them and the Archon and Viracocha passed out of the study into the corridor. The double doors of the laboratory were to their left at the very end of the hall. Viracocha slowed his own pace to walk alongside the Archon; he listened to the sandal striking the stones, the long dragging sound of the Archon's right foot, and then the click as his cane came forward to bear his weight before he extended his left foot again. The Archon did not bother to speak but put all his thoughts to making the effort to walk. Finally, they came to the double doors; the page moved quickly inside and then held the door open as the Archon passed through.

As Viracocha entered the room he saw a semicircle of about twelve examining tables. The huge stomachs of the naked women struck him first, and then he noticed that all of the women were in the last month of pregnancy. Two wires were connected to each of their heads, one at the forehead and another behind at the base of the skull. Viracocha felt uncomfortable at the sight of all this vulnerable raw flesh, and did not know where to put the focus of his attention, since from where

he was standing he seemed to be looking straight into the vulva of the woman in the center.

The temperature in the room was unusually warm but the atmosphere and tone of the place was even more immediately thick and oppressive. The page removed His Excellency's outer cape and then stood by waiting for Viracocha to release the gold chain at his neck so that he could hang the garments up with the others by the door. Viracocha released the clasp and then moved quickly over to the woman in the center to examine the wires fastened to her head. His memories of staring down the thighs of the concubines returned to mock him, and he felt embarrassed and confused.

The Archon seemed pleased with Viracocha's immediate attention to the details of the technology and began to speak to him in a friendlier tone:

"Very good, Your Grace. I see the fast Viracocha goes right to the heart of the matter. You see, we have found a way to reproduce music through the pulsed oscillation of emanating crystals. We can't, of course, repeat the exact sound and timbre of musical instruments, but we can take the pattern of the music and duplicate it. We induce this abstract pattern directly into the lattice of the skeleton, so that the bones themselves become the crystalline-receptors. We eliminate the orchestra, as it were, and make the subject both the performer and the audience."

"What on earth for?" Viracocha asked, as he tried to imagine what the sensation would feel like to have an abstraction of music vibrating in his bones.

"It eliminates the deleterious effects of prolonged coma, Your Grace. Observe the healthy skin tone and the firmness of the muscles. Although these women have been in a deep coma for several months, their organs . . . ."

"Months!" Viracocha exclaimed, "but the fetus . . ."

"No, the fetus doesn't die, but it is born mentally retarded. We haven't solved that problem yet, but we have succeeded in being able to slow down the metabolism enormously. But you're ignoring the most important point, Your Grace. You see, the music we have induced into the lattice of their bones is your own 'Canticle to the Sun.' It has absolutely amazing mathemati-

cal properties, and these translate rather nicely into the abstracted patterns. We don't quite understand all of it yet, and, no doubt, you could be of help to us here, but the resonance does seem to arrest the decay of the cells and alters the timing of the rate of growth and decay."

It wasn't rage, though it felt like that, it was knowledge, pure, immediate, and utterly without thought: Viracocha knew that he could kill the Archon. He saw in the purity of image himself twisting the neck around to see just how much twisting the neck could take until it snapped. The murder would be poetry. Nothing else would matter beyond the appropriate perfection of the act itself. Viracocha began to take a deep and slow breath, preparing to spring, but the breath, like a receding wave that exposed the ocean floor before it hurled itself beyond the limits of the shore, exposed something inside him. It took the form of a voice, but he knew it wasn't a person. "If you kill him now he kills your soul by succeeding in turning you into a twisting killer." The withdrawing tidal wave of his breath stopped, and with its force spent, everything inside Viracocha stopped. With action canceled out, a paralysis of soul took hold of him.

The Archon looked at Viracocha to see what effect his words were having on him, and then continued his lecture in that artificially friendly and avuncular tone:

"We choose to use pregnant women in this study because, you see, the etheric body is much stronger. And with the growth of the fetus we have a marvelous opportunity to observe the formation of the etheric body itself. Even if the mother or the fetus does die, we have the opportunity of slowing the rate of death down so that we can study the various stages of the process of disengagement in which the etheric body slips out of resonance with the physical organism. For example," the Archon patted the medallion on his chest, "we have developed a new lens that permits us to observe the form and radiance of the etheric body directly."

The Archon raised the lens, which before had seemed merely to be a jewel inside the medallion, and looked at Viracocha:

"I must say Your Grace is in exceptionally good health. Why, I have never seen . . . hmmm, interesting."

The Archon dropped the medallion to his chest and looked at Viracocha with a puzzled expression, but Viracocha simply stared down at the wires attached to the woman's head. Everything seemed to be so far away, as if he were looking down at her from a great height.

The Archon waited a moment to see if Viracocha had any questions, and then resumed his lecture as Viracocha stared down at the woman's face.

"The two etheric bodies of mother and fetus provide us with an enormous level of vitality, but eventually we had to move beyond this stage of research, for we were not able to detach the etheric body of either mother or fetus. The instinct for self-preservation is too strong, and so both etheric bodies stay locked into the physical. What you see here is actually the phasing out of this aspect of our work. Now we are simply monitoring the metabolic conditions of the deep coma."

"Then why not revive them and send them back before it is too late?" Viracocha asked, and it began to dawn on him that whatever horrible sight he was confronting, he was only at the beginning of his tour of the Institute.

"Oh, at first we did try to return them, but the mother and infant were always mentally retarded. We found out that by returning them we stirred up a bit of trouble in the villages of Anahuac. It was better to let them simply be taken away in the garudas, that way they could think that they were being carried off by the gods to heaven."

Viracocha remembered the Captain of the garuda that he had talked with that morning. Now he understood the tightness around his eyes as they had both watched the children pass through the gate.

"And what will become of these women?" Viracocha asked in a flat and steady voice.

"We will slowly deepen the coma by infinitesimal degrees to study, with meticulous attention, the moment of death and the behavior of the two etheric bodies. But, come, to appreciate the significance of what you are seeing here, you must see the results of our work in the next room."

As the Archon walked to the door that led to the adjacent

laboratory, Viracocha turned his back on him and pretended to survey the entire scene. He was stalling to try to gain some idea of what to do. He wanted to be anywhere but where he was, and was disgusted to think that all of this had been going on while he had wandered around the peninsula in long walks of contemplative solitude. Small wonder that Brigita had been disgusted with his artistic powers of concentration. But the thought of Brigita only contributed to his confusion and disorientation. If Brigita was right, could that mean that she was right in all the rest?

"Your Grace?" the Archon said to interrupt Viracocha's reflections. "There will be time to return if you wish to study the effects of your music more closely, but I would like to give you a higher perspective of the entire project."

Viracocha took one more look at the women, and their great rounded stomachs reminded him of the clay ovens the Anahuacan women used to bake their bread. As he turned round to follow the Archon into the next room, he saw that the Archon was once again looking at him through the lens in his medallion. He dropped the medallion to his chest, smiled as much as his stricken face would allow, and then waited by the door as Viracocha passed slowly by him into the room.

Once again Viracocha encountered the semicircle of a dozen tables, but this time the women were all quite old. Their withered breasts sagged off their chests and the flesh on their thighs listed to the side of their bones. And then he discovered in horror that they were all pregnant. He could see the corpse of the fetus inside their collapsed stomachs: a withered raisin inside a withered prune. As soon as he realized what he was seeing, the sense of disorientation came back with renewed intensity and altered the dimensions of the room. Everything was small and far away, not truly in his space, not touching his personal sense of presence.

"Although each of these women is only around eighteen," the Archon commented, "the drug without the musical analogue seems to trigger the aging process. But we have been able to remedy the situation somewhat, for now we are able to slow down the flow of time in their bodies. You see, what we have to

do is to find a way to keep a person exquisitely poised between life and death, neither alive nor dead. It has been quite a challenge, for all too often they move past the desired point and simply die. Then we lose the etheric body, and we can't have that. You can see how important your music is, for it arrests the aging process. The glandular system is not confused in its sense of timing. But there was a negative side to your music, and that was that it seemed to increase the inseparability of the physical and etheric bodies. Beauty, it would seem, is too harmonizing, if you will excuse the pun."

Viracocha knew the Archon was watching him closely, so he turned slightly to the side to pretend to be interested in something the Archon had not mentioned. He looked at a leg of one of the tables, for there was one spot where he could focus and not be able to see either the wrinkled faces of the young women or the Archon's half-stricken, rigid face. He needed that neutral place to orient himself, for he was becoming confused and dizzy. Three different commands kept rising inside his being, but they were locked and tangled in a knot of conflict, with no one emerging as victor. One was a command of rage to kill the Archon, to stop that irritating, lecturing voice by twisting his neck, not just once after the first snapping sound, but again until the head would wobble on its empty neck like a puppet from which the supporting hand had been removed. The other command was to flee from his own delight in murder, to go deep inside himself where they could never get him, no matter what they did with his body. There he would be safe and silent and would never have to talk to anyone ever again. The third command seemed coming from some place on the other side of himself, above the other two commands locked in a wrestler's embrace. It was a mind without a body, a voice without a mouth, and it said simply: "Do nothing now. You will be able to end all this later." But the counsel of doing nothing seemed only to enrage the violence inside him, and thus strengthen his desire to escape the intolerable situation by fleeing from consciousness completely. Viracocha reached out to find some human support and took the wrinkled, withered hand of one of the ravaged young women, and held it. And that helped, for he

thought he could feel the spirit of the woman respond, to accept his hand in gratitude.

"I am impressed, Your Grace," the Archon commented. "Most people who have had no training in biology are a little squeamish and are afraid to touch them. It is good to see that you are basically a man of science. As you will observe, the flesh is quite cool, but not yet deadly cold. Now, as I was saying, music was effective, but too harmonizing. But when I considered the power of sound, it occurred to me that if beauty could be so effective, then perhaps an equally intense, and sometimes even related, experience might serve as well. In other words, terror. We know from our research in sound that there are some screeches that are so horrible, so irritating, that the organism cannot tolerate them. As it turned out, that was an inspired intuition, if I may be permitted to say so, for the screech of terror immediately separates the etheric body from the physical. Come, let me show you the results of this work in the next room."

Viracocha knew at once who would be in the next room. The children would be there. The image of the twelve children passing by him in the morning fog returned. He did not want to go into that room. He did not want to see or even imagine in what condition they were. The struggle inside himself increased in intensity. His rage now knew no bounds, for he no longer wanted simply to kill the Archon, he wanted to kill the entire civilization: all of it, every temple, every garuda, every poem, even every piece of his own music that had found its way inside the Institute. It was all guilty. While he had been writing music, supported in luxury, less than a mile away, all of this had been going on. His music was like an exotic, beautiful flower that they grew because it produced a poison for their darts. The flower thought it was admired for its beauty, but did not know for what reason they could reach down into its center to come out with the perfection of death. But how could he destroy the whole civilization? Snapping the Archon's neck could not accomplish that.

As his imagination began to race through scenarios of destruction, his raging self quieted down, appeased for the moment

with the fascinations of revenge; as the rage quieted, so did the balancing pull into absolute immobility. The strong buzzing sensation in his abdomen returned and he began to feel that he had the power to break the neck of the Tower itself. But the power was not there where he stood, but in the next room; he needed to go in there, deeper into the spine of the Tower where he had only to reach out to crack it and sink the whole civilization into the sea.

Viracocha walked with new vigor and speed into the third room. They were there all right, they, the children, just as he had expected. Once again there was that semicircle of a dozen tables. Once again there was that exposure of the complete vulnerability of the bodies that contrasted with the cleanliness of all the instruments. The children were stretched out in all the naked awkwardness of early puberty: the beginnings of pubic hair and the puffiness within the nipples of the girls. But unlike the pregnant women, their eyes were all open. They all seemed to be staring at the ceiling in a state of shock.

"I think you will appreciate the delicacy and precision of this operation," the Archon said as he moved over to one of the boys closest to him. "The drug is introduced into their bloodstreams through the tubes here in the right arm, but once they are all deeply comatose, we then remove all their blood, collect it in the large quartz vessel in the center of the room, and then sensitize it by vibrating it ultrasonically. We use a stabilizing fluid to maintain the bloodstreams for the few moments that their own blood is outside the body, then we return the transformed blood back into them, and slowly bring them up to the level of dreaming consciousness. It is at that point that we induce the screech of terror into the lattice of their bones and skull. The shock causes their eyes to open, and if the eyes do not open we know immediately that we have not been successful. We then maintain the screech, but lower their metabolic processes, and then, when we have them stabilized in terror, the etheric body flies off. But because we have confused their bloodstreams we are able to collect the twelve etheric bodies, if we simply drain half of the bloodstreams into the quartz vessel, there in the center of the room. If you were clairvoyant, Your

Grace, you could see the twelve etheric bodies now within the vessel. Unfortunately, we have found that if the screech of terror is stopped even momentarily, they die, and we lose the dozen etheric bodies, which, of course, we can't have. But everything is fine as long as they remain suspended, neither alive nor dead, in this state of terror."

The Archon turned away from Viracocha for a moment to survey his work. With both hands on his cane, he leaned forward like a commander surveying the battlefield that had just put victory into his hands. Then with a backward glance, he turned to see how Viracocha was taking his initiation into a knowledge beyond life and death. Viracocha stood motionless and simply stared ahead at the vessel of blood in the center of the semicircle.

"You see, Your Grace," the Archon said, resuming his lecture, "the difficulty with the pregnant women was that the protective instinct of the mother was too strong. We could not detach the etheric body of either mother or fetus. But children at the early stages of puberty were a totally different case . . ."

The Archon signaled to the page to bring him a chair. He leaned over with his hands cupped on the top of his cane and surveyed his work with a nod of satisfaction. Then slowly he lowered himself into the chair, and continued his lecture, as if the audience were not Viracocha alone, but a posterity looking in over his shoulder:

"You, no doubt, know, Your Grace, that sometimes pubescent children—girls more often than boys—will cause peculiar disturbances around them. Things will go flying around their homes without anyone touching them. This comes from the transformation of the etheric body at the onset of puberty. As soon as the girls menstruate, or the boys have their first ejaculation, the phenomenon disappears. Puberty also generates a psychic form of detachment. It is as if the soul resents what is happening to it. I suppose it is some form of racial memory of the entrapment into the animal body, so that the development of the genitals seems to suggest a further descent into animality, a loss of spiritual innocence, and so some souls resent it and pull away from the body. Girls are often moody and daydreaming in

their detachment. Boys are simply clumsy, not truly into their bodies. Well, this detachment, this pulling away from the body was just what we needed, for it enabled us to detach the etheric body without terminating the physical body. By maintaining the screech of terror inside the physical body, we ensure that the etheric body will not return but will maintain its new association in the cluster. As long as we can prevent the child from dying, we can ensure that the etheric body will not disintegrate. The key to it all was in collecting and mixing the blood. Strange, isn't it? Savages often emphasize blood in their disgusting rituals, but they were on to something after all. Blood is the critical element, the true amorphous liquid crystal. Would Your Grace prefer to sit?"

"No, I prefer to stand." Viracocha blurted out the words and did not bother to look at the Archon. "I feel strangely energized after my month away. Please continue, I'm simply concentrating on all the possibilities."

"Ah, yes, they are so many, are they not?" the Archon said, as he took in a deep breath before resuming his narration.

The image of twisting the Archon's neck was still a pleasant thought, and Viracocha had not entirely put it out of his mind, but the thought of doing more cosmic damage was far more appealing. He went over all the possibilities, and as all the images of destruction flashed through his mind, he felt a strange sense of exhilaration that seemed somehow to be localized within his body. The peculiar buzzing had returned to his lower abdomen, except now it had become so loud that he felt as if he could hear it, but in his right ear only. He leaned his head slightly to the side to pay more attention, but the buzzing did not seem to become melodic as much as it grew louder and more powerful in an unsettling way.

"Your attention that is so sharply focused on the quartz vessel that contains one half of the united bloodstreams is well placed, Your Grace, for this is the receptacle for the dozen etheric bodies. The difficulties we now face come from the fact that the psychic dream bodies are still individuated, but, fortunately, terror is most effective on the psychic level, so we are continuing along these lines. Here we found that we had need of some-

one who could project the psychic body from its association with the physical to create a thought-form that could begin to unite the etheric bodies to a common dream body. We found a shaman from the Eastern Continent who was quite skillful at this. Strange isn't it, this relevance of the primitive to advanced science, but there it is. When the Tower is working again, we should be able to create an opening in the etheric range so that one of the Old Ones will be able to get through."

The buzzing was growing intolerable within Viracocha, for it seemed to have shifted from his abdomen to his spine, but still he could not ignore this reference to the Old Ones, for it meant that they were not simply trying to ascend, but were trying to bring something down.

"And who or what are the Old Ones?"

"We're not exactly certain yet," the Archon said. "They may be something precipitated from the collective unconscious of the children, something precipitated by the protracted state of terror in them. It is the shaman who sees and takes messages from these Old Ones, for they seem only to operate in the psychic dimensions and are not able to penetrate through the etheric into the physical. The voices themselves say that when the solar crystal is activated that they will be able to project with the power of the etheric-cluster into human sperm. We will certainly look into this, for conception is the next phase of the work here, at least a slightly more immaculate form of conception than the one we've been habituated to in our animal bodies."

Viracocha began to feel as if his own body were being taken from him, and he stared wildly ahead. His gaze was so distracted that the Archon shifted round to see what was disturbing Viracocha.

One of the naked young boys on the table just to the left of center caught the Archon's attention as the child's penis began to lift in an erection.

"Oh, is that all that is disturbing you?" the Archon asked with a bemused tone. "That sometimes happens. It means that he is too far advanced into puberty to be of any use to us. He'll ejaculate in a minute, which will actually indicate that he's

dying. It is interesting, is it not, this continual association of sexuality and death. The ejaculation of the dying man whose spinal cord is snapped. It's much the same here when the screech carries the sexually developed over into death. They get this erection and then this last ejaculation, this last pathetic little spurt of life in the face of death. Ah, you see, here it comes."

The little penis waved back and forth in the futility of its search for a womb, and then it coughed up a small jet of semen, strained for a moment, and then collapsed. As it fell over with the last stream of life oozing out of it, Viracocha felt as if his own spinal column had been snapped and he doubled over and fell to his knees in sharp pain.

It felt as if something had kicked him in the stomach and then ripped him open with an electric knife that went up from his abdomen to his heart. The pain in his chest was so unbearable that he felt as if his heart were being ripped out. Uncontrollable spasms of vomiting took hold of him and he felt as if he were vomiting his own entrails onto the floor. Even when his stomach was empty, he could not stop but continued heaving. He was on all fours, like an animal, staring down into the brown pool of his own vomit. Finally the spasms subsided and he was left gasping for breath. As he gazed down into his own reflection in the liquid mass, he felt a peculiar sense of peace.

The Archon was utterly astonished and yet somehow seemed pleased with this unexpected display of weakness on Viracocha's part. With a new sense of dominance in his voice, he took on a more openly patronizing tone:

"Please help His *Grace* up," the Archon said to his page. "I had no idea Your Grace would be so affected by the sight of a boy ejaculating. Perhaps it means that you have some attraction to young boys that you have been ignoring. There's no need to be so ashamed that you have to expose your insides in this symbolic manner. The temple concubines are really not appropriate for a man in your high position, and when you have had a chance to take a new assessment of yourself, I suggest you come round. We have more young boys here than we need for the work. But, perhaps, I am being overly analytical. Perhaps, you

have simply picked up some infirmity in the primitive huts of Eiru."

The page lifted Viracocha up and held onto his arm for a moment, but Viracocha did not take his eyes off the pool of vomit on the floor. Somehow in his mind the pool of vomit and the sea in his dream had become one. He had found something, but he still could not say just what it was. All that he knew was that the pain in his abdomen and chest had gone.

"I hope Your Grace will be feeling better by this evening," the Archon said. "It would be a pity to miss the great honor of a private dinner with His Holiness. Perhaps, if you lie down for an hour or two, you will feel better. Please escort His Grace back to his quarters, and send in one of the servants to take care of this."

Viracocha did not bother to speak or to look up, for the Archon's petty jibes seemed far removed from the important things going on inside himself. He could hear the sandal strike the floor, hear the long drag of his right foot, and then the click of his cane on the stones. He waited for several moments until all the sounds of the Archon's movements had disappeared; then he stood up straight, removed his arm from the page's grasp, and took command of the situation:

"I can walk back by myself. You need only go for the servant. I will return to my quarters by myself."

He turned quickly away from the page and left him to his bewilderment as to whom he should obey. Viracocha kept his eyes straight ahead as he passed through the three laboratories into the main corridor, and he kept his eyes fixed straight ahead as he passed the guards at the control point and at the copper doors. He was not interested in them, for his anger had become pure and more refined. Revenge against individuals or the entire civilization now seemed petty and irrelevant. He knew the Plates he had given them for the Tower would destroy the City, perhaps even the three Great Islands. He was not interested in them anymore; they were mere tools. It was God who was the real enemy; it was God that he needed to kill.

6

He did not want to look up at the Tower. He did not want to look down at the sea. He did not want to be in the College, the City, the Great Island. He did not even know if he wanted to be, but the only thing that still held him in the grip of life was his rage at God.

He kept his eyes to the ground so that he would not have to look at anything, but there was little chance that he would see anyone. They would all be inside chanting the evening prayers to the setting sun. But he could see the pale blue hem of his priestly robes that they still had on him and that sickened him. He wished he could vomit out everything they had ever put in him or on him. He was sick of priestcraft, sick of this religious analysis of misery that only seemed to produce more misery. The religious study of suffering only gave them a license to prolong it. Their sophisticated doctrines of reincarnation gave them the illusion of power over life and death with a transcendent dispensation to kill. What did one life matter in the great cycles of reincarnation, in that vulgar profligacy of verminous, abundant life after life?

But what did a hundred lives matter, if one did not? A hundred zeros still added up to zero. Nothing could add up unless there was a value to *one.* But he no longer wanted to be one. He no longer wanted to share existence with Evil, to say this side of the bed is mine, but you stay over there.

He had looked down into the ground of being and found that there was this filthy secret to existence. Evil was not simply a disease of being, it was Being. The Institute was not some kind of infection in an otherwise good civilization; it was the very essence of the whole civilization. Beauty and goodness were simply allowed a little season to flourish to make the ultimate

desecration more delicious: like someone tenderly nurturing a pregnant cow, waiting patiently to slaughter her calf for the delicacy of veal. Nothing could stand in the face of Evil because it was the face of everything. Evil would always win in the battle of good and bad, for in fighting the bad one had to use the methods of Evil. He knew that if he had exploded in rage and killed the Archon and his dumb submissive attendant, he would simply have added one more murderer to history's lengthening list. One could not be alive to fight Evil, for life itself was Evil. Men were not locked into some prison by a geometrizing god, they were penned in like animals. No, worse than that. There was a malevolence of consciousness at work here, not simply slaughter. There was a Mind seeking to grow Evil like some dark flower in a pot. When "good" armies went off to fight "bad" armies, that malevolence steamed above the battlefields and enjoyed much more the hot pleasure of the "good" man killing the bad than the obvious, cold, dry, and prosaic evil of some dumb brute lopping off a hundred heads. And how that malevolent consciousness loved being prayed to as God, loved the pretense of kindness that only made the final revelation of cruelty more perfectly Evil.

Viracocha lifted up his head as he passed through the gate into the College. The City of Knowledge, he thought, was well named, for behind all the arts and sciences of civilization was the hidden truth of the Institute, and the Institute was the universe in miniature. It was that malevolence made manifest. The children in the Institute were the open expression of what all suffered. Not even death could deliver them from suffering. There was no escape, unless somehow one could find a way to undo existence. It would not be enough simply to die; one had to find a way to eliminate death along with life, and the only way to do that was to reach up and pull down this maniacal geometrizing God, the cunning author of all these restraining traps of love and beauty.

Viracocha passed through the last doorway and came into the columnal arcade that surrounded the residential quadrangle. He noticed someone sitting on the low wall in front of his door. The figure had his back turned toward Viracocha and his head

was hidden behind one of the columns. As Viracocha approached, the man turned round, stood to attention, and then gave a formal bow. As Viracocha recognized the Captain, he also remembered in anger the appointment he had made.

The last person Viracocha wanted to see at that moment was another member of the staff of the Institute. His rage at everything flashed out at the intruder and he decided that he was not going to bother with courtesy any longer.

"Good Evening, Your Grace."

"Considering that I have just returned from a tour of your Institute, Captain, I do not consider it to be a good evening at all, and I would thank you to leave me at once so that I may be alone with my thoughts."

The Captain looked shocked and nervously glanced around to see if they were being overheard:

"Then that is all the more reason why I have to talk with Your Grace. It is as important for you as it is for me, *especially now,*" Bran said, as his voice trailed off into a whisper.

Viracocha stared at the Captain in silent anger for a moment, but the more he did look at him, the more he realized that he was not angry at him:

"All right, Captain," Viracocha said with an assenting tone of defiance, "but don't expect it to be a pleasant social visit."

Viracocha opened the door to his quarters and left it open for Bran to follow behind. He did not turn back but moved across the room to his cabinet and took out a crystal decanter of brandy and two goblets.

"I can't offer you any of your native Eireanne dry mead," Viracocha said as he turned to glower at Bran, "though I wish I could, for its power to take your mind away would be most welcome now. Unfortunately, we'll have to make do with a coarse brandy."

Bran moved into the center of the room and looked around to see if there were servants present:

"Thank you, Your Grace, but I didn't come to get drunk. Are we alone?"

"You are alone, but unfortunately I'm not," Viracocha said as he took a drink. "No, there are no servants present. I prefer that

they clean when I'm not around. I don't like to be disturbed when I'm working, so over the three years here, I've been able to frighten them all away."

Viracocha moved into the center of the room, set the decanter down on the table, filled the goblets, and then sat down without bothering to remove his cape. As he took another drink, he looked up at Bran over the rim of his goblet.

Bran looked down awkwardly for a moment, then he picked up the goblet and sat down on the chair on the other side of the table.

"Why are you so dressed up?" Viracocha asked. "There isn't going to be another Solemn Assembly. As a matter of fact, I don't think there will ever be another Solemn Assembly."

"I thought it would be easier to be inconspicuous if I were more conspicuous, as if I were on an errand of official importance."

"You sound positively conspiratorial," Viracocha said. "I hope you've come to tell me that you're going to destroy the Empire, for I'll drink to that."

"I have come to tell you that I have done as you commanded and that the vessel will be ready to take us at dawn the day after tomorrow."

"As I have commanded?" Viracocha remarked with a laugh. "I haven't commanded anything, Captain. I'm afraid you're not making much sense."

"When you gave me the official password this morning, I contacted the 144 to prepare for the escape flight to Tiahuanaco. Did you make any contact with Abaris through Brigita when you were on Eiru?"

"Wait just a moment, Captain," Viracocha said in annoyance as well as surprise. "First of all, I spent only one day with Abaris when he was alive. Second of all, I was with Brigita, but I don't see how or why you should know about that. And third, I did not contact the ghost of Abaris, for that's not exactly my style. Now, what 'password' are you talking about?"

" 'It had no arms to fight / But only wings for flight.' That is the password that Abaris said the leader of the colony would give me to identify himself and to let me know that I was to prepare

the 144 for escape to Tiahuanaco just before the Great Islands were to sink in the sea."

Viracocha took another large drink of brandy and then folded his arms across his chest as he considered Bran in a new perspective:

"Tiahuanaco *is* a beautiful place. I lived there once for over a year. If I were a leader of a colony of refugees, it is certainly the place I would choose. I hate to end this little fantasy, Captain, but I made up the end to your poem on the spot. Abaris did not give me any password, and he never told me about any colony of 144 for Tiahuanaco. As for Brigita, she never said a word about it either. I went to Eiru for the pleasures of cohabitation. I'd never slept with an Abbess before, and it held out a certain sexual mystique that I found exciting . . ."

Bran stiffened in his chair, lowered his goblet to the table, and looked at Viracocha in disbelief.

"Ah, I see I've shocked you. Well, you see it just proves that I am not a leader, and certainly not a good disciple of Abaris, so where does that leave us, Captain?"

"I'm not sure. You certainly wouldn't be my choice of a leader, that's true. But the facts remain: Abaris put the password into your mind, Brigita singled you out for the initiation with the dagger and the chalice, and you are familiar with Tiahuanaco and its native peoples. I guess it means that you've been kept in ignorance so that you could more effectively penetrate into the hierarchy here, for Abaris' prophecy said that the new leader would sit by the side of the High Priest himself. I guess the test would be to ask if you had a vision of the winged serpent when you lived in Tiahuanaco?"

"Shit!" Viracocha exclaimed in a feeling that he was being conscripted into an army against his will.

"So, there you have it, Your Grace," Bran said with an air of one taking command of the situation. "I would hazard a guess that your ostentatiously sacrilegious manner is simply a convenient blind that your higher Self uses to keep you out of the way while it goes about its business. You may have thought you were going off for a sexual frolic with Brigita, but you can be certain

that she knew just exactly who you really were and just exactly what she was doing in preparing you."

"Then what do you mean by 'the initiation of the dagger and the chalice'?" Viracocha asked, beginning to suspect that Brigita was indeed not a witch and that he just might have failed in some important way.

"The dagger and the chalice are very ancient symbols in our religion. They go back to the time before farming. Originally, I think the dagger was a spear. Anyway, they are symbols of the male and female forces in the universe. The chalice is the vulva and the cup of the crescent moon. It is also a sacred wound. As the vulva bleeds every month with the moon, it is seen as wound that heals itself, the crescent moon that fills up with light. In the dark of the moon, all the women bleed, then the moon grows full. The dagger is the shaft of light from the sun that opened up the earth to light. It is the phallus, and the dagger that creates a wound to open darkness to new life. In the vulva, the phallus causes the bleeding to stop, to flow into the womb for the child. The child is also the higher self born into the world. All of these meanings are seen at once in these images, but didn't Brigita explain any of this to you?"

"I suppose she was trying to in her own way," Viracocha said in a feeling of embarrassment that was beginning to make him feel very tired. "I guess because she's a dancer, she believes in showing more than talking. Anyway, we were . . . meditating. The moon came through the window. It made me open my eyes, then I saw the chalice and the dagger on the altar. I became taken over by this vision of Brigita as a witch and that in some primitive rite in your Eireanne religion, I was about to be sacrificed to fill her chalice."

"That's really interesting," Bran said with an air of serious and sober concern. "In your state of psychic sensitivity brought on by meditation what you picked up was part of the religion of the past on Eiru. Abaris was able to drive the witches out of Eiru, but they still exist on the Eastern Continent. In their dark rites they prefer boys in puberty to men like yourself. They drug them and then hypnotize them to make them see themselves as beautiful. They stimulate them for a long time, and

then they castrate them to fill the chalice with the blood and semen, which they drink in the delusion that the concoction will grant them eternal youth and magic powers. It's all a mistaken literalism, taking something etheric and trying to make it simply physical. It is sort of a demonic parody for something much deeper, and that is why Abaris was so angry and drove them out of Eiru. It's also the reason why Abaris was so secretive and would not give out the teachings to very many."

"Well, the ritual you describe is exactly what I thought Brigita had in mind for me, and I ran from her house in terror, all the way to the port. There I intimidated the crew into flying me back here before the old witch could wake from her trance."

"And just in perfect timing for you to accidentally bump into me, and to just happen to give the passwords," Bran said as he laughed at Viracocha's bewilderment. "Were you running from, or running toward?"

Viracocha said nothing. After a good long minute of silence, he lifted his goblet and took another drink, and then folded his arms across his chest and looked at Bran:

"And I am still going to keep on running *from.* I'm not a leader, Captain. You seem like a nice fellow, you take the 144, and I'll stay here to sink into the brandy before we all sink into the sea. I'm not very good with people, as you should be able to see by now."

"What we're good at, Your Grace, is not always the source from which our inner strength springs. People have been destroyed as often by their strengths as by their weaknesses. Look at your High Priest."

"You're a veritable fountain of wisdom, Captain. You're wasted in the Institute. It's a pity you weren't one of my colleagues in our blessed Sacred College here. I could have used a good drinking partner these last three years."

"Someone was needed who could fly, and even more than that, someone was needed who could fly the largest garuda in the Empire. I can assure you that if it weren't with the hope of being the only one who could get the people out, I wouldn't have continued, but would have gone off to a remote island to work on my poetry. But this last year has been the worst. You

saw the children this morning, so now you know what it has been like waiting for you to show up and give the damn password so that we could get the hell out of here."

"Hell is not an inappropriate word, Captain, for not only have I seen the children with you this morning, but I have followed their career all the way into the heart, if you will excuse the word, of our blessed Laboratory of Incarnation. Have you ever been inside to see our noble work in perfecting the evolution of the human race?"

Bran looked down at his feet and spoke without lifting his eyes to Viracocha:

"No. But I saw what the women and children were like when we used to take them back. Bran raised his eyes to look at Viracocha, imploringly, but also accusingly: "Why did all this have to take so long? Why didn't you show up sooner?"

"Don't you start accusing me!" Viracocha exclaimed in anger. "Brigita did that too. Listen, Captain of the great conspiracy: I am not one of your 144. I am not a leader. I don't give a damn about anything, not you, or Abaris, or Brigita, or this whole rotten civilization, or God. You've been had, Captain. God is simply the Archon, writ large. This whole fantasy of escape of yours is his little cat and mouse game. He uses me to set up the Tower so that it destroys this whole wretched city, maybe even all the Great Islands. Then he plays a joke on me and uses me to give you some magic password to carry the 144 to safety so that a few dumb followers of Abaris will think that Good has triumphed over Evil. Then slowly at first, but with increasing haste as the new civilization gets going down there, he will begin again to grow a lovely little garden of Evil. And if we're all too primitive down there and have forgotten how to build our great towers, why then we can improvise with stones and rip out people's hearts and stand on our tiptoes to lift them up in prayers to the sun."

Bran stared at Viracocha in disbelief and showed in his eyes a fear that the plan he had based his life on was about to be ruined. He looked at Viracocha in a state of shock that began to modulate into anger, as Viracocha gazed back at him in defiance.

"You see," Viracocha said, "you've got the wrong man. It's God's little perverted joke on you. You thought I was going to be some holy and devout disciple of Abaris, full of faith in the Great Plan, in the great battle of Good and Evil. But the illusion of battle is just a trick to keep you fighting, for Evil always wins. I don't want to escape, and I am not thrilled by visions of survival, because we would simply carry sweet Evil with us to Tiahuanaco, and the whole thing would start over again. I'm sick of Evil, Captain, sick of playing its game, so I'm kicking over the table and the pieces and saying to hell with you all."

"You're sick of Evil!" Bran shouted. "You act as if you've just discovered it. Where have you been your whole life? No, I know the answer to that. You've been here, a pampered favorite of His Holiness's private collection of artistic geniuses. You've been playing here in luxury a little more than a mile away from the Institute. And now you've discovered that there are bad men in your private universe, and so you sulk. You feel sorry for your poor self whose private party has been ruined. Then you come up with some great dramatic anger at God to avoid your own guilt. This God of yours sounds more like an inflated image of yourself, a being who tolerates Evil in order to compose without disturbance. Who divided up your life with Evil here and Good there? Your great anger at God is just more of your own theatrical vanity. But even in your grand self-righteous anger, you indulge yourself and carry on as if Evil were a personal insult and affront to you. God! You would have thought Abaris would have taught you more than that!"

"Abaris spent a little more than an hour with me. He told me a few things about the winged serpent, gave me a few meditation techniques, told me to wait six months before starting kriya, and then he left. I never saw him again."

"You saw him every night," Bran said. "Especially if you were given kriya. Though how you can do kriya and not be aware of your psychic life amazes me."

"My, how you do sound like Brigita. Even like the High Priest, for that matter."

"Well, perhaps there was a purpose even in your being so self-indulgent. You don't seem to be very patient, so I can't imagine

that if you knew about the conspiracy, you would wait for the right minute to move. You probably would have been rash and impulsive and moved too soon; God knows it's been hard enough for me to wait for you to show up."

Viracocha was silent for a moment and slowly sipped his drink as he considered Bran:

"If you had escaped before the elevation of the Tower and my revision of the design of the foundation vault, the Great Island would not explode and the garudas of the Empire could follow you down to Tiahuanaco at their will."

"So, there you have it," Bran said, beginning to relax and feel less threatened about the safety of the plan. "Abaris said that 'A grain of sand is more effective as a weapon in the eye than on the foot.' He prophesied that the leader would come from the hierarchy itself. Every time I came to the City of Knowledge, I kept looking to see if I could find him. And every time I had to fly back to Anahuac, I would try to think things out. I kept telling myself that if things were twisted, then they needed something straight to twist. That if there were some depraved being looking for the perfect evil, then he needed some idea of perfection. That if I were the Archon and wanted to have an army of torturers obedient to my will, then I would need them to be obedient, loyal, and truthful with one another. That if they became evil in every instant of their lives, they would kill one another and cease to be an effective force for Evil. I began to see that in order to *do* Evil one had to *be* Good, that Goodness was more basic, more fundamental to the nature of existence, that even Evil had to invoke it to try to thwart it. The greater the Evil, the longer the time criminals must spend in Goodness to effect their crime. Eventually, I think humanity will see the inherent contradiction within Evil and transcend it."

"Or is it simply," Viracocha asked, "that Goodness exists to make Evil more enjoyable? After all, you have to have straight things around for the fun of twisting them."

"But the pleasure of twisting, as you say, comes from the prior existence of the straight thing and the freedom that exists. If God had created the universe as a machine in which nothing was allowed to deviate from His will, He would not have to

bother to create it, because He would know how it would turn out. But in an open and free universe, you must have souls who are free to deviate from the will of God. They can only express love if they are free, but if they are free, you open up the possibility for Evil to deny existence. The question then becomes: Do you risk being for love, or do you play it safe in nonbeing?"

"And so for you," Viracocha said, "the fact that we're having a philosophical debate while the world goes to hell means that our 'Beloved Creator' chose to risk it?"

"In a word, Yes. God emanated the Conscious, but it was the Conscious who dreamed the universe and became caught in the projections of their own dreams. God didn't create the universe, we did. So if the universe appears to be Evil, it is a description of our own nature, that we like a story with Evil and danger in it. Evil is a projection, an expression of humanity's inability to express Being without Evil. Humanity is fascinated with Evil. It has discovered this dark and secret thing that comes out of itself. It is like a child playing with its feces, enjoying its expressive power in the patterns it can create by smearing it around. Eventually, the infant loses interest and discovers other forms of expression, and humanity will too, someday."

"But in the meantime the mother comes in and washes the infant and sets him down again in a clean bed. Who is going to mother this infantile humanity if God doesn't come in and clean up the mess?"

"And just how could God do that?" Bran asked, enjoying his new found status of authority. "Actually, you, both of us, are in a better position to *act*, since we are within this dream called the universe. God can whisper to us in our sleep and try gently to lead us out and away from our self-inflicted torment, but if he destroys the dream and the universe ends, we'd probably project another universe with the same problems. Actually, it could be much worse. We have Abaris who has one foot in the world and one foot out of it. The rest is up to us, and now that you have finally shown up, I am more than ever ready to get on with it."

"You know, when I listen to you," Viracocha said sardonically,

"I get the impression that you've memorized your lines at the feet of Abaris, and that now you're thrilled to play Teacher and to show that hidden in the uniform of a Captain lurks the soul of a poet and a prophet yet unborn."

"Don't patronize me, Your *Grace!*" Bran said in a flash of anger. "You claim not to respect your own Sacred College, but you still don't respect anybody outside your little world. You think I have to be stupid because I am in the Military. Well even if I don't succeed in becoming a poet in this life, I'll keep coming back to one culture or another until I do succeed. And long after your music is forgotten, my story about you will be remembered."

"Well, it's great to see a little vanity and passion in Captain Perfect," Viracocha said, lifting his goblet in a toast. "Here's to your fame and my oblivion. Now, since you have taken over my role as artist, I humbly suggest you take over my role as leader."

"I'm sorry," Bran said, trying to regain self-mastery. "I didn't want to lose my temper. Look, it's silly for us to argue. All right, I got angry because what you said is true. I have memorized my lines and little speeches from Abaris. I studied with him for twelve years. You still remember him after only a meeting of one day. Well, I had twelve years of instructions, and it's the only thing that got me through this last year. I'm a follower, not a leader. Abaris said that the leader would sit at the side of the High Priest, that he and Brigita would be able to work with the angels and elementals to free the children and sink the Islands. I can't do any of that. All I have been able to do is go over Abaris' talks, again and again. Last month when I had to fly the Governor back to Anahuac, I sat there watching him drool, watching the idiot spittle drip from his lips, and I kept telling him in my mind that you were coming, that soon his people would be free again.

Viracocha's face became deadly serious. He lowered his drink and looked questioningly into Bran's eyes.

Bran looked back at Viracocha and was surprised that he did not seem to know what he was talking about:

"I take it you don't know. The Archon has a contempt for death. He prefers to maintain control. He prefers to return

leaders so that the people can see what happens to those who try to oppose them. I'm told that he says that the death of a leader inspires his people, but his method completely demoralizes them. He has invented or discovered some drug that when placed into the bloodstream turns the person into a slobbering idiot. And it has had its effect. Anahuac is numb. He has crippled their will."

Neither man spoke. Viracocha set the goblet down and stood up. He turned away from Bran and walked silently to his desk. He looked down at the thin metal sheets of the transcriptions for his music. It all seemed so irrelevant. It was too soon for that kind of thing, another age, another civilization, perhaps; but not now, not in this world. He picked up the sheets of his music, folded them neatly into a leather portfolio, and tied the ribbon around them. He picked it up, but let the portfolio hang down by his side as he walked back to the chair where Bran was seated.

"All right, Captain. I've packed my things. Here's a souvenir to take with you in the vessel. I'm not certain what we do next."

Bran stood up and took the portfolio into his hands:

"I guess you do whatever is next. You seem to have been on course even when you thought you weren't. What is next here for you?"

"I have been invited to a private dinner with His Holiness, the Archon, and the Master of the Sacred College."

"Well, there you have it," Bran said. "The prophecy said the leader would sit at the side of the High Priest, that he would come from the summit of the hierarchy. Since it has been prophesied, you might as well continue to walk in the footsteps of prophecy. Perhaps there is more leverage to effect a shift of the whole society at that exalted level. Perhaps that is what was meant when Abaris said that a grain of sand is more effective as a weapon in the eye than on the foot."

"It doesn't seem that there is anything else to do. I really don't have much interest in escape or survival, but, I suppose, if we do make it out of here alive, I would like you sometime to take me to Eiru to see Brigita. I didn't exactly leave in the best of spirits, and I would like to see her again."

"She's been placed under arrest," Bran said quietly. "Her Captain has been ordered to bring her to the City. She'll be here tomorrow morning. So, we will have to work fast, otherwise she is going to end up in the Archon's hands. And I promise you that I will not let that happen."

Neither man said anything. As Viracocha looked at Bran and realized how much Bran was in love with Brigita, he felt his cynicism dissolve in the complexity of his hate for the Archon, his own confused feelings about Brigita, and his own amazement at the self-mastery and control of Bran. All these years he had been a disciple of Abaris as a symbolic way of making love to the virgin disciple. All these months, he had been ferrying victims to their dismemberment, waiting to be faithful to a prophecy. And here all the time, Viracocha had cared about no one, had lived in complete ignorance and privilege, and even been privileged to make love to Brigita, and he had failed everybody, especially Brigita.

"All right," Viracocha exhaled in an agreement that was more of an exhaustion of his defenses than an affirmation. "I don't know who is leader or follower in any of this. And I certainly don't know what I'm doing. But go on, get the garuda ready, and I will go sit by the side of the High Priest. After that, I don't know what to do. I don't have your self-control. I just may indulge myself and kill the Archon so that, at least, he won't get his hands on Brigita."

"If you don't I will," Bran said with quiet finality. "I'll gladly go to hell if I can send him there." Bran walked to the door. As he opened it, he turned back, bowed, and said in a louder and more official voice:

"Thank you for the transcriptions of your music, Your Grace. Your music has been an important inspiration to my service in the Empire."

The door closed and as Viracocha stared ahead he thought it was time for the leader to follow. He would take Bran's example and dress to perfection for dinner. The High Priest had said full ceremonial dress would not be required, but for this occasion, he wanted the irony of imperial disguise.

Viracocha walked into the bedroom, removed his cape, and

fell onto the bed. The brandy and everything that Bran had told him about Brigita made him feel deeply tired. He knew that he did not have time to sleep, that it would not be long before the page came to escort him to the dinner, and yet he had no energy to move. He closed his eyes momentarily, telling himself that he would get up in a moment, but a thick heavy darkness seemed to crush down on him, rendering him completely paralyzed. He knew in the darkness that was his shame and guilt, he had somehow failed Brigita and now she was under arrest. He had failed the children and now they were caught, as he was caught, in this thick, impenetrable darkness. If he could only move, he knew that he might be able to save them all, but the darkness crushed every muscle and the pain in the back of his neck held him in some electric paralysis. He tried to scream, but he could only grunt like a moron without articulate voice. He could hear his own voice, but still he could not move. Then into the darkness from another world he felt Brigita's hands reach down to shake him, to move him, to break the grip of the paralysis.

"Your Grace? Your Grace!" The page shook Viracocha more vigorously by the shoulders. "Wake up. I have come to escort you to His Holiness's presence."

Viracocha's eyes snapped open as he stared at the page in a state of disorientation. He could feel a buzzing in the back of his neck, and he realized that he had stuffed the cushion behind his neck in such a way that he had cut off the circulation to his head. As soon as he gained command of his body, he jumped up.

"I'll need one minute while I change robes. Wait outside for me."

The page bowed obediently, and Viracocha ran to his wardrobe and slid back the panel. He quickly removed his daily robe and threw it onto the bed; then as he took out his formal robe and raised his arms to enter it, he felt as if the robe descending on his body had become one of the Old Ones coming down within the Tower to settle on the human form. Who were these Old Ones? he asked himself as he brushed out the snarls in his hair. Would the High Priest know? Thrusting his feet into his

gold slippers, he reached up for his midnight blue silk cape, and decided that he would bring the matter up at dinner.

Viracocha turned from the mirror and moved hurriedly across the room and out into the colonnade where the page stood in attendance. The page bowed and then turned round to lead the way to the Palace. As Viracocha looked down at his gold slippers, and midnight blue cape with its magenta lining, he felt as if he were dressed in an aura and not a priestly costume. The High Priest's ceremonial robe at the Solemn Assembly, with its broad peacock-feathered headdress, certainly seemed to be a rendering of the human aura. Perhaps, he thought, there was some kind of historical progression here. We start out naked with extended auras and no priests; then we move into a society of priests with simple white smocks like Brigita's, but with light pastel auras; and then the priesthood decays, grows richly elaborate as the gorgeous robes begin to cover the darkening auras with midnight blue, purple, and magenta. Thus camouflaged, the priest was safe and no seer could discern the true colors of his aura to see in what a depraved condition he truly was.

Viracocha felt not only disguised but armored as he walked with his hands hidden behind his silk cape. He had the feeling that he was indeed going into battle, but he had absolutely no sense of what he was to do, or what he could do; and yet he realized that if everything was true that Bran had told him, then he had never *known* what he was doing.

What a strange form of knowledge it is, Viracocha thought as they came out of the College Gate onto the Broad Street, to know that you don't know what you're doing. It made him feel as if his whole personality, his whole identity as Viracocha, were a convenient fiction. Some other being was playing him the way he played his own harp; even his own fear had been used by this being that saw over the curvature of the earth to send him running in terror to be on time for an appointment with Bran. If that were true, then what was he doing now, he wondered, as he headed toward the Palace for an unknown encounter?

It was easy not to believe in the substantial reality of the great City of Knowledge as he moved down the street toward the

hypostyle hall. The whole architecture was nothing but a few metaphors of power that were truly powerless in their shallow life on the surface of the volcanic earth. The city was already dead and all those in it were simply ghosts who had not yet realized that they were no longer among the living but were actually dreaming the world that surrounded them.

The unreality of it all gave Viracocha a new sense of freedom and fearlessness. He did not care to escape to Tiahuanaco. Bran could worry about that. He had no other purpose than to shatter the Tower and to release the hold the Archon had on every soul kept from death and life within the Institute. How a mere dinner party could serve that end he did not know, but since he did not know anything, he would take it one step at a time.

One step at a time, he went up the stairs from the Broad Street to the terrace above the hypostyle hall. One step at a time, he went up the stairs in the reception hall to the private dining room opposite the High Priest's study. But where were all the guards? The squadrons that had stood in the reception hall this morning were gone, and merely two guards stood at either side of the door to the small dining room as the page led him into the presence of the hierarchy. Had the crisis been resolved? Was the High Priest celebrating some victory with this dinner?

As Viracocha passed the two guards and entered the room, he saw the three of them standing by the portals to the south. He could sense at once that they had been talking about him and his behavior within the Institute that afternoon. As the High Priest turned round to greet him the last of the evening light touched only half of his face, and half of the white robe whose central band of color was dyed reddish purple from the cochineal of Anahuac.

"And here he is!" the High Priest said as he walked away from the Master and the Archon to greet him. "We were just talking about Your Grace and were hoping that you were not too sick to join us this evening for our small celebration."

"Thank you, Your Holiness," Viracocha said as he bowed first to the High Priest, and then to the Master and the Archon. "I think the sudden change from the northern climate of Eiru to

our warmer climate here made me feel ill. But I have rested, and feel well, if still a little tired from the journey."

"Good. Come, take your drink and join us," the High Priest said as he signaled to the servants to bring Viracocha a crystal cup of wine. "We were enjoying the evening light and its beautiful reflection on the copper doors of the Institute. By the way, Your Grace, you didn't have to go to the trouble of ceremonial dress. This is to be an informal celebration."

"Since this is the first occasion of my dining with Your Holiness, and with Your Excellencies, I didn't feel quite right in my ordinary robes."

"Well, no matter," the High Priest said with a smile. "Perhaps, it adds to our festive air. You see, this is a victory celebration. The revolt in Anahuac has been crushed, and the attempt by the Commander of the Military to join with the insurgents against the Theocracy has been dealt with appropriately. When the Commander was escorted to the meeting of the conspirators, they found him to be a smiling gentle fellow who wouldn't want to hurt an ant. He sang a few little children's songs to them, and the other officers decided that revolutions were silly, so they all joined the party and decided to be good boys. Now we all can get back to the real work."

"It would seem, Your Holiness, that much has happened in the month of my absence." Viracocha took his cup from the servant's tray and wondered what drug might be in it to make him at one with the party.

The High Priest watched Viracocha as he took his first drink, and then smiled, as if to set him at ease.

Viracocha was surprised to discover that the drink was dry mead:

"Unless I am mistaken, this is the dry mead of Eiru. I am surprised to find this in the Palace, Your Holiness."

"Yes, I confess that I prefer it to our brandy. The herbs combined with the honey have a soothing effect on the subtle bodies. But come, join us in our little corner over here."

Viracocha was startled to feel the High Priest lead him over to the Archon and the Master by placing his arm on his shoulder. He felt as if the High Priest were signaling by this gesture that

he was to be the successor. The Master and the Archon both showed by the momentary change of expression in their eyes that the gesture was not lost on them.

"I was just telling His Holiness when you came in," the Master said, "that I think we have all been too busy in these last three years. Here you are a Member of the Sacred College, and yet I hardly know you. The ceremonial dinners at high table are all well and good, but they really don't provide the occasion like this one for developing a true friendship, so we are indebted to His Holiness for this evening. I certainly hope it won't be the last."

"Not at all," the High Priest said. "Now that the political matters have been taken care of, we can all get on with the true business of our society. After all, that is why I had this City of Knowledge built in the first place."

"Yes, I quite agree," the Archon said. "Now that His Grace has been made the College's representative to the Institute for the work of redesigning the Tower, I think we can make progress more quickly. This specious opposition between tradition and innovation has been obstructive in the past, but now that we have the most innovative Member of the Sacred College working within the Institute, I predict that our progress will become astounding."

"Why then I think you owe the College a favor, Your Excellency," the Master said. "You should send us the most traditional of your team to work on a project with us."

"And what kind of project might that be?" the Archon asked.

"Astronomy, of course!" the Master exclaimed in a slightly artificial tone of clubbiness.

Viracocha looked at the flushed red cheeks of the Master, and the tiny beads of sweat on his bald head, and found it hard to understand how the Master had ever come to be an astronomer. Of course, that was a long time ago; now the Master did little work except run the College and dine with his cronies. It was no surprise to Viracocha that the Master did not know him. There wasn't anybody under sixty in the circle in which he hid himself.

The Master must have felt Viracocha's penetrating stare, for

he turned to Viracocha and tried to include him in his imaginary little circle of clubby intimacy:

"I must say, Your Grace, that I was very impressed with the Plates you drew up today. The system of correspondences between the Hyperborean Solids and the subtle bodies was fascinating; but the correspondence to the planetary intervals was brilliant. I am ashamed that I never thought of it. By the way, Your Holiness, are we permitted to discuss the project, or should we avoid discussions of work this evening?"

"How could we avoid the subject?" the High Priest said. "A cycle of 25,920 years ends in a few days in the coming feast of Sauwyn, what other subject is so compelling? When I said this dinner was to be informal, I meant that we could avoid formal protocol and polite, inane chatter. So, feel free to indulge your obsessions."

"Excellent!" the Master exclaimed obsequiously. "Well then, Your Grace, as I understand it there are only, in terms of General Vibration Theory, three possible frequency-signatures that can be induced into the solar crystal. Which one do you think appropriate?"

Viracocha turned from the Master to consider the High Priest, and without taking his eyes off him, he spoke to the Master:

"Knowing His Holiness's sense of imagination and daring, I would say that there is no question about it. The third, the highest and most dangerous one, is the only one that will provide the geometrical power to strain the galactic alignment and pull apart the fabric of space and time."

"Oh, come now," the Master scoffed, "it can't be that dangerous. We're only talking about setting up a restricted singularity within the foundation vault of the Tower."

"The point is, Your Excellency," Viracocha said as he turned his gaze from the High Priest to the Master, "that it may not prove that simple to restrict the singularity *to* the foundation vault of the Tower."

"Viracocha is right, Your Excellency," the High Priest said in appreciation. "There are always risks commensurate with the

scope of any project. But the civilization that stops taking risks dies long before its institutions collapse."

"But surely, Your Grace, you can't think the danger that great?" the Master said. "Just what do you think could happen?"

"An earthquake, if the singularity creates merely a corona of force around the vault, but if the singularity extends to the center of the earth, then the earth's crust will crack, and the islands will disappear."

"All three Great Islands!" the Master scoffed. "Really, I find that hard to swallow."

"Let's hope the ocean does too, Your Excellency."

The Master appreciated Viracocha's witticism and laughed more strongly than the remark deserved, as if to make certain that lighter spirits would rule the evening.

"Well, let's put off the end of the world until at least after dinner," the High Priest said as he stepped back from the circle to lead them to the table.

The emanating crystals were set to a low ambient light which was reflected in the quartz goblets. Viracocha waited for the others to take their places, but the seating arrangement would have been immediately obvious from the size of the chairs around the table. The throne of the High Priest was flanked on either side with large chairs that were handsomely carved but had no gemstones. Viracocha's chair had no gemstones, but was composed of inlaid woods. It was, however, directly opposite the throne of the High Priest. As the four servants pulled out the chairs from the round table, there was a polite murmuring between the Master and the High Priest. The Archon was to Viracocha's left, and when they all were seated, there was a moment of silent prayer before the first dish was set down before them.

"Since this is the first time Your Grace has been to this table, I should warn you that all the food will be vegetarian. When I moved into the new Palace and began sleeping directly under the Tower, I found meat to be too heavy. It caused unpleasant dreams."

"That does not surprise me, Your Holiness," Viracocha said. "I

have had no meat for a month, so I'm almost growing used to the diet."

"So, our witch of Eiru is a vegetarian. At another time, Your Grace, I will have to learn more from you about this curious young woman."

The High Priest began the first course, and the others followed. It was the custom to eat in silence, to make a meal an exercise in contemplation. Conversation was accepted only between the courses, but because the intervals between each course were long, the social nature of the feast was still preserved. As the old wine was cleared with the dish and new wine was set on the table, it was the custom to sit back in the chairs, sip the wine slowly, and savor the conversation.

And so they ate their soup in silence, a silence made slightly more contemplative by the performance of a harpist whose music came out of the darkness at the end of the room. Out of this darkness the servants would appear, carrying their dishes and decanters of wine; out of this darkness, the music would come, and disappear appropriately during the intervals of conversation.

As they ate the soup in silence, Viracocha thought how severe the table seemed compared with Brigita's. In Eiru they ate from potter's vessels on wooden boards, but there were always flowers, different settings of flowers for breakfast and dinner. Here they ate off glass plates and bowls, crystal goblets, and anything that could take up the light from the emanating crystals and continue it, endlessly. The table itself was a glossy white marble that was so highly polished that one could see one's own reflection as in a mirror. Only the facets of light and the shadowy reflections seemed to move, for otherwise the table was set in a geometrical perfection that seemed to mock spontaneity.

And yet the play of light within the crystal and glass did have its fascination. Viracocha gazed down into the ocean of his soup and imagined that the sprinkling of three leaves on the top were the Great Islands, and as he pushed them down under the soup, he imagined that he was playing avenging god. But because the music was soft and sweet, he destroyed the islands without malice. It was then that he realized that the dry mead

was indeed made from psychedelic herbs as well as honey. He studied his mind for a moment, but found no distortion, no hallucination; it was simply that whatever thought he had normally seemed more heightened, more emphasized in its purity of sensation. The mead certainly did seem to enhance his sense of taste, for the soup had not one taste, but half a dozen following in succession like ocean waves rolling across his tongue. He remembered how the mead had enhanced his sense of smell before and how he had loved the odor of Brigita's arms and thighs. But with the image of Brigita passing through his mind, there also passed the thought she had to be now in the garuda on her way here to the City. Suddenly his emotions took a wild turn into rage and he looked down at the knife by his right hand and was sorry that it was a dull, round vegetarian implement, for he saw in an instant how quickly he could kill the Archon before any of the servants could stop him.

Viracocha struggled to contain himself, to force his mind to think of some ruse through which he could talk to Brigita to work out a plan for her escape with the others. He held his mind close to the thought of her and could feel just how much he wanted to see her again, wanted to atone for whatever failings within himself that had made him turn away from her in projected terror. He held on to the image of Brigita's love of flowers at the table. Here there was nothing but the cold precision of clear glass and crystal on white marble. Better the flowers, the clay vessels on bare boards of Eiru than this costly sterility.

As his emotions tossed back and forth between love and rage, Viracocha suddenly realized why the mead was chosen and why the High Priest had given it to him. It was some kind of truth-elixir that made you more of what you were and forced your personality out into the open. The dinner was to be a ritual of truth, a rite of initiation.

Viracocha put all his mind to tasting his soup as a way of avoiding his rage at the Archon, or his fear of the High Priest. But once his mind was steadied on sensation, he began to take an inventory of his other senses. The drug was not overpowering; he could feel that his will was still intact. It was only that tendencies and moods were intensified and exaggerated. Pro-

vided that there were no Anahuacan mushrooms in the soup, he should be able to observe his emotions without acting on them.

The High Priest leaned back from the table to indicate that conversation could resume, and the music of the harp in the darkness instantly stopped. It was the Master who signaled his desire to speak first by turning toward Viracocha and wrinkling his bald head in an attitude of serious inquiry:

"The music of the harp has set me to thinking that since Your Grace is also a musician, you could perhaps answer a question that has been troubling me for some time."

"By all means, Your Excellency," Viracocha said as he sat back in his chair and decided to go very lightly on the wine.

"Well, I have heard that you recently returned from a stay in the Remnants of Hyperborea. While you were there did you hear of any legends concerning 'The Singers from the Pleiades'?"

"I have heard some legends, Your Excellency, but it was not during my stay on Eiru. There used to be an old teacher who lived on Eiru, but often traveled around the world. His name was Abaris. He taught that the planets had their resonances within the centers along the spinal column, and that if one plugged up one's eyes and ears to listen to these centers, he could then hear this other celestial music. But I confess, it was not a practice that I ever mastered."

"So much the better for you," the High Priest said, and then turned to the Master. "You'll notice, Your Excellency, that the practice requires blocking up the ears and the eyes with the fingers. The whole teaching is nothing but a rejection of the physical world. Every technique is aimed at disassociating the psychic body from the personality, so that rather than transforming the world, the person sits in primitive squalor, but goes off, out of his body, to imaginary palaces in shimmering dimensions. Well, this Palace here is not imaginary, and we have built our temples out of mammoth stones to declare to the world that the mysteries of the spirit are to be found in matter, not in the shifting, intermediate realms."

Viracocha looked at the High Priest in surprise. He himself had said nothing about using the fingers to block up the eyes

and ears. And why was there such unusual intensity to his words? Could it be that the High Priest, that they all were drinking this dry mead together? Was this truth-telling some kind of social ritual for the hierarchy?

The Master was about to speak, but Viracocha interrupted:

"Your Holiness, I said nothing about stopping up the eyes and ears with the fingers. It could have been with cloth. How is it that you know about the practice in such detail?"

"When you were still a child, thirty years ago, I was a student of Abaris. Yes, even I. Like you, I had been bored with life on the Great Islands. His wandering all over the world fascinated me, and so I followed him until I coaxed him into giving me some of his 'secret teachings.' Then, again like you, I went off by myself and practiced with furious intensity. Nightly, I would fly out of my body and struggle with demons and witches. I would draw secret sounds up and down the inside of my spine, enjoying the energy, the private, secret ecstasy. Each day I would lift up the sun, and at night I moved the stars about. I became the Ruler of all the universes. Then one morning I decided I did not want the sun to rise in the east, but in the west instead. When it did not, I became enraged, and I stood up to strike it. But clouds came up suddenly between me and the sun, and it began to rain. It was then that I realized that I was mad and that Abaris was an old fool, playing with phantoms. Like little boys who play with their penises, but have no understanding of sexuality, Abaris played with his psyche, but had no real understanding of spirituality. The secrets of the spirit are in matter, not in the psyche."

"That is most certainly true, Your Holiness," the Archon said, "and yet I must say that I am amazed at the capacity of the dream-body to change the etheric conductivity of the blood. The blood does seem to be a physical medium that interacts with the etheric and psychic dimensions in the most amazing ways."

"I am not denying the existence of the dream-body and its imagery," the High Priest answered. "I am simply denying the wisdom of constructing a whole religion, or a whole civilization, on so limited a foundation. Eiru is an example of what happens

when you do that. It is a culture of poetry and thatched huts. But let me put it another way."

The High Priest set down his crystal cup and picked up his white linen napkin and held it taut between his two hands:

"One end is the physical world, the other end is the archetypal world of causative patterning. All the middle is the intermediate world, the world of the psyche. It is a shifting gaseous atmosphere of clouds and wind, not the world of the stars beyond, or the solid earth underneath." Then he overlapped one end on the other, and holding it by the corners, let the middle sag down.

"Of course, I agree with Your Holiness," the Archon said, "for this has been the basis of my work. But I am finding new things out every day. To use your own metaphor, the shifting and unstable atmosphere is the means by which the solar radiation is rendered usable by living things on the solid earth. Perhaps this pyschic world takes the causal patterning and renders it into organic form through the power of psychic imagery. I must say that our shaman from the East seems to be having a profound effect on the psyches of the children. He's an amateur, of course, compared to Abaris, and I think it a pity that Abaris refused to work with us, for our progress would have been much faster."

"You asked Abaris to work in the Institute?" Viracocha asked quietly, forcing himself to stay calm, as he felt again his violent rage whenever he looked at the Archon.

"He didn't, but I did," the High Priest said. "I had been informed that Abaris' travels had been increasing, and that his spider web had become world-wide. I was curious about him and we had a discussion concerning him in the Supreme Council. Then one day he simply showed up. I am told that he just sailed into the bay, docked at the quai of the Sacred College, and walked up the processional ramp, wearing that old medallion of his as if it were some kind of protection. He appeared in the throne room unannounced and said directly that unless I dismissed the Archon and closed down the Temple Institute that we would create, what he with his love of ghosts, called a civil war among the elemental spirits of the earth. I found it

amusing to look at this sorcerer standing in the middle of the City of Knowledge and making pronouncements about the future of civilization on the basis of his traffic with the dead."

"Yes, I remember that day very well," the Archon said, turning toward the High Priest and so making only his paralyzed right side visible to Viracocha. "You asked him to help us, that we had need of someone who was so skilled in projecting out of the physical body and manipulating psychic images, but he, unfortunately, refused."

"Yes," the High Priest laughed, "he made one of those dramatic gestures he was so fond of, and the Palace guard took it as a threatening gesture and shot him with a fatal dart. I will give the old man credit though, he did go out with style. He looked at me as he pulled out the dart, and said: 'You wish to take hold of this body? Here, take it, but the body of the earth will take you.' Then he rolled up his eyes and his soul went out through the top of his skull, and he became the ghost he always wanted to be. His way is not for us. No, Your Grace, your music interests me much more than his magic, for your music is the wedding of cosmic, archetypal structures and physical sounds. Abaris' soul is still out there floating in the psychic sea, but your music goes directly up from the ground of the flesh to the archetypal realm."

"I confess, Your Holiness," the Master said, looking down into his cup, "that when you first proposed His Grace here to me as a candidate for the Sacred College, I was shocked. He was so young, and, well, unknown, outside the inner circle, so to speak. But now I can see what you are driving at. Altering the body through sound and music, accelerating the whole rate of evolution, it is certainly more exciting than jabbering with the dead."

"I am glad that Your Excellency sees the pattern now, for you notice," the High Priest said, looking at Viracocha, "that it is a question of seeing the *whole* pattern. It is one thing to run a College of the traditional forms of knowledge, another thing to run an Institute devoted to the new sciences, and yet another to compose a great piece of music. But what a High Priest must do is create a civilizational form in which all of these are possible.

But now to honor the art of cooking, let us move on to the contemplation of the main dish."

The High Priest extended his hands to left and right and immediately four servants removed the wine vessels, and four more servants set down the new dish before them, and brought with it a decanter of a clear white wine.

Viracocha studied the lovely pattern and colors of the spiral terrine before him. Each layer was a different food that faded into the color of the next turn of the spiral. The pastry shell that held it together was almost translucent, but it was impossible to tell what each layer was composed of, nuts or marinated mushrooms, and yet each turn of the spiral was distinctly different in taste, progressing from mild to spicy. At the center was a fruit that he had never tasted before; it was peppery and sweet, and yet seemed as if it had been marinated in dry mead. The basic idea of the dish seemed to be one of opposites held in unresolved tension: hard and soft, sweet and sour, savory and bland, hot and cool. And there in the very core of it all was that exotic fruit, the burning sweetness of the truth the dry mead was sure to bring out.

They ate in silence throughout the main course as once again the harpist played softly, but this time he had altered the tuning to play in the Hyperborean semitones. As Viracocha listened, the five layers of the spiral in the terrine, the five Hyperborean Solids, and the five modes all began to dance in his imagination, and he found himself thinking about the five levels of being, the worlds associated with each of the subtle bodies. The worlds spiraled around one another in Viracocha's imagination, and he could see each world feeding off the world below, just as he fed off this world on his plate.

As he considered all the images in a grand fugue in his imagination: the dish, the subtle bodies, the Hyperborean Solids, the five modes, and the five levels of being, Viracocha realized just how saturated with mead the small fruit of truth really was. But he no longer cared. He wanted to know about these other levels of being. As he looked up from his empty glass plate when the servant took it away, he noticed that the High Priest was eyeing him intently. Very well, he thought to himself, let us find out:

"Who are the Old Ones, Your Holiness?" Viracocha asked as he sat back in his chair and returned the High Priest's penetrating gaze with one of his own.

The High Priest regarded Viracocha with a curiously serious stare. He was silent for a moment, and then he relaxed slightly and began to answer the question as if he had come to some sort of judgment:

"Without giving you a long lecture on the history of the earth, I'm afraid that I need to know just a little more about the context in which you heard about these Old Ones. Was it on Eiru?"

"Not at all," Viracocha answered. "Quite the contrary, it was in the Institute this afternoon. His Excellency told me this afternoon that the new configuration would enable one of the Old Ones to join a psychic, dream-body to the new etheric body cluster. Considering Your Holiness's dismissal of the intermediate realms, this providing of a body for a psychic wraith does not strike me as the joining of the physical and the archetypal worlds."

"So now I must turn to you, Your Excellency," the High Priest said, leaning into the corner of his chair as he turned toward the Archon, "to provide me with a little more of the context of your remarks about the Old Ones."

"Yes, by all means, Your Holiness. Actually, this seems like the perfect moment. I was going to ask Your Holiness about the Old Ones, but our attention was so taken up with these military matters the last month, that we haven't had the time to explore these more theoretical matters of research. Also, I'm not quite sure that we have enough information yet, but what we seem to be finding is that as we energize the blood with sound, we seem to create a heightened receptivity to the intermediate realms. In our evolutionary ascent, we seem to be effecting a recapitulation of history, and the Old Ones are irresistibly drawn to our work."

The High Priest became strangely silent and it seemed to Viracocha as if some fascinating division or fault line had just opened up between them. He hadn't a clue as to what it all

meant, but he saw the opening and he knew that he wanted to explore it further:

"I must confess my ignorance of history or mythology, or whatever it is we are talking about here, but the subject seems to be becoming even more mysterious."

The High Priest continued to stare at the Archon but clearly addressed his remarks to Viracocha:

"Mythology is simply a garbled form of the prehistory of earth. When our ancestors came to this world, it was not uninhabited."

The High Priest relaxed his hold on the Archon and turned to face Viracocha:

"On the physical plane there were two different species of hominids, but the dominant presence was on the etheric plane. When the solar system took its present form, after the explosion of the binary star, these beings were constrained to this world. Not wishing to disrupt the terms of 'the war in heaven,' when our ancestors came they made an agreement with these Old Ones that no attempt would be made to disturb their dominance on the planet. At first, this agreement was kept, but then after a considerable length of time, one of the species of hominids began to reach an evolutionary threshold, and this caught the interest of one faction of our ancestors. This group could not resist the temptation to facilitate the hominids in crossing that threshold, so they violated the agreement. They called their project 'Teaching the Serpent How to Fly,' using as a metaphor for their work the fact that both the snake and the bird had evolved from a common genus . . ."

"And is that why Abaris wore a medallion of the winged serpent?" Viracocha asked, beginning also to wonder about the sources for his vision on the island in the lake at Tiahuanaco.

"Abaris would have liked to think that his Order of the Winged Serpent goes as far back as all that, but that is an act of vanity, religious fantasy, and mythically self-created history. We're talking about millions of years, Your Grace, and there are no human institutions that can span that length of time. But, as I was saying, these Old Ones were furious at this interference with life on earth, for when the hominids crossed over the

evolutionary threshold, the balance of energy was upset, and a great deal of energy began to pass over into the physical world. To use a metaphor, the Old Ones fed off the subtle energies, both etheric and psychic, that emanated from the animals. When the hominids crossed over into consciousness, then that etheric and psychic energy was no longer available to them. Instead, the group-soul of the hominids began to be its own collector to take its subtle energies directly from the sun, so that . . ."

"In other words," Viracocha interjected, "they did what we are trying to do again on another turn of the spiral, and so the Old Ones are drawn back into the scheme."

"So it would seem, but there is more that I need to know from His Excellency before I can determine the nature of their involvement. But to finish the story, so that we can choose our own appropriate response, the Old Ones became the enemies of the new hominids and constantly tried to pull them back. Basically, they used religious forms of inspiration, establishing various rituals of blood-sacrifice that enabled them to feed again off the subtle energies, and as well to pull the hominids back into the ancient collective animal psyche.

"That was when one of our ancestors, in this case, a female, took the forbidden step of having actual sexual relations with one of the most advanced of the hominids. From this union our race came into being. But with this intrusion, hostility broke into outright war, and the Old Ones, exercising their control over the etheric forces of the earth, pulled the entire Lemeurian Continent down into the sea. Scarcely any of the two species survived. Our ancestors, these 'Singers from the Pleiades' that His Excellency the Master asked you about, came to our aid and came between ourselves and the furious Old Ones. The Old Ones, having lost their evolutionary lead, were beginning to degenerate. This, by the way, is what I fear for us, if we don't press on and escape our planetary confinement. The new race was removed to Hyperborea, which was a continent then, and a treaty was established whereby the humans were restricted to the physical plane, and the Old Ones were restricted to the intermediate realms. The direct solar access was

not taken away from the humans, but the Old Ones had the mineral and plant domains for their fields of energy. One faction agreed to cooperate with the Singers from the Pleiades, and they lifted this continent out of the sea to become our future home. But the other faction refused. There was civil war among them. At the end of that war, a new arrangement was established whereby the Pleiadean side of the Old Ones had dominance over the etheric, and the losers withdrew completely into the subtler intermediate realms. With their powers reduced, they became homeless exiles, filled with jealousy and an impotent rage. The Singers from the Pleiades remained for a while, until life on this continent was well established, and then they withdrew. The Hyperboreans insist that there were continual sexual relations between the Pleiadeans and themselves; in fact, they made sexuality into some sort of ritual, much as the Old Ones had turned blood and death into an atavistic ritual. But this is all superstitious nonsense. The way to alter the genetic linkage is not through clumsy forms of sexual reproduction, but through music and sound. The immaculate conception of the future will be with music. Eventually, we will have bodies of pure musical geometry; then we will have purged these hominids out of us entirely. So, you see, Your Grace, now you know why your music is so important, and just how in you the Sacred College and the Temple Institute can become united. Or should we say, 'Your Excellency' rather than 'Your Grace'?"

Both the Archon and the Master were shocked and felt uncomfortable as they realized that the ritual of the Meal of Truth was moving rapidly in a direction they had not expected.

Viracocha said nothing for a moment, drank down his cup of wine in a single move, and then slowly returned it to the marble table.

"Not yet, Your Holiness, not yet. You move too fast even for me, for I still do not know if *His* Excellency our esteemed Archon is working for one faction of the Old Ones, or the other."

"I'm sure I have no idea," the Archon said with an aloofness that tried to float at a level above the atmosphere of political intrigue the High Priest had introduced into the discussion. "I

shall need to ask the shaman tomorrow when he goes into trance and starts channeling these Old Ones. By tomorrow your metallic rendering of the Hyperborean Solids should be in place within the foundation vault, so tomorrow should be time enough to find out what this is all about. Biology, much more than history, is my field. I confess that most of this is rather new to me. In fact, some of this I am hearing for the first time."

"One aspect of the period of initiation," the High Priest said, "is that the old High Priest must take the new one into training. In that period, the successor is given the true and complete history of this planet. There is more than I have presented here, but only the successor himself is permitted to know. Human beings, it would seem, cannot stand too much reality. Now, Your Excellencies, the time has come for us all to share the cup."

Once again the High Priest extended his arms out, and once again the four servants cleared the table of all the dishes and goblets of wine. When the table had been cleared and wiped clean, one servant set a crystal goblet of the golden dry mead before the High Priest. His Holiness waited for the servant to disappear into the darkness, and then, taking the cup, he lifted it with both hands. Through the facets of the crystal, it seemed to Viracocha as if the High Priest had a dozen eyes. Then he lowered the cup and drank a fourth of it, and passed it to the Archon.

The Archon took it with his one good hand, drank the second fourth, and passed it on to Viracocha. Viracocha took the cup, rotated it so that the place the Archon had touched would be away from his own lips, and drank half of the remaining mead before he passed it to the Master. The Master held the cup aloft with both hands trembling, and then drained it. As he set the cup down, he closed his eyes and looked down. He seemed visibly shaken, and the High Priest did not wait for him to return the cup, but reached over to place it perfectly in the center of the table.

And there it sat, an empty vessel of invocation on a table that had become an altar in some ceremony unknown to Viracocha. The Archon closed his eyes, and the Master kept his head down, but he seemed to be clasping his hands tightly together under

the table. Very slowly, the High Priest sat back and touched his fingertips together and looked at Viracocha, who did not try to avoid his stare or follow the example of the others to close his eyes.

Viracocha watched the eyes of the High Priest go out of focus, and he might have grown nervous at the thought of him examining his aura were it not for the fact that he began to see the High Priest's emanation. A deep dark and midnight blue seemed wedded to a powerful red creating an evening purple more vivid than any sky he had ever seen before. Then the High Priest's faces began to change, slowly at first, then with an increasing speed as face after face flickered and then disappeared. Were they all the faces of his previous incarnations, Viracocha wondered, or the images of all his possibilities?

Viracocha would have continued to study the High Priest's face, except that his attention was now drawn to the Master, for he had begun to sob silently, and tears were streaming down his face.

"I can't take it anymore. The stars, they're too frightening. I don't want to hear anymore about these spirits and these demonic wars. The stars, the stars frighten me. I don't want the Tower. I don't want the Tower. I don't want any of it. Leave the stars alone! Leave me alone! I don't want any of it."

Viracocha looked from the sobbing figure of the Master, slumped down in his chair, to the High Priest, whose eyes were still set in their unfocused gaze. He could feel the presence of the mead working inside himself, and yet he felt nothing except disgust at the sight of the Master and anger at the Archon, who had closed his eyes, seeming to be bored with the whole ritual.

And yet, Viracocha thought, the Archon may be right after all; the whole ritual was a farce, for nothing could be changed with this moment of mead-induced truth. They would all go on just as before: the Master would blubber, but return to hide from the stars with his studies of astronomy, the Archon would go on torturing children, and the High Priest would go on babbling about the secret thread that ties College to Institute, and City to Civilization. The whole thing was a stupid farce, for what difference would the Truth make to these people?

Viracocha reached out and seized the cup, turned it upside down, and smashed it down on the table. The High Priest's eyes snapped back into focus as he looked at Viracocha, and then at the cup, which had cracked but not yet broken apart.

"I have had enough of this stupid, meaningless ritual," Viracocha said as everyone stared at him in shock. "In fact, I have had just about enough of everything. Here we are building a Tower that is aimed at cracking open the heavens and stealing primacy from the sun, and the Chief Astronomer is afraid of the stars, and the Archon is sucking the blood out of children to draw down devils that have hated the human race since the foundation of the world. Good God! What on earth do you two demented fools *think* you are doing?"

A wry smile crossed the lips of the Archon, as if to say that he was glad that his suspicions about Viracocha had been confirmed, but the Master stared at Viracocha in disbelief and his tear-reddened eyes appeared as pink and bulbous as his bald head. The High Priest looked at Viracocha for a moment, and no trace of emotion or judgment escaped him. Then slowly he turned to the Archon:

"He has a point, Your Excellency. You do seem to have gotten in over your head. You should have come to me at once as soon as the shaman began to channel the Old Ones."

"Your Holiness!" the Archon exhaled in quiet but still expressed exasperation. "We have not exactly been idle these last months, and in spite of illness and an aborted revolution, we have still managed to keep this project on schedule for the feast of Sauwyn. I was fully intending to bring this matter to you, but I needed more information. And so I decided to complete the test with the shaman and the dozen subjects. Then His Grace's proposal for the use of the Five Hyperborean Solids was sent over, and so I decided to test it all together, and then report to you. If I reported to Your Holiness every day, I think neither one of us would get much work finished."

"The principle of what you say is true," the High Priest said, "but you need to develop greater sensitivity to the whole context in which you work. When you shifted your studies from captives and criminals to pregnant women from Anahuac, you

did not give me sufficient notice, and we had an insurrection on our hands, a native insurrection that quickly promised to develop into a military coup. Then you went on with the studies of children, and now this working directly with the Old Ones."

"I wouldn't call it 'working with' the Old Ones, Your Holiness," the Archon said. "When the shaman in trance spoke about conception, and providing a body for the Old Ones, I was more interested in his words about conception than Old Ones, simply because we had not yet progressed from our studies of the children to conception itself. Since the shaman seemed to be anticipating our work and your instructions about 'immaculate conception,' I became interested. I would have informed Your Holiness sooner or later."

"And I am telling Your Excellency that it should have been sooner rather than later," the High Priest said with finality.

The Archon bowed his head in silent submission, but even before he could raise it, Viracocha blurted out:

"And I am telling both of you that it should have been *never* with the whole project. This rage for transcendence or escape, or whatever it is, is insane. To escape death we seem to be inflicting it all around us."

"Stop!" the Master screamed with his pink eyes bulging in a rage against disorder. "This is against all the rules of our society! You can't do this! I can hardly call you, 'Your Grace.' You're a disgrace! Your Holiness, are you going to permit this outrage in our presence!"

The High Priest kept his eyes on Viracocha in complete fascination with the move he had just taken, but with a casual and contemptuous wave of the hand at the Master, he dismissed him:

"You are excused, Your Excellency. Leave these matters to your superiors and go back to your cloister. You will find nothing outrageous there."

The Master stood up quickly, looked at Viracocha as if to say, "Now you are going to learn your lesson," and then he bowed quickly and turned to leave as he snapped his fingers impatiently at one of the pages to open the door and so honor his exit with ceremonial dignity.

The High Priest never took his eyes off Viracocha as he waited for the Master to leave and the door to close. The only sound was of the fingers of the Archon lightly tapping on the marble table. Once again the High Priest regarded Viracocha's aura in that diffuse, examining stare. After a moment, he took a deep breath, and returned his eyes to normal focus. Almost as if he had come to some final judgment, he looked without anger or malice and spoke sympathetically to Viracocha:

"I am sorry, truly sorry, Viracocha, that you have made your decision against us. I had hoped for more from you, but human beings are curious animals, and sometimes when they come near power or success, they run from it in a fear they like to think of as innocence. Since you choose to remain within the prison of your ignorance, you give me no choice but to place you under house arrest and confine you to the guest rooms within the Institute. If you are right about the danger of the project, then your term of arrest will be short, as catastrophe overtakes us all. But if, as I think, you are wrong, then perhaps you will come to your senses to realize that you are not omniscient.

"And since you are not omniscient, let me tell you a few things you do not understand, things you can think about when you are in your new quarters.

"Let us suppose we stop the project. What then: The kairos of our time is lost, and we fall down into another cycle of 25,920 years. I doubt very much whether our civilization, or even any priesthood with knowledge, will be intact when we come to the opportunity of the next opening. You seem to have a particular sensitivity to death in the illusion that the individual organism is some sort of solid foundation on which to build a civilization. Well then, consider the future. War has a way of stimulating more wars. Consider a time when the High Priest is weak, and the military class succeeds in displacing the Priesthood. Consider a cycle of 25,920 years in which dreary militaristic empire follows dreary militaristic empire. All music will be marching songs, all science simply the cunning craft of weapons. When the next opportunity comes along for humanity, what traditional body of knowledge will be intact to enable us to move

beyond this dull eternal round? No, by then humanity will have sunk so deeply into warfare that there will be no spiritual knowledge surviving anywhere, and this planet will become the place where the human race is buried alive.

"You object to the people who have died in the Institute, though it can't be more than a hundred, perhaps two. Well, add up all the deaths from wars that are going to come in the next cycle. Whole cities burned to the ground. Pregnant women with their bellies ripped open and the fetus put to their breasts by laughing soldiers. And perhaps someday the military mind will learn how to destroy continents, or the whole planet. Then humanity will have to wait in suffering for another solar system to evolve and carry them to the point where we are now. For what great destiny are you saving humanity, you young, ignorant fool?"

"I don't see the difference," Viracocha said in defiance, "between the soldiers disemboweling pregnant women and what is now going on in the Institute. Except for the fact that the women and children of the future will at least be able to die. The soldiers won't be able to go after them to continue the torture in the space between life and death, but we will. And perhaps soon with the help of the Archon's colleagues, the Old Ones, we will know how to continue the torture long after death, so the wretch is prevented from escaping in death or life. Of course, that is all assuming that our Lord the Archon is an infallible genius. But let's assume he's not. Let's assume he and I are both wrong: no opening and no earthquake. Do you think he will release his addiction to human blood? Do you think he'll say, 'No more experiments for 25,920 years'? I doubt it! No, His Excellency has seen a more powerful reflection of himself in the eyes of the children glazed over with terror."

"All life is supported by terror," the Archon said with cold contempt for Viracocha's hot passions. "Your self-deluding sentimentality is disgusting, disgusting because under the pretense of knowledge you refuse to see. Had you the sensitivity of a savage, you would hear the tree scream as it is cut down to make wood for your harp. Personally, I would prefer it if beauty were the foundation of life, if all beings did not eat one another to

live, that stars did not periodically explode, wiping out whole solar systems, or that galaxies in their epochal swing did not sometimes collide with other galaxies. But I was not consulted on the design of the universe, but the mocking archangel who built life on a foundation of terror seems somehow to manage to keep the universe in some kind of order. And so I accept that, even learn from that to use death and terror as merely the tools of Order that they truly are."

"You don't accept death," Viracocha answered with his anger fully released. "You hate it. You take it as a personal insult. The mocking archangel you describe is your own self-image thrown up in the sky. You inflict torture on others because you think that if you, great as you are, have to be subjected to this slow and rotting death that takes bits and pieces of your body and removes them from the control of your will, that lesser beings have no right to object if they become one with your half-dead body." Viracocha turned from the Archon to look at the High Priest:

"Here we sit eating the delicate vegetarian cuisine of adepts. Well, why don't we all be honest with ourselves and serve up the tender fetus? We sit here in elegance, but really we should be sitting around the roasting spit, waiting for the great belly to boil and burst to drop the fetus onto our plates."

"My, what an imagination Your Grace seems to possess!" the Archon exclaimed in a mockingly jovial tone. "But you don't seem to understand civilization. Everything in the Pyramid of Life feeds off the creatures underneath it, as savages feed off animals. When we moved up to become civilized, it did mean that we were entitled to feed off savages, but not literally in the way you suggest. Because we are now beings of knowledge, we feed off them through science, just as you as an artist feed off society to produce your music. And just as the artist sentimentalizes his perception of mental suffering to make his work sad or tragic, so we rationalize pain to put it to the higher use of knowledge."

As if the word "pain" were a signal, Viracocha brushed his thoughts of restraint aside, seized the fork at his side, and lunged toward the Archon, who moved quickly enough to avoid

the stab touching his head or neck, but could not move fast enough out of the chair to avoid the thrust striking his shoulder. The High Priest touched his ruby ring to the receptive-crystal in the arm of his chair and Viracocha could hear the guards running from out of the darkness at the end of the hall. But the sound he paid much more attention to was the scream of the Archon.

"There! Rationalize that!" Viracocha said as he twisted the tines around to enlarge the wound.

The guards grabbed Viracocha by the arms, but the High Priest held up his hand to stop them. Viracocha did not take his eyes off the Archon but watched the blood soak through the silk of his robe.

"There," Viracocha said with satisfaction, "you see the pain is real, and its reality takes place *in* and only *in* the individual. So much for your theories of the abstract human race, or your Pyramid of Life."

The Archon leaned back for support into the corner of his chair, reached up with his single good arm, pulled the fork out, and set it down on the table. Slowly and with dignity, he rose, bowed to the High Priest, and then picking up the napkin to stanch his wound, he turned to face Viracocha:

"I see that when our passionate artist loses the argument, he also loses his mind and resorts to force. Interesting revelation, don't you think? I, for one, will gladly accept this pain and rationalize it, for in revealing yourself tonight you have saved me from the pain of having to work with you in the Institute, and, perhaps, you have even saved our society from the pain of having you in a position of power.

"Your Holiness, may I be excused?"

The High Priest said nothing, but with a wave of the hand, he dismissed the Archon. Then with a deep breath of acceptance he folded his fingertips together and looked over his hands to consider Viracocha:

"You didn't have to blurt out your feelings here," the High Priest said calmly. "You could have observed your feelings without acting on them, for that is the training for which the mead is used. But you are so self-important in your moral outrage that

you think that if you have a feeling, why then the whole world should have it too. And that may be all well and good for an artist, but when you walked out of the ranks of the College, you showed me that you were interested in Power, in what holds a civilization together and not simply the voices of a choir. But you have failed, Viracocha. Think about that during the long days of your confinement. You, the brilliant and fast Viracocha could not get beyond your own emotions to understand the impersonal nature of Power. And as you think about your own emotional life, ask yourself just where your moral indignation is coming from, and why it can so easily flip over into cruelty. There is no cruelty in the Archon. He uses terror the way the sculptor uses a hammer and chisel, not to 'hurt' the rock, but to make Order out of chaos. But his cold and impersonal use of terror, like your hot and personal use of emotion, is unstable. Just as you flip over from dramatic moralism to cruelty, so he could flip into a self-dramatizing saintliness to become something like old Abaris. You see, the reason that I am the High Priest, and that neither you nor the Archon could be, is that I know another level. You are fond of beauty and self-expression, His Excellency is fond of terror and the illusion of impersonal control it gives him, but the universe is neither wholly beautiful nor terrible. His Excellency wishes to extend his range of control from life into death. You simply want to extend the theater of your personal expression from music into governance. Ironic, isn't it? You can't even govern your tongue. Ruled by your feelings, and deceived as you are with your own perception of yourself as a 'good' man, you would make a very dangerous High Priest indeed. The Theocracy was ruled by 'good' men before. They didn't do much good.

"Well, your training is over, Viracocha. You didn't even last a single day. You should have stayed a garden flower in the College, for now you're going to become a potted plant. Perhaps you can turn the sadness of your confinement into great music composed under tragic circumstances. If not, and if I see that you are degenerating, I will honor your gifts and mercifully order your execution."

"You're sending me to prison," Viracocha said in a struggle

against the wave of exhaustion he felt coming over him, "but I think you'll find that you've built only a slightly larger cell for yourself. Even you must see from this evening that neither the Archon nor the Master serves your vision. You're all locked into your cells of private fears and obsessions. You alone rule, and you rule alone, and no one will ever succeed you. You have nothing to pass on, small wonder that your obsession is of solitary escape."

The High Priest stared at Viracocha, but Viracocha did not try to avert his gaze with the customary bow. Instead, he kept his eyes locked onto the High Priest as he took the three steps back and turned to be taken away by the guards. But even as he passed into the corridor, Viracocha still continued to imagine the High Priest seated upon his throne and enclosed within a geometrical lattice of light.

7

ΔΔΔΔΔΔΔΔΔΔΔΔΔΔΔΔΔΔΔΔΔΔΔΔΔΔΔΔΔΔΔΔΔΔΔΔΔΔΔΔΔΔΔΔΔΔΔΔΔΔΔΔΔΔΔΔΔΔΔΔΔ

The touch of failure or misfortune for Viracocha was always a physical sensation, as if someone had pulled out the stop that held his blood in place. He could feel precisely the point in his abdomen where the life-force would pour out of him. Within seconds exhaustion would overwhelm him and his mind would drop into sleep as if he had been drugged. Mercifully, some higher part of his being knew better than to let his consciousness continue its collapse into depression or to let his powerful imagination loose in unending morbidity, and so it simply took his mind away for safekeeping.

So habitually supported was his personality by the joys of confidence and creativity that when they were not there, he did not know how to make do with life. If he was not filled, he was empty. Life had no meaning by itself for him; it had to be given meaning in the act of creation. Failure, therefore, was all the more threatening, and deep and abject depression was the shadow to all his creativity. Fortunately for Viracocha, he had not failed very often, had not felt more than once or twice in his life that heavy crush of annihilation as that black shadow became the form and measure of himself.

But just as he did not know where his sudden intuitions came from, so he did not know where the rescuing oblivion of sleep sprang from. All that he knew was that it was his salvation, for whenever he awakened out of that sleep of defeat, he did so with renewed dedication and formidable strength that had been somehow smuggled into his body under the cover of night.

As the guards had led him out of the Palace, it was the irony of his situation that had first struck him. On the same day that the High Priest had suggested that he might become the successor

to the throne, he was being led away to imprisonment in the Institute. But then as he began to reflect on all the other events of that single day, his sense of irony faded into exhaustion. The High Priest had been right: he did not need to speak out, to try to triumph over them all as if it were some collegial debate. He could have remained silent and cunning, waiting for the right moment to spring to attack. But he was habituated to exhibiting himself and the contents of his own mind, so he had accepted the pull of the mead and blurted out everything.

The guards must have thought that Viracocha had passed out in terror as they approached the copper doors of the Institute; perhaps they even grumbled about the weakling as they carried him through the doors, up the staircase to the left, and down the long corridor to the apartment built into the living rock of the cliffs above the sea. And perhaps it was with some degree of satisfaction as well as relief, that they had thrown the body onto the bed and pulled the curtains around his imprisonment.

When Viracocha awoke at dawn he thought the bed itself was his cell. It was not until his fingers touched the curtains that he realized he could part the walls and step out onto the balcony that overlooked the sea and the neighboring island.

His prison was really a guest apartment for dignitaries visiting the Institute. It had a large living room and a dining table, and a bedroom separated from the sitting room by a series of sliding panels of translucent wafers of thinly sliced and polished rocks. Both the bed and the bath faced out to the balcony that looked on to the sea. But it was the bath that seemed to proclaim the luxury of the space. With its hot-spring-fed waterfall that cascaded down the enclosing rocks to fill a pool large enough for a sybarite with more than one attendant concubine, it showed that it had been built to impress its guests with the fact that the Institute was a good deal more than a laboratory.

After Viracocha had explored his new surroundings, he began to explore his own feelings. He knew that he should be afraid and he could not understand why he was not. He could remember no dream to account for any change of mood, for he had not awakened slowly out of sleep into drifting thoughts. His eyes had snapped open and his mind was instantly and completely

alert. Once he had determined the shape of his imprisonment, he began to set his mind to the feelings of new resolve concerning his future. But resolve to do what? he wondered, and the more he tried to think about this peculiar feeling of resolve, the more confused he became.

Viracocha paced back and forth on the balcony that ran along the entire length of the apartment, but as the day began to lighten, his thoughts began to darken. An oppressive feeling of shame began to creep over him in the brightening dawn as he realized that his impulsive trip to Eiru had endangered Brigita, and with Brigita in custody, Bran might not be so eager to fly off, but might try some foolish effort of rescue that could endanger the escape of the 144. As his mood began to shift he found himself even staring over the wall of the balcony in consideration of the certain death that lay far below on the surf striking the broken rocks. In moments like that he knew he was indulging himself in histrionics of shame and despair and his inner voice seemed to ridicule his need to think or to feel, when clearly neither thought nor moods were of any use at all in that particular moment. What was of use was patience, but the fast Viracocha simply was not a patient man.

And so he paced back and forth in useless exercises of useless thoughts and feelings until at last he began to grow bored with the endless circles and decided that he would meditate to get above the confusion of rational investigation and irrational moods. It was then, as he turned away from the balcony, that he heard the knock on the door and the shuffle of several feet and went into the bedroom to see what was going on.

He noticed first the soldier carrying the stretcher and saw the other soldier before he saw the long hair trailing over the side and almost touching the floor. When the soldier turned to move to the bed, he recognized Brigita at exactly the same moment that he caught sight of the Archon coming through the door:

"My God! What have you done to her?"

"On the contrary," the Archon said, smiling, "we have to ask you what *you* have done to her. Her servant told the pilot of her garuda that she has been like this ever since you left. I don't

know what the two of you were up to on Eiru, but it would seem that she is in some kind of deep ecstatic trance."

As the soldiers lifted her onto the bed from the stretcher, Viracocha began to smell a strong odor of wild roses, and as soon as the soldiers had moved away, he leaned over the bed to determine whether it was really emanating from Brigita.

The Archon walked slowly over to the stairs but did not make an effort to come up into the bedroom. Instead, he lifted his medallion to his eye and began to examine Viracocha as he leaned over Brigita:

"Yes, it's curious, isn't it?" he said in his studiedly artificial manner. "She seems to be giving off a strong emanation of a floral scent, and if you observe her closely you will notice that she only takes a breath once a minute. Hmmmm. Fascinating. Yes, I can see that I was right in my intuition to bring her here."

Viracocha turned momentarily to see the lens-distorted eye of the Archon staring at him, but he quickly returned his glance to Brigita to consider the strangely beatific smile that rested on her full lips.

"No, there seems to be little doubt," the Archon said as he dropped his medallion to his chest. "Your etheric aura is twelve times stronger in intensity than what would be normal for a man your age, and hers practically fills the room. But the most curious thing is to observe what happens at the edges where your two auras meet. Very curious indeed."

"What are you talking about?" Viracocha said in annoyance as he turned away from Brigita to glower down on the Archon, who seemed to be smiling up at him like some doting parent.

"Yes, I think I shall have to do a few tests with the two of you later when I have finished the attunement of the crystal to the foundation vault. You two do seem to make such a lovely and interesting couple. Later, you must tell me all about your stay on Eiru."

"Will you please stop this little game and tell me what you are talking about?" Viracocha said as he noticed his anger and hatred of the Archon returning.

"Sometimes the slowness of the fast Viracocha amazes me," the Archon said in clear enjoyment of Viracocha's confusion. "It

is obvious, isn't it? Our esteemed Abbess of Eiru was initiating you into some Pleiadean ritual, the results of which are that she is in a deep ecstasy in which she has slowed down the rate of time in her body, and you have increased the intensity of your etheric body enormously. As you will recall, both the metabolic rate and the etheric body are keen interests of mine. So, you can understand why I thought it was more appropriate, and perhaps even more productive of discovery, if we kept you here together. Besides, her servant says that the two of you were together for a month, so please make yourselves at home. Who knows. perhaps she will awaken from your magic kiss."

"You're disgusting!" Viracocha exclaimed quietly but intensely. "Just what do you plan to do with us?"

"Why, make you my guests, of course! Rest assured that I will not stick a fork into you. Perhaps a tube or two, but only for a few tests. In the meantime, enjoy your comfortable surroundings. Your Hyperborean Solids have given me a little work to do, but when I have finished with your helpful suggestions I will come back to bring you news of how your invention is coming along. By tonight the children should be all nicely arranged within the vault."

Viracocha moved from the bed toward the Archon in anger, but the guard lifted his catapult and aimed it straight at his chest.

"That would be useless, Your Grace," he said as he turned to walk away. He walked slowly toward the door as the two soldiers moved to come between Viracocha and the Archon.

"Oh yes, I forgot." The Archon turned back and rested both hands on his cane. "Feel free to slide the bolt to ensure your privacy. It is a little bolt, but it will ensure that the guards knock before they enter with your meals. So make this your home and feel free to resume your Eireanne way of life. Good day, Your Grace."

The Archon turned and walked out of the room, and it seemed to Viracocha that his step was lighter and less pained, as if he were indeed happy to have the two of them under his control.

Viracocha waited until the guards had left before he walked

down the stairs to bolt the door. For a moment, he leaned against the door and looked back into the bedroom, as if that little distance could give him the perspective that he needed, but from that angle he could not see her; he could only detect the odor of wild roses that was slowly filling the room. He took a deep breath and felt the odor itself entering his body. It brought knowledge, as if scent were the equal of light in bringing forms and ideas to his body, and it was his body and not his mind that responded as his penis began to lift in an erection.

With the awakening of the body came an awakening of soul, as if from opposite sides the mind was being taken in hand and lifted up to some other level. He could hear again the voice of secret instruction, less primal than the smell that excited him, it seemed to be from a world beyond smell or sight, a world of disincarnate knowing. "She is trapped in paradise," it seemed to be saying. "Only you can bring her back, but you must go back to the point at which you abandoned her." For a moment it seemed as if his mind would rise up to come between body and soul to offer its objections and interpretations, but the primal, worldly wisdom of that erection took hold of his body.

Viracocha moved away from the door and walked toward the bed. Yesterday there had been no moment in which the rush of events had allowed him to find his casual release in the embrace of a concubine. Now there was nothing in his awareness but that feeling of engorgement that the ritual of the month with her had created in him. And yet there was a difference that he could feel inside himself. The fullness was there, but the pressure was gone. There were no thoughts as to how that could be, but as if he were dreaming with his eyes open he saw the semen flowing up the spinal column while a stream of light came down from his brain. When they had touched it was as if a star was created in white light, and it was that feeling that he had misinterpreted as having his heart ripped out of his chest; it was the shock of that dismemberment and that agony of love for the children that had torn him apart and sent him puking to the floor to find his own reflection in his vomit.

But now he could feel that his body had changed. While his mind had been occupied with the politics of the Theocracy, his

body had gone on about its own business, little caring whether he *knew* what was going on or not. Now he could feel with clear intensity that that pool of seminal light had been transformed by rage and pain into another essence of life. The love that he had not been able to find in the paradise of bliss where Brigita remained, he had found in hell. And now deep in the center of this hell, inside the Institute itself, he would have to return to that terror.

Viracocha came to the edge of the bed and looked down on the smile of bliss that still rested on Brigita's lips and remembered in astonishment just how he had read that beatific expression as an evil, ecstatic leer. The odor of wild roses was now very strong, but he noticed an emanation of pastel light that seemed to shift from pink to a soft amethyst whenever she would take the shallowest of breaths. He climbed up onto the bed and sat at her feet for a moment and could feel her emanation enveloping him, as if he had already entered her body.

The simple act of reaching up to pull the curtains closed around the bed not only dampened the light, but also increased his sensitivity to the slight pulsation of light that hovered a few feet above her body. Enclosed within the bed and within her aura, he felt as if he were no longer within the Institute but had entered a vessel every bit as powerful as a garuda. Secure in that space, he clasped his hands together and bowed to her, not merely to honor her but to ask permission; and then he lifted his robe over his shoulders and set it to the side.

As if it were an act of slow, walking meditation, he took hold of her ankles and very gently pulled her toward him, lifting her silk robe above her waist. Unhurriedly, reverently, like a priest folding an altar cloth, he removed her robe entirely, and then sat down cross-legged at her feet. He waited a moment until his breath had quieted, and as he gazed down at her vulva he understood why the savages of Anahuac chose to bless an object by inscribing a triangle with a line in the middle of it. The savages knew what the priests of the Theocracy did not, in their domestication of women into servants and concubines, that this softest of forms was both origin and altar, and not until he had made a vulva of his own heart and had felt it break open to give

birth to a love he had always felt to be the embarrassing, illegitimate bastard of his secret life, did he dare approach this other altar of the immediate, intimate God. With her permission, he touched her legs to allow her thighs to encircle him. With her permission, he lifted her back to allow her arms to encircle his neck; and with her permission, he lifted her hips and entered her entirely. With arms, thighs, and vulva all about him, he closed his eyes in meditation and allowed the slowness of her breath and the slightest pulse of her heart to draw him into the time and rhythm of her own being. In less than an hour, their pulse was one.

Whether she awakened or he joined her in trance, he could not say, but slowly he felt her body open and take him up. Once again he began to feel the strange vibration and the strong, vortex-like sensation that felt as if it were trying to pull his entire being into hers. He remembered the time before, remembered how his mind had interpreted this sensation first in fear, then in panic, and finally in complete terror, but now he knew that it had been his own self that had been the source of that terror, the fearful author of that inadequate interpretation. This time he would not interpret, not raise his defenses against her. He would surrender and allow his thoughts to become merely clouds passing through the sky. He was the sky and not the clouds.

As he relaxed she widened enormously, unbelievably, taking him all in: legs, hips, half of him had already disappeared into her. He saw a vision of binary stars in space, alternately consuming one another in complete intercourse, but he felt no fear in this vision of being utterly consumed. He felt her body move up along his stomach and back, and then peacefully, gently, he was completely taken in. The moistness of her vagina was soft on his face and the taste of her moved past his lips into his mouth and tranquilized him, as if he were sipping some richly sweet opiate. The physical pleasure of a close and intimately embracing peace allowed him to stretch out in relaxation to feel the caress of her body everywhere at once all over his skin. He did not linger there but slowly continued as he was eased up further into the ultimate embrace of her womb. Space widened to

receive him and he began to float. The smell, the taste, the touch all over his body were so intensely enjoyable that he floated on a perfect concentration of effortless pleasure.

The sea grew enormous as he grew small, and then the sea became all space. Suspended in that infinite space, he felt no need to do anything and he knew that he could rest there forever in that absolute bliss. But no sooner had he felt the wish to rest in dark eternity that he felt the pull of motion drawing him through space as if he were moving in approach to some distant planet. And then he saw it. It must have been the earth of long ago, for it was red and molten and slowly turning. He saw it struck by a shower of meteors and he knew that these meteors and comets brought strange metals from outer space to startle this world with the possibilities of others. The red molten earth cracked open, and light met light, but not alone, not without sound. There was a music to it all that he had never heard before. There was a theme and then a complex variation on the theme as the earth underwent so many changes before his sight. He had entered down into the sky of this world and now overhead he could hear the vault of heaven respond with variations of its own. Like a primitive drumbeat moving toward some climax, the monotony of the double beat began to push him toward a shore he could not make out. He wanted more space. He wanted to float in the sea, but now there was nothing but the muffled dullness of distant sounds. Sound began to thicken but as it did, taste returned and he felt again the drugged tranquility it induced in him to allow him to relax to enjoy this moist kissing all over his body. As his head emerged into the light and he felt her lips glide down the back of his neck, he remembered the first time one of the temple concubines had taken his penis into her mouth to glide her moist lips up and down the entire length of it: this was like that but better because these erotic lips could glide down the entire length of his body at once. It felt so unimaginably good as his head emerged and her lips slid down his neck, and then down his back and legs. He hung there suspended in the light that announced that he had eyes and that she had a face, a face that drew him toward her as new lips began to kiss again. The taste

of her lower lips was still lingering on his mouth as she began to kiss him as if the taste of her own life was an offering that he had brought up to her in gratitude. How lovely was this gentle tongue in his mouth, these warm living arms around his neck, and the soft words of surprise breathed into his ear:

"It's morning! We must have meditated all night!"

Viracocha opened his eyes and saw Brigita's green eyes smiling at him as she leaned back to look at him while she clasped her hands behind his neck.

"Where are we? Are we in the body, or have you magically transported me to the heaven of lovers?"

Brigita looked down in amazement as she began to notice the feeling of Viracocha completely inside of her. She shifted her hips from side to side, enjoying the new sensation of his presence within her.

"I guess we really are in the body. I've never experienced this feeling in the ethereal worlds."

She took her hands from around his neck and leaned back, pushing her hips tight against his. Her long reddish-brown hair fell like a waterfall and collected in a deep pool on the bed, and as she stretched her head back, her little breasts lifted high, and Viracocha admired again the full, soft, pink areolas whose color seemed to match the odor of wild roses that still hovered around her. As he looked at her he wondered how he could ever have seen her as plain, or how he could ever have seen her as ugly to run from her in terror. How could he not have seen the beauty that was so completely revealed to him then?

Brigita straightened her back and returned her hands to clasp them behind his neck as she looked into his eyes with a depth of feeling that needed no other expression:

"Thank you," Brigita whispered. "I'm glad that I've never experienced any of this before now. And I am glad we've both survived. I would never have come out of it without you diving down and lifting me up. I feel rather ashamed. I collapsed into myself completely and could have floated in that trap of paradise forever."

Brigita looked into Viracocha's eyes and began to comb his

hair away from his forehead with her fingers. They were silent for a moment, then Viracocha took in a breath and said:

"Well, since this is a time for naked confessions, you should know that paradise wasn't my trap. You may have collapsed in on yourself in bliss, but I exploded and ran from you in terror, completely possessed and completely out of my mind in fear. I had better tell you now where . . ."

"Not now," Brigita said as she placed her fingers across his lips. "Not now. I can feel the changes in both of our bodies. Now we need to put aside all of the restraints of the last month. Now it's time to let go."

Brigita fell back and pulled Viracocha on top of her. She smiled and kept brushing his hair back as he responded with surprise to her and struggled to unlock his legs and stretch them out against the length of hers. She quickly wrapped her legs around him and became a different woman: her thighs lifted high, her hand reaching under her legs to fondle him, her tongue fluttering in his ear. The more excited she became, the more lifted up he was. He could see and feel every center along his spine open and tremble with hers so that he could no longer tell who was male or female to the other.

Brigita's head rocked back and forth on the bed, and then her eyes opened wide in uncomprehending wonder as her body became the being she had never known in her eagerness to escape it. Like a woman completely taken over in the shock of sibylline possession, she stared at miracle and then reached up to take his head with her hands and cover his ear with her mouth. She began murmuring oracles, wild images, and then thrust these teachings of the god into his mind with her tongue. If Brigita had told him before, he might have questioned revelation, but she breathed her visions into him precisely at the moment of his own release, and so he followed her as word, image, and moving bodies became a dance.

They rolled over, as soon the continent would. And then she towered above him and as he looked up from underneath her high floating breasts, he understood that she was the only tower given to man for the return to the stars, that the "Singers from the Pleiades" had placed in both of them the true crystals that

could be touched by light. With wild astonished eyes she looked down at his face between her thighs and blessed him, and then like a mother in graceful birth she reached down to lift him up to her. Face to face, kneeling in complete communion, they drank from the light-transfigured chalice of their bodies and became, not mystic vision or beatific scent, but the incarnate taste of each in the other: tasted, the way an anemone tastes the sea and opens.

Gently, with a slight downward pressure of her fingertips on his shoulders, she led him back into the seated posture of meditation, and then encircling him once more with her thighs, she sat with her arms around his neck. Lightly with her fingertips she closed his eyes, and then with one finger she touched the crown of his head. It felt as if she were opening a door, for immediately he could feel a shaft of light descending into it. She took his fingertip and placed it onto the crown of her own head for a moment only, and then she moved his hand to her waist. He could feel her heart beat against his chest and soon the beat of their hearts and the lifting tide of their breaths became one.

He could not tell whether time slowed down, or whether it took a long time for their heartbeats to disappear and their breath to stop, but he could feel that this time all was effortless and easy. As if they had removed the impediments that before had made the vibrations seem shattering, they now could simply accept and not resist the disappearance of their flesh. Viracocha saw with a colorful intensity beyond physical vision the bright centers along Brigita's spine, saw the blue light that shot out from the base of her spine into his. Back and forth, as if it were a shuttle weaving their two beings together, the light shot from center to center, from the base of the spine to the crown of the head. When the blue light reached the crown, he felt himself being pulled up out of his body completely into the light. As he emerged out of his body and hovered in the ethereal light, Brigita reached over, took his hand, and the feeling of ascent continued.

They seemed to pass into a world not built of the stark contrasts of light and shadow, but simply different qualities of light. It seemed as if the sun were shining in the sea rather than on it,

and if Viracocha looked closely he could see that this ocean was composed of numberless crystalline beings, just as a field of snow is composed of numberless crystalline snowflakes. The light was never still or monotonous but was constantly changing in color and sound. How primitive and limited his lists of earthly colors now seemed to him, for he would need the names of over a hundred different kinds of white and gold. As he listened to the sounds the colors made he realized by some inner form of direct knowledge that this ocean of light was a membrane all around the earth that took in the energy of the sun, stepped it down to a safer level for animals and plants, and spread it throughout the world in all the innumerable variations of climate and weather.

The feeling of ascent continued and they passed above the ocean of light into an ambience of radiant indigo, and as he moved out of the earthly orbit he could feel a burden being lifted from him and he began to be able to think purely, no longer in words, but in the crystals he had seen when composing. As these forms moved, they rotated through more than three dimensions, and the friction of their passing through a higher dimension created a sound that was also a meaning, a glyph. The unpacking of all the information in a single crystal would take ages, for the language they generated from the crystal's unfoldment was what humans experienced as time. These crystals did not sit in space, anymore than they were located in time, but space seemed to be an emanation that came out of them. As the crystals sounded and rotated in and out of their several dimensions, Viracocha could see that in some baffling way the inside of one crystal became the outside of the other.

Still they moved higher, or deeper, for there was no longer an up or a down. The only frame of reference was the vestigial sense of holding Brigita's hand, but Viracocha knew that that was merely a habitual thought-form. Although he could "see" in a complete circle, and was no longer limited to what was ahead, still there was a horizon to his perception. And from the edge of that horizon he could see a being appear moving toward them.

He knew that something was before him, but he could not

make it out because it had absolutely no relationship to anything he had ever known. He could feel the being reach into his mind to find some frame of reference or image with which to establish a relationship, and he could sense its frustration at all the limitations his consciousness imposed. It seemed to settle for a collection of images, as if they were hints or metaphors, but not its own substantial nature. First, it became one of the multidimensional crystals of blue ice, then it shifted that image slightly so that every one of the thousand facets of the crystal had a pair of eyes staring at him in the most amazing, a-human way. He looked back at this creature of the thousand eyes in puzzlement, and then he understood the hint. His own being was polar; he had eyes at one end of his spine, organs of generation at the other, and he was more located in his head than in the other parts of his body. But this being was omnipresent in itself; it was not a polarized creature at all. Then all the crystalline facets with the thousand eyes were set into motion and it gave him the feeling of wings, of the speed of hummingbird wings. Accepting the limitations of images and perceptions, the creature moved and passed into Viracocha, and then drew Brigita inside as well. As soon as they were all "inside" one another, Viracocha lost all sense of his subtle body with its imprinted image taken from the physical body. What had been Viracocha became a geometrical lattice of light and enormous energy. What had been Brigita became the music vibrating and emanating from and around the lattice.

They became a single being. Before the foundation of the world, they were. They had only parted in an act of offering and sacrifice that made entry into the limited world of dense matter possible. Before the adventure of the world they had been with this creature of the thousand eyes in another universe entirely, and now that they had come out of the world, it had come to welcome them in the joy of return, of homecoming. In an act of recognition and communion, Brigita and Viracocha offered up to this being from another universe their own new consciousness of matter and flesh that they had brought with them out of the world. But there arose this felt difference and the communion was touched by their differences. The world of matter and

flesh was inconceivable to it, and with a tone of peace and acceptance, the being moved out of them and the atmosphere began to lighten from vibrant indigo to a pearl-white iridescent with many colors.

Into this color-scape that seemed to hover at the edge of becoming a transfigured earthly landscape, the sun behind a beautiful cloud began to speak to them.

I am Melchizedek.
I formed this earth as a vessel
To receive the fall of light
In which humanity descended.
From Godhead humanity came,
To Godhood humanity will come
When time is finished.

But you have come out of time
As a twin-being only,
While all the others
Who went down with you
Remain below in suffering
And dark forgetfulness.
The great being that is Humanity
Has not gathered itself back
Into its future form.
Why do you abandon time
Before it is time?

I offered this earth as a cup,
But can a cup be filled
With a pulp of grapes
That has not turned to wine?
I offered this earth as bread
For nourishment of Spirit,
But a mass of dough
Is neither grains nor bread.
Why do you abandon time
Before its time?

You are now what you were
When you chose to descend,
But have you forgotten
Why you chose to fall?
Spirit had become a vacant sphere
Whose inner surface was
A vacant reflecting mirror.
The love you now hold
In your bright duality
You did not hold before
In your vacant solitude.
How much more will there be then,
When each is in All
And humanity reflects
Mind to Spirit, earth to sun,
And bright ecstatic carnal love
To the innocence of virgin stars?
Why then do you stand
Here two as one,
But not All as One?

But greater than the Plan
Is freedom, novelty, and creative joy.
Now if you should choose
The return to the Pleiades,
Then the Plan for earth
Must change of Necessity.
The vessel will be cracked
Where you came out of it.
The unremembering souls
Will spill and fall
Out of form to nothingness.
Not until another sun
Can shape another world
Will they be able to return.
Out of ignorance and fear
They will create hell-worlds
from their broken, aborted minds.

One who is beyond our sun,
Even beyond our galaxy,
Is preparing to descend,
First in the form of
Origin and Emergence,
Mother and Child,
Then into the life of death.
Only at the end of time
When Humanity was One
And Earth another
Would they become lovers,
As you have become,
To consume the world
In the last copulation
Of matter and of light.
What you have invoked
Is a future far off for humanity.
But all this can change,
For eternity is in love with time.
You are gods
And freedom is your divinity.
You are free
And now must choose.

The archetypal pulse and tone of the words was in no earthly language, and both Brigita and Viracocha realized that what they were experiencing was being translated from a higher life into a familiar religious form, but just as the timbre and tone of this prehistoric language seemed to suggest a liturgy of sacred utterance, so did it seem to recall what they were before the world began. The power of this sacred language seemed to be capable of harmonizing dissonances and in its presence the creature of a thousand eyes was able to become completely one with Brigita and Viracocha. But once they were a Unity, they felt the formation of a higher pattern that Melchizedek himself could enter, and then as an unimaginably powerful quaternity they could feel the intimate emergence of a Being so cosmic that it could take form only in precisely that Unity of hitherto

separate streams of universal existence. There was no thought whatsoever except the return that would make the emergence of this inconceivable level of consciousness possible in the nature of humanity. As soon as the emotion of willing their return cried out in affirmation, the cloud passed away from the sun, and Melchizedek stood before them as the Archetype of Humanity in its sublime and future form, a being in whom evil had been taken in and transformed in the erotic transfiguration of matter into incarnate light. Now in a new voice and in the natural language of Eiru, the being spoke with a voice of love and compassion that was at once everywhere in space and in the deepest recesses of themselves:

> "Before there was Mind, Being was entranced with its own reflection. Love did not exist for there was nothing other to itself. When Being broke apart into beings, love became possible. The risks that you are now about to take will not be yours alone. In the opening you have made in space and time, the one you know as the creature of a thousand eyes will accompany you back to the edges of matter. And Abaris shall return to earth from the sun, and here I promise that the Fellowship of the Sun will never be absent from the earth. From here to the end of the world, it will remain on earth to guide it through its crises and seizures of hate and forgetfulness. Go now, you are a priest forever after the Order of Melchizedek."

The cloud returned to cover the sun, and then slowly, beautifully, the sun set, and they were again in the deep indigo of planetary night. A sensation of falling came over them, and as they fell they could feel their unity as a single being come apart. Now alone in a denser body, Viracocha could feel the wrenching sense of loss.

The descent continued and the spherical sense of global consciousness, in which the inside and the outside of a hypersphere were a single surface, collapsed into a merely circular field of consciousness. He felt as if he were at the edge of time and as he passed through the barrier, like someone falling into the sea,

there was a splash that was more like an explosion. In that explosion, with its fragments or sparks flung out, he could see all his incarnations at once. It was not a line of life following life; rather, the fragments were the effects of the impact of his spirit with time. He knew that he was none of these incarnations truly and entirely, that they were not so much beings but relationships between his being and time. Only with the ending of time could he gather up these sparks, these fragments to make a being such as he recognized himself to be. Although each of the sparks was flung outward by the impact, yet each still had the possibility of tracing itself back to its origin, its original spirit.

But then as he fell more deeply into time, the fragments took on a more substantial reality; they seemed more solid, more like trees in a forest than sparks from a disintegrating meteor. The quality of each life seemed to be directly related to the quality of the culture in which it was embedded. In some cases, the lives were strong and flourishing; in others, weak and stunted. In some cultures, the Spirit was known and life could grow and reach up; in others, everything was forgotten or denied and the stunted beings gasped in a poisoned air.

As he looked down at the circle of his "lives," four seemed to stand out, like the points of orientation on a compass. There was the life of Viracocha, then another as a monk ten thousand years later in the Remnants of Hyperborea. There again Brigita appeared as a mate, but in that culture sexuality was forbidden as sinful and they chose not to meet in their personal lives, but simply spiritually join their different monasteries while their bodies were asleep. Then, in a later time, they were together again in Anahuac; there they struggled to remember in chastity the rituals they had experienced as Brigita and Viracocha. But they could not achieve, either in chastity or in the shame of their falling into carnal love, anything remotely close to their previous transfiguration, for the culture was hopelessly broken, and eventually the priesthood was overrun by warriors. Viracocha could see himself dying amid flames in the splendor of his priestly robes as the morning star rose over the sea. The last life was at the end of the world when they all were reunited in the consummation of time itself.

He fell, entering the circle which contracted to a point in which he became Viracocha only. There he could see Brigita once more, and he reached out in human space to take her hand. Together they continued their descent until they hovered over their physical bodies. With a renewed sense of loss, he released her hand, was drawn thickly down, and returned to a sense of the almost unbearably dense and heavy animal body.

Viracocha opened his eyes and stared into the crying eyes of Brigita. He reached over and kissed her tears, tasting the salt, and accepting it. Brigita took a deep breath, and then slowly unwound her thighs from around Viracocha's hips. She moved a little away from him and then sat back on her feet and looked down on her thighs. As Viracocha followed her eyes, he thought that her thighs looked like the iridescent coloring of an abalone shell, for the glazing of semen and blood had flown together to create a new quality of light.

"I haven't menstruated in three years," Brigita said quietly. "I really have returned to the body. Strange, isn't it, this unconscious sense of timing? That too was written on the pages I saw in my vision of the mind of the future. Along with the instructions for this ritual of love, there was a poem:

"Let the moon and sun
become
A single light.
Let the red and the white
Be taken in delight.
When One became two,
Heaven broke apart.
When two become One
Humanity will start
Returning to the sun."

"I wonder," Viracocha said, "if you saw the writings in a vision of the future, or, if in a life to come, you will look back on our lovemaking to write that poem?"

"Well, whatever this has been," Brigita said sadly, "it is over."

"In more ways than one," Viracocha added as he pulled back the curtains to reveal the balcony looking out to the sea. "We are back on earth in a prison all right. We were both arrested and brought back to the Institute."

Brigita looked up at Viracocha with a seriously determined look. For a moment she probed his mind to orient herself amid the danger, and then with a look of formidable resolution that revealed to Viracocha just how strong this passionate lover could be when called upon to play the role of warrior, she stepped up off the bed and extended her hand back to him:

"Come, let's bathe together before we go into battle; the fools don't know what they have made possible by bringing the two of us back together."

8

△△

Perfectly at sunset, when the light behind the mountain let it take growth from its shadows, the volcano on the other island exploded. They heard the crack and then the deep percussive boom and ran from the table out to the balcony to see a sunset that was no sweet farewell to day, but a war between light and dark. Intense orange, raging scarlet, and demonic purple were more vividly present in the sky than in any sunset Viracocha had ever seen before. Light and dark fought openly as they had not since they argued over the inhabitants of the world, when the clouds were ripped apart to let in the stars and when a new celestial quiet was forced down upon the heaving earth. Now the old debate raged again, and the cloud rising up to darken the sky seemed the largest, most powerful being there could be. The claim of man to rule the earth seemed about to be hurled aside.

"It looks as if we are going to have some help," Brigita said as she gazed across the sea to the island.

"Do you think it can take care of this island to do any good soon enough?" Viracocha asked.

Brigita gave Viracocha a smile of exasperation: "I don't expect you to think like an Eireanne, but it might help you to see into the volcano if you didn't call her 'it.' "

"It's hard to break old habits," Viracocha answered. "But why a 'her' and not a 'him'?"

"We like to think of volcanoes as breasts of the Great Mother."

"I don't think that I would like to nurse on that," Viracocha said as he nodded toward the smoking cone.

"Infants don't nurse on hot blood; they nurse on the milk the blood helps to produce. It's the same for the volcano. The lava

produces soil, the plants nurse on the soil, the air nurses on the plants."

There was another cracking rumble and a bright red stream of lava found the sea and sent up a great cloud of steam.

"She obviously agrees with you. What else is she telling you?"

"That your work on the Tower has helped them speed up the process of change. This explosion is a result of today's tests."

"If that explosion is coming from the Archon's tests in the foundation vault, then I have a lot to learn about how etheric waves are propagated through the earth. That explosion should have been on this island."

"It will be tomorrow, or whenever the sun strikes the solar crystal when the opening is made. This volcano is simply the beginning. All the three Great Islands will disappear. The elemental Spirit whose outer physical body is the volcano is telling me that there is war among themselves, among those you called the Old Ones. One group is using the Archon to break into the human body. He thinks he is advancing human evolution, but actually they are trying to take it over, the way a parasite takes over a host. But what this volcano is saying is that they are not going to let things go that far. They take their energy purely from the sun, and if the Tower were to succeed in displacing the primacy of the sun's position in the larger configuration, the whole solar system would be destroyed."

Viracocha imagined the new geometry; he could see the new galactic pattern disturb the inner core of the sun, could see it explode in a supernova that wiped out the entire solar system, could see the multitude of souls wandering in dreams of agony, bodiless and unevolving for the eons until another world could spin into form.

"I begin to understand what Melchizedek was talking about, if I can use the word 'talking' for that kind of communication."

"The side of the elemental Spirits that is aligned with the Hyperboreans will not allow things to go that far. They will work to destroy the Tower as soon as the sun touches the crystal. They will not allow enough time for the opening to establish itself in relationship with the distant sun."

"So, we need to get the children out of the vault tonight, and

Bran needs to leave before sunrise. It's going to be a busy night."

"Bran probably won't be able to make his move until the middle of the night," Brigita said. "And I don't think we should wait that long. The children must be in unimaginable agony. I think we should project down there now."

"Projecting down into the Tower vault in your psychic body won't help the children, Brigita. That screech of terror is operating in the etheric range. I would have to get down there physically to dismantle the thing."

Brigita was silent for a moment. She closed her eyes and tried to feel her way into the vault. "But I can eliminate the thought-form the shaman is creating with the Old One."

"You've lost me there."

"The Old Ones feed off fear. The more frightened a person is, the more power they can suck away from him. The shaman is terrifying the children to create a thought-form that the Old Ones can use as a temporary vehicle, until they can enter the physical body itself."

"It sounds like a nightmare," Viracocha said.

"That is what nightmares are. But if I project down there, I can eliminate the shaman's thought-form and that will force the Old One to withdraw out of the etheric."

"But what if the Old One goes after you, Brigita?"

"They can't, unless of course, one is afraid. If one has no fear, one can walk right through them and they disperse like a mirage in hot air. I'll deal with the Old Ones, but I'm going to have to leave the screech of terror to you. Since I don't understand how it's being produced, I don't know how to annul it."

"It would be easy enough to annul it on the physical plane, but to annul it on the etheric, one would have to hear it even before you could get near it, and I certainly don't look forward to that. But what makes you think I can just pop out of my body and follow you down there? You know I'm no good at that sort of thing."

"You didn't seem to have any difficulty this morning, and that voyage was far more elevated than anything I have ever experienced before. We went far above the etheric and psychic

realms. Perhaps if we come together again and align the spinal centers, we could become a single being with the angel and shatter the crystals that are generating the vibration."

"That might work," Viracocha said as he tried to imagine how to produce a counter sound to cancel out the frequency of the oscillating crystal. "And if we fail, there isn't a nicer way to die. But what if we fail, what then?"

"As we all die in the morning when the volcano explodes, I'll take the children to Abaris' school, but I guess you'll have to try to find a life in Tiahuanaco. It would certainly be better if that doesn't happen."

"Well, then, I guess that's the plan," Viracocha said with a smile as he reached out to take Brigita's hand. But before he could turn to lead her back to the bed, there was a loud knock on the door.

"It's the Archon," Brigita said quietly as Viracocha looked toward the door. "We'll have to work alone to start with."

The knock sounded again, but this time more loudly. Viracocha moved toward the door, but he did not hurry. There was a third knock by the time he came to the door and opened it.

It was the Archon, the Archon and four guards. Two of the guards entered before him and held their catapults aimed at Viracocha.

"Good evening, Your Grace. The servants who brought you your dinner passed on the good news that Her Reverence, our esteemed Abbess, was up and about. You do seem to have many talents. Ah, and there she is."

Brigita came in from the balcony and stood at the top of the stairs between the bedroom and the living room. She stood there silently looking at the Archon in that unfocused way that told Viracocha that she was trying to find out as much about her enemy as she could. The Archon returned the compliment by lifting his medallion to stare at Brigita through its crystal.

"Fascinating!" the Archon said. "The two of you do make such an interesting couple. Good evening, Your Reverence. It is an honor to have you here as our guest. It is not often that we have a guest with such amazing faculties, and so I hope you won't

mind if we request your presence in the examining room. It's only a matter of some routine tests to measure the intensity of the etheric radiance."

Viracocha flinched and was ready to jump, but the guard immediately placed a restraining hand on him. As soon as one guard moved, the other grabbed him while the other two pointed their catapults at him.

"Don't be alarmed, Your Grace. I will return for you in an hour or so. You will, no doubt, be glad to know that your model works beautifully. We induced the frequencies into the metallic rods and tested the first two of the frequency-signatures. Tomorrow at dawn we shall be able to test the third. You might be interested to know that our shaman has been in touch with the Old Ones who interest you so much, and he tells us that this volcanic eruption was to relieve the strain and the pressure on the crust, as well as to provide a release passage for tomorrow's activities. So, you may put aside your anxieties about an earthquake. Right now, all is prepared. The children are sleeping comfortably in the foundation vault with all your lovely Hyperborean geometry around them. I find it hard to sleep with all this excitement, so I thought I would take an hour or two to have a closer look at the two of you."

The Archon signaled to the guards to lead Brigita out, but Brigita did not wait for them to set hands on her and walked quickly to the door and out into the corridor.

"Do stay up, won't you, Your Grace?" the Archon said as he turned from watching Brigita leave to look at Viracocha. "Since the two of you have lived together, it might be interesting to examine you together, for your auras interlock in the most extraordinary way. It should be only an hour or two before I send for you. Until then, Your Grace."

The Archon smiled in that absurd grimace of his that was more a twisting of half of his mouth than an expression of ease, and then turned his back and walked out the door. Two of the guards followed him, but two kept their catapults aimed at Viracocha until the Archon had passed out into the corridor. As the door closed after them, Viracocha felt his hatred of the

Archon begin to take over and he struggled to remain collected in his search for a new strategy.

The images of the Archon putting his stricken hands on Brigita to examine her sprang up in his mind, and he found his rage boiling up again. He struggled to gain distance, to rediscover that connection inside himself with that other presence, as he paced in circles in the bedroom; but the more he looked at the bed, the more disturbed he became. He went out to the balcony, but there the red streams flowing down the volcano in the black night called to his anger to come forth. He paced back and forth on the balcony and tried chanting the *Ome ulubi ra, se tata mak*, but the melodic call of that work now seemed to be a faint memory of a past life. The suffusing presence in him now was not a lifting up into the spaces of detachment and concentration, but an unknown and secret power that he could feel singing in his muscles. It wanted to have the time and place for its own recital. Viracocha closed his eyes to try to see into this undiscovered part of his own being.

The first thing that he felt was that there had been something false about his image of himself as a composer. It was not what he was, but what he wanted to be precisely because it covered up something he did not want to see in himself. He had never been self-indulgent; he had been self-escaping into his little dramatization of the artist. The surprising thing was just how good the music had been, or just how plastic any being was and how it could be molded into artist and lover, lunatic and warrior.

But it was the warrior he felt inside himself now. The solitary walker lost in artistic contemplation had allowed him to ignore this other being of muscular power and instantaneous violence. And yet the artist had taught this other being something in the process of disguising it. The force he could feel inside now was not inchoate rage, but an empowering presence that was as forceful and direct as his music had been. More than anything, it loathed the perverse cruelties of the Archon. The Archon hated the body and was waging war against it, but Viracocha could feel that he loved the feeling of this presence inside his

marrow and muscles and that he was about to fight with the body and not against it.

He was ready. Viracocha knew just how he would do it: he would fall to the ground, and as the two guards struggled to lift him, he would move quickly to strike their heads against one another, take their catapults, and rush downstairs to the laboratory. With a clear resolve that the time for action had come, Viracocha moved from the balcony back into the bedroom, but before he was past the three stairs and into the sitting room there was a loud perfunctory knock on the door and four guards entered, two with their catapults already in hand.

The vision of his strategy flashed through his mind, and Viracocha could see that although it would work with two easily, and maybe with the right kind of kick, with three, it would not work with four. Why had the Archon sent so many guards for simply one man?

But he knew the answer to that question. The Archon's medallion had told him that if Viracocha's etheric aura was that intense, his phsyical strength would also be doubled as that swirling vortex in his abdomen sent its empowering light into all his muscles. The Archon knew Viracocha's secret even before he himself had known it.

The guard gestured with contempt for Viracocha to move out the door, and as Viracocha passed by him into the corridor he tried to think of a way to isolate two of the guards while he grabbed one of the catapults. But nothing seemed to work out in his imagination. Two guards walked beside him, and two guards walked behind. He would have to wait for a moment of opening, a moment when they did not expect him to strike.

Viracocha walked down the corridor in a hurry; he was glad that the waiting was over. As he came down the stairs and passed the sentry desk, he noticed for future escape that there were only two guards posted, with probably another two outside at the front of the copper doors. He would need to take a catapult away from one of the guards. They passed the sentry desk and moved into the main corridor; he could see that there were no guards posted in front of the Archon's study, or in front

of the double doors to the laboratory itself, so that meant that he had only four guards to deal with immediately.

The guard to his right opened the door in front of him and as they went into the first laboratory, Viracocha saw that the pregnant women had been removed. Where were they now, he wondered, and an image of a long slide flashed into his mind, a long slide that carried round body after body down into the green sea.

They entered the second laboratory, and once again Viracocha observed that the withered bodies of the young women had been removed. But the Archon was there in the very center, with his cape thrown back over the edges of one of the tables. As Viracocha came closer he wondered why he had exposed the stricken arm that was frozen to his side. Always before he had kept that arm hidden within his robes.

"Ah, there you are," the Archon said in a state of obvious pleasure. "Sorry to keep you waiting. Again, I find that I must congratulate you. Very much like your own, Her Reverence has an etheric radiance twelve times what would be normal for a woman her age. It seems that we have been working with people at the wrong stage of life. Pregnant women and pubescent children, why they're nothing compared to the two of you. We will, of course, need to know more about these meditational practices that the two of you have been using. They must be rather original, I would say, considering that I found traces of semen in her vagina."

The Archon moved from the end of the table and his cape fell away to reveal Brigita lying naked on the slab. The tube inserted in her vagina was directly in front of him and he could see the menstrual blood being collected in a vessel. There was no thought or premeditation in his mind, but simply as if lightning had flashed, Viracocha moved with a speed that yet seemed slow and precise to him. With his right foot he kicked out to the side of the guard's knee at a point where he knew he could drive it out of its socket and immobilize him with pain, and at the same time he reached out to his left and seizing the guard by the neck, he snapped the cervical vertebrae and then pulled him off his feet to strike the third guard, whose catapult

discharged into the swinging body that Viracocha released to send him hurtling against the other as he jumped onto the fourth. He could tell that he had the force to stop him from reaching for his catapult when he felt the sharp electric wasp sting in his back. He had both of the guard's hands pinned and a knee in position to give him the leverage to snap both of the wrists when he realized that he couldn't move.

The guard spun his own wrists out of Viracocha's loosening grasp, and seizing him by the upper arms, he spun him round to see the Archon smiling at him as he held up a slender white cylinder:

"Actually, these are medical darts. We haven't used this particular drug on the battlefield, but we soon will. You will notice, Your Grace, that it takes effect instantly, much more quickly than the old poisonous darts. They, of course, paralyzed for life, but these are more subtle and precise. In this dosage, the effects of paralysis are temporary. Please put His Grace on this table here.

"You will observe, Your Grace, that with this drug, you can see everything, hear everything, but you can't move any of those powerful muscles of yours. By the way, thank you for that demonstration. I had rather thought that the intensity of the etheric radiance would mean that you would also be much stronger physically than normal."

Viracocha tried to turn his head to keep his eyes on the Archon, but all he could do was stare helplessly ahead or up at the ceiling as the guard lifted him and stretched him out on the slab. He could hear the moaning of one guard and the shuffling of feet, but there was no way to move his head to the side to observe them.

After a moment the Archon intruded his head into his field of vision, and Viracocha felt that the face was even more disgusting when seen upside down. The Archon smiled down at him and patted him approvingly on the shoulder:

"Most heroic, Your Grace. You've actually managed to kill one of the guards and break the other's leg. You see, it's not all that difficult to kill. It is, however, much more difficult to keep the balance at the point between life and death."

The Archon stood back so that Viracocha could only see the middle of his robes.

"Take him away. And you go into the infirmary. Post two guards at the door and then leave us."

Viracocha listened to the shuffle of feet and the noise of the departing guards. To his right he saw the blue cloth and the head of the Archon's assistant peering down at him.

"Remove his robe. I want to do the same blood tests," he commanded.

"This evening, Your Excellency?" the assistant asked in surprise.

"Yes, tonight. I feel intuitively that we are close to a great discovery. I don't think either one of us will sleep much tonight. But you, Your Grace," the Archon said, patting him on the shoulder, "you shall have a good rest. But before you do, since you are a man of knowledge, let me explain our process of examination here. First, we will introduce this tube into your right arm. This will strengthen the dosage of the dart and begin to take you down into a deep coma. Then we introduce this tube into your left arm so that we may remove some of your blood to subject it to glyphic analysis. By the way, should you find yourself hovering above your body here, I would recommend that you don't go into the Tower. The screech of terror is activated, and though I do find that an amusing touch for a musician, your sensitive soul might be disturbed. Now, have a pleasant rest. When you awaken tomorrow morning, we will share the results of our tests with you."

Viracocha felt as if he were falling backward, but falling slowly back through darkness into warm sunlit water, water that was pleasant on his skin. The water was rich, bright aquamarine, and lovely in the thick liquid softness of its silence. There was no need to breathe. There was no need to move; he had only to float there suspended in peace. Somewhere from above he began to hear a voice. It was distant and unclear, mere vowels mumbled in oil. Still, his body seemed to lift and float toward the sound. Now that he could hear it better, he could make out that the voice was calling his name. It had a familiar tone to it, this voice, and it kept telling him to wake up. But

why? He wasn't asleep, simply comfortable. Then his face emerged above the surface, and he could hear the voice more distinctly. It was Brigita. But why did she just stand there in her white robe, why didn't she take off her clothes to drift naked in this bright, warm water?

"No, come up, Viracocha. Don't sink back. We've lost a few hours already. Viracocha! Wake up! Come up out of your body."

Brigita held out her hand and touched Viracocha's forehead, and instantly he came awake. He moved up out of his body and looked around in confusion through the grayish, indistinct light.

"Are you clear?" Brigita asked. "We shouldn't move if you're still disoriented."

"Yes, I'm not dreaming anymore, I'm just trying to figure out this light, and why you're wearing a white robe when your body is naked on the table."

"We're on the etheric plane. You won't be able to soar. You'll move in leaps and float back down like a sinking feather. We can pass through the walls though, so now we will be able to get into the Tower."

"But why the clothes?" Viracocha asked.

"All perceptions are habits of mind, interpretations. Dispensing with the interpretations is transcendence, but come on, we can discuss philosophy later. We need to move, I think it must be after midnight."

Brigita set off in the direction of the doors at the opposite end of the laboratory. The first thing that Viracocha noticed was the intensely physical sensation of flying. He could float, with the energy perfectly balanced like a garuda, if he stayed about three feet above the floor. If he jumped, or tried to fly higher, he would always settle down to that level. He stretched out, floating on that stream of energy and found that he could move forward simply by willing himself in that direction. Brigita led the way and he passed through closed doors and walls like a swimmer moving through colored water. Within a few moments they were out of the series of laboratories and into the underground tunnel that connected the laboratory with the foundation vault of the Tower.

They passed through the enormous megalithic stones of the

foundation and moved from the gray indistinct light into total darkness. Viracocha lost sight of Brigita as a force pulled him into the center of a viscous blackness that increased its compressive hold on him. He lost all sense of orientation, could see nothing, hear nothing, but could feel this mass crushing him with a black suffocating heat. He couldn't even move his fingers. The compression began to increase, and the pain became excruciating, as if each finger were being smashed in a vise, and then when the pain was unbearable, he heard it, and the pain became inconceivable.

Sound returned to his void in the form of a shrill, high-pitched screech that kept pulsing back and forth between two notes. One note produced a sensation of extreme irritability, as if all the nerves were being confused in their impulses and were reporting every painful sensation they knew: burning, freezing, tearing, breaking. He was being flayed alive and disemboweled at the same time. Then the note oscillated to its demonic twin and the extreme irritability concentrated itself into an unbearable itching under the skin. If he could have moved his hands he would have gladly torn his skin to pieces in a frenzy. The rage against his own body was so intense as he hung there suspended like a fly in black amber that after he had heard the screech only twice, some power took over and he found that he had flung himself out of his etheric body and was now flying in some kind of dirty black tornado.

He was not alone. Others had been sucked up by this psychic tornado and had been pulled from their sleep by the power of this nightmare that hung over their world. Some of the others tried to grab onto his ankles or hair as if he could save them from the force that was pulling them down to the center of the vortex. His speed increased as he drew nearer to the center, and then he saw it. And he knew at once that it was an Old One.

At the end of the vortex was a cauldron of blood in which the Old One stood as if it were bathing, but it wasn't bathing as much as it was kneading itself into shape. The creature was a conglomerate mass of the twelve children's bodies. It would seethe like a pool of lava in which bubbles would come to the surface and explode, but the bubbles were the children's faces.

From one moment to another, one of the children's faces would emerge and scream in agony the single syllable of "God!" and then the creature would hear and thrust its head back into the molten mass of its slowly congealing body.

The closer he came to the Old One, the more horrible the smell became. Every foul odor he had ever encountered in his life seemed gathered into the thing. Then the Old One saw him and reached up to grasp his ankle to pull him in. As he stopped stuffing the children's heads into his body, all their faces came out at once and screamed "God!" over and over again in a dissonant chorus of despair. The touch of the Old One felt as if a leech had gelatinously stuck itself to his skin, but the odor that came up from his hands was the smell of black, oozingly wet and rotten potatoes.

Then a rational thought shot through the emotional stream of his panic like a salmon leaping up a waterfall: "Why rotten potatoes in hell?" And then he remembered the experience in Tiahuanaco and knew that the thing was into his mind, into his own memory, and was playing back at him every fear and revolting thing he could think of. And with a great liberating laugh at the gross banality of this aggregate monster's mind, he pulled his arms up and dived head-first into the misshapen, screaming mass.

It popped like a bubble. Then there was only an enormous cauldron of black obsidian and a bent little old man standing in the center. He looked totally ridiculous in his animal skin and horns on his head. His spindly legs were wrinkled and from hopping about in his little dance, his testicles had slipped out of the dirty skin that served for a loin cloth and they now sagged and bounced against his withered thighs in rhythm with the drum and stick he beat with his hands. All around him, tightened up in fetal knots, were the twelve children. One of them looked out at nothing with insane eyes.

The shaman stopped beating his drum and looked at Viracocha with a puzzled expression. He mumbled a few incoherent syllables, beat his drum in Viracocha's face, and then backed up a pace when the magic did not have its intended effect. Fear began to form in the old man's eyes, and as he lifted

his glance to look beyond Viracocha's shoulder, the eyes widened in terror, then panic. With a scream, he threw down his drum and stick and flew out of the cauldron in the shape of a startled owl. Viracocha turned around to see Brigita floating above him.

"Anyone who rules by fear is always afraid of something," Brigita said as she came down to stand next to Viracocha. "I simply played his own magic back at him."

"And the Old One?" Viracocha asked.

"As soon as you broke the connection between it and the shaman, it lost its point of contact with our level. The souls in the whirlwind have been flung back to their sleeping bodies."

Brigita knelt down to look at the children. Each one of them still seemed locked into its individual nightmare.

"The terror has been sustained too long," Brigita said. "If we brought them back to their bodies now, they would be insane for the rest of their lives. The best that I can do for them now is to carry them safely over into death. They can rest with Abaris until they are ready to move back into life."

"How are you going to do that?" Viracocha asked.

"If you can get rid of that screech, then the dream body can join the etheric. Together they will have enough force of will to sever the connection with the physical body to experience death peacefully. But as long as that screech is sounding, they'll be paralyzed in terror."

"The sound is vibrating in the First Hyperborean Solid," Viracocha said, feeling once again the paralytic grip of his own fear. "To annul the oscillation, I would have to be able to function on the physical or the etheric level, both of which are impossible. So where does that leave us?"

"We'll have to go back into the sound," Brigita said as she reached out to grasp Viracocha's hands.

"No! That's impossible!" Viracocha shouted. "I can't do anything inside it. Why can't we just wait? We'll all be dead in a few hours anyway."

"If the children die in the state they are in now, they will attract the Old One back to them and he will be able to take them down into his level. Then I would have to go down there

alone to try to get them out, and I just might not be able to do that. Look, I lost you last time because I didn't know what I was getting into. Even in my worst nightmares as a young girl, I've never experienced anything like it. But now that I know it, I think that I can stay with you. We have to do it. There's no other way."

Viracocha became silent for a moment, searching for the courage to will himself back, but he couldn't find it; all he could find was the feeling of panic at the thought of being buried alive in that blackness. There was no courage in him at all, but suddenly there was understanding. The calculations flashed into his mind in a complete and detailed analysis of the oscillatory pattern.

"That's it!" Viracocha exclaimed in excitement. "If we can produce the right overtones from *hoomi* singing, the screech will be annulled and will leak right out of the metal rods as the wave pattern is flattened out."

Brigita saw the vision in Viracocha's mind and felt a wave of joy and gratitude sweep through her:

"Abaris comes from the Eastern Continent where that singing originates. He taught me the woman's part when I was also learning how to dance. Now I know why."

"It will be stronger and the effect will take place sooner, if we can do it together. Here, follow me. Try to sound the occult reciprocal of the overtone. Don't overpower it. It's a question of geometry and not force."

Viracocha began the low twang that at first sounded absurd, like a ridiculous emphasis on sound in nasal passages and skull cavities, but then, like a sacrament of the mystery of incarnation, all the physical absurdities of the ridiculous body began to gather into form until high above, this clear, crystalline angel came down to hover in beauty above it. When Brigita heard the overtone, she closed her eyes and began to intone the occult reciprocal, and no sooner had the two sounds married, that the creature of the thousand eyes manifested in the resonant space between them. Brigita's eyes opened wide in the same expression of astonishment that took her over in her first experience of orgasm. The ecstatic resonance of the angelic being inside them

was so erotic that for as long as they could produce the sound they remained a single being of three in one.

"I don't think I'm afraid anymore," Viracocha said as he steadied his breath. "But to be certain, hold my wrists tightly, and I will lock mine on yours. I don't want to loose you again. All right. Remember, sound the music in the interval, in the trough between the two waves of that screech. Let's go then."

With an act of resolution, they willed themselves back, but as soon as Viracocha heard the first note, he lost all resolve. He did not sing. He screamed. Brigita collapsed into herself in mute silence as the black compression became a suffocating coffin. Then the second note sounded and Viracocha became hysterical and if he could have moved a finger he would have ripped his face off his skull to get the underside of his skin away from his body. He could not observe his pain with spiritual detachment. He could not spiritually transcend that maddening itch of perversity, but he could gather his being together in anger, and in the concentration of his being in anger, he found a new strength. His whole being rose up in one united and enormous burst of anger at the Archon. The musician in him was in a state of rage that anyone could take sound and put it to so perverse a use. The rage lifted him up and the anger ennobled him as his spirit rose in revolt to the torture that was being inflicted on it. He sang out his note as the second note ended, and the sound released Brigita from the crush of annihilation and she joined him.

The overtone was even more beautifully present than before, gaining power as it approached closer to the world of matter in the turbulent shore of the etheric. As the overtone hovered in the ether above them, they felt again the creature of the thousand eyes take on a body from the sound, and as it took its body in the space between them, the hot liquid blackness froze into a dark crystal of obsidian and then shattered as light reentered hell.

Space returned with its familiar dimensions and they saw themselves standing within the complex alignment of the five interlocking Solids. Within the space of enclosure were the twelve tables with the bodies of the children, connected by

tubes to the large basin of blood in the center. Slumped over, dead in the basin, was the body of the shaman.

Brigita and Viracocha moved to the center by the basin, and to ensure their reconstitution of that space, they held one another's hands, and sang out the overtones once again. With the emergence of the creature again, the Solids became a singularity of exquisite sensitivity and Viracocha knew that with the first touch of sunlight at dawn on the crystal far above, the crust of the earth would begin to resonate and part to allow volcanic fire to mate with sunlight.

Like the touch of a mother's hand on a child tossing in nightmare, the music completely transformed the atmosphere of the vault. For a few lingering moments, the three of them continued to sing, not in triumph, but for the love of the sound. Its beauty was so strong, so completely antithetical to everything in that world of torture, that the children heard it and awakened from their nightmares. They rushed to Brigita and shouted out in Anahuacan the name of their mother goddess and reached out to touch her skirts.

The mother of none became the mother of all, and as Viracocha stepped back to look at Brigita with each of the twelve children kneeling and touching the hem of her skirt, it seemed as if Brigita had become the sun and each of the children the sign through which the coming age would pass. The aeon of Atlantis was indeed over, and a new cycle of 25,920 years was about to begin at dawn with the sound that would soon come from the solar crystal high overhead.

"Thank you," Brigita said as she looked at Viracocha, wanting to say more, but not knowing how to say it. "There is something more I want to say, but after being melted down, body and soul into one another, I don't know why I'm still reaching after words. You'd better go back now. I'll take the children to Abaris and then come back to be with you. I can't see how Bran can rescue us now. It must be getting close to dawn. Anyway, I'll come back to be with you at the moment of death."

Viracocha turned away and began to flow on that stream of energy that was a membrane between the physical and etheric planes. He passed through the tunnel that connected the Insti-

tute to the vault, but found no one at work in any of the inner laboratories. When he came back into the main laboratory, he found the Archon and his assistant seated in front of a table of a dozen Mercury Tablets. He moved behind them to try to understand what they were doing, when behind him he heard a shout in a strangely distorted voice. As he turned round to question the source of the strange call, he saw the High Priest enter with two guards on either side, one of whom he recognized as the man who held him when the Archon had shot him with the dart.

"Your Excellency!" Again the command rang out, but this time Viracocha realized the distortion was coming from the interference between the physical and etheric. Evidently, the Archon was in the process of slowly bringing him up out of his coma and his consciousness was being pulled through a buzzing thickness back into the waking world. But Viracocha did not want to return to his body; he wanted to hold onto that spectral freedom long enough to find out why the High Priest had suddenly appeared. Had he discovered Bran's conspiracy to escape?

The Archon and his assistant rose immediately and bowed as the High Priest approached. He stopped momentarily at the tables where Brigita and Viracocha lay stretched out amid a tangle of wires and tubes, and then he looked up at the Archon with anger in his eyes:

"Who gave you permission for this? I placed them under house arrest for good reason. I did not give you permission to make them subjects of midnight research."

"One moment, Your Holiness. It is much better than it would appear. Tonight has been a time of unimaginable discovery."

"I gave you permission to examine their etheric auras," the High Priest said, and reached up to fling the Archon's medallion into his face. "I said nothing about subjecting them to this kind of mutilation until I had determined what Pleiadean rituals of transformation they had uncovered."

"But, Your Holiness, that is precisely what I have determined. What we have learned tonight equals almost everything we

have learned in the whole project. I have not injured either one of them."

"One guard is dead, two are wounded. That is injury. You have acted without permission once again, and that is injury. You have taken a Member of the Sacred College, as well as the hereditary custodian of the Treaty with Hyperborea, and subjected them to research as if they were common criminals. You have done all this flagrantly in the presence of the Palace Guards, and then act as if this were not the most serious affront to my authority."

"With what I have learned tonight, Your Holiness, I may have given you authority over death."

"*You* have no authority to give *me!*" the High Priest said with a deadliness that unsettled the Archon. "I begin to see now. In a way, Viracocha was right. I had thought that you men of knowledge would be different, but you are just the military in different uniforms."

"Please, Your Holiness, when you hear me out, you will see that what I have discovered is of the most use to you. After this morning's tests, I was going to come with the full report, not just of the new attunement of the solar crystal, but of the discoveries with His Grace and Her Reverence. Hear me out now, and if afterward you wish to order my execution, so be it. But at least let me show you what is in the Mercury Tablets. They are right here for you to see. Nothing that we have discovered here in the Institute equals what we have discovered tonight."

At that proclamation, Viracocha's interest shifted from the High Priest to the series of Mercury Tablets and he moved over to the table to study the glyphs shimmering in the liquid crystals.

The High Priest signaled to the guards to remain by the doors and then he turned and walked to the table where the assistant quickly pulled out one of the chairs for him to sit on. The Archon picked up the Emerald Stylus and pointed to the first Tablet:

"We analyze the blood by inducing a sound into it when it is inside the quartz vessel. This then produces a three-tone melodic sequence, which is in turn transposed into its mathemati-

cal isomorph. Then, through the normal procedures of glyphic resolution, we generate the associated hieroglyph, and as these all come in pairs, we take the melodic sequence associated with the twin, and reintroduce it into vibration of the blood within the vessel. This produces a four-tone melodic sequence, which can be resolved into its associated glyph. So far, this is what we have tried on the women and children, but always with uninteresting results. But tonight I noticed that Her Reverence was menstruating, and when I did an analysis of the menstrual blood, it produced a new and completely different melodic sequence, and this glyphic resolution."

The Archon pointed to the first of the Mercury Tablets with his Emerald Stylus. As he activated the Tablet, a four-note melodic sequence sounded and a spiraling mandala appeared in the liquid metal. The Archon watched the High Priest closely and then smiled as he noticed his attention quicken.

"Precisely. His Holiness immediately recognizes the chant still used in the Remnants of Hyperborea. Now, the curious fact is that there were traces of semen in the blood, which, of course, leads me to suspect that their practices were not at all what we know of as meditation, but some form of intensification of incarnation through sexual practice, or, perhaps, concentrations in trance upon the blood and semen. It would seem that in our work on pregnant women and pubescent children, we were working on the wrong stage of life. They evidently chose to concentrate on the essence of the sexual act itself. You see, I think that I now know that what is being expressed in this glyph is the effect of the female blood and the male semen, not in conception, of course, but in self-conception, as it were. Later we shall need to study the blood and semen separately, but this initial analysis led me to think that it might be worthwhile to do a tonal analysis of their bloodstreams together. And it certainly was."

"You mean to tell me," the High Priest said with returning anger, "that you removed their entire bloodstreams without even asking for permission for this kind of extreme activity?"

"Your Holiness, there is a momentum and an excitement to scientific discovery. If I stopped in the middle of it, I would lose

touch with my own intuition, and it is that intuition tonight that has produced these, if I may say so, inspired results. Besides, there is no danger to them. As long as they are in deep coma and I keep the blood flowing through the vessel and back into them, they cannot be harmed. I was in the process of finishing the return of their blood when Your Holiness arrived. They will revive shortly and should have no effects more than a mere headache."

"*Should* have no effect," the High Priest corrected, "but the point is, you don't really know. And the more that you argue how unusual they are, the more you indicate that you cannot possibly know what you are doing."

"I believe Your Holiness said that we must take risks or risk becoming a decadent civilization. Well, men of knowledge must take risks as well if they are to make new discoveries."

"The point, Your Excellency, is that *I* am to decide which risks to take by looking at the whole situation, something you are incapable of doing."

"Your Holiness, you must choose your leaders and then trust them. If you are going to lead on the battlefield, do research in the laboratory, and write poetry and music in the College, then you will have to be a civilization by yourself."

"I observe that Your Excellency is now trying to tell me what I 'must' do and what I can be. You really are like the military. In your ignorance you try to make the eyes into the brain. The military wants a society of warriors. You want a society of scientists. You're all the same, the same blindness, the same ignorance."

"In my ignorance, Your Holiness, I discovered that the glyphic resolution for the mixture of their bloodstreams together was the Pleiades."

The Archon struck the second Mercury Tablet with his stylus and a four-note melody sounded and a new geometrical figure shimmered in the crystal. The High Priest looked at the figure with intense interest as he heard the melodic sequence.

"This glyph is extremely ancient," the Archon said. "I had to go through the Temple Archives to identify it, but I see that Your Holiness recognizes it. We see this glyph as the one for the

Morning Star, but the most ancient records show that before the last reversal of the poles, this was the glyph of the Pleiades. What I think this means, this signature of the stars that they carry in their blood . . ."

Viracocha moved from the edge of the group to come closer to the Tablet as soon as he saw the glyph. He wished he could activate the other three Tablets and turned impatiently to the Archon, as if to command him, when he noticed Brigita's subtle body descend and hover over her physical form before she turned to look where he was standing. She moved within an instant to his side, and Viracocha pointed to the Tablet. Brigita nodded, removed a ring from her finger, and showed him the same glyph ingraved on the inside. Viracocha did not need to ask who gave her the ring, but realized at once that Abaris must have given it to her at some level in her initiation.

". . . is that when the Hyperborean Custodians defeated the Old Ones, they must have introduced the lattice of the subtle bodies and spinal centers into the upwardly evolving animal body of the mixed race. Perhaps they saw this as some sort of rescue mission to make up for the bungling of the earlier forbidden experiments. Into the geometry of these spinal centers, they constructed various mathematical and musical patterns. In effect, they signed their work with the musical traditions of the Pleiades."

It flashed into Viracocha's mind all at once. The Tower, like the black magic of the witches' brew of blood and semen, was a crude literalism, and that was why the solar crystal would always rebound in dense matter to create a volcano. They had not been locked in by some demonic geometrizing demiurge; they were misunderstanding the message of the geometry. The centers along the spine were the solar crystals that resounded behind space to tell the Pleiadeans when humans had reached the next level. As soon as he and Brigita had ascended through the inner Tower of their own spines, it was not a volcano that appeared, but Melchizedek. It was the dense, literal, cruel materialism of human misunderstanding that was dooming their civilization and not the traps of some demented god.

"What Viracocha and Brigita seem to have been able to do,"

the Archon continued, "is to carry this mathematical-musical structure from the psychic, through the etheric, and directly into the physical bloodstream. They have turned the bloodstream into a liquid crystal of enormous, and I truly mean, enormous power and renewed etheric radiation. The most obvious and immediate ramification of this is that their blood could have tremendous healing powers. And, as His Holiness can appreciate, for someone like myself, dying of a blood-clotting disease that is daily taking bits and pieces of my body from the control of my will, such healing properties are of truly *vital* interest."

The High Priest did not say anything for a moment, but simply stared blankly at the glyph. Finally, he took a deep breath and looked up at the Archon:

"Once again, I see that you are not looking at the whole situation, though I suppose with your sickness you could not be expected to do so. Can't you see that if what you say is true, then the Priesthood is ruined? It means that women are not simply the old animal bodies of our entrapment, but that they carry a transformative power of the spirit. Can't you see, this would make a sacrament out of the grossest form of sexual intercourse, make a joke out of the ceremony of transference of power from the old High Priest to the new one, and tear apart all our institutions, from the Tower down to the temple concubines themselves."

"With all due respect, Your Holiness," the Archon said with the slightest of smiles touching the left corner of his mouth, "that is only true if the emphasis of our institutions is on escape *from* incarnation, rather than manipulation of it. Perhaps, it is time to set aside these archaic institutions of the Priesthood to have a new priesthood of science. If we placed Brigita in the Tower and controlled her conception, one of the Old Ones would be able to descend into her and we could begin to create a new generation, not for the stars, but here for earth."

"You still don't understand, do you?" the High Priest said in a tone of weariness. "The reason I came here myself in the middle of the night is because I could trust no one else either to ask or to know if what I feared was true. You fool! Brigita is the

hereditary custodian of the Treaty with Hyperborea! In manipulating her like this, you may just have sounded an alarm that will bring the Custodians back to reestablish Hyperborea. That was what Abaris was all about. Do you begin to understand now? The Empire is finished. And it is all because you did not come to me immediately when you contacted the Old Ones. They are not the future of the human race, you idiot, they are a race of psychic parasites who would like nothing better than to attach themselves to the human race. You see, you don't know everything in this science of yours, and that is why the Priesthood can never be replaced with the likes of you. Now revive these two immediately. Tomorrow, after the morning's test, you are to report for a meeting with the Supreme Council. Now that the military is well in hand, it is time to take care of all of you. Guards!"

The Palace Guards came running from their station to the side of the High Priest.

"Now, revive them. I wish to know exactly how much damage you have done tonight."

The Archon gestured in silence to his assistant to remove the wires and tubes.

"Their bloodstreams were returning to normal while we were speaking. My assistant will revive them now with the antidote. I hope that when Your Holiness sees the beneficial results of tomorrow's test, that you will have renewed confidence in my ability to provide you with all that you need for the opening of the new cosmic cycle. I can assure you that I am not interested in challenging your authority, or in any kind of political power. Power over life and death, yes. Power to take the ovum of Brigita and project myself at the moment of death into her womb, yes. And power to create a new etheric body so that one does not have to go through the inanity of childhood all over again, yes. Power to flush out my blood to replace it with Viracocha's, yes; but power to rule, no, not at all."

Brigita shivered in disgust as the Archon spoke of projecting himself into her womb, but Viracocha reached out to take her hand as a plan of escape began to form in his mind. He looked at Brigita to question her willingness to try it, and she nodded her

assent just as they began to feel the strong pull back to their bodies.

Viracocha stretched out above his body and began to sense the realignment of the two energy fields pull him down into sleep. He did not want to go to sleep. He did not want to move through that swamp of confused dreams that would, in its garbled jumble of images, produce a confused memory and interpretation of what he had experienced in the clarity of this other state of mind. But, for all his resistance, he sank into unconsciousness.

He saw himself standing in front of a mirror. It was confusing, because he could see clearly that he now had Brigita's breasts. He still had a penis, but he could sense that behind the scrotum there was this secret, hidden vagina. He began to reason with himself, telling himself that he remembered the single body of crystalline light that he shared with Brigita when they were with Melchizedek, but he should not have come down from there with her body confused with his. He opened his eyes to question the ceiling about his descent, and then he woke up, then he remembered his plan.

"My breasts! My breasts come from the sky!" Viracocha shouted in a childish voice.

The High Priest, the Archon, and the two guards came closer to the table.

"My breasts! My breasts come from the sky!" Viracocha shouted once again.

"Is this what you call normal?" the High Priest asked in disgust.

"He is merely dreaming, Your Holiness," the Archon answered. "He is simply creating dream images out of his perception of the presence of her blood in his. His mind will clear in a moment."

Viracocha turned his head to look at the High Priest:

"Please, Daddy, don't let that bad man hurt me again. Please take me home with you. I've come back from the stars to be with you. Please take me away from here."

"Marvelous piece of work!" the High Priest said. "You've found yet another way to turn a genius into a babbling idiot. But

notice, that even in his babbling, he's talking about the stars. They have made contact."

Brigita began to shout in a little girl's voice:

"I have to pee. I have to pee bad!"

Viracocha picked up on Brigita's suggestion and the High Priest's fear and began to chant:

"Peepee, peepee,
Please, please.
Peepee on the Pleiades."

"I have had enough of this. Guards, take them back with me to the Palace and keep them under guard in the guest rooms."

"But Your Holiness, I *need* them here! Even a small amount of their blood might be able to cure my disease."

"The only thing you need right now is to learn how to obey. You will return to your quarters, and you will remain there under guard, until I summon you to a meeting of the Supreme Council. Now, pick these two up, and let us all go to the Captain of the guard together."

The guards moved quickly to slip their arms under the neck and thighs of Brigita and Viracocha, but as the two of them were lifted from the table, Brigita began to urinate on the hands and arms of the guard, who dropped her feet to the floor and stared at the High Priest in confusion as the stream of urine spilled to the floor. And all the time, Viracocha clapped his hands and kept up his little chant:

"Peepee, peepee,
Please, please,
Peepee from the Pleiades."

As the second guard struggled to hold the giggling Viracocha in his arms, Viracocha shouted "Now!" in his mind to Brigita and both sprang into action at the same instant.

With a single maneuver they lifted the dart-catapults out of their sleeves and shot each of the guards in the thigh. The assistant saw what was happening first and ran to the table to

seize one of the white cylinders that contained the medical darts, but Brigita shot him in the back before he reached them. Viracocha kept his catapult aimed at the High Priest and the Archon, but ran over to the table to pick up one of the cylinders.

"Now, Your Holiness," Brigita said as she moved closer to him, "I feel uncomfortable walking about in the nude, so if you will please slowly release the chain at your neck and drop your cloak to the floor and step back, I will put on the robes, but not the office. No, I wouldn't try that. If you swing the robe into my face, I'll shoot you now."

"So, our esteemed Abbess reads minds," the High Priest said as he released the thin chain at his neck to let his purple cloak fall to the floor. "I suppose that is all part of the training with Abaris."

Viracocha approached the High Priest and the Archon as Brigita fastened the cloak about her neck.

"Surely, Viracocha, you don't think you can get very far with this escape. How long do you think it will be before the garudas of the Empire will follow you?"

"This is much more than an escape, Your Holiness," Viracocha said. "Can't you see? Your Empire is a fantasy in your own mind. The Military has its agenda, the scientists have theirs, the Old Ones have another, and the Society of 144 waiting now with the largest garuda in the Empire to take us away have more than an agenda in mind. They have a whole new era. This is over, Your Holiness. The Empire has broken into pieces even before these islands break apart and disappear into the sea."

"Not quite, Your Grace," the Archon said with an air of superiority. "If the two of you leave now, before you receive the antidote which only I know, now that you have killed my assistant . . ."

"What?" Brigita exclaimed. "I thought these darts were only to paralyze temporarily."

"The medical darts stun," the Archon said, "but the Palace Guards are not part of the medical force, Your Reverence."

Brigita picked up one of the white cylinders from the table,

but Viracocha still kept the catapult aimed at the High Priest and the Archon.

"Your ruse won't work. I stood by you as you gave His Holiness a lecture on the magical properties of our bloodstreams."

"Be reasonable, Viracocha," the Archon continued. "You have discovered some new practices, but we had the knowledge to join your bloodstreams. What you are now, you owe to us."

Viracocha shot the Archon with the medical dart and spoke to him as he slumped to the floor:

"What we are, we certainly do not owe to you. You will excuse me if I have difficulty hearing you out, for ever since I heard your screech in the Tower, I find your voice harder to take."

Viracocha looked up from the Archon, collapsed on the floor, to the High Priest, as if in his eyes he could find an answer to it all:

"I wonder if *you* had experienced it, whether you still could inflict it on anyone. Do you even begin to realize what you have created in your 'City of Knowledge'? Why did you let him do all this? It's all so completely unnecessary. Had you meditated inside the aligned Solids, you could have ascended out of the physical world all the way to the archetypal realm."

"You know as well as I do, Viracocha, that the archetypal realm is simply the bars of our cage, the grid of cause and effect that lock us in. You've rescued a dozen children for what? For a thousand more lives of torture and unending misery. I'm not interested in mystical experiences anymore. I'm interested in ending this whole wheel of torture called existence."

"And you think by torture you're going to end torture? You admit that the Old Ones are parasites. Well, what kind of torture would there be if they had attached themselves to humanity, the way a lamprey eel attaches itself to a fish? In a little while at sunrise, you are going to have a few moments to gain understanding. I am going to put both you and the Archon in the center of the vault, staring up through the hollow spine into the solar crystal. If I am wrong, you can come to get me and we can have civil war. If I am right, you will have a few seconds when all the human dimensions, etheric, psychic, mental, and

archetypal, implode into you. But it will be only a few seconds before the whole city explodes and drops in pieces into the sea."

"I observe, Viracocha, that you have become my successor after all. In struggling to defeat me you have changed enormously. Think about it. You are much closer to me than you were before. Now that you have taken power from me, rather than accepting it, you have killed three or four people. You weighed their deaths against the whole and made a decision to execute them in order to execute your own plan of escape. And I have done no differently. And so, Your Holiness, I pass on the succession to you and offer you my medallion of office."

Before the High Priest could lift his hands to his chest, Brigita shot him with the paralyzing dart from the white cylinder she had exchanged for the more deadly catapult. As he slumped down to the floor, she spoke to Viracocha without taking her eyes off the High Priest.

"He was going to shoot you. There's a dart-catapult hidden to the side of the large emerald."

Viracocha stared down at the High Priest on the floor and then bent down to pick him up to place him on the table. Very gently and with great respect he removed his white sleeveless robe and medallion and chain. He set them to the side, folded the High Priest's hands over his chest, and then put rolls of cloth to support his head so that it would not turn to the side and away from the point of the Tower.

Very slowly, Viracocha lifted the High Priest's robe over his own naked body and let it fall down over his arms and shoulders. He stood for a moment, looking down at the robe, and then spoke quietly to Brigita:

"He's right, you know. I'll take this robe with me, and when I forget, I'll take it out."

Viracocha turned to look questioningly at Brigita, took a breath, looked around at the four bodies sprawled on the floor, and took command of the situation:

"We had better put them all in the vault, so that everything goes on as usual in the morning when the work begins."

"But if the Archon is not here, won't they become suspicious?" Brigita asked.

"I'll leave new instructions in the Mercury Tablets and warn them all to stay out of the foundation vault. I can sign the order with the imprint of his ring. We'd better put them on the tables now and try to straighten up in here."

Brigita and Viracocha worked together picking up the dead and setting them on the tables on which their victims had lain before. As they set the last guard down and folded his hands over his chest, Viracocha recognized him to be the one who had reported to the High Priest.

"We owe our lives to this one," Viracocha said. "If he hadn't reported to the High Priest, we might still be stretched out on these tables. Not exactly a fitting reward, is it?"

Brigita said nothing for a moment, then she looked up at Viracocha as she still held on to the ankles of the dead man:

"He is the one I shot."

"Would it make you feel any better," Viracocha asked, "if I told you that he was the one who shot Abaris with that same catapult?"

"Yes, but that only makes me feel worse. I asked Abaris once about this problem of becoming evil to fight evil. He said: 'Those who swim in dirty water will get wet, but those who drink it will die.' We're not innocent, and we will have to accept our purification, however it comes."

Brigita moved away from the body to pick up a cloth to clean the puddle of urine she had made on the floor. As she bent down, she could see her own reflection, hesitated for a moment, and then spoke to it:

"The virgin has become a lover, the Abbess has become a warrior. Love and Death."

"And if they had carried catapults that only stunned, they would not be dead, but would be in a few hours. We won't be able to sort this out for a long time. Come on, let's take them into the Tower."

Brigita stood up, placed the cloth under the table, and then looked with confusion at the wheelless slabs:

"How do these move?"

"They're like tiny garudas. Touch the amethyst there, and then they will hover weightlessly above the floor. Then you

simply push or pull them wherever you want. We'd better take one at a time to leave a hand free for the catapult, just in case we run into anybody else. But at least this time we can use the paralyzing darts."

As Viracocha set the white catapult down between the feet of the High Priest, he heard the sound of running feet, and picked up the cylinder as he pulled Brigita down behind the table. They listened intently for a moment.

"It's all right," Brigita said. "I can feel that it is Bran."

"I know," Viracocha answered. "They're running in bare feet: invaders, not defenders."

The doors burst open and Bran and six men rushed into the room, with catapults in hand. Viracocha did not take any chances with a nervously discharged dart and shouted before he stood up:

"It's all right, Bran. We're over here."

"Thank God both of you are not drugged," Bran exclaimed. "We've got to hurry." Bran ran over to them, noticed the High Priest and the Archon, and then the dead assistant and guards:

"Good God! How did you do all this?"

"Since you were a little tardy in rescuing us, we had to rescue ourselves. Great bladdered Brigita pissed on the Empire, and it went under. By the way, Brigita, that was a brilliant tactic."

"It wasn't all that clever," Brigita said with an embarrassed smile, "I really had to pee."

"Well, it was a godsend, however it came out," Viracocha said. "Bran, we have to hide them all in the Tower. The High Priest and the Archon are alive and only temporarily paralyzed. I want them both to see what they have done when the sun strikes the crystal in the morning."

Bran looked down at the Archon:

"I will move him in. After all the people I have had to bring to this wretched place, it will be a great satisfaction to make him my last."

"Let me take his ring off first to put the instructions into the Mercury Tablets for the staff in the morning," Viracocha said as he took off the ring from the tightly clenched fist of the Archon.

"Now, he's all yours, but I suppose, according to protocol, the High Priest should go first."

They moved down the aisle to the end of the laboratory where Viracocha erased the Tablets, set in the new instructions, and then impressed the ring into the template of the living metal. Meticulously, he placed the ring back on the withered finger of the Archon and then took up his position at the table bearing the High Priest.

Viracocha led the way down the underground tunnel to the vault, and clothed as he was in the robe of the High Priest, it seemed as if he were leading a funeral procession down into the great megalithic chamber. No one made a sound. There was no wailing or chanting, but only the slight serpentine hiss of the irradiating crystals that lit the way.

When they came into the vault, Brigita went over immediately to the bodies of the children, crossed their arms over their chests, and began to chant a prayer in Eireanne. Viracocha placed the High Priest and the Archon side by side, staring up the hollow central axis to the golden pyramid that held the solar crystal at the very peak of the Tower. Then Bran placed the two guards and the assistant between them and the children, and when all were in place, he waited nervously while Brigita completed her circuit around the twelve children.

When she had finished, she bowed to the children, turned to Bran, and said:

"We can go now."

"And we had better go quickly, on the run," Bran said as he signaled the others to go ahead.

They started to run with a nervous energy at first, but when they had come out of the vault into the underground tunnel, the power of the deep quiet that had descended onto the city settled on them, and their run became a more rhythmic and steadily beating trot. As he listened to the repetitive beat, it seemed to Viracocha that he was hearing the heartbeat of the whole civilization moving out from the Tower to pulse its last in the arteries of the streets and passageways.

The fog was thick and conspiring with them as they came out of the entrance to the Institute onto the Broad Street. It was first

light, but no bird sang. No person stirred, waking in his bed. The only sound that could be heard was that of their bare feet striking the stones and scattering the few grains of sand and dirt that the winds had brought into the city from the wild.

No guard stirred in his post, and no resistance sprang up in surprise to threaten their escape. As they passed the last of the guard posts, they came out onto the quai of the Institute where the two garudas lay covered by the thick fog.

Brigita stopped for a moment to catch her breath, and Viracocha turned round in a flash of anxiety to look back at her:

"Are you all right?" he asked. "We're almost there, it's just a few more yards to the vessel."

Brigita placed her palms on her knees, bent over to take a deep breath, and shook her head to say yes before she stood up straight to look at Viracocha.

Bran signaled the others to go on ahead into the ship, and then walked back to Brigita:

"I've changed your crew, Brigita. I thought you would want to go back," Bran said as he struggled to catch his breath and quiet his own beating heart.

"Thank you," Brigita said, and then she looked at Viracocha in surprise. "I'm sorry. I thought it would be obvious to you. It has to be this way. You have your work. I have mine. And now more than ever we're going to be needed on opposite sides of the world. I'm no man's wife, and you certainly would make a poor wife for me on Eiru."

Viracocha was stunned into silence. He had never really thought of what they would do, or could do in the future. He had only some vague romantic notion of their flying off in the dawn together.

"Good-bye, Bran. Although I know I'll see you again. You're not likely to stay in one place very long."

Brigita embraced Bran in gratitude, and Viracocha could see that part of her wanted to be a normal woman, free of the burden of her own psychic gifts, and if there were any part of her that could be a wife, that normal woman in her would be wife to Bran.

Brigita took a step back into the fog and looked from Bran to Viracocha:

"More time. We all need more time. Maybe only after we've loved everybody once, can we love everybody all at once."

Brigita backed into the fog and disappeared. Viracocha heard the creak of the footbridge to her vessel, heard the rustle of material and then its passage through the air. As he looked down at the cloak of the High Priest, which lay crumpled at his feet, he thought of Brigita standing naked at the doorway to her ship, and he remembered when she stood naked in defiance of his lust, and when she had stood naked in front of her meditation chamber to approach him and carry him out of time and space. What more did he want from her? And why did he stand there feeling so sorry for himself? With a gesture of sadness and self-disgust, he kicked the High Priest's cloak over the edge and down into the water, and then turned to follow Bran to the mooring of the larger vessel.

They passed in silence up the ramp and through the main chamber into the control room. Viracocha hardly looked up to examine all the faces of those with whom he would spend the rest of his life. Instead he stared out the window and watched as Bran slowly followed Brigita's vessel out into the center of the bay. They held their position for a few moments, waiting as the small garuda lifted into the air; then they began their own slow ascent. As soon as they were up a few hundred feet, they came above the clouds. The entire city was hidden by the fog, and only the Tower stood out alone in the sky.

In the distance Viracocha could see Brigita's vessel moving off into the lightening horizon.

"Do you wish to stay here?" Bran asked. "I mean, we can hover here if you wish to see the first beams of the sun touch the crystal."

Viracocha did not look up at Bran. Instead, he stared out the portal at the Tower and he began to speak quietly as if he were seeing a vision or reading from a history book:

"When the first beam of light touches the crystal, nothing very unusual will seem to have happened. One will only wonder why the birds have stopped singing. And then the animals

will begin to howl, but that will quickly pass. Then the greatest silence on earth will come over the whole island. Those who are asleep will bolt upright in bed, fully awake, fully alert as they listen to nothing that they can hear. It will seem to them for a moment that a sound was coming to them without direction. No man will recognize sounds that no man has ever heard before. Here and there will be a trickle of sand, a gentle dripping of gravel, as a seam appears in the Tower. When the sides pull apart, the crystal will seem to hang there for a moment, supported on a shaft of light. And then it will drop, not very far, for by then the cone will have been drawn up to meet it. The buildings will not so much fall as slip to the side in their irrelevance to the cone's attraction to the crystal. Perhaps a few farmers far up on the peninsula will notice a peculiar smell soaking up through the grass. They will have a moment to wonder about the little jets of steam in the earth, but only a moment, for with the not very heavy touch of the crystal on the cone, the volcano will explode and the entire city will pass in fragments through the air where we are now. And that will simply be the beginning. By the end of the day, all the islands will have sunk into the sea."

Bran turned back to the controls and began to move the vessel round to face toward the south and the island where the others waited for their appointed rendezvous. Viracocha did not take his eyes from the portal, but simply accepted in silence the part of the sky that was being given to him as they turned, the part of the sky in which the Morning Star appeared leading the sun into the last Atlantean dawn.

Where catastrophe still haunts the brittle edges of our land, there the mind dwells; no, not mind as we know it exactly, but that atavistic faculty of dream, vision, and spontaneous images sprung out of time. The skeptical mind is its own proud tower that resists the pull of the darkening sea. Not forgotten, nor yet accepted, is that other morning when the usual return to the body slipped and glanced off into dimensions of future time: there upon the palisades above the Hudson were seen the crowd of towers in midtown, but everywhere the water covered the streets, although not entirely the buildings, which

seemed to sway like reeds in a tidal marsh. New York was at peace with nature in death as it had never been in life. Like an ancient and romantic ruin, the Empire State Building rose gloriously above the water which covered only six of its numberless stories. More beneficent than its ancestor had been, it did not need to be obliterated from the face of the earth. Its fall would stand for ages.